3.2 SLOPE OF A NONVERTICAL LINE

If $P(x_1, y_1)$ and $Q(x_2, y_2)$ lie on a nonvertical line l, then

Slope of line $l = m = \dfrac{y_2 - y_1}{x_2 - x_1}$

Lines with equal slopes are parallel.

Lines with slopes that are negative reciprocals are perpendicular.

3.3 EQUATIONS OF LINES

If $P(x_1, y_1)$ is a point on a line l with a slope of m, then the equation of the line can be written as

$$y - y_1 = m(x - x_1) \qquad \text{Point-slope form}$$
$$y = mx + b \qquad \text{Slope-intercept form}$$
$$Ax + By = C \qquad \text{General form}$$

3.4 FUNCTIONS AND FUNCTION NOTATION

A function defined by an equation of the form

$$y = f(x) = mx + b$$

is called a **linear function**.

3.5 POLYNOMIAL FUNCTIONS

A function defined by an equation of the form

$$y = f(x) = ax^2 + bx + c \qquad (a \neq 0)$$

is called a **quadratic function**.

The graph of

$$y - k = a(x - h)^2 \qquad (a \neq 0)$$

is a parabola with vertex at (h, k).

Tests for Symmetry:
If $(-x, y)$ lies on a graph whenever (x, y) does, the graph is symmetric about the y-axis.

If $(x, -y)$ lies on a graph whenever (x, y) does, the graph is symmetric about the x-axis.

If $(-x, -y)$ lies on a graph whenever (x, y) does, the graph is symmetric about the origin.

The graph of $\begin{cases} y = f(x) + |k| \\ y = f(x) - |k| \end{cases}$ is identical to the graph of

$y = f(x)$, except that it is translated $|k|$ units $\begin{cases} \text{upward} \\ \text{downward} \end{cases}$.

3.5 POLYNOMIAL FUNCTIONS (continued)

The graph of $\begin{cases} y = f(x - |k|) \\ y = f(x + |k|) \end{cases}$ is identical to the graph of

$y = f(x)$, except that it is translated $|k|$ units to the $\begin{cases} \text{right} \\ \text{left} \end{cases}$

3.7 COMPOSITION OF FUNCTIONS

$$(f \circ g)(x) = f(g(x))$$

3.9 PROPORTION AND VARIATION

$y = kx \qquad y$ varies directly with x

$y = \dfrac{k}{x} \qquad y$ varies inversely with x

$y = kxz \qquad y$ varies jointly with x and z

4.1 LINEAR INEQUALITIES

If $a < b$, then

$$a + c < b + c$$
$$a - c < b - c$$
$$\left. \begin{array}{c} ac < bc \\ \dfrac{a}{c} < \dfrac{b}{c} \end{array} \right\} \text{if } c > 0$$
$$\left. \begin{array}{c} ac > bc \\ \dfrac{a}{c} > \dfrac{b}{c} \end{array} \right\} \text{if } c < 0$$

If $a < b$ and $b < c$, then $a < c$.

4.2 ABSOLUTE VALUE

If $a > 0$, then

$|a| = \sqrt{a^2}$

$|x| = a$ is equivalent to $x = a$ or $x = -a$

$|x| < a$ is equivalent to $-a < x < a$

$|x| > a$ is equivalent to $x < -a$ or $x > a$

For real numbers a and b

$$|ab| = |a||b| \qquad \left| \dfrac{a}{b} \right| = \dfrac{|a|}{|b|} \quad (b \neq 0)$$

$$|a + b| \leq |a| + |b|$$

Books in the Gustafson and Frisk Series

4

TH EDITION
College Algebra

R. David Gustafson
Peter D. Frisk

Rock Valley College

Brooks/Cole Publishing Company
Pacific Grove, California

To our wives, Carol and Martha,
and our children, Kristy and Steven;
Sarah, Heidi, and David

Consulting Editor: *Robert Wisner*

Brooks/Cole Publishing Company
A Division of Wadsworth, Inc.

Printed in the United States of America

10   9   8   7   6   5   4   3   2

Library of Congress Cataloging-in-Publication Data

Gustafson, R. David (Roy David), [date]
    College algebra / R. David Gustafson, Peter D. Frisk, — 4th ed.
      p.   cm.
    Includes index.
    ISBN 0-534-10380-4
    1. Algebra.   I. Frisk, Peter D., [date].   II. Title.
QA154.2.G87   1989
512.9—dc20   89-31887
                                                                                    CIP

Sponsoring Editor: *Sue Ewing, Paula-Christy Heighton*
Editorial Assistant: *Sarah Wilson*
Production Editor: *Ellen Brownstein*
Manuscript Editor: *David Hoyt*
Permissions Editor: *Carline Haga*
Interior and Cover Design: *Kelly Shoemaker*
Cover Photo: *Lee Hocker*
Art Coordinator: *Lisa Torri*
Interior Illustration: *Lori Heckelman*
Typesetting: *Jonathan Peck Typographers, Ltd.*
Cover Printing: *The Lehigh Press Company*
Printing and Binding: *R. R. Donnelley & Sons*

Preface

TO THE INSTRUCTOR

A wave of educational reform continues to sweep the country. Educators and legislators, alarmed by what they perceive to be an erosion in the quality of education, are pressing for pervasive changes in the educational system. At the heart of the matter is the desire to promote greater student achievement through higher academic standards.

College Algebra, Fourth Edition, answers the challenge of this reform by providing a thorough, no-nonsense approach to the topics of college algebra. It is designed to be easy for students to use and understand, and it will demonstrate that algebra can be useful and interesting.

Because of our classroom teaching and helpful suggestions from many instructors who use *College Algebra*, we have made several improvements in the fourth edition:

1. Chapter 1 now includes more set terminology, more work with interval notation, and more work with rational exponents and radicals. The FOIL method of multiplying two binomials is now discussed.
2. Chapter 2 includes more work with radical equations and more applied problems.
3. Sections 3.4 and 3.5 explain more clearly the fundamental principles of graphing. The discussion includes piecewise functions, vertical and horizontal translations of graphs, and vertical and horizontal stretchings of graphs. More applications are also included.
4. Rational inequalities are now solved by the test-point and sign-graph methods, with emphasis on interval notation.
5. Chapter 8 now includes techniques of translation to aid in graphing exponential and logarithmic functions. The material on logarithms is divided into two sections, one stressing properties of logarithms and one stressing applications of logarithms. More work with exponential and logarithmic equations strengthens the chapter.
6. The binomial theorem now appears earlier in Chapter 9. The treatment of probability is expanded.

In this new edition, we have been able to refine many of the following features of the previous editions.

Solid Mathematics   The treatment of college algebra is direct and straightforward. Although the treatment is mathematically sound, it is not so rigorous that it will confuse students. Every effort has been made to ensure the accuracy of the mathematics and of the answers to the exercises. The text has been critiqued by dozens of reviewers. Each exercise has been worked by both authors and by a problem checker. Although the exercise sets are designed primarily to provide practice and drill, they also contain problems that will challenge the best students. The text contains nearly 4000 exercises.

Accessibility to Students   The text is written for students to read and understand. On the Fry readability test, the writing is at the tenth-grade level. The numerous problems within each exercise set are carefully keyed to over 400 worked examples, in which the author's notes explain many of the steps used in the problem-solving process. The answers to the odd-numbered exercises appear in an appendix.

Students will like the chapter summaries, the review exercises, the functional use of second color, and the end papers that list, in order of presentation, the important formulas developed in the text. They will also appreciate the *Student Solutions Manual*, which contains the solutions to half the even-numbered exercises.

Emphasis of Applications   To show that mathematics is useful, we have included a large number of word problems and applications throughout the text.

Built-in Redundancy   Because skills taught in early chapters are used throughout the text, students have several opportunities to review or relearn material. This constant review helps to improve student retention.

Supplementary Materials   A test manual containing three tests for each chapter is available in printed format. For compiling examinations, a bank of test items with EXP-TEST®, a full-featured test-generating system for the IBM-PC, is available to those who adopt the text. Also available is a teacher's manual that gives answers to all the even-numbered problems.

ORGANIZATION AND COVERAGE

The text can be used in a variety of ways. To maintain optimum flexibility, several chapters are sufficiently independent to allow you to pick and choose topics that are relevant to your students' needs. The following diagram shows how the chapters are interrelated.

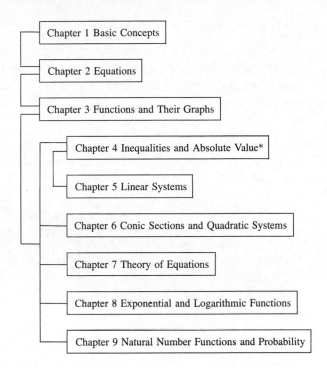

CALCULATORS AND COMPUTERS

The use of calculators is assumed throughout the text. We believe that students should learn calculator skills in the mathematics classroom. They will then be prepared to use calculators in science and business classes and for nonacademic purposes. The directions within exercise sets indicate which exercises require calculators.

Computer exercises have been incorporated at various places within the text. A copy of the software to solve these exercises is included with each teacher's edition. This software, called *Algebra Pak*, may be duplicated without permission and distributed to students free of charge. If computers are not available, the computer exercises can easily be omitted without interrupting the flow of topics.

TOPICS COVERED

Review

Chapters 1 and 2 primarily review basic algebra—the real number system and its properties, exponents and radicals, polynomial arithmetic, and solutions of equations. Complex numbers are introduced early, so they are available whenever needed in later chapters.

---

*The first three sections of Chapter 4 can be taught immediately after Chapter 2 if so desired.

Functions

The concept of function is introduced in Chapter 3, with emphasis on function notation and graphing polynomial functions. Also included is a section on variation and a thorough treatment of inverse functions, with discussions of domain and range.

Inequalities

Chapter 4 discusses inequalities in one and two variables, along with their graphical interpretations. Rational inequalities are solved by both the test-point and sign-graph methods.

Systems of Equations and Inequalities

Chapter 5 includes techniques for solving systems of linear equations. Matrix methods are developed, and some matrix algebra is presented. Linear programming in two variables is discussed. The topic of partial fractions is introduced as an application of systems of linear equations. There are several optional problems that require *Algebra Pak* and a computer.

Conic Sections

Chapter 6 develops the basic forms of the equations for conic sections and provides opportunities for graphing these equations. Solutions of simultaneous second-degree equations are obtained both graphically and algebraically.

Theory of Equations

Chapter 7 discusses methods for finding roots of polynomial equations. Several examples illustrate the interplay between the fundamental theorem of algebra, Descartes' rule of signs, the remainder and factor theorems, the rational-root theorem, and the conjugate-pairs result. Binary chopping is used to approximate irrational roots. There are several optional problems that require *Algebra Pak* and a computer.

Exponential and Logarithmic Functions

Chapter 8 discusses exponential functions, logarithms, and many of their applications. The use of a calculator is emphasized in this chapter. Computing with logarithms is included as an optional section.

Natural Number Functions and Probability

Chapter 9 begins with an introduction to mathematical induction. The binomial theorem, permutations, and combinations lead into a presentation of simple and compound probabilities, odds, and mathematical expectation. A proof of the binomial theorem for natural number exponents is given in an appendix.

TO THE STUDENT

We have tried to write a text that you can read and understand. We have provided an extensive number of worked examples that parallel the problems within each exercise set. Be sure to read the explanations carefully, since they contain much of the value of the text.

A *Student Solutions Manual* is available to you, which contains the answers to half the even-numbered exercises. Also available is a *Study Guide*, designed to enhance your study techniques.

Many of the exercises within the text require the use of a computer and *Algebra Pak*, a piece of computer software. If you would like a copy of this software, check with your instructor.

The material presented in *College Algebra*, Fourth Edition, will be of value to you in later years. Therefore, we suggest you keep this text after completing the course. It will be a good source of reference and will keep at your fingertips the material that you have learned here.

We wish you well.

ACKNOWLEDGMENTS

We are grateful to the following people who reviewed the manuscript in its various stages. They all had valuable suggestions that have been incorporated into the text.

The reviewers include Richard Andrews, University of Wisconsin; James Arnold, University of Wisconsin; Wilson Banks, Illinois State University; Jerry Bloomberg, Essex Community College; Dale Boye, Schoolcraft College; Lee R. Clancy, Golden West College; Jan Collins, Embry Riddle College; Cecilia Cooper, William & Harper College; Romae J. Cormier, Northern Illinois University; John S. Cross, University of Northern Iowa; Grace DeVelbiss, Sinclair Community College; Lena Dexter, Faulkner State Junior College; Emily Dickinson, University of Arkansas; Robert E. Eicken, Illinois Central College; Eric Ellis, Essex Community College; Dale Ewen, Parkland College; Ronald J. Fischer, Evergreen Valley College; Mary Jane Gates, University of Arkansas at Little Rock; Jerry Gustafson, Beloit College; Jerome Hahn, Bradley University; Douglas Hall, Michigan State University; Robert Hall, University of Wisconsin; David Hansen, Monterey Peninsula College; Kevin Hastings, University of Delaware; William Hinrichs, Rock Valley College; Arthur M. Hobbs, Texas A & M University; Jack E. Hofer, California Polytechnic State University; Ingrid Holzner, University of Wisconsin; Warren Jaech, Tacoma Community College; Nancy Johnson, Broward Community College; William B. Jones, University of Colorado; David Kinsey, University of Southern Indiana; Helen Kriegsman, Pittsburg State University; Judy McKinney, California Polytechnic Institute at Pomona; Sandra McLaurin, University of North Carolina; Marcus McWaters, University of Southern Florida; James W. Mettler, Pennsylvania State University; Eldon L. Miller, University of Mississippi; Stuart E. Mills, Louisiana State University, Shreveport; Gilbert W. Nelson, North Dakota State; Marie Neuberth, Catonsville City College; Anthony Peressini, University of Illinois; David L. Phillips, University of Southern Colorado; William H. Price, Middle Tennessee State University; Janet P. Ray, Seattle Central Community College; Barbara Riggs, Tennessee Technological University; Paul Schaefer, SUNY, Geneseo; Vincent P. Schielack, Jr., Texas A & M University; Robert Sharpton, Miami Dade Community College; L. Thomas Shiflett, Southwest Missouri State University; Richard Slinkman, Bemidji State University; Merreline Smith, California Polytechnic Institute at Pomona; John Snyder, Sinclair Community College; Warren Strickland, Del Mar College; Ray Tebbetts, San Antonio College; Faye Thames, Lamar State University; Carroll G. Wells, Western Kentucky University; Charles R. Williams, Midwestern State University; Harry Wolff, University of Wisconsin; Clifton Whyburn, University of Houston; Albert Zechmann, University of Nebraska.

We wish to thank the staff at Brooks/Cole, especially Craig Barth, Jeremy Hayhurst, Sue Ewing, Paula-Christy Heighton, Ellen Brownstein, Kelly Shoemaker, and Lisa Torri for their competent work and support.

R. David Gustafson
Peter D. Frisk

Contents

8 Exponential and Logarithmic Functions

345

9 Natural Number Functions and Probability

388

1 Basic Concepts

The concept of number is fundamental to mathematics. For this reason, we begin by discussing various sets of numbers and their properties.

1.1 THE SET OF RATIONAL NUMBERS AND ITS SUBSETS

A **set** is any collection of objects. Each object is called a **member** or an **element** of the set. To denote a set, we often use braces to enclose the list of its elements. For example, the notation $\{a, b, c\}$ means the set whose elements are the letters a, b, and c. To indicate that b is an element of this set, we write

$$b \in \{a, b, c\} \qquad \text{Read as "b is an element of the set containing a, b, and c."}$$

The expression

$$d \notin \{a, b, c\}$$

indicates that d is not an element of $\{a, b, c\}$.

Capital letters are often used to name sets. For example, the expression

$$\mathbf{A} = \{a, e, i, o, u\}$$

means that $\mathbf{A}$ is the set containing the vowels a, e, i, o, and u.

In **set-builder notation**, a rule is given that establishes membership in a set. The set of vowels in the English alphabet, for example, can be denoted as

$$\mathbf{V} = \{x : x \text{ is a vowel of the English alphabet.}\}$$

The statement above is read as "$\mathbf{V}$ is the set of all letters x such that x represents a vowel of the English alphabet." Because x can represent many different elements of the set, x is called a **variable**.

When two sets such as $\mathbf{A}$ and $\mathbf{V}$ have the same elements, we say that they are equal, and we write $\mathbf{A} = \mathbf{V}$. If two sets $\mathbf{A}$ and $\mathbf{B}$ do not have the same elements, they are not equal, and we write

$$\mathbf{A} \neq \mathbf{B}$$

If $\mathbf{B} = \{a, c, e\}$ and $\mathbf{A} = \{a, b, c, d, e\}$, each element of set $\mathbf{B}$ is also an element of set $\mathbf{A}$. When this is so, we say that $\mathbf{B}$ is a **subset** of $\mathbf{A}$. In symbols, we write

$$\mathbf{B} \subseteq \mathbf{A} \qquad \text{Read as "set $\mathbf{B}$ is a subset of set $\mathbf{A}$."}$$

1

Because every element in set **A** is an element in set **A**, set **A** is a subset of itself. In symbols,

$$\mathbf{A} \subseteq \mathbf{A}$$

In general, any set is a subset of itself. The expression

$$\mathbf{A} \not\subseteq \mathbf{B}$$

indicates that **A** is not a subset of **B**. If **A** $\not\subseteq$ **B**, then there is at least one element of **A** that is not an element of **B**.

A set with no elements is called the **empty set** or the **null set** and it is denoted by the symbol $\emptyset$. Thus,

$$\emptyset = \{ \qquad \}$$

The empty set is considered to be a subset of every set.

If the elements of some set **A** are united with the elements of some set **B**, the **union** of set **A** and set **B** is formed. The union of set **A** and set **B** is denoted as

$$\mathbf{A} \cup \mathbf{B} \qquad \text{Read as ``the union of set } \mathbf{A} \text{ and set } \mathbf{B}\text{.''}$$

The elements in **A** $\cup$ **B** are *either* elements of set **A**, *or* elements of set **B**, *or* elements of *both* set **A** and set **B**.

The set of elements that are common to set **A** and set **B** is called the **intersection** of **A** and **B**. The intersection of set **A** and set **B** is denoted as

$$\mathbf{A} \cap \mathbf{B} \qquad \text{Read as ``the intersection of set } \mathbf{A} \text{ and set } \mathbf{B}\text{.''}$$

The elements of **A** $\cap$ **B** are those elements that are in *both* set **A** and set **B**. If **A** and **B** have no elements in common, then **A** $\cap$ **B** = $\emptyset$. When the intersection of two sets is the empty set, we say that the two sets are **disjoint**.

Example 1   If **A** = $\{a, b, c, d, e\}$ and **B** = $\{a, d, g\}$, find   **a. A** $\cup$ **B**,   **b. A** $\cap$ **B**, and **c. A** $\cup$ (**B** $\cap$ $\emptyset$).

Solution   **a. A** $\cup$ **B** = $\{a, b, c, d, e, g\}$

b. A $\cap$ **B** = $\{a, d\}$

c. A $\cup$ (**B** $\cap$ $\emptyset$) = **A** $\cup$ $\emptyset$ Do the work in parentheses first.
    = **A**   ■

Sets of Numbers

In mathematics, the most basic set of numbers is the set of **natural numbers**. These are the numbers that we use for counting.

$$\mathbf{N} = \{1, 2, 3, 4, 5, 6, 7, 8, 9, 10, 11, \ldots\}$$

The three dots, called the **ellipsis**, following the natural number 11 indicate that the list of natural numbers continues forever. Because we can always add 1 to any chosen natural number to obtain a larger one, there is no largest natural number.

If a set, such as the set of natural numbers, has an unlimited number of elements, it is called an **infinite set**. If a set has a limited number of elements, it is called a **finite set**.

Certain natural numbers are called **prime numbers**. These are the natural numbers that are greater than 1 and that are divisible with a remainder of 0 only by 1 and by the number itself. The prime numbers are the set

$$\mathbf{P} = \{2, 3, 5, 7, 11, 13, 17, 19, \ldots\}$$

Because every prime number is also a natural number, the set of prime numbers is a subset of the set of natural numbers:

$$\mathbf{P} \subseteq \mathbf{N}$$

If a natural number is greater than 1 and can be divided with a remainder of 0 by some number other than itself and 1, the number is called a **composite number**. The set of composite numbers is the set

$$\mathbf{C} = \{4, 6, 8, 9, 10, 12, 14, 15, 16, \ldots\}$$

Because every composite number is also a natural number, we have

$$\mathbf{C} \subseteq \mathbf{N}$$

Because no composite numbers are prime, we have

$$\mathbf{P} \cap \mathbf{C} = \varnothing$$

If we include 0 with the set of natural numbers, we have the set of **whole numbers**:

$$\mathbf{W} = \{0, 1, 2, 3, 4, 5, 6, 7, 8, 9, \ldots\}$$

If we form the union of the set $\{-1, -2, -3, \ldots\}$ and $\mathbf{W}$, we have the set $\mathbf{J}$ of **integers**.

$$\mathbf{J} = \{\ldots, -6, -5, -4, -3, -2, -1, 0, 1, 2, 3, 4, 5, 6, \ldots\}$$

Those integers that are divisible by 2 are called **even integers**, and those that are not are called **odd integers**. If $\mathbf{E}$ is the set of even integers and $\mathbf{O}$ is the set of odd integers, then

$$\mathbf{E} = \{\ldots, -8, -6, -4, -2, 0, 2, 4, 6, 8, \ldots\}$$
$$\mathbf{O} = \{\ldots, -7, -5, -3, -1, 1, 3, 5, 7, \ldots\}$$

If a number can be written as a fraction with an integer for its numerator and a nonzero integer for its denominator, it is called a **rational number**. To denote the set $\mathbf{Q}$ of rational numbers, we use set-builder notation and write

$$\mathbf{Q} = \{x : x \text{ is a number that can be written in the form } \frac{a}{b}, \text{ where } a \text{ and } b \text{ are integers and } b \neq 0.\}$$

The previous statement is read as "$\mathbf{Q}$ is the set of numbers x such that x is a number that can be written in the form $\frac{a}{b}$, where a and b are integers and b is not equal to 0."

The rational numbers include numbers such as

$$\frac{3}{4}, \frac{-1}{3}, \frac{5}{1}, -\frac{8}{4}, \frac{0}{5}, \text{ and } \frac{99}{113}$$

Numbers such as 3, 0, −0.25, and 0.333 . . . are rational numbers because each one can be written in the required form:

$$3 = \frac{3}{1}, \quad 0 = \frac{0}{7}, \quad -0.25 = -\frac{1}{4}, \quad \text{and} \quad 0.333 \ldots = \frac{1}{3}$$

It is important to remember that the denominator of a fraction cannot be 0. A symbol such as $\frac{8}{0}$ is undefined, because there is no number that when multiplied by 0 gives 8. The symbol $\frac{0}{0}$ is also undefined, because all numbers when multiplied by 0 give 0.

Every rational number in fractional form can be changed to decimal form. For example, to change $\frac{3}{4}$ to a decimal fraction, we divide 3 by 4 to obtain 0.75:

$$
\begin{array}{r}
0.75 \\
4\overline{)3.00} \\
2\ 8 \\
\hline
20 \\
20 \\
\hline
0
\end{array}
$$

Because the division leaves a remainder of 0, the quotient is a **terminating decimal**. If we change a fraction such as $\frac{4}{15}$ to a decimal fraction, we obtain a **repeating decimal**:

$$
\begin{array}{r}
0.266 \ldots \\
15\overline{)4.000} \\
3\ 0 \\
\hline
1\ 00 \\
90 \\
\hline
100 \\
90 \\
\hline
10
\end{array}
$$

It can be shown that the decimal forms of all rational numbers are either terminating decimals or repeating decimals. It is also true that any decimal fraction that is a terminating decimal or a repeating decimal can be changed to fractional form.

Example 2   Change   **a.** 0.75 and   **b.** 0.3 876 876 876 . . . to fractional form.

Solution   **a.** Because the *terminating decimal* 0.75 means 75 hundredths, write 0.75 as the fraction $\frac{75}{100}$, and simplify it to obtain $\frac{3}{4}$.

b. The decimal 0.3 876 876 876 . . . has a repeating block of three digits, 876. Form an equation by setting x equal to the decimal.

1. $x = 0.3\,876\,876\,876$ . . .

Then form another equation by multiplying both sides of Equation 1 by 10^3, which is 1000.

2. $1000x = 387.6\,876\,876\,876$ . . .

Subtract each side of Equation 1 from the corresponding side of Equation 2 to obtain $999x = 387.3$.

$$
\begin{aligned}
1000x &= 387.6\,876\,876\,876 \ldots \\
x &= 0.3\,876\,876\,876 \ldots \\
\hline
999x &= 387.3\,000\,000\,000 \ldots
\end{aligned}
$$

Finally, solve this equation for x and simplify the fraction.

$999x = 387.3$

$x = \dfrac{387.3}{999}$ Divide both sides by 999.

$= \dfrac{3873}{9990}$ Multiply both the numerator and the denominator by 10.

$= \dfrac{1291}{3330}$ Simplify.

Use a calculator to show that the decimal representation of $\frac{1291}{3330}$ is

0.3 876 876 876 . . .

The key step in this example was multiplying both sides of Equation 1 by 10^3. If there had been n digits in the repeating block of the decimal, you would have multiplied both sides of Equation 1 by 10^n.   ∎

A repeating decimal such as 0.3 876 876 876 . . . is often written in the form $0.3\overline{876}$, where the overbar indicates the repeating block of digits.

Because it is always possible to write rational numbers in fractional form as terminating or repeating decimals and to write terminating or repeating decimals as rational numbers in fractional form, these two sets of numbers are one and the same. Thus, the set of rational numbers can be described as the set of all decimals that either terminate or repeat.

$\mathbf{Q} = \{x : x$ is a terminating or a repeating decimal.$\}$

Graphs of Sets of Rational Numbers

Sets of numbers can be represented geometrically as points on a **number line**. To construct a number line, we draw a line, choose some arbitrary point on it and label the point 0. We then mark off equal distances to the right and to the left of

0, labeling the points as in Figure 1-1. The line and the number labels on it continue forever in both directions.

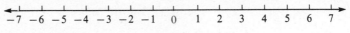

Figure 1-1

The numbers that represent points to the right of 0 are **positive numbers**, and the numbers that represent points to the left of 0 are **negative numbers**. The number 0 is neither a positive number nor a negative number.

To graph the set of natural numbers from 1 to 10 on the number line, we place large dots on the points represented by 1, 2, 3, 4, 5, 6, 7, 8, 9, and 10, as shown in Figure 1-2**a**. The point associated with each number is called the **graph** of that number. The number is called the **coordinate** of its corresponding point. The graph of the prime numbers less than 12 is shown in Figure 1-2**b**, and the graph of the whole numbers less than 10 is shown in Figure 1-2**c**.

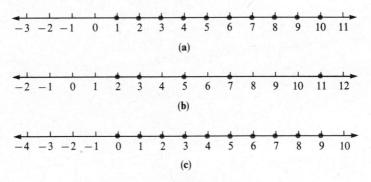

Figure 1-2

The graph of the integers is shown in Figure 1-3. Both the line and the points marked on it continue forever in both directions.

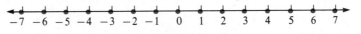

Figure 1-3

It is also possible to graph rational numbers on a number line. For example, the point with a coordinate of $\frac{3}{2}$ is at a distance midway between the points with coordinates of 1 and 2. The point with a coordinate of $\frac{1}{3}$ is at a distance one-third of the way from 0 to 1. These points and others are shown in Figure 1-4.

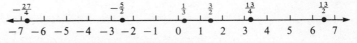

Figure 1-4

EXERCISE 1.1

In Exercises 1–6, **A** = {2, 4, 6, 8}, **B** = {2, 4, 6, 7}, *and* **C** = {x : x *is an even integer.*}. *Insert either an* ∈ *or a* ⊆ *symbol to make a true statement.*

1. **A** _____ **C**   **2.** 4 _____ **A**   **3.** 7 _____ **B**   **4.** {4, 6} _____ **C**

5. ∅ _____ **B**   **6.** **B** _____ **B**

In Exercises 7–14, **A** = {2, 3, 5, 7}, **B** = {1, 3, 5, 7, 9}, *and* **C** = {2, 4, 6, 8, 10}. *Find each set.*

7. **A** ∩ **B**   **8.** **A** ∪ **C**   **9.** **B** ∪ **C**   **10.** **A** ∩ **C**

11. (**A** ∩ ∅) ∪ **B**   **12.** **B** ∩ (∅ ∪ **C**)   **13.** (**A** ∩ **B**) ∪ **C**   **14.** (**A** ∪ **C**) ∩ **B**

In Exercises 15–18, list the elements in each set.

15. {x : x is a composite number between 9 and 19.}   **16.** {x : x is a prime number between 10 and 20.}

17. {x : x is a number that is both even and prime.}   **18.** {x : x is a natural number that is neither prime nor composite.}

In Exercises 19–24, tell whether each set is a finite or an infinite set.

19. {x : x is a composite number.}   **20.** {x : x is an even integer.}

21. {1, 2, 3, 4, 5, 6, 7, 8, 9, 10}   **22.** {x : x is a rational number.}

23. {x : x is a country in the world.}   **24.** {x : x is an integer that is both prime and composite.}

25. Find the first prime larger than 50.   **26.** Find the largest prime less than 100.

In Exercises 27–36, indicate whether each statement is true. If a statement is false, explain why,

27. The number 27 is a prime number.   **28.** The number 2 is a prime number.

29. The set of composite numbers is a subset of the set of integers.   **30.** The set of prime numbers is a subset of the set of odd integers.

31. The number 0 is not a rational number.   **32.** The symbol $\frac{16}{0}$ represents a rational number.

33. All integers are rational numbers.   **34.** Some integers are greater than 1.

35. The number 1 is a prime number.   **36.** The number 0 is neither even nor odd.

In Exercises 37–50, indicate whether each statement is true. If a statement is false, give an example to show that it is false.

37. The product of two prime numbers is a prime number.   **38.** The sum of two prime numbers is a prime number.

39. The square of a prime number is a prime number.   **40.** The sum of two prime numbers can be a composite number.

41. The sum of two composite numbers is a prime number.   **42.** All natural numbers are integers.

43. No even integers are prime numbers.   **44.** All odd integers are prime numbers.

45. The sum of an even integer and an odd integer is an odd integer.   **46.** The sum of three odd integers must be another odd integer.

47. If the product of several integers is an even integer, then at least one of the integers is an even integer.

48. If the product of several integers is an odd integer, then each of the integers must be an odd integer.

49. If the sum of three integers is an even integer, then each of the integers must be an even integer.

50. The union of the set of even integers and the set of odd integers is the set of integers.

In Exercises 51–56, change each fraction into an equivalent decimal fraction.

51. $\dfrac{1}{4}$ **52.** $\dfrac{7}{2}$ **53.** $\dfrac{2}{9}$ **54.** $\dfrac{3}{11}$ **55.** $-\dfrac{5}{12}$ **56.** $-\dfrac{1}{7}$

In Exercises 57–68, change each decimal fraction into an equivalent common fraction.

57. 0.3 **58.** 0.375 **59.** $-0.\overline{75}$ **60.** $-0.\overline{28}$

61. $0.\overline{123}$ **62.** $0.8\overline{41}$ **63.** $0.3\overline{456}$ **64.** $0.9\overline{245}$

65. $-8.\overline{61}$ **66.** $-4.3\overline{21}$ **67.** $1.6\overline{17}$ **68.** $2.3\overline{51}$

69. Change the decimal $1.\overline{9}$ to a common fraction and thereby show that

$$1.999 \ldots = 2$$

70. Change the decimal $4.\overline{9}$ to a common fraction and thereby show that

$$4.999 \ldots = 5$$

71. Does $0.999 = 1$? Explain.

72. Determine the pattern in the decimal

$$0.1212212221 \ldots$$

Does this decimal represent a rational number? Explain.

In Exercises 73–80, graph each set on a number line, if possible.

73. The set of even integers between 19 and 31.

74. The set of integers from -5 to -1.

75. The set of prime numbers between 10 and 20.

76. The set of even composite numbers between 20 and 30.

77. The set of natural numbers that are neither prime nor composite.

78. The set of integers that are both even and prime.

79. The set of numbers that are both even and odd.

80. The set of numbers that are odd and composite and are less than 10.

1.2 THE REAL NUMBERS AND THEIR PROPERTIES

Numbers whose decimal representations neither terminate nor repeat are called **irrational numbers**. Numbers such as

$$\sqrt{2} = 1.414213562 \ldots \quad \text{and} \quad \pi = 3.141592653 \ldots$$

are examples of irrational numbers because their decimal representations are non-terminating, nonrepeating decimals. To express the set of irrational numbers, we write

$\mathbf{H}$ = {$x : x$ is a nonterminating, nonrepeating decimal.}

The union of the set of rational numbers $\mathbf{Q}$ (the set of terminating or repeating decimals) and the set of irrational numbers $\mathbf{H}$ (the set of nonterminating, nonrepeating decimals) is the set of all decimals. This set is called the set of **real numbers**, and it is denoted by the symbol $\mathcal{R}$:

$\mathcal{R}$ = {$x : x$ is a decimal number.}

In symbols,

$$\mathcal{R} = \mathbf{Q} \cup \mathbf{H}$$

The set of rational numbers and the set of irrational numbers are both subsets of the set of real numbers.

The graphs of many subsets of the real numbers are portions of the number line called **intervals**. Figure 1-5 shows the graph of the real numbers x that are between -2 and 4. We describe this set with the expression

$\{x : -2 < x < 4\}$ or $\{x : -2 < x$ and $x < 4\}$

or more simply by

$-2 < x < 4$ or $-2 < x$ and $x < 4$

where the symbol $<$ is read as "is less than."

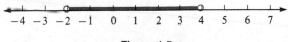

Figure 1-5

The open circles at points with coordinates of -2 and 4 indicate that the endpoints are not included in the graph. When neither endpoint is included, the interval is called an **open interval**. The open interval between -2 and 4 is often denoted as $(-2, 4)$. The parentheses in this notation indicate that the endpoints are not included.

The graph of the set of real numbers from 1 to 6 is shown in Figure 1-6 on page 10. We describe this set with the expressions

$\{x : 1 \le x \le 6\}$ or $\{x : 1 \le x$ and $x \le 6)$

or more simply as

$1 \le x \le 6$ or $1 \le x$ and $x \le 6$

where the symbol $\le$ is read as "is less than or equal to." The closed circles at points with coordinates of 1 and 6 indicate that these points are included in the graph. When both endpoints are included, the interval is called a **closed interval**. The closed interval from 1 to 6 can be denoted as $[1, 6]$. The brackets in this notation indicate that the endpoints are included.

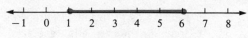

Figure 1-6

Sometimes an interval is half-open. For example, Figure 1-7 shows the graph of the half-open interval $(-6, -1]$. This interval can also be described with the expressions

$$-6 < x \le -1 \qquad \text{or} \qquad -6 < x \text{ and } x \le -1$$

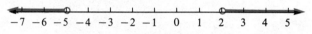

Figure 1-7

The graph of all real numbers x such that $x < -5$ or $x > 2$ (read the symbol > as "is greater than") is shown in Figure 1-8.

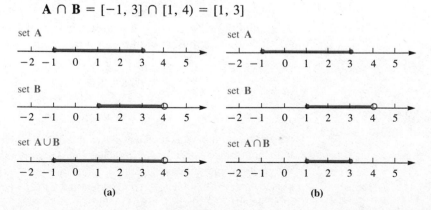

Figure 1-8

Example 1   If $\mathbf{A} = [-1, 3]$ and $\mathbf{B} = [1, 4)$, find the graph of   **a.** $\mathbf{A} \cup \mathbf{B}$ and   **b.** $\mathbf{A} \cap \mathbf{B}$.

Solution   **a.** The union of intervals **A** and **B** is the set of all real numbers that are elements of either set **A** or set **B** or both. Numbers from -1 to 3 are in set **A**, and numbers from 1 to 4, not including 4, are in set **B**. Numbers from -1 to 4, not including 4, are in at least one of these sets. To see this, refer to Figure 1-9**a**. Thus,

$$\mathbf{A} \cup \mathbf{B} = [-1, 3] \cup [1, 4) = [-1, 4)$$

b. The intersection of intervals **A** and **B** is the set of all real numbers that are elements of both set **A** and set **B**. The numbers that are in both of these sets are the numbers from 1 to 3. To see this, refer to Figure 1-9**b**. Thus,

$$\mathbf{A} \cap \mathbf{B} = [-1, 3] \cap [1, 4) = [1, 3]$$

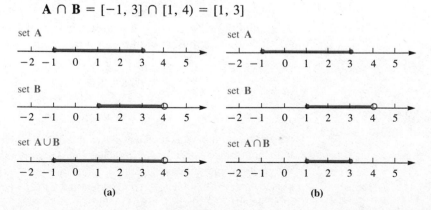

(a)   (b)

Figure 1-9

There is a distinction between the words *and* and *or* as used in the previous discussion. The statement $-2 < x$ *and* $x < 4$ means that both conditions on x must be true at the same time. The statement $x < -5$ *or* $x > 2$ means that only one of the conditions on x needs to be true.

If a and b are real numbers and $a < b$, then the graph of a is to the left of the graph of b. If $a > b$, then the graph of a is to the right of the graph of b.

Two real numbers are called **negatives** of each other if their graphs on a number line lie at equal distances from the origin on opposite sides of the origin. For example, the negative of 2 is negative 2, denoted as -2, and the negative of -2, denoted as $-(-2)$, is 2. See Figure 1-10. In general, for any real number x,

$$-(-x) = x$$

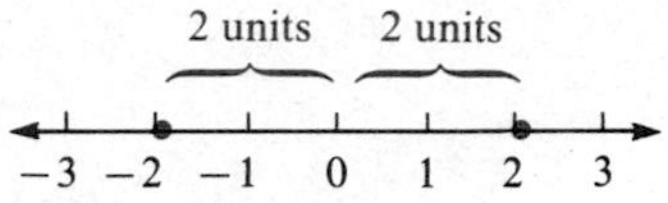

Figure 1-10

This property of negatives and others are summarized as follows:

Properties of Negatives.

1. $-(-x) = x$

2. $(-x) + (-y) = -(x + y)$

3. $x - y = x + (-y)$

4. $-(x - y) = y - x$

5. $(-1)x = -x$
    $(-x)y = x(-y) = -xy$
    $(-x)(-y) = xy$

6. $\dfrac{-x}{y} = \dfrac{x}{-y} = -\dfrac{x}{y}$
    $\dfrac{-x}{-y} = \dfrac{x}{y}$
    provided that $y \neq 0$

The **absolute value** of a real number a, denoted as $|a|$, is the distance on a number line between 0 and the point with the coordinate of a. Because points with coordinates of 4 and -4 both lie 4 units from 0, it follows that $|4| = |-4| = 4$. Likewise, because points with coordinates of -17 and 17 both lie the same distance from 0, it follows that $|-17| = |17| = 17$. In general,

$$|-a| = |a|$$

for any real number a.

The absolute value of a number can be defined more formally. In the following definition, read the symbol $\geq$ as "is greater than or equal to."

Definition. $|x| = x$ if $x \geq 0$.
   $|x| = -x$ if $x < 0$.

The definition of the absolute value of x indicates that if x is a positive number or 0, then x is its own absolute value. However, if x is a negative number, then

$-x$ (which is positive) is its absolute value. Thus, $|x|$ always represents a non-negative number, and we have, for all real numbers x,

$$|x| \geq 0$$

Example 2   Find   **a.** $|3|$,   **b.** $|-4|$,   **c.** $|0|$, and   **d.** $|1 - \sqrt{2}|$.

Solution   **a.** $|3| = 3$ **b.** $|-4| = -(-4)$
$$= 4$$

c. $|0| = 0$

d. $|1 - \sqrt{2}| = -(1 - \sqrt{2})$ Because $1 < \sqrt{2}$, the number $1 - \sqrt{2}$ is negative.
$$= \sqrt{2} - 1 \qquad \blacksquare$$

Example 3   If $x = -2$, $y = 3$, and $z = -5$, find the value of   **a.** $y|x + z|$ and
b. $\dfrac{|z - x|}{y}$.

Solution   **a.** $y|x + z| = 3|-2 + (-5)|$ **b.** $\dfrac{|z - x|}{y} = \dfrac{|-5 - (-2)|}{3}$
$$= 3|-7|$$
$$= 3(7) \qquad\qquad\qquad\qquad = \dfrac{|-5 + 2|}{3}$$
$$= 21$$
$$= \dfrac{|-3|}{3}$$
$$= \dfrac{3}{3}$$
$$= 1 \qquad \blacksquare$$

Consider the number line in Figure 1-11. The distance between the points with coordinates of 1 and 4 is $4 - 1$, or 3 units. However, if the subtraction were done in the other order, the resulting number would be $1 - 4$, or -3 units. To guarantee that the distance between two points is always positive, we use absolute value symbols. Thus, the distance d between points with coordinates of 1 and 4 is

$$d = |4 - 1| = |1 - 4| = 3$$

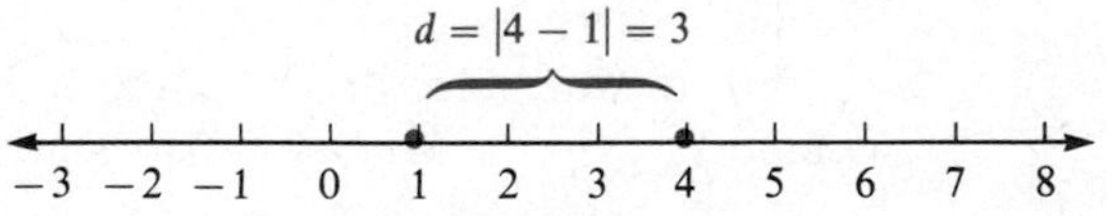

Figure 1-11

The previous discussion suggests that we can define the distance d between two points with coordinates of a and b on the number line by using the concept of absolute value.

> **Definition.**   If a and b are the coordinates of two points on a number line, then the distance between those points is given by the formula
>
> $$d = |a - b|$$

Example 4   Find the distance d on a number line between points with coordinates of   **a.** 3 and 5,   **b.** -2 and 3, and   **c.** -5 and -1.

Solution   **a.** $d = |5 - 3| = |2| = 2$ or $d = |3 - 5| = |-2| = 2$

b. $d = |3 - (-2)| = |5| = 5$ or $d = |-2 - 3| = |-5| = 5$

c. $d = |-5 - (-1)| = |-4| = 4$ or $d = |-1 - (-5)| = |4| = 4$ ∎

Any two numbers a and b can be compared. Either a and b are equal, or they are not. If they are not equal, then one or the other must be the larger. The possibilities are summed up in the **trichotomy property**.

The Trichotomy Property.   For any two real numbers a and b, *exactly one* of the following relationships must hold:

1. a and b are equal $(a = b)$

2. a is less than b $(a < b)$

3. a is greater than b $(a > b)$

To indicate that x is not equal to y, we use the notation $x \neq y$. If x is not less than y, then x must be greater than or equal to y, which is denoted as $x \geq y$. Similarly, $x \leq y$ denotes that x is less than or equal to y.

Three properties of equality are basic in algebra.

Properties of Equality.   If a, b, and c are real numbers, then

$a = a$ (Reflexive property)

If $a = b$, then $b = a$.   (Symmetric property)

If $a = b$ and $b = c$, then $a = c$.   (Transitive property)

Example 5   Determine if the relation $<$ is   **a.** reflexive,   **b.** symmetric, and   **c.** transitive.

Solution   **a.** To determine if the relation $<$ is reflexive, ask the question "Is a real number less than itself?" Because the answer is no, the relation is *not* reflexive.

b. To determine if the relation $<$ is symmetric, ask the question "If one real number is less than a second, is it also true that the second number is less than the first?" Because the answer is no, the relation is *not* symmetric.

c. To determine if the relation $<$ is transitive, ask the question "If one real number is less than a second, and if that second real number is less than a third, does this mean that the first is less than the third?" Because the answer is yes, the relation *is* transitive.   ■

There are many assumptions, called the **field properties**, that we must use when working with real numbers.

If a, b, and c are real numbers, then

The Closure Properties of Real Numbers.

$a + b$ is a real number

$a - b$ is a real number

ab is a real number

$\dfrac{a}{b}$ is a real number   (provided $b \neq 0$)

The Associative Properties for Addition and Multiplication.

$(a + b) + c = a + (b + c)$ $(ab)c = a(bc)$

The Commutative Properties for Addition and Multiplication.

$a + b = b + a$ $ab = ba$

The Distributive Property of Multiplication over Addition.

$a(b + c) = ab + ac$

The Identity Elements.   The real number 0 has the property that for any real number a,

$a + 0 = 0 + a = a$

The real number 1 has the property that for any real number a,

$a \cdot 1 = 1 \cdot a = a$

The number 0 is called the **additive identity element** and the number 1 is called the **multiplicative identity element**.

The Inverse Elements.   For every real number a, there exists a real number $-a$ such that

$a + (-a) = -a + a = 0$

(continued)

(*continued*)

There also exists a real number $\frac{1}{a}$ $(a \neq 0)$ such that

$$\frac{1}{a} \cdot a = a \cdot \frac{1}{a} = 1$$

The number $-a$ is called the **additive inverse** or the **negative** of a. The number $\frac{1}{a}$ is called the **multiplicative inverse** or the **reciprocal** of a.

Example 6   The statements in the left column are true because of the property of equality or the property of real numbers listed in the right column.

$a < b$ or $a = b$ or $a > b$.	trichotomy property
$2 + 7$ is a real number.	closure property for addition.
$2(7)$ is a real number.	closure property for multiplication
$9 + 3 = 3 + 9$	commutative property for addition
$8 \cdot 3 = 3 \cdot 8$	commutative property for multiplication
$9 + (2 + 3) = (9 + 2) + 3$	associative property for addition
$2(xy) = (2x)y$	associative property for multiplication
$2(x + 3) = 2x + 2 \cdot 3$	distributive property
$(a + b) + c = c + (a + b)$	commutative property for addition
$37 + 0 = 37$	identity property for addition
If $a = 37$ and $37 = b$, then $a = b$.	transitive property of equality
$\frac{3}{7} + \left(-\frac{3}{7}\right) = 0$	additive inverse property
If $a = b + c$, then $b + c = a$.	symmetric property of equality
$\frac{3}{7} \cdot \frac{7}{3} = 1$	multiplicative inverse property
$32 = 32$	reflexive property of equality

EXERCISE 1.2

In Exercises 1–8, indicate whether each statement is true. If a statement is false, explain why.

1. The rational numbers are a subset of the real numbers.
2. Every real number is an irrational number.

3. Some irrational numbers are integers.
4. Some real numbers are integers.

5. There are real numbers that are neither rational nor irrational.

6. The smallest irrational number is $\sqrt{2}$.

7. The intersection of the set of real numbers and the set of rational numbers is the set of decimals that either terminate or repeat.

8. The union of the set of rational numbers and the set of irrational numbers is the set of real numbers.

In Exercises 9–30, graph each subset of the real numbers on the number line, if possible.

9. $3 < x < 10$ **10.** $-2 \leq x \leq 3$ **11.** $-4 \leq x < 0$ **12.** $5 < x \leq 15$

13. $3 < x$ and $x \leq 7$ **14.** $x \leq -4$ or $x > 8$

15. $x < -2$ or $x > 10$ **16.** $x \leq 0$ and $x \geq 5$

17. $x \geq 0$ and $x \leq 0$ **18.** $x > 0$ or $x < 0$

19. $(-2, 3)$ **20.** $[-5, 5]$

21. $[4, 8)$ **22.** $(-10, -2]$

23. $(2, 6) \cup (1, 5)$ **24.** $(3, 5] \cup [4, 6]$

25. $[-2, 2) \cap [1, 2]$ **26.** $(-5, -1) \cap (-3, 0)$

27. $(-5, -3] \cup [-3, -2)$ **28.** $[-5, 2) \cap (-2, 0]$

29. $\{x : x \neq 0\} \cap (-3, 2]$ **30.** $\{x : x > 4\} \cup [0, 4)$

In Exercises 31–36, find the value of each expression.

31. $|-7|$ **32.** $-|-6|$ **33.** $|7| - |-6|$ **34.** $|8 - 5| + |5 - 8|$

35. $|\sqrt{2} - 1|$ **36.** $|1 - \pi|$

In Exercises 37–48, assume that $x = -5$, $y = -2$, and $z = 3$. Find the value of each expression.

37. $|xy|$ **38.** $|x + y| + z$ **39.** $x|yz|$ **40.** $x|y - z|$

41. $|x - y|z$ **42.** $|y| + |xz|$ **43.** $\dfrac{|y - z|}{x}$ **44.** $\dfrac{|x + 2y|}{z}$

45. $\dfrac{x|yz|}{yz}$ **46.** $\dfrac{3y|2x + 4z|}{yz}$ **47.** $\dfrac{|x + z| \neg |y - z|}{|x + y|}$ **48.** $\dfrac{|x + y + z|}{|xy| - |yz| + |xz|}$

In Exercises 49–50, assume that $x = -5$, $y = -2$, and $z = -3$.

49. Find the distance on the number line between two points with coordinates of x and y.

50. Find the distance on the number line between two points with coordinates of x and z.

In Exercises 51–58, tell if the given relation is reflexive, if it is symmetric, and if it is transitive.

51. $>$ **52.** $\geq$ **53.** $\leq$ **54.** $\neq$

55. $\cong$ ("is congruent to," from geometry)   **56.** "is divisible by" on the set of natural numbers

57. "is the same color as"   **58.** "is taller than"

In Exercises 59–74, indicate which field property or property of equality justifies each given expression.

59. $(8 + a) + b = b + (8 + a)$ **60.** $3(x + y) = 3x + 3y$

61. $\dfrac{2}{3}\left(-\dfrac{3}{4}\right)$ is a real number.   **62.** $2 + 0 = 2$

63. If $3 = a$ and $a = c$, then $3 = c$.   **64.** $\dfrac{2}{5}\left(\dfrac{5}{2}\right) = 1$

65. $5 = 5$ **66.** $(a + b)c = c(a + b)$

67. $-5 + (-8)$ is a real number.   **68.** $3(4 + t) = 3 \cdot 4 + 3t$

69. $a + (b + c) = a + (c + b)$ **70.** $a(bc) = (bc)a$

71. If $a = b$, then $b = a$.   **72.** $3(1) = 3$

73. $[a(b + c)]d = a[(b + c)d]$ **74.** $5 + 4 + (2 + 1) = 5 + (4 + 2) + 1$

1.3   INTEGER EXPONENTS AND SCIENTIFIC NOTATION

When two or more quantities are multiplied together, each quantity is called a **factor** of the product. The expression x^4 indicates that x is to be used as a factor four times. Thus,

$$x^4 = x \cdot x \cdot x \cdot x$$

In general,

Definition.   For any natural number n,

$$x^n = \overbrace{x \cdot x \cdot x \cdot \cdots \cdot x}^{n \text{ factors of } x}$$

The number x in the **exponential expression** x^n is called the **base** and the number n is called the **exponent** or the **power** to which the base is raised. The expression x^n is also called a **power of x**. A natural number exponent tells how many times the base of an exponential expression is to be used as a factor in a product.

Example 1   **a.** $4^2 = 4 \cdot 4 = 16$ Read 4^2 as "four squared."

b. $5^3 = 5 \cdot 5 \cdot 5 = 125$ Read 5^3 as "five cubed."

c. $(-5)^3 = -5(-5)(-5) = -125$

d. $3x^4 = 3 \cdot x \cdot x \cdot x \cdot x$ Read x^4 as "x to the fourth power."

e. $(3x)^4 = 3x(3x)(3x)(3x) = 81x^4$ ■

It is important to note the distinction between ax^n and $(ax)^n$, and between $-x^n$ and $(-x)^n$:

$$ax^n = a \cdot \overbrace{x \cdot x \cdot x \cdot \cdots \cdot x}^{n \text{ factors of } x} \qquad -x^n = -\overbrace{(x \cdot x \cdot x \cdot \cdots \cdot x)}^{n \text{ factors of } x}$$

$$(ax)^n = \overbrace{(ax)(ax)(ax) \cdot \cdots \cdot (ax)}^{n \text{ factors of } ax} \qquad (-x)^n = \overbrace{(-x)(-x)(-x) \cdot \cdots \cdot (-x)}^{n \text{ factors of } -x}$$

We begin to develop the rules of exponents by considering the product $x^m x^n$. Because x^m indicates that x is to be used as a factor m times and because x^n indicates that x is to be used as a factor n times, there are $m + n$ factors of x in the product $x^m x^n$.

$$x^m x^n = \overbrace{\underbrace{x \cdot x \cdot x \cdot \cdots \cdot x}_{m \text{ factors of } x} \underbrace{x \cdot x \cdot x \cdot \cdots \cdot x}_{n \text{ factors of } x}}^{m + n \text{ factors of } x} = x^{m+n}$$

Thus, to multiply exponential expressions with the same base, we keep the same base and add the exponents.

The Product Rule for Exponential Expressions.   If m and n are natural numbers, then

$$x^m x^n = x^{m+n}$$

The product rule applies only to exponential expressions with the same base. A product of two powers with different bases, such as $x^4 y^3$, cannot be simplified.

To find another property of exponents, we consider the exponential expression $(x^m)^n$. The exponent n indicates that x^m is to be used as a factor n times. This implies that x is to be used as a factor mn times.

$$\overbrace{(x^m)^n = \underbrace{(x^m)(x^m)(x^m) \cdots\cdots (x^m)}_{n \text{ factors of } x^m} = x^{mn}}^{mn \text{ factors of } x}$$

Thus, to raise an exponential expression to a power, we keep the same base and multiply the exponents.

To raise a product to a power, we raise each factor to that power.

$$\overbrace{(xy)^n = \underbrace{(xy)(xy)(xy)\cdots\cdots(xy)}_{n \text{ factors of } xy}}^{n \text{ factors of } xy} = \underbrace{(x\cdot x\cdot x\cdots\cdots x)}_{n \text{ factors of } x}\underbrace{(y\cdot y\cdot y\cdots\cdots y)}_{n \text{ factors of } y} = x^n y^n$$

To raise a fraction to a power, we raise both the numerator and the denominator to that power. If $y \neq 0$, then

$$\left(\frac{x}{y}\right)^n = \overbrace{\left(\frac{x}{y}\right)\left(\frac{x}{y}\right)\left(\frac{x}{y}\right)\cdots\cdots\left(\frac{x}{y}\right)}^{n \text{ factors of } \frac{x}{y}}$$

$$= \frac{\overbrace{x\,x\,x\cdots\cdots x}^{n \text{ factors of } x}}{\underbrace{y\,y\,y\cdots\cdots y}_{n \text{ factors of } y}}$$

$$= \frac{x^n}{y^n}$$

The previous three results are often called **power rules of exponents.**

The Power Rules of Exponents.   If m and n are natural numbers, then

$$(x^m)^n = x^{mn} \qquad (xy)^n = x^n y^n \qquad \left(\frac{x}{y}\right)^n = \frac{x^n}{y^n} \quad (y \neq 0)$$

Example 2   **a.** $x^5 x^7 = x^{5+7} = x^{12}$ **b.** $x^2 y^3 x^5 y^1 = x^{2+5} y^{3+1} = x^7 y^4$

c. $(x^4)^9 = x^{4 \cdot 9} = x^{36}$ **d.** $(x^2 x^5 y^5)^3 = (x^7 y^5)^3 = (x^7)^3 (y^5)^3$

$$= x^{21} y^{15}$$

e. $\left(\dfrac{x}{y^2}\right)^5 = \dfrac{x^5}{(y^2)^5}$ **f.** $\left(\dfrac{5x^2 y}{z^3}\right)^2 = \dfrac{5^2 (x^2)^2 y^2}{(z^3)^2}$

$$= \dfrac{x^5}{y^{10}} \quad (y \neq 0)$$ $$= \dfrac{25 x^4 y^2}{z^6} \quad (z \neq 0)$$ ∎

Until now we have defined only natural-number exponents. We can, however, extend the definition to include other exponents. For example, if we assume that the rules for natural-number exponents hold for exponents of 0, we can write

$$x^0 x^n = x^{0+n} = x^n = 1 x^n$$

Because $x^0 x^n = 1 x^n$, it follows that if $x \neq 0$, then $x^0 = 1$. Thus, we make the following definition:

Definition.   If $x \neq 0$, then

$$x^0 = 1$$

Furthermore, if we assume that the rules for natural-number exponents hold for exponents that are negative integers, we can write

$$x^{-n} x^n = x^{-n+n} = x^0 = 1$$

provided that $x \neq 0$. However, because $\dfrac{1}{x^n} \cdot x^n = \dfrac{x^n}{x^n}$ and any nonzero number divided by itself is 1, we have

$$\frac{1}{x^n} \cdot x^n = 1 \qquad (x \neq 0)$$

Because $x^{-n} x^n$ and $\dfrac{1}{x^n} \cdot x^n$ both equal 1, we have $x^{-n} x^n = \dfrac{1}{x^n} \cdot x^n$ and it follows that if $x \neq 0$, then $x^{-n} = \dfrac{1}{x^n}$. Thus, we make the following definition.

Definition.   If n is an integer and $x \neq 0$, then

$$x^{-n} = \frac{1}{x^n} \qquad \text{and} \qquad \frac{1}{x^{-n}} = x^n$$

Because of the two previous definitions, all of the rules for natural-number exponents hold for integer exponents.

Example 3   **a.** $(3x)^0 = 1$ **b.** $3(x^0) = 3(1) = 3$

c. $x^{-4} = \dfrac{1}{x^4}$ **d.** $\dfrac{1}{x^{-6}} = x^6$

e. $x^{-3}x = x^{-3+1}$ **f.** $(x^{-4}x^8)^{-5} = (x^4)^{-5}$

$\qquad\quad = x^{-2}$ $\qquad\qquad\quad = x^{-20}$

$\qquad\quad = \dfrac{1}{x^2}$ $\qquad\qquad\quad = \dfrac{1}{x^{20}}$ ■

To develop the quotient rule for exponents, we proceed as follows:

$$\frac{x^m}{x^n} = x^m\left(\frac{1}{x^n}\right) = x^m x^{-n} = x^{m+(-n)} = x^{m-n} \qquad (x \neq 0)$$

Thus, to divide two exponential expressions with the same nonzero base, we keep the base and subtract the exponents.

The Quotient Rule for Exponential Expressions.   If m and n are integers and $x \neq 0$, then

$$\frac{x^m}{x^n} = x^{m-n}$$

Example 4   **a.** $\dfrac{x^8}{x^5} = x^{8-5}$ **b.** $\dfrac{x^{-6}}{x^2} = x^{-6-2}$

$\qquad\qquad = x^3$ $\qquad\qquad = x^{-8}$

$\qquad\qquad\qquad\qquad\qquad\qquad\qquad = \dfrac{1}{x^8}$

c. $\dfrac{x^2 x^4}{x^{-5}} = \dfrac{x^6}{x^{-5}}$ **d.** $\dfrac{x^m x^2}{x^3} = \dfrac{x^{m+2}}{x^3}$

$\qquad\quad = x^{6-(-5)}$ $\qquad\quad = x^{m+2-3}$

$\qquad\quad = x^{11}$ $\qquad\quad = x^{m-1}$

e. $\left(\dfrac{x^3 y^{-2}}{x^{-2} y^3}\right)^{-2} = (x^{3-(-2)} y^{-2-3})^{-2}$ **f.** $\left(\dfrac{x}{y}\right)^{-n} = \dfrac{x^{-n}}{y^{-n}}$

$\qquad\qquad\qquad = (x^5 y^{-5})^{-2}$ $= \dfrac{x^{-n} x^n y^n}{y^{-n} x^n y^n}$

$\qquad\qquad\qquad = x^{-10} y^{10}$ $= \dfrac{x^0 y^n}{y^0 x^n}$

$\qquad\qquad\qquad = \dfrac{y^{10}}{x^{10}}$ $= \dfrac{y^n}{x^n}$

$\qquad\qquad\qquad\qquad\qquad\qquad\qquad\qquad = \left(\dfrac{y}{x}\right)^n$ ■

Part **f** of Example 4 establishes this useful theorem.

> **Theorem.**   If n is a natural number and neither x nor y is 0, then
> $$\left(\frac{x}{y}\right)^{-n} = \left(\frac{y}{x}\right)^{n}$$

If several mathematical operations occur in one expression, the order in which those operations are performed might change the result. To avoid conflicting answers, arithmetic operations should be performed in the following manner:

> **Order of Operations.**   If an expression does not contain grouping symbols such as parentheses or brackets, follow these steps:
>
> **1.**   Find the values of any exponential expressions.
> **2.**   Do all multiplications and/or divisions as they are encountered while working from left to right.
> **3.**   Do all additions and/or subtractions as they are encountered while working from left to right.
>
> If an expression contains grouping symbols, use the rules above to perform the calculations within each pair of grouping symbols, working from the innermost pair to the outermost pair.
>
> In a fraction, simplify the numerator and the denominator separately. Then simplify the fraction, whenever possible.

Example 5   If $x = -2$, $y = 3$, and $z = -4$, find the value of   **a.** $-x^2 + y^2 z$ and **b.** $\dfrac{2z^3 - 3y^2}{5x^2}$.

Solution   **a.** $-x^2 + y^2 z = -(-2)^2 + 3^2(-4)$

$$\begin{aligned}
&= -(4) + 9(-4) && \text{Evaluate the powers.}\\
&= -4 + (-36) && \text{Do the multiplication.}\\
&= -40 && \text{Do the addition.}
\end{aligned}$$

b. $\dfrac{2z^3 - 3y^2}{5x^2} = \dfrac{2(-4)^3 - 3(3)^2}{5(-2)^2}$

$$\begin{aligned}
&= \frac{2(-64) - 3(9)}{5(4)} && \text{Evaluate the powers.}\\[4pt]
&= \frac{-128 - 27}{20} && \text{Do the multiplications.}\\[4pt]
&= \frac{-155}{20} && \text{Do the subtraction.}\\[4pt]
&= -\frac{31}{4} && \text{Simplify the fraction.}
\end{aligned}$$

∎

Scientific Notation

Scientists and engineers often work with numbers that are either very large or very small. These numbers can be written compactly by expressing them in **scientific notation**.

> **Definition.**   A number is written in **scientific notation** if it is written as the product of a number between 1(including 1) and 10, and an appropriate power of 10.

Example 6   Light travels 29,980,000,000 centimeters per second. Express this number in scientific notation.

Solution   To write the number 29,980,000,000 in scientific notation, you must express it as a product of a number between 1 and 10 and some power of 10. The number 2.998 lies between 1 and 10. To get the number 29,980,000,000, the decimal point in 2.998 must be moved ten places to the right. This is accomplished by multiplying 2.998 by 10^{10}. The number 29,980,000,000 written in scientific notation is 2.998×10^{10}.   ∎

Example 7   One meter is approximately 0.00062 miles. Express this number in scientific notation.

Solution   To write the number 0.00062 in scientific notation, express it as a product of a number between 1 and 10, and some power of 10. The number 0.00062 may be obtained by moving the decimal point in 6.2 four places to the left. This is accomplished by multiplying 6.2 by 10^{-4}, because

$$6.2 \times 10^{-4} = 6.2 \times \frac{1}{10^4} = \frac{6.2}{10,000} = 0.00062$$

The number 0.00062 written in scientific notation is 6.2×10^{-4}.   ∎

Example 8   Express 3.27×10^{-5} in standard notation.

Solution   The factor of 10^{-5} indicates that 3.27 is divided by 5 factors of 10. Because each division by 10 moves the decimal point one place to the left, you must move the decimal point in the number 3.27 five places to the left. In standard notation, 3.27×10^{-5} is 0.0000327.   ∎

Study each of the following numbers written in both scientific and standard notation. In each case, note that the exponent gives the number of places the decimal point moves, and the sign of the exponent indicates the direction in which it moves.

a. $3.72 \times 10^5 = 3\,7\,2\,0\,0\,0.$ 5 places to the right

b. $9.93 \times 10^9 = 9\,9\,3\,0\,0\,0\,0\,0\,0\,0.$ 9 places to the right

c. $5.37 \times 10^{-4} = 0.\underset{\smile}{0\,0\,0\,5\,3\,7}$ 4 places to the left

d. $1.529 \times 10^{-1} = 0.\underset{\smile}{1\,5\,2\,9}$ 1 place to the left

e. $7.36 \times 10^{0} = 7.36$ No movement of the decimal point

Example 9   Use scientific notation to calculate $\dfrac{(3,400,000)(0.00002)}{170,000,000}$.

Solution
$$\frac{(3,400,000)(0.00002)}{170,000,000} = \frac{(3.4 \times 10^{6})(2.0 \times 10^{-5})}{1.7 \times 10^{8}}$$

$$= \frac{6.8}{1.7} \times 10^{6+(-5)-8}$$

$$= 4.0 \times 10^{-7}$$

$$= 0.0000004 \qquad \blacksquare$$

EXERCISE 1.3

In Exercises 1–56, simplify each expression. If the base is a variable, express all answers with positive exponents only. Assume that all variables are restricted to those numbers for which the expression is defined.

1. 13^2

2. 10^3

3. 5^0

4. 0^0

5. $4x^0$

6. $(4x)^0$

7. $-x^0$

8. $(-x)^0$

9. x^2x^3

10. y^3y^4

11. $(z^2)^3$

12. $(t^6)^7$

13. $(y^5y^2)^3$

14. $(a^3a^6)a^4$

15. $(z^2)^3(z^4)^5$

16. $(t^3)^4(t^5)^2$

17. $(3x)^3$

18. $(-2y)^4$

19. $(x^2y)^3$

20. $(x^3z^4)^6$

21. $\left(\dfrac{a^2}{b}\right)^3$

22. $\left(\dfrac{x}{y^3}\right)^4$

23. z^{-4}

24. $\dfrac{1}{t^{-2}}$

25. $y^{-2}y^{-3}$

26. $-m^{-2}m^3$

27. $(x^3x^{-4})^{-2}$

28. $(y^{-2}y^3)^{-4}$

29. $\dfrac{(a^3)^{-2}}{aa^2}$

30. $\dfrac{r^9r^{-3}}{(r^{-2})^3}$

31. $\left(\dfrac{a^{-3}}{b^{-1}}\right)^{-4}$

32. $\left(\dfrac{t^{-4}}{t^{-3}}\right)^{-2}$

33. $\left(\dfrac{r^4r^{-6}}{r^3r^{-3}}\right)^2$

34. $\dfrac{(x^{-3}x^2)^2}{(x^2x^{-5})^{-3}}$

35. $\left(\dfrac{x^5y^{-2}}{x^{-3}y^2}\right)^4$

36. $\left(\dfrac{2x^{-7}y^5}{x^7y^{-4}}\right)^3$

37. $\left(\dfrac{5x^{-3}y^{-2}}{3x^2y^{-2}}\right)^{-2}$

38. $\left(\dfrac{3x^{-2}y^{-5}}{2x^{-2}y^{-6}}\right)^{-3}$

39. $\left(\dfrac{x^2+3x+y}{2x^5-y}\right)^0$

40. $\left(\dfrac{x^0+y^0-z^0}{x^0-y^0+z^0}\right)^6$

41. $\left(\dfrac{3x^5y^{-3}}{6x^{-5}y^3}\right)^{-2}$

42. $\left(\dfrac{12x^{-4}y^3z^{-5}}{4x^4y^{-3}z^5}\right)^3$

43. $(-x^2y^x)^5$

44. $(-x^2y^y)^6$

45. $-(x^2y^3)^{4xy}$

46. $-(x^2y)^{3xy}$

47. $x^{m+1}x^{3-m}$

48. $y^{n-3}y^{3-n}$

49. $(x^{2n}x^{3n})^n$

50. $(a^{-3m}a^m)^{-m}$

51. $\dfrac{x^{3m+5}}{x^{3m}}$

52. $\dfrac{12x^{m+3}}{3x^m}$

53. $\dfrac{(8^{-2}z^{-3}y)^{-1}}{(5y^2z^{-2})^3(5yz^{-2})^{-1}}$

54. $\dfrac{(m^{-2}n^3p^4)^{-2}(mn^{-2}p^3)^4}{(mn^{-2}p^3)^{-4}(mn^2p)^{-1}}$

55. $\left[\dfrac{(m^{-2}n^{-1}p^3)^{-2}}{(mn^2)^{-3}(p^{-3})^4}\right]^{-2}$

56. $\left[\dfrac{(3x^2)^{-4}(3p)^2}{(x^{-2}p^4)^{-3}(3x^{-3})^{-2}}\right]^{-3}$

In Exercises 57–74, let $x = -2$, $y = 0$, and $z = 3$, and evaluate each expression.

57. x^2

58. $-x^2$

59. x^3

60. $-x^3$

61. $(-xz)^3$

62. $-xz^3$

63. $\dfrac{-(x^2z^3)}{z^2 - y^2}$

64. $\dfrac{z^2(x^2 - y^2)}{x^3z}$

65. $5x^2 - 3y^3z$

66. $3(x - z)^2 + 2(y - z)^3$

67. $3x^3y^7z^{15}$

68. $5x^{-2}z^3 + y^3$

69. $\dfrac{-3x^{-3}z^{-2}}{6x^2z^{-3}}$

70. $\dfrac{(-5x^2z^{-3})^2}{3xz^2}$

71. $x^zz^xz^y$

72. $(x^yz^y)^{-z}$

73. $\dfrac{x^xz^zx^y}{-x^2z^2}$

74. $\dfrac{x^yz^x}{x^{x+z}}$

In Exercises 75–88, express each number in scientific notation.

75. 372,000

76. 89,500

77. 177,000,000

78. 23,470,000,000

79. 0.007

80. 0.00052

81. 0.000000693

82. 0.000000089

83. one trillion

84. sixty-three billion

85. one trillionth

86. forty-three billionths

87. 99.7×10^{-4}

88. 0.0085×10^5

In Exercises 89–94, express each number in standard notation.

89. 9.37×10^5

90. 4.26×10^9

91. 2.21×10^{-5}

92. 2.774×10^{-2}

93. 0.00032×10^4

94. 9300.0×10^{-4}

In Exercises 95–98, use the method of Example 9 to perform each calculation. Write all answers in scientific notation.

95. $\dfrac{(65,000)(45,000)}{250,000}$

96. $\dfrac{(0.000000045)(0.00000012)}{45,000,000}$

97. $\dfrac{(0.00000035)(170,000)}{0.00000085}$

98. $\dfrac{(0.0000000144)(12,000)}{600,000}$

99. The speed of sound (in air) is 3.31×10^4 centimeters per second. Use scientific notation to compute the speed of sound in meters per minute.

100. Calculate the volume of a box that has dimensions of 6000 by 9700 by 4700 millimeters. Use scientific notation to perform the calculation, and express the answer in scientific notation.

101. The mass of one proton is 0.0000000000000000000000000167248 gram. Find the mass of one billion protons.

102. The speed of light (in a vacuum) is approximately 30,000,000,000 centimeters per second. Find the speed of light in miles per hour, and express your answer in scientific notation. There are 160,934.4 centimeters in one mile.

In Exercises 103–108, use the calculator to work each problem.

103. Show that $4.57^0 = 1$.

104. Show that $(1.2)^3(3.2)^3 = [(1.2)(3.2)]^3$.

105. Show that $(4.1)^2 + (5.2)^2 \neq (4.1 + 5.2)^2$.

106. Show that $[(3.7)^2]^3 = (3.7)^6$.

107. Show that $(3.2)^{1.2}(3.2)^{2.1} = (3.2)^{3.3}$.

108. Show that $4.75^{-0.2} = \dfrac{1}{4.75^{0.2}}$.

1.4 FRACTIONAL EXPONENTS AND RADICALS

The rules for integral exponents can be extended to include rational (fractional) exponents if we assume that some base a is nonnegative and define the exponential expression $a^{1/n}$ in the following way:

> **Definition.**   If n is a natural number and $a \ge 0$, then $a^{1/n}$ is the nonnegative number whose nth power is a. In symbols,
>
> $$(a^{1/n})^n = a$$
>
> The nonnegative number $a^{1/n}$ is called the **principal nth root of a**.

Example 1   **a.** $16^{1/2} = 4$ because $4^2 = 16$.   Read $16^{1/2}$ as "the square root of 16."

b. $27^{1/3} = 3$ because $3^3 = 27$.   Read $27^{1/3}$ as "the cube root of 27."

c. $\left(\dfrac{1}{81}\right)^{1/4} = \dfrac{1}{3}$ because $\left(\dfrac{1}{3}\right)^4 = \dfrac{1}{81}$.   Read $(\frac{1}{81})^{1/4}$ as "the fourth root of $\frac{1}{81}$."

d. $(64a^6)^{1/3} = 4a^2$ because $(4a^2)^3 = 64a^6$.

e. $-32^{1/5} = -(32^{1/5}) = -(2) = -2$ Read $32^{1/5}$ as "the fifth root of 32."

f. $(-25)^{1/2}$ is not a real number because the square of no real number is -25. This illustrates that if n is even and a is negative in the exponential expression $a^{1/n}$, there is no real number that is the nth root of a.   ∎

If n is an odd natural number in the expression $a^{1/n}$, we can remove the restriction that $a \ge 0$.

> **Definition.**   If n is an odd natural number and a is any real number, then $a^{1/n}$ is the real number whose nth power is a. In symbols,
>
> $$(a^{1/n})^n = a$$

Example 2   **a.** $(-8)^{1/3} = -2$ because $(-2)^3 = -8$.

b. $15{,}625^{1/6} = 5$ because $5^6 = 15{,}625$.

c. $(-32x^5)^{1/5} = -2x$ because $(-2x)^5 = -32x^5$.

d. $(256x^{16})^{1/4} = 4x^4$ because $(4x^4)^4 = 256x^{16}$.   ∎

The following chart summarizes the definitions concerning the exponential expression $a^{1/n}$.

If n is a natural number and a is a real number, then

If $a > 0$, then $a^{1/n}$ is the positive number such that $(a^{1/n})^n = a$.

If $a = 0$, then $a^{1/n} = 0$.

If $a < 0$ $\begin{cases} \text{and } n \text{ is odd, then } a^{1/n} \text{ is the real number such that} \\ \quad (a^{1/n})^n = a. \\ \text{and } n \text{ is even, then } a^{1/n} \text{ is not a real number.} \end{cases}$

All positive numbers have two square roots. For example, the two square roots of 121 are 11 and -11, because 11^2 and $(-11)^2$ both equal 121. The symbol $121^{1/2}$ represents the positive square root of 121:

$$121^{1/2} = 11$$

and the symbol $-121^{1/2}$ represents the negative square root of 121:

$$-121^{1/2} = -(121^{1/2}) = -11$$

In general, if $a \neq 0$, the number a^2 has a and $-a$ as its two square roots. One of these square roots is positive and the other is negative. Because the expression $(a^2)^{1/2}$ represents the positive square root, we cannot write $(a^2)^{1/2} = a$ if a might be negative. Instead, we must write

$$(a^2)^{1/2} = |a|$$

The absolute value symbols are needed to guarantee that the square root is positive. For example, if $a = -5$, we have

$$[(-5)^2]^{1/2} = |-5| = 5$$

Example 3   **a.** If $x \geq 0$, find $(25x^2)^{1/2}$,   **b.** If $x < 0$, find $(36x^2)^{1/2}$, and   **c.** If x is unrestricted, find $(49x^2)^{1/2}$.

Solution   **a.** If $x \geq 0$, then

$$(25x^2)^{1/2} = 5x$$

because $(5x)^2 = 25x^2$, and because $5x$ is positive, no absolute value symbols are needed.

b. If $x < 0$, then

$$(36x^2)^{1/2} = -6x$$

because $(-6x)^2 = 36x^2$, and because x is negative, $-6x$ is positive.

c. If x is unrestricted, then

$$(49x^2)^{1/2} = 7|x|$$

Because x is unrestricted, $7x$ could be negative. In this case, you must use absolute value symbols to guarantee that the square root is positive. ∎

The definition of $a^{1/n}$ can be extended to include rational exponents whose numerators are not 1. For example, $4^{3/2}$ can be written as either

$$(4^{1/2})^3 \qquad \text{or} \qquad (4^3)^{1/2}$$

In general, we have the following rule.

If m and n are positive integers and the fraction $\frac{m}{n}$ has been simplified to lowest terms, then

$$a^{m/n} = (a^{1/n})^m = (a^m)^{1/n}$$

Thus, we can interpret the expression $a^{m/n}$ in two different ways:

1. $a^{m/n}$ means the mth power of the nth root of a.
2. $a^{m/n}$ means the nth root of the mth power of a.

For example, $16^{3/4}$ and $(-27)^{2/3}$ can be interpreted as

$$16^{3/4} = (16^{1/4})^3 = 2^3 = 8 \qquad\qquad (-27)^{2/3} = [(-27)^{1/3}]^2 = (-3)^2 = 9$$

or as

$$16^{3/4} = (16^3)^{1/4} = (4096)^{1/4} = 8 \qquad (-27)^{2/3} = [(-27)^2]^{1/3} = (729)^{1/3} = 9$$

Either way, the results are the same. As the previous examples suggest, however, it is usually easier to take the root of the base first in order to avoid large numbers.

Because of the definition of negative exponents, we define $a^{-m/n}$ as follows:

Definition.   If m and n are positive integers, the fraction $\frac{m}{n}$ has been simplified to lowest terms, and $a \neq 0$, then

$$a^{-m/n} = \frac{1}{a^{m/n}} \qquad \text{and} \qquad \frac{1}{a^{-m/n}} = a^{m/n}$$

Example 4   **a.** $25^{3/2} = (25^{1/2})^3 = 5^3 = 125$

b. $\left(-\dfrac{x^6}{1000}\right)^{2/3} = \left[\left(-\dfrac{x^6}{1000}\right)^{1/3}\right]^2 = \left(-\dfrac{x^2}{10}\right)^2 = \dfrac{x^4}{100}$

c. $32^{-2/5} = \dfrac{1}{32^{2/5}} = \dfrac{1}{(32^{1/5})^2} = \dfrac{1}{2^2} = \dfrac{1}{4}$

d. $\dfrac{1}{81^{-3/4}} = 81^{3/4} = (81^{1/4})^3 = 3^3 = 27$

∎

The rules of integral exponents also hold for rational exponents. The following list illustrates the use of each rule.

a. $a^m a^n = a^{m+n}$ $5^{2/7} 5^{3/7} = 5^{2/7+3/7} = 5^{5/7}$

b. $(a^m)^n = a^{mn}$ $(7^{2/5})^6 = 7^{(2/5)6} = 7^{12/5}$

c. $(ab)^n = a^n b^n$ $(7 \cdot 3)^{5/2} = 7^{5/2} 3^{5/2}$

d. $\dfrac{a^m}{a^n} = a^{m-n}$ $\dfrac{5^{5/7}}{5^{3/7}} = 5^{5/7-3/7} = 5^{2/7}$

e. $\left(\dfrac{a}{b}\right)^n = \dfrac{a^n}{b^n}$ $\left(\dfrac{3}{5}\right)^{2/7} = \dfrac{3^{2/7}}{5^{2/7}}$

f. $a^{-n} = \dfrac{1}{a^n}$ $9^{-2/5} = \dfrac{1}{9^{2/5}}$

g. $\dfrac{1}{a^{-n}} = a^n$ $\dfrac{1}{3^{-2/5}} = 3^{2/5}$

h. $a^0 = 1$ $(3.75)^0 = 1$

The next example shows how to use the rules of exponents to simplify exponential expressions with variables in their bases.

Example 5 Assume that all variables represent positive numbers. Write all answers without using negative exponents.

a. $(36x)^{1/2} = 36^{1/2} x^{1/2}$
$$= 6x^{1/2}$$

b. $\dfrac{(a^{1/3} b^{2/3})^6}{(y^3)^2} = \dfrac{a^{6/3} b^{12/3}}{y^6}$
$$= \dfrac{a^2 b^4}{y^6}$$

c. $\dfrac{b^{3/7} b^{2/7}}{b^{4/7}} = b^{3/7+2/7-4/7}$
$$= b^{1/7}$$

d. $\dfrac{a^{x/2} a^{x/4}}{a^{x/6}} = a^{x/2+x/4-x/6}$
$$= a^{6x/12+3x/12-2x/12}$$
$$= a^{7x/12}$$

e. $\left[\dfrac{-c^{-2/5}}{c^{4/5}}\right]^{5/3} = (-c^{-2/5-4/5})^{5/3}$
$$= (-c^{-6/5})^{5/3}$$
$$= (-c)^{-30/15}$$
$$= (-c)^{-2}$$
$$= \dfrac{1}{(-c)^2}$$
$$= \dfrac{1}{c^2}$$

f. $\dfrac{(9r^2 s)^{1/2}}{rs^{-3/2}} = \dfrac{(9r^2 s)^{1/2} s^{3/2}}{r}$
$$= \dfrac{9^{1/2}(r^2)^{1/2} s^{1/2} s^{3/2}}{r}$$
$$= \dfrac{3rs^{1/2+3/2}}{r}$$
$$= 3r^{1-1} s^{4/2}$$
$$= 3r^0 s^2$$
$$= 3s^2$$

Radical Expressions

Another notation for expressing roots of numbers uses a symbol called the *radical sign*.

Definition.   If n is a natural number greater than 1 and if $a^{1/n}$ is a real number, then

$$\sqrt[n]{a} = a^{1/n}$$

The symbol $\sqrt[n]{a}$ is called a **radical expression** (or **radical**). In this expression, the symbol $\sqrt{}$ is called the **radical sign**, the number a is called the **radicand**, and n is called the **index** (or the **order**) of the radical. If the order of a radical is 2, the expression is called a **square root**, and we do not write the index. Thus,

$$\sqrt{a} = \sqrt[2]{a}$$

If the index of a radical is 3, the radical is called a **cube root**.

We now restate the definition for the nth root of a number, using radical notation.

Definition.   If n is a natural number greater than 1 and $a \geq 0$, then $\sqrt[n]{a}$ is the nonnegative number whose nth power is a. In symbols,

$$(\sqrt[n]{a})^n = a$$

The nonnegative number $\sqrt[n]{a}$ is called the **principal nth root of a**.

If 2 is substituted for n in the equation $(\sqrt[n]{a})^n = a$, we have

$$(\sqrt[2]{a})^2 = (\sqrt{a})^2 = \sqrt{a}\sqrt{a} = a$$

Thus, if a number a can be factored into two equal factors, either of those factors is a square root of a. Likewise, if a can be factored into n equal factors, any of those factors is an nth root of a.

In the expression $\sqrt[n]{a}$, if n is an odd natural number greater than 1, we can remove the restriction that $a \geq 0$.

Definition.   If n is an odd natural number greater than 1 and if a is any real number, then $\sqrt[n]{a}$ is the real number whose nth power is a. In symbols,

$$(\sqrt[n]{a})^n = a$$

Example 6   **a.** $\sqrt{81} = 9$ because $9^2 = 81$.

b. $\sqrt[3]{-125} = -5$ because $(-5)^3 = -125$.

c. $-\sqrt[4]{256} = -(\sqrt[4]{256}) = -(4) = -4$

d. $-\sqrt[5]{-243} = -(\sqrt[5]{-243}) = -(-3) = 3$ ■

The following box summarizes the definitions concerning the expression $\sqrt[n]{a}$.

If n is a natural number greater than 1 and a is a real number, then

If $a > 0$, then $\sqrt[n]{a}$ is the positive number such that $(\sqrt[n]{a})^n = a$.

If $a = 0$, then $\sqrt[n]{a} = 0$.

If $a < 0$ $\begin{cases} \text{and } n \text{ is odd, then } \sqrt[n]{a} \text{ is the real number such that } (\sqrt[n]{a})^n = a. \\ \text{and } n \text{ is even, then } \sqrt[n]{a} \text{ is not a real number.} \end{cases}$

We have seen that $a^{m/n} = (a^{1/n})^m = (a^m)^{1/n}$. This same fact, stated in radical notation, is

$$a^{m/n} = (\sqrt[n]{a})^m = \sqrt[n]{a^m} \qquad \text{provided } \sqrt[n]{a} \text{ is a real number}$$

Thus, the mth power of the nth root of a is the same as the nth root of mth power of a. For example, to find $\sqrt[3]{27^2}$, we can proceed in one of two ways:

$$\sqrt[3]{27^2} = (\sqrt[3]{27})^2 = 3^2 = 9 \qquad \text{or} \qquad \sqrt[3]{27^2} = \sqrt[3]{729} = 9$$

Either way, the result is the same.

Because $\sqrt{a^2} = (a^2)^{1/2}$, the symbol $\sqrt{a^2}$ represents a nonnegative number. Thus, if a is not restricted,

$$\sqrt{a^2} = |a|$$

A similar argument holds when the index is any even natural number. The symbol $\sqrt[4]{a^4}$, for example, means the *nonnegative* fourth root of a^4. Thus, if a is not restricted,

$$\sqrt[4]{a^4} = |a|$$

Example 7   If x could be any real number, simplify   **a.** $\sqrt[6]{x^6}$,   **b.** $\sqrt[3]{x^3}$, and   **c.** $\sqrt{x^8}$.

Solution   **a.** Because the symbol $\sqrt[6]{x^6}$ represents the positive sixth root of x^6, the sixth root cannot be negative. Thus, you must use absolute value symbols to guarantee that the result is nonnegative.

$$\sqrt[6]{x^6} = |x|$$

b. Because the index is odd, no absolute value symbols are needed.

$$\sqrt[3]{x^3} = x$$

c. Because $\sqrt{x^8} = \sqrt{(x^4)^2}$ and x^4 is always positive, no absolute value symbols are needed.

$$\sqrt{x^8} = x^4$$

■

Simplifying Radicals

Many properties of exponents have counterparts in radical notation. For example, because $a^{1/n}b^{1/n} = (ab)^{1/n}$, we have

$$\sqrt[n]{a}\sqrt[n]{b} = \sqrt[n]{ab}$$

Thus, as long as all expressions represent real numbers, the product of the nth roots of two numbers is equal to the nth root of their product.

Furthermore, because

$$\frac{a^{1/n}}{b^{1/n}} = \left(\frac{a}{b}\right)^{1/n} \qquad (b \neq 0)$$

it follows that

$$\frac{\sqrt[n]{a}}{\sqrt[n]{b}} = \sqrt[n]{\frac{a}{b}} \qquad (b \neq 0)$$

Thus, as long as all expressions represent real numbers, the quotient of the nth roots of two numbers is equal to the nth root of their quotient.

Properties of Radicals. If $\sqrt[n]{a}$ and $\sqrt[n]{b}$ are real numbers, then

1. $\sqrt[n]{ab} = \sqrt[n]{a}\sqrt[n]{b}$

2. $\sqrt[n]{\frac{a}{b}} = \frac{\sqrt[n]{a}}{\sqrt[n]{b}} \qquad (b \neq 0)$

Properties 1 and 2 of radicals involve either the nth root of the product of two numbers or the nth root of the quotient of two numbers. There is no such property for sums or differences. For example,

$$\sqrt{9 + 4} \neq \sqrt{9} + \sqrt{4}$$

because

$$\sqrt{9 + 4} = \sqrt{13} \qquad \text{but} \qquad \sqrt{9} + \sqrt{4} = 3 + 2 = 5$$

and $\sqrt{13} \neq 5$. In general,

$$\sqrt{a + b} \neq \sqrt{a} + \sqrt{b} \qquad \text{and} \qquad \sqrt{a - b} \neq \sqrt{a} - \sqrt{b}$$

Numbers such as 1, 4, 9, 16, 25, and 36 that are squares of positive integers are called **perfect squares**. Numbers such as 1, 8, 27, 64, 125, and 216 that are cubes of positive integers are called **perfect cubes**. Sometimes expressions such as x^2 and x^4 are called perfect squares because each is the square of another expression. Likewise, x^3 and x^9 are perfect cubes. There are also perfect fourth powers, perfect fifth powers, and so on.

We can use these perfect powers and Property 1 of radicals to simplify many radical expressions. For example, to simplify the radical $\sqrt{12x^5}$, we factor $12x^5$

so that one factor is the largest perfect square that divides $12x^5$. In this case, the largest perfect square is $4x^4$. We then rewrite $12x^5$ as $4x^4 \cdot 3x$, use Property 1 of radicals, and simplify.

$$\sqrt{12x^5} = \sqrt{4x^4 \cdot 3x} = \sqrt{4x^4}\sqrt{3x} = 2x^2\sqrt{3x}$$

To simplify the radical $\sqrt[3]{432x^9y}$, we find the largest perfect cube factor of $432x^9y$, which is $216x^9$, and proceed as follows:

$$\sqrt[3]{432x^9y} = \sqrt[3]{216x^9 \cdot 2y} = \sqrt[3]{216x^9}\sqrt[3]{2y} = 6x^3\sqrt[3]{2y}$$

Two radical expressions with the same index and the same radicand are called **like** or **similar radicals** and can be combined. For example, to combine the like radicals in the expression $3\sqrt{2} + 2\sqrt{2}$, we use the distributive property, as follows:

$$3\sqrt{2} + 2\sqrt{2} = (3 + 2)\sqrt{2} = 5\sqrt{2}$$

Thus, to combine like radicals, we simply add their coefficients and keep the same radical.

If two radical expressions have the same index but different radicands, we can often adjust them so that they have the same radicand. They can then be combined. For example, to simplify the expression $\sqrt{27} - \sqrt{12}$, we simplify both radicals and combine like radicals.

$$\begin{aligned}
\sqrt{27} - \sqrt{12} &= \sqrt{9 \cdot 3} - \sqrt{4 \cdot 3} \\
&= \sqrt{9}\sqrt{3} - \sqrt{4}\sqrt{3} \\
&= 3\sqrt{3} - 2\sqrt{3} \\
&= \sqrt{3}
\end{aligned}$$

Example 8   Simplify   **a.** $\sqrt{50} + \sqrt{200}$,   **b.** $3z\sqrt[5]{64z} - 2\sqrt[5]{2z^6}$, and   **c.** $\dfrac{a\sqrt{16a} - \sqrt{a^3}}{\sqrt{9a} + \sqrt{36a}}$.

Assume that all variables represent positive numbers.

Solution   **a.** $\begin{aligned}[t]
\sqrt{50} + \sqrt{200} &= \sqrt{25 \cdot 2} + \sqrt{100 \cdot 2} \\
&= \sqrt{25}\sqrt{2} + \sqrt{100}\sqrt{2} \\
&= 5\sqrt{2} + 10\sqrt{2} \\
&= 15\sqrt{2}
\end{aligned}$

b. $\begin{aligned}[t]
3z\sqrt[5]{64z} - 2\sqrt[5]{2z^6} &= 3z\sqrt[5]{32 \cdot 2z} - 2\sqrt[5]{z^5 \cdot 2z} \\
&= 3z\sqrt[5]{32}\sqrt[5]{2z} - 2\sqrt[5]{z^5}\sqrt[5]{2z} \\
&= 3z(2)\sqrt[5]{2z} - 2z\sqrt[5]{2z} \\
&= 6z\sqrt[5]{2z} - 2z\sqrt[5]{2z} \\
&= 4z\sqrt[5]{2z}
\end{aligned}$

c. $\begin{aligned}[t]
\frac{a\sqrt{16a} - \sqrt{a^3}}{\sqrt{9a} + \sqrt{36a}} &= \frac{a\sqrt{16}\sqrt{a} - \sqrt{a^2}\sqrt{a}}{\sqrt{9}\sqrt{a} + \sqrt{36}\sqrt{a}} \\
&= \frac{4a\sqrt{a} - a\sqrt{a}}{3\sqrt{a} + 6\sqrt{a}}
\end{aligned}$

$$= \frac{3a\sqrt{a}}{9\sqrt{a}}$$

$$= \frac{a \cdot 3\sqrt{a}}{3 \cdot 3\sqrt{a}}$$

$$= \frac{a}{3} \qquad \frac{3\sqrt{a}}{3\sqrt{a}} = 1 \text{ because any nonzero}$$
number divided by itself is 1. ∎

We now consider radicals such as $\sqrt{\frac{5}{3}}$, whose radicands are fractions. Because of Property 2 of radicals, this radical can be written in the form $\sqrt{5}/\sqrt{3}$. There is a process, called **rationalizing the denominator**, that enables us to write this fraction as a fraction with a rational number in the denominator. All that we must do is multiply both the numerator and the denominator of the fraction by $\sqrt{3}$ to make the denominator a rational number.

$$\sqrt{\frac{5}{3}} = \frac{\sqrt{5}}{\sqrt{3}} = \frac{\sqrt{5}\sqrt{3}}{\sqrt{3}\sqrt{3}} = \frac{\sqrt{15}}{3}$$

Example 9 Rationalize each denominator and simplify, whenever possible. Assume that all variables represent positive numbers.

a. $\dfrac{1}{\sqrt{7}}$, **b.** $\sqrt{\dfrac{3}{x}}$, **c.** $\sqrt{\dfrac{3a^3}{5x^5}}$, **d.** $\sqrt{\dfrac{1}{2}} + \sqrt{\dfrac{1}{8}}$, **e.** $\sqrt[3]{\dfrac{1}{y^2}} - \sqrt[3]{\dfrac{16}{x^4}}$

Solution **a.** $\dfrac{1}{\sqrt{7}} = \dfrac{1\sqrt{7}}{\sqrt{7}\sqrt{7}} = \dfrac{\sqrt{7}}{7}$

b. $\sqrt{\dfrac{3}{x}} = \dfrac{\sqrt{3}}{\sqrt{x}} = \dfrac{\sqrt{3}\sqrt{x}}{\sqrt{x}\sqrt{x}} = \dfrac{\sqrt{3x}}{x}$

c. $\sqrt{\dfrac{3a^3}{5x^5}} = \dfrac{\sqrt{3a^3}}{\sqrt{5x^5}} = \dfrac{\sqrt{3a^3}\sqrt{5x}}{\sqrt{5x^5}\sqrt{5x}} = \dfrac{\sqrt{15a^3x}}{\sqrt{25x^6}} = \dfrac{\sqrt{a^2}\sqrt{15ax}}{5x^3} = \dfrac{a\sqrt{15ax}}{5x^3}$

d. $\sqrt{\dfrac{1}{2}} + \sqrt{\dfrac{1}{8}} = \dfrac{1}{\sqrt{2}} + \dfrac{1}{\sqrt{8}} = \dfrac{1\sqrt{2}}{\sqrt{2}\sqrt{2}} + \dfrac{1\sqrt{2}}{\sqrt{8}\sqrt{2}}$

$$= \dfrac{\sqrt{2}}{2} + \dfrac{\sqrt{2}}{\sqrt{16}} = \dfrac{\sqrt{2}}{2} + \dfrac{\sqrt{2}}{4} = \dfrac{3\sqrt{2}}{4}$$

e. $\sqrt[3]{\dfrac{1}{y^2}} - \sqrt[3]{\dfrac{16}{x^4}} = \dfrac{\sqrt[3]{1}}{\sqrt[3]{y^2}} - \dfrac{\sqrt[3]{16}}{\sqrt[3]{x^4}}$

$$= \dfrac{1}{\sqrt[3]{y^2}} - \dfrac{2\sqrt[3]{2}}{\sqrt[3]{x^4}} \qquad \sqrt[3]{16} = \sqrt[3]{8}\sqrt[3]{2} = 2\sqrt[3]{2}$$

$$= \dfrac{1\sqrt[3]{y}}{\sqrt[3]{y^2}\sqrt[3]{y}} - \dfrac{2\sqrt[3]{2}\sqrt[3]{x^2}}{\sqrt[3]{x^4}\sqrt[3]{x^2}}$$

$$= \frac{\sqrt[3]{y}}{\sqrt[3]{y^3}} - \frac{2\sqrt[3]{2x^2}}{\sqrt[3]{x^6}}$$

$$= \frac{\sqrt[3]{y}}{y} - \frac{2\sqrt[3]{2x^2}}{x^2} \qquad\qquad \blacksquare$$

Another property of radicals can be derived from the properties of exponents. If all of the expressions represent real numbers, then

$$\sqrt[n]{\sqrt[m]{x}} = \sqrt[n]{x^{1/m}} = (x^{1/m})^{1/n} = x^{1/(mn)} = \sqrt[mn]{x}$$

Similarly,

$$\sqrt[m]{\sqrt[n]{x}} = \sqrt[mn]{x}$$

These results are summarized in the following theorem.

Theorem.   If all of the expressions involved represent real numbers, then

$$\sqrt[m]{\sqrt[n]{x}} = \sqrt[n]{\sqrt[m]{x}} = \sqrt[mn]{x}$$

We can use the previous theorem to simplify many radicals. For example,

$$\sqrt[3]{\sqrt{8}} = \sqrt{\sqrt[3]{8}} = \sqrt{2}$$

Fractional exponents can be used to simplify many radical expressions, as shown in the following example.

Example 10   Assume that x and y are positive numbers and simplify   **a.** $\sqrt[6]{4}$,   **b.** $\sqrt[12]{x^3}$, and   **c.** $\sqrt[9]{8y^3}$.

Solution   **a.** $\sqrt[6]{4} = 4^{1/6} = (2^2)^{1/6} = 2^{2/6} = 2^{1/3} = \sqrt[3]{2}$

b. $\sqrt[12]{x^3} = x^{3/12} = x^{1/4} = \sqrt[4]{x}$

c. $\sqrt[9]{8y^3} = (2^3y^3)^{1/9} = (2y)^{3/9} = (2y)^{1/3} = \sqrt[3]{2y}$ $\blacksquare$

Multiplying and Dividing Radicals with Different Indexes

To multiply radicals with different indexes, we first write the factors as radicals with a common index. To do so, we write each radical as an exponential expression, change the exponents to equivalent fractions with a common denominator, and change the resulting exponential expression back to radical notation.

$$\sqrt{3} = 3^{1/2} = 3^{3/6} = (3^3)^{1/6} = \sqrt[6]{3^3} = \sqrt[6]{27}$$

$$\sqrt[3]{5} = 5^{1/3} = 5^{2/6} = (5^2)^{1/6} = \sqrt[6]{5^2} = \sqrt[6]{25}$$

To multiply $\sqrt{3}$ and $\sqrt[3]{5}$, we multiply the sixth roots.

$$\sqrt{3}\sqrt[3]{5} = \sqrt[6]{27}\sqrt[6]{25} = \sqrt[6]{(27)(25)} = \sqrt[6]{675}$$

Example 11 Simplify $\dfrac{\sqrt[3]{3}}{\sqrt{5}}$.

Solution Proceed as follows:

$$\frac{\sqrt[3]{3}}{\sqrt{5}} = \frac{\sqrt[3]{3}\sqrt{5}}{\sqrt{5}\sqrt{5}} = \frac{3^{1/3}5^{1/2}}{5} = \frac{3^{2/6}5^{3/6}}{5} = \frac{\sqrt[6]{(3^2)(5^3)}}{5} = \frac{\sqrt[6]{1125}}{5}$$

■

EXERCISE 1.4

In Exercises 1–20, simplify each expression. Write all answers without using negative exponents.

1. $9^{1/2}$ **2.** $-16^{1/2}$ **3.** $0^{1/7}$ **4.** $625^{1/4}$

5. $\left(\dfrac{1}{25}\right)^{1/2}$ **6.** $\left(-\dfrac{8}{27}\right)^{1/3}$ **7.** $4^{3/2}$ **8.** $(-8)^{2/3}$

9. $-1000^{2/3}$ **10.** $100^{3/2}$ **11.** $64^{-1/2}$ **12.** $25^{-1/2}$

13. $64^{-3/2}$ **14.** $49^{-3/2}$ **15.** $-9^{-3/2}$ **16.** $(-27)^{-2/3}$

17. $\left(\dfrac{4}{9}\right)^{5/2}$ **18.** $\left(\dfrac{25}{81}\right)^{3/2}$ **19.** $\left(-\dfrac{27}{64}\right)^{-2/3}$ **20.** $\left(\dfrac{125}{8}\right)^{-4/3}$

In Exercises 21–40, simplify each expression. Write all answers without using negative exponents. Assume that all variables represent positive numbers.

21. $(100s^4)^{1/2}$ **22.** $(64u^6v^3)^{1/3}$ **23.** $(32y^{10}z^5)^{-1/5}$ **24.** $(625a^4b^8)^{-1/4}$

25. $(x^{10}y^5)^{3/5}$ **26.** $(64a^6b^{12})^{5/6}$ **27.** $(r^8s^{16})^{-3/4}$ **28.** $(-8x^9y^{12})^{-2/3}$

29. $\left(-\dfrac{8a^6}{125b^9}\right)^{2/3}$ **30.** $\left(\dfrac{16x^4}{625y^8}\right)^{3/4}$ **31.** $\left(\dfrac{27r^6}{1000s^{12}}\right)^{-2/3}$ **32.** $\left(-\dfrac{32m^{10}}{243n^{15}}\right)^{-2/5}$

33. $\left(\dfrac{a^8c^{24}}{b^{20}}\right)^{0.25}$ **34.** $\left(\dfrac{u^{10}v^0}{z^5}\right)^{0.2}$ **35.** $\dfrac{a^{2/5}a^{4/5}}{a^{1/5}}$ **36.** $\dfrac{x^{6/7}x^{3/7}}{x^{2/7}x^{5/7}}$

37. $\dfrac{(16a^{-6}s^2)^{1/4}}{a^{3/2}s^{-1/2}}$ **38.** $\dfrac{(64x^{11/3}y^{-10/3})^{1/3}}{(27x^{-2/3}y^{1/3})^{-1/3}}$ **39.** $\dfrac{b^{x/3}b^{x/4}}{b^{x/8}}$ **40.** $\dfrac{c^{x/3}c^{x/6}}{c^{x/12}}$

In Exercises 41–56, simplify each radical. Assume that all variables represent positive numbers.

41. $\sqrt{49}$ **42.** $\sqrt{81}$ **43.** $\sqrt[3]{125}$ **44.** $\sqrt[3]{-64}$

45. $-\sqrt[4]{81}$ **46.** $\sqrt[5]{-243}$ **47.** $\sqrt[5]{-\dfrac{32}{100,000}}$ **48.** $\sqrt[4]{\dfrac{256}{625}}$

49. $\sqrt{36x^2}$ **50.** $-\sqrt{25y^6}$ **51.** $\sqrt{x^2y^4}$ **52.** $\sqrt{a^4b^8}$

53. $\sqrt[3]{8y^3}$ **54.** $\sqrt[3]{-27z^9}$ **55.** $\sqrt[4]{\dfrac{x^4y^8}{z^{12}}}$ **56.** $\sqrt[5]{\dfrac{a^{10}b^5}{c^{15}}}$

In Exercises 57–68, assume that all variables are unrestricted. Simplify each expression and use absolute value symbols when appropriate. Write answers without using negative exponents.

57. $(16a^2)^{1/2}$ **58.** $(25a^4)^{1/2}$ **59.** $(16a^{12})^{1/4}$ **60.** $(-64a^3)^{1/3}$

61. $(-32a^5)^{1/5}$ **62.** $(64a^6)^{1/6}$ **63.** $\sqrt{16b^4}$ **64.** $\sqrt{4a^2}$

65. $\sqrt[3]{27x^3}$ **66.** $\sqrt[4]{16x^4}$ **67.** $\sqrt[6]{x^6y^{12}}$ **68.** $\sqrt[8]{a^{16}b^{32}c^{64}}$

In Exercises 69–84, simplify each expression. Assume that all variables represent positive numbers.

69. $\sqrt{8} - \sqrt{2}$ **70.** $\sqrt{75} - 2\sqrt{27}$

71. $\sqrt{200x^2} + \sqrt{98x^2}$ **72.** $\sqrt{128a^3} - a\sqrt{162a}$

73. $2\sqrt{48y^5} - 3y\sqrt{12y^3}$ **74.** $y\sqrt{112y} + 4\sqrt{175y^3}$

75. $2\sqrt[3]{81} + 3\sqrt[3]{24}$ **76.** $3\sqrt[4]{32} - 2\sqrt[4]{162}$

77. $\sqrt[4]{768z^5} + \sqrt[4]{48z^5}$ **78.** $-2\sqrt[5]{64y^2} + 3\sqrt[5]{486y^2}$

79. $\sqrt{8x^2y} - x\sqrt{2y} + \sqrt{50x^2y}$ **80.** $3x\sqrt{18x} + 2\sqrt{2x^3} - \sqrt{72x^3}$

81. $\sqrt[3]{16xy^4} + y\sqrt[3]{2xy} - \sqrt[3]{54xy^4}$ **82.** $\sqrt[4]{512x^5} - \sqrt[4]{32x^5} + \sqrt[4]{1250x^5}$

83. $\dfrac{3x\sqrt{2x} - 2\sqrt{2x^3}}{\sqrt{18x} - \sqrt{2x}}$ **84.** $\dfrac{4x\sqrt{3x} + \sqrt{12x^3}}{\sqrt{27x} + \sqrt{12x}}$

In Exercises 85–102, assume that all variables are positive numbers. Rationalize each denominator and simplify.

85. $\dfrac{3}{\sqrt{3}}$ **86.** $\dfrac{10}{\sqrt{5}}$ **87.** $\dfrac{2}{\sqrt{x}}$ **88.** $\dfrac{3}{\sqrt{y}}$

89. $\dfrac{2}{\sqrt[3]{2}}$ **90.** $\dfrac{3}{\sqrt[3]{9}}$ **91.** $\dfrac{5a}{\sqrt[3]{25a}}$ **92.** $\dfrac{16}{\sqrt[3]{16b^2}}$

93. $\dfrac{2b}{\sqrt[4]{3a^2}}$ **94.** $\dfrac{64a}{\sqrt[5]{16b^3}}$ **95.** $\sqrt{\dfrac{x}{2y}}$ **96.** $\sqrt{\dfrac{57}{3x}}$

97. $\sqrt[3]{\dfrac{2u^4}{9v}}$ **98.** $\sqrt[3]{-\dfrac{3s^5}{4r^2}}$ **99.** $\sqrt{\dfrac{1}{3}} - \sqrt{\dfrac{1}{27}}$ **100.** $\sqrt{\dfrac{1}{5}} + \sqrt{\dfrac{2}{5}}$

101. $\sqrt{\dfrac{x}{8}} - \sqrt{\dfrac{x}{2}} + \sqrt{\dfrac{x}{32}}$ **102.** $\sqrt[3]{\dfrac{y}{4}} + \sqrt[3]{\dfrac{y}{32}} - \sqrt[3]{\dfrac{y}{500}}$

In Exercises 103–108, simplify each expression. Assume that all variables represent positive numbers.

103. $\dfrac{\sqrt[3]{x^6y^3}}{\sqrt{9x^2y}}$ **104.** $\dfrac{\sqrt[6]{128x^{-9}y^8z^7}}{\sqrt[6]{x^6yz^4}}$ **105.** $\dfrac{\sqrt[4]{81x^{-6}y^8}}{\sqrt[4]{x^{-2}y^4}}$ **106.** $\dfrac{\sqrt[4]{625x^8y^4}}{\sqrt{16x^3y^2}}$

107. $\dfrac{\sqrt{x^{12}y^{12}}}{\sqrt[5]{32x^6y^{11}}}$ **108.** $\dfrac{\sqrt[3]{-27x^{10}y^{14}}}{\sqrt[3]{64x^7y^2}}$

In Exercises 109–112, simplify each radical.

109. $\sqrt[4]{9}$ **110.** $\sqrt[6]{27}$ **111.** $\sqrt[10]{16x^6}$ **112.** $\sqrt[6]{27x^9}$

In Exercises 113–116, write each expression as a single radical.

113. $\sqrt{2}\sqrt[3]{3}$ **114.** $\sqrt[3]{3}\sqrt{5}$ **115.** $\dfrac{\sqrt[4]{3}}{\sqrt{2}}$ **116.** $\dfrac{\sqrt[3]{2}}{\sqrt{5}}$

117.   For what values of x does $\sqrt{x^2} = x$?

118.   For what values of x does $\sqrt[3]{x^3} = x$?

119.   For what values of x does $\sqrt[4]{x^4} = -x$?

120. If all of the radicals involved represent real numbers and $y \neq 0$, prove that

$$\sqrt[n]{\frac{x}{y}} = \frac{\sqrt[n]{x}}{\sqrt[n]{y}}$$

121. If all of the radicals involved represent real numbers and there is no division by 0, prove that

$$\left(\frac{x}{y}\right)^{-m/n} = \sqrt[n]{\frac{y^m}{x^m}}$$

122. The definition of $x^{m/n}$ requires that $\sqrt[n]{x}$ be a real number. Why is this important? (*Hint:* Consider what happens when n is even, m is odd, and x is negative.)

In Exercises 123–128, let $x = 3.5$, $y = 1.2$, and $z = 1.4$. Use a calculator to verify each statement.

123. $(xy)^z = x^z y^z$.

124. $\left(\dfrac{x}{y}\right)^z = \dfrac{x^z}{y^z}$

125. $(x^y)^z = (x^z)^y$

126. $(x + y)^z \neq x^z + y^z$

127. $\sqrt[x]{\sqrt[y]{x}} = x^{1/(xy)}$

128. $(x^y)^z \neq x^{(yz)}$

1.5 ARITHMETIC OF POLYNOMIALS

A **monomial** is either a number or the product of a number and one or more variables with whole-number exponents. The number is called the **numerical coefficient**, or just the **coefficient**. Some examples of monomials are

$$3x, \quad 7ab^2, \quad -5ab^2c^4, \quad x^3, \quad \text{and} \quad -12$$

with coefficients of 3, 7, −5, 1, and −12, respectively.

The **degree** of a monomial is the sum of the exponents of its variables. The degree of $3x$ is 1, the degree of $7ab^2$ is 3, the degree of $-5ab^2c^4$ is 7, the degree of x^3 is 3, and the degree of -12 is 0 (because $-12 = -12x^0$). All nonzero constants have a degree of 0. However, the constant 0 does not have a defined degree.

A single monomial or a finite sum of monomials is called a **polynomial**. Each of the monomials in that sum is called a **term** of the polynomial. A polynomial with two terms is called a **binomial**, and a polynomial with three terms is called a **trinomial**. The **degree of a polynomial** is the degree of the term in the polynomial with highest degree. The only polynomial with no defined degree is 0, which is called the **zero polynomial**. Some examples of polynomials are

$3x^2y^3 + 5xy^2 + 7$ is a trinomial. It is of degree 5 because its term of highest degree (the first term) is of degree 5.

$3ab + 5a^2b$ is a binomial of degree 3.

$5x + 3y^2 + \sqrt[3]{3}z^4 - \sqrt{7}$ is a polynomial because its variables have whole-number exponents. It is of degree 4.

$-7y^{1/2} + 3y^2 + \sqrt[5]{3}z$ is not a polynomial because one of its variables does not have a whole-number exponent.

If two terms of a polynomial have the same variables with the same exponents, they are called **like** or **similar terms**. The distributive property enables us to combine such terms. For example, to combine the like terms in the binomial $3x^2y + 5x^2y$, we proceed as follows:

$$3x^2y + 5x^2y = (3 + 5)x^2y = 8x^2y$$

Thus, to combine like terms, we add their numerical values and keep the same variables with the same exponents.

Adding and Subtracting Polynomials

To add and subtract polynomials, we often use the **extended distributive property**:

$$a(b + c + d + e + \cdots) = ab + ac + ad + ae + \cdots$$

We can use this property to remove parentheses enclosing several terms of a polynomial. If the sign preceding the parentheses is $+$, we simply drop the parentheses:

$$+(a + b - c) = +1(a + b - c) = 1a + 1b - 1c = a + b - c$$

Polynomials are added by removing parentheses, if necessary, and combining any like terms that are contained within the polynomials.

Example 1   Perform the addition $(3x^3y + 5x^2 - 2y) + (2x^3y - 5x^2 + 3x)$.

Solution   To add the polynomials, remove parentheses and combine like terms. Because the terms in each polynomial represent real numbers, you can use the commutative and associative properties of real numbers to rearrange and regroup terms.

$$
\begin{aligned}
&(3x^3y + 5x^2 - 2y) + (2x^3y - 5x^2 + 3x) \\
&= 3x^3y + 5x^2 - 2y + 2x^3y - 5x^2 + 3x \\
&= 3x^3y + 2x^3y + 5x^2 - 5x^2 - 2y + 3x \\
&= 5x^3y - 2y + 3x
\end{aligned}
$$

■

To remove parentheses enclosing several terms of a polynomial when the sign preceding the parentheses is $-$, we simply drop the parentheses and the $-$ sign and *change the sign of each term within the parentheses.*

$$
\begin{aligned}
-(a + b - c) &= -1(a + b - c) \\
&= -1a + (-1)b - (-1)c \\
&= -a - b + c
\end{aligned}
$$

This suggests that the way to subtract polynomials is to remove parentheses and combine like terms.

Example 2   Perform the subtraction $(2x^2 + 3y^2) - (x^2 - 2y^2 + 7)$.

Solution   To subtract the polynomials, remove parentheses and combine like terms.

$$(2x^2 + 3y^2) - (x^2 - 2y^2 + 7) = 2x^2 + 3y^2 - x^2 + 2y^2 - 7$$
$$= 2x^2 - x^2 + 3y^2 + 2y^2 - 7$$
$$= x^2 + 5y^2 - 7 \qquad \blacksquare$$

When an expression in parentheses is multiplied by a constant, we use the distributive property to remove the parentheses. We multiply each term within the parentheses by the constant and drop the parentheses. For example,

$$4(3x^2 - 2x + 6) = 4(3x^2) - 4(2x) + 4(6) = 12x^2 - 8x + 24$$

This suggests that to add multiples of one polynomial to another, or to subtract multiples of one polynomial from another, we remove parentheses and combine like terms.

Example 3   Simplify the expression $7x(2y^2 + 13x^2) - 5(xy^2 - 13x^3)$.

Solution   Remove parentheses and combine like terms.

$$7x(2y^2 + 13x^2) - 5(xy^2 - 13x^3) = 14xy^2 + 91x^3 - 5xy^2 + 65x^3$$
$$= 14xy^2 - 5xy^2 + 91x^3 + 65x^3$$
$$= 9xy^2 + 156x^3 \qquad \blacksquare$$

Multiplying Polynomials

To find the product of two monomials such as $3x^2y^3z$ and $5xyz^2$, we proceed as follows:

$$(3x^2y^3z)(5xyz^2) = 3 \cdot x^2 \cdot y^3 \cdot z \cdot 5 \cdot x \cdot y \cdot z^2$$
$$= 3 \cdot 5 \cdot x^2 \cdot x \cdot y^3 \cdot y \cdot z \cdot z^2$$
$$= 15x^3y^4z^3$$

Thus, to multiply two monomials, we first multiply the numerical factors and then multiply the variable factors.

To find the product of a monomial and a polynomial, we use the distributive property or the extended distributive property. For example,

$$3xy^2(2xy + x^2 - 7yz) = 3xy^2(2xy) + (3xy^2)(x^2) - (3xy^2)(7yz)$$
$$= 6x^2y^3 + 3x^3y^2 - 21xy^3z$$

Thus, to multiply a polynomial by a monomial, we multiply each term of the polynomial by the monomial.

To multiply one binomial by another, we use the distributive property repeatedly.

Example 4   Find each of the products   **a.** $(x + y)(x + y)$, **b.** $(x - y)(x - y)$, and **c.** $(x + y)(x - y)$.

Solution   **a.** $(x + y)(x + y) = (x + y)x + (x + y)y$
$$= x^2 + xy + xy + y^2$$
$$= x^2 + 2xy + y^2$$

b. $(x - y)(x - y) = (x - y)x - (x - y)y$
$$= x^2 - xy - xy + y^2$$
$$= x^2 - 2xy + y^2$$

c. $(x + y)(x - y) = (x + y)x - (x + y)y$
$$= x^2 + xy - xy - y^2$$
$$= x^2 - y^2$$ ∎

The products found in Example 4 are called **special products**. Because they occur so often, it is worthwhile to learn their forms.

Special Product Formulas.

$$(x + y)^2 = (x + y)(x + y) = x^2 + 2xy + y^2$$
$$(x - y)^2 = (x - y)(x - y) = x^2 - 2xy + y^2$$
$$(x + y)(x - y) = x^2 - y^2$$

We must remember that $(x + y)^2$ and $(x - y)^2$ have trinomials for their products and that

$$(x + y)^2 \neq x^2 + y^2 \qquad \text{and} \qquad (x - y)^2 \neq x^2 - y^2$$

The results of Example 4 suggest that to multiply one binomial by another, we multiply each term of one binomial by each term of the other binomial and combine like terms, whenever possible.

There is a shortcut method, called the **FOIL method**, that we can use to multiply one binomial by another. The word FOIL is an acronym for First terms, Outer terms, Inner terms, and Last terms. In this method, we draw curves to show the indicated products. To multiply binomials such as $(3x - 4)(2x + 5)$, we can write

First terms   Last terms

$$(3x - 4)(2x + 5) = 3x(2x) + 3x(5) - 4(2x) - 4(5)$$
$$= 6x^2 + 15x - 8x - 20$$
$$= 6x^2 + 7x - 20$$

Inner terms

Outer terms

In this example, the product of the first terms is $6x^2$, the product of the outer terms is $15x$, the product of the inner terms is $-8x$, and the product of the last terms is -20. The resulting terms of the product are then combined.

Example 5   Use the FOIL method to find the product $(\sqrt{3} + x)(2 - \sqrt{3}x)$.

Solution

$$
\begin{aligned}
(\sqrt{3} + x)(2 - \sqrt{3}x) &= 2\sqrt{3} - \sqrt{3}\sqrt{3}x + 2x - x\sqrt{3}x \\
&= 2\sqrt{3} - 3x + 2x - \sqrt{3}x^2 \\
&= 2\sqrt{3} - x - \sqrt{3}x^2
\end{aligned}
$$

∎

To multiply a polynomial with more than two terms by another polynomial, we multiply each term of one polynomial by each term of the other polynomial and combine like terms whenever possible.

Example 6   Find the products   **a.** $(x + y)(x^2 - xy + y^2)$ and   **b.** $(x + 3)^3$.

Solution   **a.** $(x + y)(x^2 - xy + y^2) = x^3 - x^2y + xy^2 + yx^2 - xy^2 + y^3$
$$
= x^3 + y^3
$$

b. $(x + 3)^3 = (x + 3)(x + 3)^2$
$$
\begin{aligned}
&= (x + 3)(x^2 + 6x + 9) \\
&= x^3 + 6x^2 + 9x + 3x^2 + 18x + 27 \\
&= x^3 + 9x^2 + 27x + 27
\end{aligned}
$$

∎

If n is a whole number, the expressions $a^n + 1$ and $2a^n - 3$ are binomials, and we can use the FOIL method to multiply them together.

$$
\begin{aligned}
(a^n + 1)(2a^n - 3) &= 2a^{2n} - 3a^n + 2a^n - 3 \\
&= 2a^{2n} - a^n - 3
\end{aligned}
$$

We can also use the methods previously discussed to multiply expressions such as $x^{-2} + y$ and $x^2 - y^{-1}$ that are not polynomials.

$$
\begin{aligned}
(x^{-2} + y)(x^2 - y^{-1}) &= x^{-2+2} - x^{-2}y^{-1} + x^2y - y^{1-1} \\
&= x^0 - \frac{1}{x^2y} + x^2y - y^0 \\
&= 1 - \frac{1}{x^2y} + x^2y - 1 \\
&= x^2y - \frac{1}{x^2y}
\end{aligned}
$$

If the denominator of a fraction is a binomial containing radicals, we can use the product formula $(x + y)(x - y)$ to rationalize its denominator. For example, to rationalize the denominator of the fraction $\dfrac{6}{\sqrt{7} + 2}$, we multiply both the numerator and the denominator of the fraction by $\sqrt{7} - 2$ and simplify.

$$\frac{6}{\sqrt{7} + 2} = \frac{6(\sqrt{7} - 2)}{(\sqrt{7} + 2)(\sqrt{7} - 2)}$$

$$= \frac{6(\sqrt{7} - 2)}{7 - 4}$$

$$= \frac{6(\sqrt{7} - 2)}{3}$$

$$= 2(\sqrt{7} - 2)$$

In the previous example, we multiplied both the numerator and the denominator of the given fraction by $\sqrt{7} - 2$. This binomial is the same as the denominator of the given fraction $\sqrt{7} + 2$, except for the signs between their terms. Such binomials are called **conjugates** of each other.

Definition.   **Conjugate binomials** are binomials that are the same except for the sign between their terms. The conjugate of $a + b$ is $a - b$, and the conjugate of $a - b$ is $a + b$.

Example 7   Simplify $\dfrac{\sqrt{3x} - \sqrt{2}}{\sqrt{3x} + \sqrt{2}}$ $(x > 0)$ by rationalizing the denominator.

Solution   Multiply both the numerator and the denominator by $\sqrt{3x} - \sqrt{2}$ (which is the conjugate of $\sqrt{3x} + \sqrt{2}$) and simplify.

$$\frac{\sqrt{3x} - \sqrt{2}}{\sqrt{3x} + \sqrt{2}} = \frac{(\sqrt{3x} - \sqrt{2})(\sqrt{3x} - \sqrt{2})}{(\sqrt{3x} + \sqrt{2})(\sqrt{3x} - \sqrt{2})}$$

$$= \frac{\sqrt{3x}\sqrt{3x} - \sqrt{3x}\sqrt{2} - \sqrt{2}\sqrt{3x} + \sqrt{2}\sqrt{2}}{(\sqrt{3x})^2 - (\sqrt{2})^2}$$

$$= \frac{3x - \sqrt{6x} - \sqrt{6x} + 2}{3x - 2}$$

$$= \frac{3x - 2\sqrt{6x} + 2}{3x - 2}$$

In calculus, it is sometimes necessary to rationalize a numerator.

Example 8   Rationalize the numerator of the fraction $\dfrac{\sqrt{x + h} - \sqrt{x}}{h}$.

Solution   Multiply both the numerator and the denominator by the conjugate of the numerator and simplify.

$$\frac{\sqrt{x + h} - \sqrt{x}}{h} = \frac{(\sqrt{x + h} - \sqrt{x})(\sqrt{x + h} + \sqrt{x})}{h(\sqrt{x + h} + \sqrt{x})}$$

$$= \frac{x + h - x}{h(\sqrt{x + h} + \sqrt{x})}$$

$$= \frac{h}{h(\sqrt{x + h} + \sqrt{x})}$$

$$= \frac{1}{\sqrt{x + h} + \sqrt{x}} \qquad \text{Divide out the common factor of } h. \quad \blacksquare$$

Dividing Polynomials

To divide a monomial by a monomial, we write the quotient of the monomials as a fraction and simplify the fraction by using the rules of exponents. For example,

$$\frac{6x^2y^3}{-2x^3y} = -3x^{2-3}y^{3-1} = -3x^{-1}y^2 = -\frac{3y^2}{x}$$

To divide a polynomial by a monomial, we write the quotient of the polynomial and the monomial as a fraction and write the fraction as a sum of separate fractions. For example, to divide $8x^5y^4 + 12x^2y^5 - 16x^2y^3$ by $4x^3y^4$, we first write

$$\frac{8x^5y^4 + 12x^2y^5 - 16x^2y^3}{4x^3y^4} = \frac{8x^5y^4}{4x^3y^4} + \frac{12x^2y^5}{4x^3y^4} + \frac{-16x^2y^3}{4x^3y^4}$$

and then simplify each of the fractions on the right-hand side of the equals sign to obtain

$$\frac{8x^5y^4 + 12x^2y^5 - 16x^2y^3}{4x^3y^4} = 2x^2 + \frac{3y}{x} - \frac{4}{xy}$$

To divide one polynomial by another, we use a process similar to the long division of whole numbers. To illustrate this process, we consider the division

$$\frac{2x^2 + 11x - 30}{x + 7}$$

This division can be written in long division form as

$$x + 7 \overline{)2x^2 + 11x - 30}$$

The binomial $x + 7$ is called the **divisor**; the trinomial $2x^2 + 11x - 30$ is called the **dividend**. The final answer, called the **quotient**, will appear above the long division symbol.

We begin the division process by asking "What expression when multiplied by x gives $2x^2$?" The answer to this question is $2x$, because $x \cdot 2x = 2x^2$. We place $2x$ as the first term in the quotient, multiply each term of the divisor by $2x$, subtract, and bring down the -30:

$$
\begin{array}{r}
2x \\
x + 7 \overline{)2x^2 + 11x - 30} \\
\underline{2x^2 + 14x} \\
-\ 3x - 30
\end{array}
$$

We continue the division by asking the question "What expression when multiplied by x gives $-3x$?" We place the answer, **-3**, as the second term in the quotient, multiply each term of the divisor by -3, and subtract:

$$
\begin{array}{r}
2x \quad\;\; -3 \\
x + 7\overline{)2x^2 + 11x - 30} \\
\underline{2x^2 + 14x} \\
-\;3x - 30 \\
\underline{-\;3x - 21} \\
-\;9
\end{array}
$$

Because the degree of the remainder, -9, is less than the degree of the divisor, the division process is completed, and we can express the result in *quotient* + $\dfrac{remainder}{divisor}$ form as follows:

$$
2x - 3 + \frac{-9}{x + 7}
$$

Note that we have subtracted from $2x^2 + 11x - 30$ first $2x$ times $x + 7$, then -3 times $x + 7$, leaving a remainder of -9.

Sir Isaac Newton (1642–1727) In Newton's lifetime, there was a bitter controversy over who first invented calculus. It is now believed that Newton and Leibniz invented the calculus independently.

Example 9   Divide $6x^3 - 11$ by $2x + 2$

Solution   Set up the division, leaving space for the missing powers of x in the dividend:

$$
2x + 2\overline{)6x^3 + 0x^2 + 0x - 11}
$$

The division process continues as usual: "What expression when multiplied by $2x$ gives $6x^3$?" The answer, $3x^2$, is the first term of the quotient. The division proceeds as follows:

$$
\begin{array}{r}
3x^2 \\
2x + 2\overline{)6x^3 - 11} \\
\underline{6x^3 + 6x^2} \\
-\;6x^2
\end{array}
$$

Because $2x(-3x) = -6x^2$, the second term of the quotient is $-3x$:

$$
\begin{array}{r}
3x^2\; -3x \\
2x + 2\overline{)6x^3 - 11} \\
\underline{6x^3 + 6x^2} \\
-\;6x^2 \\
\underline{-\;6x^2 - 6x} \\
+\;6x - 11
\end{array}
$$

Because $2x(3) = 6x$, the third term of the quotient is 3.

$$
\begin{array}{r}
3x^2 - 3x + 3 \\
2x + 2\overline{)6x^3 \hspace{3.5em} - 11} \\
\underline{6x^3 + 6x^2} \\
- 6x^2 \\
\underline{- 6x^2 - 6x} \\
+ 6x - 11 \\
\underline{+ 6x + 6} \\
- 17
\end{array}
$$

Hence,

$$\frac{6x^3 - 11}{2x + 2} = 3x^2 - 3x + 3 + \frac{-17}{2x + 2}$$

■

Example 10   Divide $-3x^3 - 3 + x^5 + 4x^2 - x^4$ by $x^2 - 3$.

Solution   The division process works most efficiently when the terms in both the divisor and dividend are written with their exponents in descending order. The long division is set up and accomplished as follows:

$$
\begin{array}{r}
x^3 - x^2 + 1 \\
x^2 - 3\overline{)x^5 - x^4 - 3x^3 + 4x^2 - 3} \\
\underline{x^5 \hspace{3.5em} - 3x^3} \\
- x^4 \hspace{3.5em} + 4x^2 \\
\underline{- x^4 \hspace{3.5em} + 3x^2} \\
x^2 - 3 \\
\underline{x^2 - 3} \\
0
\end{array}
$$

Hence,

$$\frac{x^5 - x^4 - 3x^3 + 4x^2 - 3}{x^2 - 3} = x^3 - x^2 + 1$$

■

EXERCISE 1.5

In Exercises 1–12, tell whether the given expression is a polynomial. If it is, give its degree and, if possible, classify it as a monomial, a binomial, or a trinomial.

1. $x^2 + 3x + 4$

2. $x^3 - 5xy$

3. $x^3 + y^{1/2}$

4. $x^{3/2} - 5y$

5. $\sqrt{5}x^3 + 4x^2$

6. x^2y^3

7. $\sqrt{15}$

8. $\dfrac{5}{x} + \dfrac{x}{5} + 5$

9. 0

10. $\sqrt{y}$

11. $4x^{-3} - 3x^{-2} + 2x$

12. $3y^3 - 4y^2 + 2y + 2$

In Exercises 13–70, perform the indicated operations and simplify.

13. $(x^3 - 3x^2) + (5x^3 - 8x)$

14. $(2x^4 - 5x^3) + (7x^3 - x^4 + 2x)$

15. $(y^5 + 2y^3 + 7) - (y^5 - 2y^3 - 7)$

16. $(3t^7 - 7t^3 + 3) - (7t^7 - 3t^3 + 7)$

17. $2(x^2 + 3x - 1) - 3(x^2 + 2x - 4) + 4$

18. $5(x^3 - 8x + 3) + 2(3x^2 + 5x) - 7$

19. $8(t^2 - 2t + 5) + 4(t^2 - 3t + 2) - 6(2t^2 - 8)$

20. $-3(x^3 - x) + 2(x^2 + x) + 3(x^3 - 2x)$

21. $y(y^2 - 1) - y^2(y + 2) - y(2y - 2)$

22. $-4a^2(a + 1) + 3a(a^2 - 4) - a^2(a + 2)$

23. $x(x^2 - 1) - x^2(x + 2) - x(2x - 2)$

24. $x(x - 4) - (x^2 + 3) + x(2x + 3)$

25. $xy(x - 4y) - y(x^2 + 3xy) + xy(2x + 3y)$

26. $3mn(m + 2n) - 6m(3mn + 1) - 2n(4mn - 1)$

27. $2x^2y^3(4xy^4)$

28. $-15a^3b(-2a^2b^3)$

29. $-3m^2n(2mn^2)(-mn)$

30. $-3r^2s^3(2r^2s)(-15rs^2)$

31. $-4rs(r^2 + s^2)$

32. $6u^2v(2uv^3 - y)$

33. $6ab^2c(2ac + 3bc^2 - 4ab^2c)$

34. $-3mn^2(4mn - 6m^2 - 8)$

35. $(a + 2)(a + 2)$

36. $(y - 5)(y - 5)$

37. $(a - 6)^2$

38. $(t + 9)^2$

39. $(x + 4)(x - 4)$

40. $(z + 7)(z - 7)$

41. $(a - 2b)^2$

42. $(4a + 5b)(4a - 5b)$

43. $(3m + 4n)(3m - 4n)$

44. $(4r + 3s)^2$

45. $(x - 3)(x + 5)$

46. $(z + 4)(z - 6)$

47. $(u + 2)(3u - 2)$

48. $(4x + 1)(2x - 3)$

49. $(2x + 3)(3x - 5)$

50. $(2a - 3)(2 + 3a)$

51. $(5x - 1)(2x + 3)$

52. $(4x - 1)(2x - 7)$

53. $(3v - 4)(4v + 3)$

54. $(4c + 5)(3c - 4)$

55. $(2y - 4x)(3y - 2x)$

56. $(-2x + 3y)(3x + y)$

57. $(9x - y)(x^2 - 3y)$

58. $(8a^2 + b)(a + 2b)$

59. $(5z + 2t)(z^2 - t)$

60. $(y - 2x^2)(x^2 + 3y)$

61. $(3x - 1)^3$

62. $(2x - 3)^3$

63. $(xy^2 - 1)(x^2y + 2)$

64. $(xy - z^2)(2xy + z^2)$

65. $(3x + 1)(2x^2 + 4x - 3)$

66. $(2x - 5)(x^2 - 3x + 2)$

67. $(3x + 2y)(2x^2 - 3xy + 4y^2)$

68. $(4r - 3s)(2r^2 + 4rs - 2s^2)$

69. $(x^2 + x + 1)(x^2 + x - 1)$

70. $(x^2 - x + 1)(x^2 + 2x + 3)$

In Exercises 71–82, multiply the expressions as you would multiply polynomials.

71. $2y^n(3y^n - 2y^2 + y^{-n})$

72. $3a^{-n}(2a^n + 3a^{n-1} - 7a^{n-2})$

73. $-5x^{2n}y^n(2x^{2n}y^{-n} + 3x^{-2n}y^n)$

74. $-2a^{3n}b^{2n}(5a^{-3n}b - ab^{-2n})$

75. $(x^n + 3)(x^n - 4)$

76. $(a^n - 5)(a^n - 3)$

77. $(2r^n - 7)(3r^n - 2)$

78. $(4z^n + 3)(3z^n + 1)$

79. $x^{1/2}(x^{1/2}y + xy^{1/2})$

80. $ab^{1/2}(a^{1/2}b^{1/2} + b^{1/2})$

81. $(a^{1/2} + b^{1/2})(a^{1/2} - b^{1/2})$

82. $(x^{3/2} + y^{1/2})^2$

In Exercises 83–98, rationalize each denominator.

83. $\dfrac{2}{\sqrt{3} - 1}$

84. $\dfrac{1}{\sqrt{5} + 2}$

85. $\dfrac{3x}{\sqrt{7} + 2}$

86. $\dfrac{14y}{\sqrt{2} - 3}$

87. $\dfrac{6}{\sqrt{5} - \sqrt{2}}$

88. $\dfrac{15}{\sqrt{7} + \sqrt{2}}$

89. $\dfrac{x}{x - \sqrt{3}}$

90. $\dfrac{y}{2y + \sqrt{7}}$

91. $\dfrac{y + \sqrt{2}}{y - \sqrt{2}}$ **92.** $\dfrac{x - \sqrt{3}}{x + \sqrt{3}}$ **93.** $\dfrac{\sqrt{5} - 2}{\sqrt{5} + 2}$ **94.** $\dfrac{\sqrt{7} + 4}{\sqrt{7} - 4}$

95. $\dfrac{\sqrt{2} - \sqrt{3}}{1 - \sqrt{3}}$ **96.** $\dfrac{\sqrt{3} - \sqrt{2}}{1 + \sqrt{2}}$ **97.** $\dfrac{\sqrt{x} - \sqrt{y}}{\sqrt{x} + \sqrt{y}}$ **98.** $\dfrac{\sqrt{2x} + y}{\sqrt{2x} - y}$

In Exercises 99–100, rationalize each numerator.

99. $\dfrac{\sqrt{x + 3} - \sqrt{x}}{3}$ **100.** $\dfrac{\sqrt{2 + h} - \sqrt{2}}{h}$

In Exercises 101–106, perform each division. Express all answers without using negative exponents.

101. $\dfrac{36a^2b^3}{18ab^6}$ **102.** $\dfrac{-45r^2s^5t^3}{27r^6s^2t^8}$ **103.** $\dfrac{5x^3y^2 + 15x^3y^4}{10x^2y^3}$ **104.** $\dfrac{9m^4n^9 - 6m^3n^4}{12m^3n^3}$

105. $\dfrac{24x^5y^7 - 36x^2y^5 + 12xy}{60x^5y^4}$ **106.** $\dfrac{9a^3b^4 + 27a^2b^4 - 18a^2b^3}{18a^2b^7}$

In Exercises 107–120, simplify each fraction by performing a long division. If there is a nonzero remainder, write the answer in the $\text{quotient} + \dfrac{remainder}{divisor}$ *form.*

107. $\dfrac{3x^2 + 11x + 6}{x + 3}$ **108.** $\dfrac{3x^2 + 11x + 6}{3x + 2}$ **109.** $\dfrac{2x^2 - 19x + 37}{2x - 5}$ **110.** $\dfrac{2x^2 - 19x + 35}{x - 7}$

111. $\dfrac{x^3 - 2x^2 - 4x + 3}{x^2 + x - 1}$ **112.** $\dfrac{x^3 - 2x^2 - 4x + 5}{x - 3}$

113. $\dfrac{x^5 - 2x^3 - 3x^2 + 9}{x^2 - 2}$ **114.** $\dfrac{x^5 - 2x^3 - 3x^2 + 9}{x^3 - 3}$

115. $\dfrac{x^5 - 32}{x - 2}$ **116.** $\dfrac{x^4 - 1}{x + 1}$

117. $\dfrac{36x^4 - 121x^2 + 120 + 72x^3 - 142x}{11x - 10 + 6x^2}$ **118.** $\dfrac{-121x^2 + 72x^3 - 142x + 120 + 36x^4}{x + 6x^2 - 12}$

119. $\dfrac{11x^5 - 9x^2 + 12x^6 + 3x^4 + 3x + 10x^3 - 6}{4x^4 - 3 + 5x^3}$ **120.** $\dfrac{3x^4 + 10x^3 + 3x - 6 - 9x^2 + 12x^6 + 11x^5}{2 + 3x^2 - x}$

In Exercises 121–128, simplify each expression.

121. $x^{3a+2}x^{2a-3}$ **122.** $x^{3-y^2}x^{3y^2+y}$ **123.** $(b^{3a})^{a+1}$ **124.** $(a^{2b^2+3})^{2b^2-3}$

125. $\dfrac{a^{2(x+1)}}{a^{3-2x}}$ **126.** $\dfrac{b^{x^2+1}}{b^{3-2x^2}}$ **127.** $\dfrac{r^{s^2+3}r^{s-1}}{r^{s^2+2}}$ **128.** $\dfrac{m^{n^2-n}m^{n+2}}{(m^{-n})^2m^{n+1}}$

129. Show that any trinomial can be squared using the formula $(a + b + c)^2 = a^2 + b^2 + c^2 + 2ab + 2bc + 2ca$.

130. Show that $(a + b + c + d)^2 = a^2 + b^2 + c^2 + d^2 + 2ab + 2ac + 2ad + 2bc + 2bd + 2cd$.

1.6 FACTORING POLYNOMIALS

When two or more polynomials are multiplied together, each of the polynomials is called a **factor** of the resulting product. The product $7(x + 2)(x + 3)$, for example, has polynomial factors of 7, $x + 2$, and $x + 3$. The process of writing a polynomial as the product of several factors is called **factoring**.

In this section, we will discuss factoring over the set of integers. If a polynomial cannot be factored by using integers only, we will call the polynomial **prime** or **irreducible** over the set of integers.

Example 1   Factor the binomial $3xy^2 + 6x$.

Solution   Note that each term contains a factor of $3x$:

$$3xy^2 + 6x = 3x(y^2) + 3x(2)$$

Then use the distributive property to factor out the common factor of $3x$:

$$3xy^2 + 6x = 3x(y^2 + 2)$$

The factored form of $3xy^2 + 6x$ is $3x(y^2 + 2)$. This type of factoring is called **factoring out a common** (monomial) **factor**.   ∎

Example 2   Factor the binomial $x^2y^2z^2 + xyz$.

Solution   The binomial can be rewritten, and the common factor xyz on the right-hand side of the following equation can be factored out:

$$x^2y^2z^2 + xyz = xyz \cdot xyz + xyz \cdot 1$$
$$= xyz(xyz + 1)$$

It is important to understand where the $+1$ comes from. The last term in the expression $x^2y^2z^2 + xyz$ has an understood coefficient of 1. When the xyz is factored out, the understood 1 must be written.   ∎

Example 3   Factor the expression $ax + bx + a + b$.

Solution   There is no factor that is common to all four terms, but the factor x is common to the first two terms. Factor out the x and enclose the last two terms with parentheses:

$$ax + bx + a + b = x(a + b) + (a + b)$$

The common factor $a + b$ is now present in each term on the right-hand side and can be factored out:

$$ax + bx + a + b = x(a + b) + (a + b)$$
$$= x(a + b) + 1(a + b)$$
$$= (a + b)(x + 1)$$

The technique used in this example is called **factoring by grouping**.   ∎

Example 4   Factor the binomial $49x^2 - 4$.

Solution   Observe that each term is a perfect square:

$$49x^2 - 4 = (7x)^2 - (2)^2$$

The difference of the squares of two quantities is the product of two factors. One factor is the sum of those quantities, and the other is the difference of those quantities. Thus, $49x^2 - 4$ factors as follows:

$$49x^2 - 4 = (7x)^2 - (2)^2$$
$$= (7x + 2)(7x - 2)$$

Verify that $(7x + 2)(7x - 2) = 49x^2 - 4$ by multiplying $(7x - 2)$ by $(7x + 2)$:

$$(7x + 2)(7x - 2) = 49x^2 - 14x + 14x - 4$$
$$= 49x^2 - 4$$

This type of problem is called factoring the **difference of two squares**. ■

Example 4 suggests the following formula for factoring the difference of two squares.

$$x^2 - y^2 = (x + y)(x - y)$$

With only integers to work with, the sum of two squares cannot be factored. For example, $x^2 + y^2$ is a prime polynomial.

Example 5 Factor the binomial $16x^4 - y^4$.

Solution The binomial $16x^4 - y^4$ can be factored as the difference of two squares:

$$16x^4 - y^4 = (4x^2)^2 - (y^2)^2$$
$$= (4x^2 + y^2)(4x^2 - y^2)$$

However, the second factor, $4x^2 - y^2$, is also a difference of two squares and can be factored:

$$16x^4 - y^4 = (4x^2 + y^2)(4x^2 - y^2)$$
$$= (4x^2 + y^2)(2x + y)(2x - y)$$

With only integers to work with, $4x^2 + y^2$ cannot be factored. ■

Example 6 Factor $18x^2 - 32$.

Solution Begin by factoring out a 2.

$$18x^2 - 32 = 2(9x^2 - 16)$$

Because $9x^2 - 16$ is the difference of two squares, it can be factored:

$$18x^2 - 32 = 2(9x^2 - 16)$$
$$= 2(3x + 4)(3x - 4)$$

■

Some trinomials are squares of binomials and can be factored according to the following formulas:

---

1. $x^2 + 2xy + y^2 = (x + y)(x + y) = (x + y)^2$
2. $x^2 - 2xy + y^2 = (x - y)(x - y) = (x - y)^2$

For example, to factor $a^2 - 6a + 9$, we note that $a^2 - 6a + 9$ can be written in the form

$$a^2 - 2(3a) + 3^2$$

and this form matches the left-hand side of Equation 2 above. Thus,

$$a^2 - 6a + 9 = a^2 - 2(3a) + 3^2 = (a - 3)(a - 3) = (a - 3)^2$$

Factoring trinomials that are not squares of binomials is more difficult. If a trinomial with no common factors is to be factorable, it must factor into the product of two binomials. Finding the binomials is called **factoring the general trinomial**.

Example 7 Factor the trinomial $x^2 + 3x - 10$.

Solution Because this trinomial has no common factors and is given to be factorable, it will factor into the product of two binomials. To factor $x^2 + 3x - 10$, you must find two binomials $x + a$ and $x + b$ such that

$$x^2 + 3x - 10 = (x + a)(x + b)$$

where the product of a and b is -10, and the sum of a and b is 3.

$$ab = -10 \quad \text{and} \quad a + b = 3$$

To find two such numbers, list the possible factorizations of -10:

$$10(-1) \quad 5(-2) \quad -10(1) \quad -5(2)$$

Only in the factorization $5(-2)$ do the factors have a sum of 3. Thus, $a = 5$ and $b = -2$, and

$$x^2 + 3x - 10 = (x + a)(x + b)$$
3. $x^2 + 3x - 10 = (x + 5)(x - 2)$

This factorization can be verified by multiplying $x + 5$ by $x - 2$ and observing that the product is $x^2 + 3x - 10$.

Because of the commutative property of multiplication, the order of the factors in Equation 3 is not important. Equation 3 can also be written as

$$x^2 + 3x - 10 = (x - 2)(x + 5)$$

■

Example 8 Factor the trinomial $2x^2 - x - 6$.

Solution Because the first term of the trinomial is $2x^2$, the first terms of the binomial factors must be $2x$ and x:

$$2x^2 - x - 6 = (2x + ?)(x + ?)$$

The product of the last terms must be -6, and the sum of the products of the outer terms and the inner terms must be $-x$.

$$2x^2 - x - 6 = (2x + ?)(x + ?)$$

$$O + I = -x$$

The only factorization of -6 that will cause this to happen is $3(-2)$. Thus,

$$2x^2 - x - 6 = (2x + 3)(x - 2)$$

Verify this factorization by multiplication. ∎

It is not easy to give specific rules for factoring trinomials, because some guesswork is often necessary. However, the following hints are helpful.

To factor a general trinomial, follow these steps:

1. Write the trinomial in descending powers of one variable.
2. Factor out any greatest common factor (including -1 if that is necessary to make the coefficient of the first term positive).
3. When the sign of the first term of a polynomial is $+$ and the sign of the third term is $+$, the sign between the terms of each binomial factor is the same as the sign of the middle term of the trinomial.

   When the sign of the first term is $+$ and the sign of the third term is $-$, one of the signs between the terms of the binomial factors is $+$ and the other is $-$.
4. Mentally, try various combinations of the first terms and last terms until you find one that works. If you exhaust all the possibilities, the trinomial does not factor with integer coefficients.
5. Check the factorization by multiplication.

Example 9 Factor the trinomial $10xy + 24y^2 - 6x^2$.

Solution Begin by writing the trinomial in descending powers of x and then factor out the common factor of -2.

$$10xy + 24y^2 - 6x^2 = -6x^2 + 10xy + 24y^2$$
$$= -2(3x^2 - 5xy - 12y^2)$$

Because the sign of the third term of $3x^2 - 5xy - 12y^2$ is $-$, the signs between the binomial factors will be opposite. Because the first term is $3x^2$, the first terms of the binomial factors must be $3x$ and x:

$$-2(3x^2 - 5xy - 12y^2) = -2(3x \qquad)(x \qquad)$$

The product of the last terms must be $-12y^2$, and the sum of the outer terms and the inner terms must be $-5xy$:

$$-2(3x^2 - 5xy - 12y^2) = -2(3x \qquad ?)(x \qquad ?)$$

$$-12y^2$$

$$O + I = -5xy$$

Of the many factorizations of $-12y^2$, only $4y(-3y)$ leads to a middle term of $-5xy$. Thus,

$$10xy + 24y^2 - 6x^2 = -6x^2 + 10xy + 24y^2$$
$$= -2(3x^2 - 5xy - 12y^2)$$
$$= -2(3x + 4y)(x - 3y)$$

Verify this result by multiplication.   ∎

We can often factor polynomials with variable exponents. For example, if n is a natural number,

$$a^{2n} - 5a^n - 6 = (a^n + 1)(a^n - 6)$$

because

$$(a^n + 1)(a^n - 6) = a^{2n} - 6a^n + a^n - 6 = a^{2n} - 5a^n - 6$$

Two other factoring types are the **sum of two cubes** and the **difference of two cubes**. Like the difference of two squares, they follow a definite pattern.

$$x^3 + y^3 = (x + y)(x^2 - xy + y^2)$$
$$x^3 - y^3 = (x - y)(x^2 + xy + y^2)$$

Example 10   Factor the binomial $x^3 - 8$.

Solution   This binomial can be written in the form $x^3 - 2^3$, which is the difference of two cubes. Substituting into the formula for the difference of two cubes gives

$$x^3 - 2^3 = (x - 2)(x^2 + 2x + 2^2)$$
$$= (x - 2)(x^2 + 2x + 4)$$

Thus,

$$x^3 - 8 = (x - 2)(x^2 + 2x + 4)$$ ∎

Example 11   Factor the binomial $27x^6 + 64y^3$.

Solution   This expression can be written in the form of the sum of two cubes and factored as follows:

$$27x^6 + 64y^3 = (3x^2)^3 + (4y)^3$$
$$= (3x^2 + 4y)[(3x^2)^2 - (3x^2)(4y) + (4y)^2]$$
$$= (3x^2 + 4y)(9x^4 - 12x^2y + 16y^2)$$ ∎

Example 12   Factor $x^2 - y^2 + 6x + 9$.

Solution   This expression can be written and factored as follows:

$$x^2 - y^2 + 6x + 9 = x^2 + 6x + 9 - y^2$$
$$= (x + 3)^2 - y^2 \qquad \text{Factor } x^2 + 6x + 9.$$
$$= (x + 3 + y)(x + 3 - y) \qquad \text{Factor the difference of two squares.}$$

Note that this is a factoring-by-grouping problem.   ∎

Example 13   Factor the trinomial $z^4 - 3z^2 + 1$.

Solution   An attempt to factor $z^4 - 3z^2 + 1$ as the product of two binomials will fail. One reasonable attempt is $(z^2 - 1)(z^2 - 1)$, but this product is $z^4 - 2z^2 + 1$. However, if you add and subtract the perfect square z^2, you can proceed as follows:

$$z^4 - 3z^2 + 1 = z^4 - 3z^2 + z^2 + 1 - z^2 \qquad \text{Add and subtract } z^2.$$
$$= (z^4 - 2z^2 + 1) - z^2$$
$$= (z^2 - 1)(z^2 - 1) - z^2$$
$$= (z^2 - 1)^2 - z^2$$

The expression $(z^2 - 1)^2 - z^2$ is the difference of two squares and can be factored as

$$z^4 - 3z^2 + 1 = (z^2 - 1 + z)(z^2 - 1 - z)$$

or

$$z^4 - 3z^2 + 1 = (z^2 + z - 1)(z^2 - z - 1)$$

Thus, the factorization of the given trinomial is

$$z^4 - 3z^2 + 1 = (z^2 + z - 1)(z^2 - z - 1)$$ ∎

Example 14   Factor the trinomial $x^4 - 6x^2 + 1$.

Solution   Again, factorization as the product of two binomials fails. The only reasonable possibility is $(x^2 - 1)(x^2 - 1)$, but that product is $x^4 - 2x^2 + 1$. However, if you add and subtract the perfect square $4x^2$, you can proceed as follows:

$$x^4 - 6x^2 + 1 = x^4 - 6x^2 + 4x^2 + 1 - 4x^2$$
$$= (x^4 - 2x^2 + 1) - 4x^2$$
$$= (x^2 - 1)^2 - (2x)^2$$
$$= (x^2 - 1 + 2x)(x^2 - 1 - 2x)$$
$$= (x^2 + 2x - 1)(x^2 - 2x - 1) \qquad \blacksquare$$

EXERCISE 1.6

In Exercises 1–114, completely factor each expression over the set of integers. If an expression is prime, so indicate.

1. $3x - 6$ **2.** $5y - 15$ **3.** $8x^2 + 4x^3$ **4.** $9y^3 + 6y^2$

5. $7x^2y^2 + 14x^3y^2$ **6.** $25y^2z - 15yz^2$

7. $3a^2bc + 6ab^2c + 9abc^2$ **8.** $5x^3y^3z^3 + 25x^2y^2z^2 - 125xyz$

9. $b(x + y) - a(x + y)$ **10.** $b(x - y) + a(x - y)$

11. $4a + b - 12a^2 - 3ab$ **12.** $x^2 + 4x + xy + 4y$

13. $3x^3 + 3x^2 - x - 1$ **14.** $4x + 6xy - 9y - 6$

15. $2txy + 2ctx - 3ty - 3ct$ **16.** $2ax + 4ay - bx - 2by$

17. $ax + bx + ay + by + az + bz$ **18.** $6x^2y^3 + 18xy + 3x^2y^2 + 9x$

19. $6xc + yd + 2dx + 3cy$ **20.** $ax + ay + az + bx + by + bz$

21. $4x^2 - 9$ **22.** $36z^2 - 49$ **23.** $4 - 9r^2$ **24.** $(x - y)^2 - 9$

25. $(x + z)^2 - 25$ **26.** $16 - 49x^2$ **27.** $1 + 25x^4$ **28.** $36x^4 + 121$

29. $x^2 - (y - z)^2$ **30.** $z^2 - (y + 3)^2$

31. $(x - y)^2 - (x + y)^2$ **32.** $(2a + 3) - (2a + 3)^2$

33. $x^4 - y^4$ **34.** $z^4 - 81$ **35.** $x^8 - 64z^4$ **36.** $1 - y^8$

37. $3x^2 - 12$ **38.** $3x^3y - 3xy$ **39.** $18xy^2 - 8x$ **40.** $27x^2 - 12$

41. $x^2 + 8x + 16$ **42.** $a^2 - 12a + 36$ **43.** $b^2 - 10b + 25$ **44.** $y^2 + 14y + 49$

45. $m^2 + 4mn + 4n^2$ **46.** $r^2 - 8rs + 16s^2$

47. $x^2 + 10x + 21$ **48.** $x^2 + 7x + 10$ **49.** $x^2 - 4x - 12$ **50.** $x^2 - 2x - 63$

51. $x^2 - 2x + 15$ **52.** $x^2 + x + 2$ **53.** $12x^2 - xy - 6y^2$ **54.** $8x^2 - 10xy - 3y^2$

55. $24y^2 + 15 - 38y$ **56.** $10x^2 - 18 + 3x$ **57.** $-15 + 2a + 24a^2$ **58.** $-32 - 68x + 9x^2$

59. $6x^2 + 29xy + 35y^2$ **60.** $10x^2 - 17xy + 6y^2$ **61.** $-35 - x + 6x^2$ **62.** $-5x - 6 + 6x^2$

63. $12y^2 - 58y - 70$ **64.** $3x^2 - 6x - 9$ **65.** $-6x^3 + 23x^2 + 35x$ **66.** $-y^3 - y^2 + 90y$

67. $6x^4 - 11x^3 - 35x^2$ **68.** $12x + 17x^2 - 7x^3$ **69.** $-35 + 47x - 6x^2$ **70.** $-14x^2 - 11x + 15$

71. $x^4 + 2x^2 - 15$ **72.** $x^4 - x^2 - 6$

73. $2x^2y^2z^2 - 16xyz + 80$ **74.** $3x^2 + 30x + 75$

75. $a^{2n} - 2a^n - 3$ **76.** $a^{2n} + 6a^n + 8$ **77.** $6x^{2n} - 7x^n + 2$ **78.** $9x^{2n} + 9x^n + 2$

79. $4x^{2n} - 9y^{2n}$ **80.** $8x^{2n} - 2x^n - 3$ **81.** $10y^{2n} - 11y^n - 6$ **82.** $16y^{4n} - 25y^{2n}$

83. $8z^3 - 27$ **84.** $125a^3 - 64$ **85.** $2x^3 + 2000$ **86.** $3y^3 + 648$

87. $(x + y)^3 - 64$ **88.** $(x - y)^3 + 27$ **89.** $1 - (x + 1)^3$ **90.** $1 + (x - 1)^3$

91. $64a^6 - y^6$ **92.** $a^6 + b^6$

93. $a^3 - b^3 + a - b$ **94.** $(a^2 - y^2) - 5(a + y)$

95. $64x^6 + y^6$ **96.** $z^2 + 6z + 9 - 225y^2$

97. $x^2 - 6x + 9 - 144y^2$ **98.** $x^2 + 2x - 9y^2 + 1$

99. $(a + b)^2 - 3(a + b) - 10$ **100.** $2(a + b)^2 - 5(a + b) - 3$

101. $6(u + v)^2 + 11u + 4 + 11v$ **102.** $8(r + s)^2 - 10(r + s) - 3$

103. $x^6 + 7x^3 - 8$ **104.** $x^6 - 13x^4 + 36x^2$

105. $a(c + d) + a + b(c + d) + b$ **106.** $c(a + b) + 3c + 3d + d(a + b)$

107. $x^4 + 3x^2 + 4$ **108.** $x^4 + x^2 + 1$ **109.** $x^4 + 7x^2 + 16$ **110.** $y^4 + 2y^2 + 9$

111. $4a^4 + 1 + 3a^2$ **112.** $x^4 + 25 + 6x^2$ **113.** $2x^4 + 8$ **114.** $3x^4 - 21x^2 + 27$

In Exercises 115–122, factor the indicated monomial from the given expression.

115. $3x + 2; 2$ **116.** $a + b; a$ **117.** $x + x^{1/2}; x^{1/2}$ **118.** $2x + \sqrt{2}y; \sqrt{2}$

119. $ab^{3/2} - a^{3/2}b; ab$ **120.** $ab^2 + b; b^{-1}$ **121.** $a^{n+2} + a^{n+3}; a^2$ **122.** $x^4 - 5x^6; x^{-2}$

1.7 ALGEBRAIC FRACTIONS

If $y \neq 0$, the number represented by the quotient $\frac{x}{y}$ is called a **fraction**. The number x is called the **numerator**, and y is called the **denominator**. Because division by zero is not defined, the denominator of a fraction cannot be 0.

Quotients of algebraic expressions are called **algebraic fractions**. If the algebraic expressions in a fraction are polynomials, the fraction is called a **rational expression**. Each of the following fractions is an algebraic fraction; the first two are rational expressions.

$$\frac{5y^2 + 2y}{y^2 - 3y - 7} \qquad \frac{8ab^2 - 16c^3}{2x + 3} \qquad \frac{x^{1/2} + 4x}{x^{3/2} - x^{1/2}}$$

As with all fractions, the denominators cannot be 0.

Many properties of fractions are summarized as follows:

If a, b, and c are real numbers and no denominators are 0, then	
Equality of Fractions:	$\dfrac{a}{b} = \dfrac{c}{d}$ if and only if $ad = bc$
The Fundamental Property of Fractions:	$\dfrac{ax}{bx} = \dfrac{a}{b}$
Multiplication of Fractions:	$\dfrac{a}{b} \cdot \dfrac{c}{d} = \dfrac{ac}{bd}$

(continued)

(continued)

Division of Fractions: $\dfrac{a}{b} \div \dfrac{c}{d} = \dfrac{a}{b} \cdot \dfrac{d}{c} = \dfrac{ad}{bc}$

Addition and Subtraction of Fractions with Like Denominators: $\dfrac{a}{b} + \dfrac{c}{b} = \dfrac{a + c}{b}$ and $\dfrac{a}{b} - \dfrac{c}{b} = \dfrac{a - c}{b}$

Each of these properties of fractions is illustrated in the following example.

Example 1   **a.** $\dfrac{2a}{3} = \dfrac{4a}{6}$ because $2a(6) = 3(4a)$

$\dfrac{3y}{5} \neq \dfrac{15z}{25}$ because $3y(25) \neq 5(15z)$

b. $\dfrac{6xy}{10xy} = \dfrac{3(2xy)}{5(2xy)} = \dfrac{3}{5}$

c. $\dfrac{2r}{7s} \cdot \dfrac{3r}{5s} = \dfrac{2r \cdot 3r}{7s \cdot 5s} = \dfrac{6r^2}{35s^2}$

d. $\dfrac{3mn}{4pq} \div \dfrac{2pq}{7mn} = \dfrac{3mn}{4pq} \cdot \dfrac{7mn}{2pq} = \dfrac{21m^2n^2}{8p^2q^2}$

e. $\dfrac{2ab}{5xy} + \dfrac{ab}{5xy} = \dfrac{2ab + ab}{5xy} = \dfrac{3ab}{5xy}$

f. $\dfrac{6uv^2}{7w^2} - \dfrac{3uv^2}{7w^2} = \dfrac{6uv^2 - 3uv^2}{7w^2} = \dfrac{3uv^2}{7w^2}$ ∎

To add or subtract fractions with unlike denominators, we can use the fundamental property of fractions and multiply both the numerator and denominator by the same quantity to write each fraction as an equivalent fraction with a common denominator. We can then add or subtract them. For example,

$$\frac{3x}{5} + \frac{2x}{7} = \frac{3x(7)}{5(7)} + \frac{2x(5)}{7(5)} = \frac{21x}{35} + \frac{10x}{35} = \frac{21x + 10x}{35} = \frac{31x}{35}$$

$$\frac{4a^2}{15} - \frac{3a^2}{10} = \frac{4a^2(2)}{15(2)} - \frac{3a^2(3)}{10(3)} = \frac{8a^2}{30} - \frac{9a^2}{30} = \frac{8a^2 - 9a^2}{30} = \frac{-a^2}{30} = -\frac{a^2}{30}$$

A fraction is said to be **reduced to lowest terms** if all factors common to the numerator and the denominator have been removed. To **simplify a fraction** means to reduce it to lowest terms.

To simplify a fraction we use the fundamental property of fractions. This property enables us to divide out all factors that are common to both the numerator and the denominator of a fraction.

Example 2 Simplify the fraction $\dfrac{x^2 - 9}{x^2 - 3x}$.

Solution Factor the difference of two squares in the numerator and factor out the common monomial factor of x in the denominator. Then divide out the resulting common factor of $x - 3$.

$$\frac{x^2 - 9}{x^2 - 3x} = \frac{(x + 3)(x - 3)}{x(x - 3)} = \frac{x + 3}{x}$$

■

We will encounter the following properties of fractions in the next examples.

$$
\text{If } a \text{ and } b \text{ represent real numbers and there are no divisions by 0, then}
$$

$$\frac{a}{1} = a \qquad\qquad \frac{a}{a} = 1$$

$$\frac{a}{b} = \frac{-a}{-b} = -\frac{a}{-b} = -\frac{-a}{b} \qquad -\frac{a}{b} = \frac{a}{-b} = \frac{-a}{b} = -\frac{-a}{-b}$$

Example 3 Simplify the fraction $\dfrac{x^2 - 2xy + y^2}{y - x}$.

Solution Factor the trinomial in the numerator and factor -1 from the denominator. Then divide out all common factors and proceed as follows:

$$\frac{x^2 - 2xy + y^2}{y - x} = \frac{(x - y)(x - y)}{-(x - y)}$$

$$= \frac{(x - y)(x - y)}{-(x - y)}$$

$$= \frac{x - y}{-1}$$

$$= -\frac{x - y}{1}$$

$$= -(x - y)$$

$$= y - x$$

■

Example 4 Simplify the fraction $\dfrac{x^2 - 3x + 2}{x^2 - x - 2}$.

Solution Factor both the numerator and denominator and simplify:

$$\frac{x^2 - 3x + 2}{x^2 - x - 2} = \frac{(x - 1)(x - 2)}{(x + 1)(x - 2)} = \frac{(x - 1)(x - 2)}{(x + 1)(x - 2)} = \frac{x - 1}{x + 1}$$

■

Example 5   Simplify $\dfrac{x^2 + 2x - 3}{x^2 - x - 2}$.

Solution   After factoring, note that there are no factors common to both the numerator and the denominator:

$$\frac{x^2 + 2x - 3}{x^2 - x - 2} = \frac{(x - 1)(x + 3)}{(x + 1)(x - 2)}$$

Because there are no common factors, this algebraic fraction is already in lowest terms and cannot be simplified.   ∎

Example 6   Find the product of $\dfrac{x^2 - x - 2}{x^2 - 1}$ and $\dfrac{x^2 + 2x - 3}{x - 2}$.

Solution   Multiply the fractions together, factor, and then divide out factors common to the numerator and denominator.

$$\frac{x^2 - x - 2}{x^2 - 1} \cdot \frac{x^2 + 2x - 3}{x - 2} = \frac{(x^2 - x - 2)(x^2 + 2x - 3)}{(x^2 - 1)(x - 2)}$$

$$= \frac{(x - 2)(x + 1)\,(x - 1)(x + 3)}{(x + 1)(x - 1)(x - 2)}$$

$$= \frac{(x + 3)}{1}$$

$$= x + 3$$ ∎

Example 7   Simplify $\dfrac{x^2 - 2x - 3}{x^2 - 4} \div \dfrac{x^2 + 2x - 15}{x^2 + 3x - 10}$.

Solution   Change the division to a multiplication, factor everything you can, and simplify:

$$\frac{x^2 - 2x - 3}{x^2 - 4} \div \frac{x^2 + 2x - 15}{x^2 + 3x - 10} = \frac{x^2 - 2x - 3}{x^2 - 4} \cdot \frac{x^2 + 3x - 10}{x^2 + 2x - 15}$$

$$= \frac{(x^2 - 2x - 3)(x^2 + 3x - 10)}{(x^2 - 4)(x^2 + 2x - 15)}$$

$$= \frac{(x - 3)(x + 1)\,(x - 2)(x + 5)}{(x + 2)(x - 2)\,(x + 5)(x - 3)}$$

$$= \frac{x + 1}{x + 2}$$ ∎

Example 8 Simplify $\dfrac{2x^2 - 5x - 3}{3x - 1} \cdot \dfrac{3x^2 + 2x - 1}{x^2 - 2x - 3} \div \dfrac{2x^2 + x}{3x}$.

Solution Change the division to a multiplication, factor, and simplify.

$$\dfrac{2x^2 - 5x - 3}{3x - 1} \cdot \dfrac{3x^2 + 2x - 1}{x^2 - 2x - 3} \div \dfrac{2x^2 + x}{3x}$$

$$= \dfrac{2x^2 - 5x - 3}{3x - 1} \cdot \dfrac{3x^2 + 2x - 1}{x^2 - 2x - 3} \cdot \dfrac{3x}{2x^2 + x}$$

$$= \dfrac{(2x^2 - 5x - 3)(3x^2 + 2x - 1)(3x)}{(3x - 1)(x^2 - 2x - 3)(2x^2 + x)}$$

$$= \dfrac{(x - 3)(2x + 1)\,(3x - 1)(x + 1)3x}{(3x - 1)(x + 1)(x - 3)\,x(2x + 1)}$$

$$= 3$$

■

Example 9 Add the fractions $\dfrac{2}{x + 5}$ and $\dfrac{3x}{x + 5}$.

Solution Because these fractions have the same denominator, add the numerators and keep the common denominator.

$$\dfrac{2}{x + 5} + \dfrac{3x}{x + 5} = \dfrac{2 + 3x}{x + 5}$$

■

To add or subtract fractions with unlike denominators, we must find a common denominator. The easiest common denominator to use is called the **least (or lowest) common denominator**, or more simply the **LCD**. We now consider how it can be found.

Suppose the unlike denominators of three fractions are 12, 20, and 35. First, we find the unique prime factorization of each number.

$$12 = 4 \cdot 3 = 2^2 \cdot 3$$
$$20 = 4 \cdot 5 = 2^2 \cdot 5$$
$$35 = 5 \cdot 7$$

Because the least common denominator is the smallest number that can be divided by 12, 20, and 35, it must contain factors of 2^2, 3, 5, and 7. The least common denominator is

$$\text{LCD} = 2^2 \cdot 3 \cdot 5 \cdot 7 = 420$$

That is, 420 is the smallest number that can be divided by 12, 20, and 35.

When finding a least common denominator, always factor each denominator and then create the LCD by using each factor the greatest number of times that it appears in any one denominator. The product of these factors is the LCD.

Example 10   Add the fractions $\dfrac{1}{x^2 - 4}$ and $\dfrac{2}{x^2 - 4x + 4}$.

Solution   Factor each denominator and find the LCD.

$$x^2 - 4 = (x + 2)(x - 2)$$
$$x^2 - 4x + 4 = (x - 2)(x - 2) = (x - 2)^2$$

The LCD is $(x + 2)(x - 2)^2$. Write each fraction with its denominator in factored form, convert each fraction into an equivalent fraction with a denominator of $(x + 2)(x - 2)^2$, add the fractions, and simplify.

$$\frac{1}{x^2 - 4} + \frac{2}{x^2 - 4x + 4} = \frac{1}{(x + 2)(x - 2)} + \frac{2}{(x - 2)(x - 2)}$$

$$= \frac{1(x - 2)}{(x + 2)(x - 2)(x - 2)} + \frac{2(x + 2)}{(x - 2)(x - 2)(x + 2)}$$

$$= \frac{1(x - 2) + 2(x + 2)}{(x + 2)(x - 2)(x - 2)}$$

$$= \frac{x - 2 + 2x + 4}{(x + 2)(x - 2)(x - 2)}$$

$$= \frac{3x + 2}{(x + 2)(x - 2)(x - 2)}$$

Always attempt to simplify the final result. In this case, the final fraction is already in lowest terms.   ■

Example 11   Combine and simplify $\dfrac{x - 2}{x^2 - 1} - \dfrac{x + 3}{x^2 + 3x + 2} + \dfrac{3}{x^2 + x - 2}$.

Solution   Factor the denominators to determine the LCD.

$$x^2 - 1 = (x + 1)(x - 1)$$
$$x^2 + 3x + 2 = (x + 2)(x + 1)$$
$$x^2 + x - 2 = (x + 2)(x - 1)$$

The LCD is $(x + 1)(x + 2)(x - 1)$. Now write each fraction as an equivalent fraction with this common denominator and proceed as follows:

$$\frac{x - 2}{x^2 - 1} - \frac{x + 3}{x^2 + 3x + 2} + \frac{3}{x^2 + x - 2}$$

$$= \frac{x - 2}{(x - 1)(x + 1)} - \frac{x + 3}{(x + 1)(x + 2)} + \frac{3}{(x - 1)(x + 2)}$$

$$= \frac{(x - 2)(x + 2)}{(x - 1)(x + 1)(x + 2)} - \frac{(x + 3)(x - 1)}{(x + 1)(x + 2)(x - 1)} + \frac{3(x + 1)}{(x - 1)(x + 2)(x + 1)}$$

$$= \frac{(x^2 - 4) - (x^2 + 2x - 3) + (3x + 3)}{(x - 1)(x + 2)(x + 1)}$$

$$= \frac{x^2 - 4 - x^2 - 2x + 3 + 3x + 3}{(x - 1)(x + 2)(x + 1)}$$

$$= \frac{x + 2}{(x - 1)(x + 2)(x + 1)}$$

$$= \frac{1}{(x - 1)(x + 1)}$$ ∎

A fraction that has a fraction in its numerator or in its denominator is called a **complex fraction**.

Example 12 Simplify the complex fraction $\dfrac{\dfrac{1}{x} + \dfrac{1}{y}}{\dfrac{x}{y}}$.

Solution 1 Determine that the lowest common denominator of the three fractions in the given complex fraction is xy. Multiply both the numerator and denominator of the given fraction by xy and simplify:

$$\frac{\dfrac{1}{x} + \dfrac{1}{y}}{\dfrac{x}{y}} = \frac{\left(\dfrac{1}{x} + \dfrac{1}{y}\right)xy}{\left(\dfrac{x}{y}\right)xy} = \frac{\dfrac{xy}{x} + \dfrac{xy}{y}}{\dfrac{xxy}{y}} = \frac{y + x}{x^2}$$

Solution 2 Combine the fractions in the numerator of the complex fraction to obtain a single fraction over a single fraction. Then use the fact that any fraction indicates a division; that is, the numerator is divided by the denominator:

$$\frac{\dfrac{1}{x} + \dfrac{1}{y}}{\dfrac{x}{y}} = \frac{\dfrac{1(y)}{x(y)} + \dfrac{1(x)}{y(x)}}{\dfrac{x}{y}} = \frac{\dfrac{y + x}{xy}}{\dfrac{x}{y}}$$

Change this result to a multiplication problem and simplify.

$$\frac{\dfrac{y + x}{xy}}{\dfrac{x}{y}} = \frac{y + x}{xy} \div \frac{x}{y} = \frac{y + x}{xy} \cdot \frac{y}{x} = \frac{(y + x)y}{xyx} = \frac{y + x}{x^2}$$ ∎

═══════ **EXERCISE 1.7** ═══════

In Exercises 1–4, tell whether the given fractions are equal.

1. $\dfrac{8x}{3y}, \dfrac{16x}{6y}$

2. $\dfrac{3x^2}{4y^2}, \dfrac{12y^2}{16x^2}$

3. $\dfrac{25xyz}{12ab^2c}, \dfrac{50a^2bc}{24xyz}$

4. $\dfrac{15r^2s^2}{4uv^2}, \dfrac{45uv^2}{10r^2s^2}$

In Exercises 5–12, perform the indicated operations and simplify, whenever possible.

5. $\dfrac{4x}{7} \cdot \dfrac{2}{5a}$

6. $\dfrac{-5y}{2z} \cdot \dfrac{4}{y^2}$

7. $\dfrac{8m}{5n} \div \dfrac{3m}{10n}$

8. $\dfrac{15p}{8q} \div \dfrac{-5p}{16q^2}$

9. $\dfrac{3z}{5c} + \dfrac{2z}{5c}$

10. $\dfrac{7a}{4b} - \dfrac{3a}{4b}$

11. $\dfrac{15x^2y}{7a^2b^3} - \dfrac{x^2y}{7a^2b^3}$

12. $\dfrac{8rst^2}{15m^4n} + \dfrac{7rst^2}{15m^4n}$

In Exercises 13–20, simplify each fraction.

13. $\dfrac{2x - 4}{x^2 - 4}$

14. $\dfrac{x^2 - 16}{x^2 - 8x + 16}$

15. $\dfrac{25 - x^2}{x^2 + 10x + 25}$

16. $\dfrac{4 - x^2}{x^2 - 5x + 6}$

17. $\dfrac{6x^3 + x^2 - 12x}{4x^3 + 4x^2 - 3x}$

18. $\dfrac{6x^4 - 5x^3 - 6x^2}{2x^3 - 7x^2 - 15x}$

19. $\dfrac{x^3 - 8}{x^2 + ax - 2x - 2a}$

20. $\dfrac{xy + 2x + 3y + 6}{x^3 + 27}$

In Exercises 21–54, perform the indicated operations and simplify, whenever possible.

21. $\dfrac{x^2 - 1}{x} \cdot \dfrac{x^2}{x^2 + 2x + 1}$

22. $\dfrac{y^2 - 2y + 1}{y} \cdot \dfrac{y + 2}{y^2 + y - 2}$

23. $\dfrac{2x^2 + 32}{8} \div \dfrac{x^2 + 16}{2}$

24. $\dfrac{x^2 + x - 6}{x^2 - 6x + 9} \div \dfrac{x^2 - 4}{x^2 - 9}$

25. $\dfrac{z^2 + z - 20}{z^2 - 4} \div \dfrac{z^2 - 25}{z - 5}$

26. $\dfrac{ax + bx + a + b}{a^2 + 2ab + b^2} \div \dfrac{x^2 - 1}{x^2 - 2x + 1}$

27. $\dfrac{3x^2 + 7x + 2}{x^2 + 2x} \cdot \dfrac{x^2 - x}{3x^2 + x}$

28. $\dfrac{x^2 + x}{2x^2 + 3x} \cdot \dfrac{2x^2 + x - 3}{x^2 - 1}$

29. $\dfrac{x^2 + x}{x - 1} \cdot \dfrac{x^2 - 1}{x + 2}$

30. $\dfrac{x^2 + 5x + 6}{x^2 + 6x + 9} \cdot \dfrac{x + 2}{x^2 - 4}$

31. $\dfrac{3x^2 + 5x - 2}{x^3 + 2x^2} \cdot \dfrac{2x^3 + 5x^2}{6x^2 + 13x - 5}$

32. $\dfrac{x^2 + 7x + 12}{x^3 - x^2 - 6x} \cdot \dfrac{x^2 - 3x - 10}{x^2 + 2x - 3} \cdot \dfrac{x^3 - 4x^2 + 3x}{x^2 - x - 20}$

33. $\dfrac{x^2 + 13x + 12}{8x^2 - 6x - 5} \div \dfrac{2x^2 - x - 3}{8x^2 - 14x + 5}$

34. $\dfrac{x^2 - 2x - 3}{21x^2 - 50x - 16} \cdot \dfrac{3x - 8}{x - 3} \div \dfrac{x^2 + 6x + 5}{7x^2 - 33x - 10}$

35. $\dfrac{x^3 + 27}{x^2 - 4} \div \left(\dfrac{x^2 + 4x + 3}{x^2 + 2x} \div \dfrac{x^2 + x - 6}{x^2 - 3x + 9} \right)$

36. $\dfrac{x(x - 2) - 3}{x(x + 7) - 3(x - 1)} \cdot \dfrac{x(x + 1) - 2}{x(x - 7) + 3(x + 1)}$

37. $\dfrac{3}{x + 3} + \dfrac{x + 2}{x + 3}$

38. $\dfrac{3}{x + 1} + \dfrac{x + 2}{x + 1}$

39. $\dfrac{4}{x - 1} - \dfrac{3x}{x + 1}$

40. $\dfrac{2}{5 - x} + \dfrac{1}{x - 5}$

41. $\dfrac{a + 3}{a^2 + 7a + 12} + \dfrac{a}{a^2 - 16}$

42. $\dfrac{x}{x^2 - 4} - \dfrac{1}{x + 2}$

43. $\dfrac{2}{y^2 - 1} + 3 + \dfrac{1}{y + 1}$

44. $2 + \dfrac{4}{t^2 - 4} - \dfrac{1}{t - 2}$

45. $\dfrac{1}{x - 2} + \dfrac{3}{x + 2} - \dfrac{3x - 2}{x^2 - 4}$

46. $\dfrac{x}{x - 3} - \dfrac{5}{x + 3} + \dfrac{3(3x - 1)}{x^2 - 9}$

47. $\left(\dfrac{1}{x - 2} + \dfrac{1}{x - 3} \right) \cdot \dfrac{x - 3}{2x}$

48. $\left(\dfrac{1}{x + 1} - \dfrac{1}{x - 2} \right) \div \dfrac{1}{x - 2}$

49. $\dfrac{3x}{x - 4} - \dfrac{x}{x + 4} - \dfrac{3x + 1}{16 - x^2}$

50. $\dfrac{7x}{x - 5} + \dfrac{3x}{5 - x} + \dfrac{3x - 1}{x^2 - 25}$

51. $\dfrac{1}{x^2 + 3x + 2} - \dfrac{2}{x^2 + 4x + 3} + \dfrac{1}{x^2 + 5x + 6}$

52. $\dfrac{-2}{x - y} + \dfrac{2}{x - z} - \dfrac{2z - 2y}{(y - x)(z - x)}$

53. $\dfrac{3x - 2}{x^2 + x - 20} - \dfrac{4x^2 + 2}{x^2 - 25} + \dfrac{3x^2}{x^2 - 16}$

54. $\dfrac{3x + 2}{8x^2 - 10x - 3} + \dfrac{x + 4}{6x^2 - 11x + 3} - \dfrac{1}{4x + 1}$

In Exercises 55–82, simplify each complex fraction.

55. $\dfrac{\dfrac{3a}{b}}{\dfrac{6ac}{b^2}}$

56. $\dfrac{\dfrac{3t^2}{9x}}{\dfrac{t}{18x}}$

57. $\dfrac{\dfrac{3a^2b}{ab}}{27}$

58. $\dfrac{\dfrac{3u^2v}{4t}}{3uv}$

59. $\dfrac{\dfrac{x - y}{ab}}{\dfrac{y - x}{ab}}$

60. $\dfrac{\dfrac{x^2 - 5x + 6}{2x^2y}}{\dfrac{x^2 - 9}{2x^2y}}$

61. $\dfrac{\dfrac{1}{x} + \dfrac{1}{y}}{xy}$

62. $\dfrac{xy}{\dfrac{11}{x} - \dfrac{11}{y}}$

63. $\dfrac{\dfrac{1}{x} + \dfrac{1}{y}}{\dfrac{1}{x} - \dfrac{1}{y}}$

64. $\dfrac{\dfrac{1}{x} - \dfrac{1}{y}}{\dfrac{1}{x} + \dfrac{1}{y}}$

65. $\dfrac{\dfrac{3a}{b} - \dfrac{4a^2}{x}}{\dfrac{1}{b} + \dfrac{1}{ax}}$

66. $\dfrac{1 - \dfrac{x}{y}}{\dfrac{x^2}{y^2} - 1}$

67. $\dfrac{x + 1 - \dfrac{6}{x}}{x + 5 + \dfrac{6}{x}}$

68. $\dfrac{2z}{1 - \dfrac{3}{z}}$

69. $\dfrac{3xy}{1 - \dfrac{1}{xy}}$

70. $\dfrac{x - 3 + \dfrac{1}{x}}{-\dfrac{1}{x} - x + 3}$

71. $\dfrac{3x}{x + \dfrac{1}{x}}$

72. $\dfrac{2x^2 + 4}{2 + \dfrac{4x}{5}}$

73. $\dfrac{\dfrac{x}{x + 2} - \dfrac{2}{x - 1}}{\dfrac{3}{x + 2} + \dfrac{x}{x - 1}}$

74. $\dfrac{\dfrac{2x}{x - 3} + \dfrac{1}{x - 2}}{\dfrac{3}{x - 3} - \dfrac{x}{x - 2}}$

75. $\dfrac{1}{1 + x^{-1}}$

76. $\dfrac{y^{-1}}{x^{-1} + y^{-1}}$

77. $\dfrac{3(x + 2)^{-1} + 2(x - 1)^{-1}}{(x + 2)^{-1}}$

78. $\dfrac{2x(x - 3)^{-1} - 3(x + 2)^{-1}}{(x - 3)^{-1}(x + 2)^{-1}}$

79. $\dfrac{x}{1 + \dfrac{1}{3x^{-1}}}$

80. $\dfrac{ab}{2 + \dfrac{3}{2a^{-1}}}$

81. $\dfrac{1}{1 + \dfrac{1}{1 + \dfrac{1}{x}}}$

82. $\dfrac{y}{2 + \dfrac{2}{2 + \dfrac{2}{y}}}$

In Exercises 83–86, simplify each expression.

83. $\dfrac{\sqrt{x^2 + 4} - \dfrac{5}{\sqrt{x^2 + 4}}}{x + 1}$

84. $\dfrac{\sqrt{y^2 - 5} + \dfrac{4}{\sqrt{y^2 - 5}}}{y - 1}$

85. $\dfrac{\dfrac{3}{\sqrt{z^2 - 7}} + \sqrt{z^2 - 7}}{z + 2}$

86. $\dfrac{\dfrac{2}{\sqrt{t^2 + 1}} - \sqrt{t^2 + 1}}{t - 1}$

87. Prove the formula: $\dfrac{a}{b} + \dfrac{c}{d} = \dfrac{ad + bc}{bd}$.

88. Prove that $\dfrac{a}{b} \div \dfrac{c}{d} = \dfrac{a}{b} \cdot \dfrac{d}{c}$.

1.8   COMPLEX NUMBERS

At one time, mathematicians believed that square roots of negative numbers, denoted by symbols such as $\sqrt{-1}$, $\sqrt{-4}$, and $\sqrt{-7}$, were meaningless. In fact, these symbols were termed **imaginary numbers** by René Descartes (1596–1650). Mathematicians no longer think of imaginary numbers as undefined. Instead, they know that imaginary numbers have important applications, such as describing the behavior of alternating current in electronics.

The imaginary number $\sqrt{-1}$ is often denoted by the letter i. Because $i = \sqrt{-1}$, it follows that

$$i^2 = -1$$

The powers of i with natural-number exponents follow a special pattern:

$$i = \sqrt{-1} = i \qquad\qquad i^5 = i^4 i = 1i = i$$
$$i^2 = \sqrt{-1}\sqrt{-1} = -1 \qquad i^6 = i^4 i^2 = 1(-1) = -1$$
$$i^3 = i^2 i = -1i = -i \qquad i^7 = i^4 i^3 = 1(-i) = -i$$
$$i^4 = i^2 i^2 = -1(-1) = 1 \qquad i^8 = i^4 i^4 = 1(1) = 1$$

The pattern continues $i, -1, -i, 1, \ldots$.

Because of the previous pattern, it is easy to find many powers of i. If i is raised to the nth power and n is a natural number, then

If n leaves a remainder of 1 when divided by 4, then $i^n = i$.
If n leaves a remainder of 2 when divided by 4, then $i^n = -1$.
If n leaves a remainder of 3 when divided by 4, then $i^n = -i$.
If n is a multiple of 4, then $i^n = 1$.

Example 1   Simplify i^{366}.

Solution   Divide 366 by 4 to obtain a quotient of 91 and a remainder of 2. Thus, $366 = 4(91) + 2$, and

$$i^{366} = i^{4(91)+2} = (i^4)^{91} i^2 = (1)^{91}(-1) = 1(-1) = -1$$

Because 366 leaves a remainder of 2 when divided by 4,

$$i^{366} = -1$$ ■

We can simplify powers of i involving negative integral exponents by rationalizing denominators. For example,

$$i^{-3} = \frac{1}{i^3} = \frac{1 \cdot i}{i^3 \cdot i} = \frac{i}{i^4} = \frac{i}{1} = i$$

In the exercises, you will be asked to show that if at least one of two numbers x and y is nonnegative, then

$$\sqrt{xy} = \sqrt{x}\sqrt{y}$$

This property enables us to write the square root of any negative number in the form bi, where b is a real number and $i = \sqrt{-1}$. For example,

$$\sqrt{-4} = \sqrt{4(-1)} = \sqrt{4}\sqrt{-1} = 2i$$
$$\sqrt{-9} = \sqrt{9(-1)} = \sqrt{9}\sqrt{-1} = 3i$$

and

$$\sqrt{-\frac{1}{16}} = \sqrt{\frac{1}{16}(-1)} = \sqrt{\frac{1}{16}}\sqrt{-1} = \frac{1}{4}i$$

Because a square root of a number such as -3 is a number whose square is -3, we have

$$(\sqrt{-3})^2 = \sqrt{-3}\sqrt{-3} = -3$$

In this case, it is incorrect to apply the rule $\sqrt{xy} = \sqrt{x}\sqrt{y}$.

$$\sqrt{-3}\sqrt{-3} = \sqrt{(-3)(-3)} = \sqrt{9} = 3$$

The previous false statement illustrates that if x and y are both negative, the square root of a product is not the product of the square roots.

Expressions such as $3 + 4i$ and $-1 - 9i$ that indicate the sum or difference of a real number and an imaginary number forms a set of numbers called the **complex numbers**.

Definition.   A **complex number** is any number that can be expressed in the form $a + bi$, where a and b are real numbers and $i = \sqrt{-1}$.

   The number a is called the **real part** of the complex number, and the number b is called the **imaginary part**.

If $b = 0$, the complex number $a + bi$ is the real number a. Thus, any real number is a complex number with a zero imaginary part. If $a = 0$ and $b \neq 0$, the complex number $a + bi$ is the imaginary number bi. Thus, any imaginary number is a complex number with a zero real part. It follows that the real-number set and the imaginary-number set are both subsets of the complex-number set. Figure 1-12, shown on page 66, illustrates how the various number sets discussed thus far are related.

   We must accept some definitions before performing arithmetic with complex numbers.

Equality of Complex Numbers.   Two complex numbers are equal if and only if their real parts are equal and their imaginary parts are equal. Thus, if $a + bi$ and $c + di$ are two complex numbers, then

$$a + bi = c + di \qquad \text{if and only if} \qquad a = c \quad \text{and} \quad b = d$$

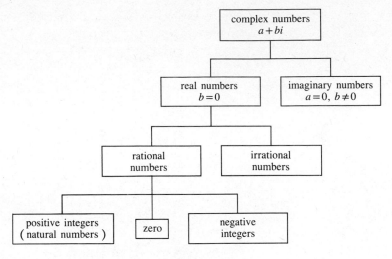

Figure 1-12

Example 2   For what real numbers x and y is $3x + 4i = (8y + x) + xi$?

Solution   Because the imaginary parts of these two complex numbers must be equal, $x = 4$. Because the real parts of the complex numbers must be equal,

$$3x = 8y + x$$

To find y, substitute 4 for x in the previous equation and solve for y:

$$3(4) = 8y + 4$$
$$12 = 8y + 4$$
$$8 = 8y$$
$$1 = y$$

Thus, $x = 4$ and $y = 1$.   ■

Complex numbers are added and subtracted as if they were binomials.

Addition and Subtraction of Complex Numbers.   Two complex numbers such as $a + bi$ and $c + di$ are added and subtracted as binomials:

$$(a + bi) + (c + di) = (a + c) + (b + d)i$$
$$(a + bi) - (c + di) = (a - c) + (b - d)i$$

Because of the preceding definition, the sum or difference of two complex numbers is another complex number.

Example 3   **a.** $(3 + 4i) + (2 + 7i) = 3 + 4i + 2 + 7i$

$$= 3 + 2 + 4i + 7i$$

$$= 5 + 11i$$

b. $(-5 + 8i) - (2 - 12i) = -5 + 8i - 2 + 12i$

$$= -5 - 2 + 8i + 12i$$

$$= -7 + 20i$$ ∎

Complex numbers are multiplied as if they were binomials.

Multiplication of Complex Numbers. Two complex numbers such as $a + bi$ and $c + di$ are multiplied as binomials, with $i^2 = -1$.

$$(a + bi)(c + di) = (ac - bd) + (ad + bc)i$$

Because of the preceding definition, the product of two complex numbers is another complex number.

Example 4   **a.** $(3 + 4i)(2 + 7i) = 6 + 21i + 8i + 28i^2$

$$= 6 + 21i + 8i + 28(-1)$$

$$= 6 - 28 + 29i$$

$$= -22 + 29i$$

b. $(5 - 7i)(1 + 3i) = 5 + 15i - 7i - 21i^2$

$$= 5 + 15i - 7i - 21(-1)$$

$$= 5 + 21 + 8i$$

$$= 26 + 8i$$ ∎

To avoid errors in determining the sign of the result, we will always express numbers in $a + bi$ form before attempting any algebraic manipulations.

Example 5   Find the product of $-2 + \sqrt{-16}$ and $4 - \sqrt{-9}$.

Solution   First change each complex number to $a + bi$ form.

$$-2 + \sqrt{-16} = -2 + \sqrt{16}\sqrt{-1} = -2 + 4i$$
$$4 - \sqrt{-9} = 4 - \sqrt{9}\sqrt{-1} = 4 - 3i$$

Then find the product of $-2 + 4i$ and $4 - 3i$.

$$(-2 + 4i)(4 - 3i) = -8 + 6i + 16i - 12i^2$$

$$= -8 + 6i + 16i - 12(-1)$$

$$= -8 + 12 + 22i$$

$$= 4 + 22i$$ ∎

The following definition is useful in finding the quotient of two complex numbers.

> **Definition.**   The complex numbers $a + bi$ and $a - bi$ are called **complex conjugates** of each other.

To find the product of $-\frac{1}{2} + 4i$ and its conjugate $-\frac{1}{2} - 4i$, for example, we proceed as follows:

$$\left(-\frac{1}{2} + 4i\right)\left(-\frac{1}{2} - 4i\right) = \frac{1}{4} + 2i - 2i - 16i^2$$

$$= \frac{1}{4} + 16$$

$$= \frac{65}{4}$$

In general, the product of the complex number $a + bi$ and its complex conjugate $a - bi$ is the real number $a^2 + b^2$, as the following work shows:

$$(a + bi)(a - bi) = a^2 - abi + abi - b^2i^2$$
$$= a^2 - b^2(-1)$$
$$= a^2 + b^2$$

This fact enables us to divide one complex number by another, as the next two examples illustrate.

Example 6   Write the number $\dfrac{3}{2 + i}$ in $a + bi$ form.

Solution   To write the number in $a + bi$ form, multiply both the numerator and the denominator by the conjugate of the denominator and simplify:

$$\frac{3}{2 + i} = \frac{3(2 - i)}{(2 + i)(2 - i)}$$

$$= \frac{6 - 3i}{4 - 2i + 2i - i^2}$$

$$= \frac{6 - 3i}{4 + 1}$$

$$= \frac{6 - 3i}{5}$$

$$= \frac{6}{5} - \frac{3}{5}i$$

$$= \frac{6}{5} + \left(-\frac{3}{5}\right)i$$

It is common practice to accept $\frac{6}{5} - \frac{3}{5}i$ as a substitute for $\frac{6}{5} + \left(-\frac{3}{5}\right)i$.   ∎

Example 7   Write the number $\dfrac{2 - \sqrt{-64}}{3 + \sqrt{-1}}$ in $a + bi$ form.

Solution $\dfrac{2 - \sqrt{-64}}{3 + \sqrt{-1}} = \dfrac{2 - 8i}{3 + i}$

$$= \frac{(2 - 8i)(3 - i)}{(3 + i)(3 - i)}$$

$$= \frac{6 - 2i - 24i + 8i^2}{9 - 3i + 3i - i^2}$$

$$= \frac{-2 - 26i}{9 + 1}$$

$$= \frac{-2}{10} - \frac{26i}{10}$$

$$= -\frac{1}{5} - \frac{13}{5}i \qquad\blacksquare$$

The results in Examples 6 and 7 illustrate that the quotient of two complex numbers is another complex number.

It is also possible to find absolute values of complex numbers.

Absolute Value of a Complex Number.   If $a + bi$ is a complex number, then

$$|a + bi| = \sqrt{a^2 + b^2}$$

Because of the previous definition, the absolute value of a complex number is a real number.

Example 8   **a.** $|3 + 4i| = \sqrt{3^2 + 4^2} = \sqrt{9 + 16} = \sqrt{25} = 5$

b. $|2 - 5i| = \sqrt{2^2 + (-5)^2} = \sqrt{4 + 25} = \sqrt{29}$ $\qquad\blacksquare$

Example 9   Find   **a.** $\left|\dfrac{2i}{3 + i}\right|$ and   **b.** $|a + 0i|$, where a is unrestricted.

Solution   **a.** To find $\left|\dfrac{2i}{3 + i}\right|$, first express $\dfrac{2i}{3 + i}$ in $a + bi$ form:

$$\frac{2i}{3 + i} = \frac{2i(3 - i)}{(3 + i)(3 - i)} = \frac{6i - 2i^2}{9 - i^2} = \frac{6i + 2}{10} = \frac{1}{5} + \frac{3}{5}i$$

Then find the absolute value of $\dfrac{1}{5} + \dfrac{3}{5}i$.

$$\left|\frac{2i}{3 + i}\right| = \left|\frac{1}{5} + \frac{3}{5}i\right| = \sqrt{\left(\frac{1}{5}\right)^2 + \left(\frac{3}{5}\right)^2} = \sqrt{\frac{10}{25}} = \frac{\sqrt{10}}{\sqrt{25}} = \frac{\sqrt{10}}{5}$$

b. If a is unrestricted, then

$$|a + 0i| = \sqrt{a^2 + 0^2} = \sqrt{a^2} = |a|$$

Note that this result is consistent with the definition of the absolute value of a real number.   ■

Because $i^2 = -1$, it is possible to factor the sum of two squares over the set of complex numbers. For example, to factor $9x^2 + 16y^2$, we proceed as follows:

$$9x^2 + 16y^2 = 9x^2 - (-1)16y^2$$
$$= 9x^2 - 16y^2i^2$$
$$= (3x + 4yi)(3x - 4yi) \qquad \text{Factor the difference of two squares.}$$

EXERCISE 1.8

In Exercises 1–4, simplify each expression.

1. i^9

2. i^{27}

3. i^{38}

4. i^{99}

In Exercises 5–8, find the values of x and y. Assume that x and y are real numbers.

5. $x + (x + y)i = 3 + 8i$

6. $x + 5i = y - yi$

7. $3x - 2yi = 2 + (x + y)i$

8. $\begin{cases} 2 + (x + y)i = 2 - i \\ x + 3i = 2 + 3i \end{cases}$

In Exercises 9–42, perform all indicated operations and express all answers in $a + bi$ form.

9. $(2 - 7i) + (3 + i)$

10. $(-7 + 2i) + (2 - 8i)$

11. $(5 - 6i) - (7 + 4i)$

12. $(11 + 2i) - (13 - 5i)$

13. $(14i + 2) + (2 - \sqrt{-16})$

14. $(5 + \sqrt{-64}) - (23i - 32)$

15. $(3 + \sqrt{-4}) - (2 + \sqrt{-9})$

16. $(7 - \sqrt{-25}) + (-8 + \sqrt{-1})$

17. $(2 + 3i)(3 + 5i)$

18. $(5 - 7i)(2 + i)$

19. $(2 + 3i)^2$

20. $(3 - 4i)^2$

21. $(11 + \sqrt{-25})(2 - \sqrt{-36})$

22. $(6 + \sqrt{-49})(6 - \sqrt{-49})$

23. $(\sqrt{-16} + 3)(2 + \sqrt{-9})$

24. $(12 - \sqrt{-4})(-7 + \sqrt{-25})$

25. $(2 - i)^3$

26. $(2 + 3i)^3$

27. $\dfrac{1}{i^3}$

28. $\dfrac{3}{i^5}$

29. $\dfrac{-7}{i^9}$

30. $\dfrac{-8i}{i^3}$

31. $\dfrac{1}{2 + i}$

32. $\dfrac{-2}{3 - i}$

33. $\dfrac{2i}{7 + i}$

34. $\dfrac{-3i}{2 + 5i}$

35. $\dfrac{2 + i}{3 - i}$

36. $\dfrac{3 - i}{1 + i}$

37. $\dfrac{4 - 5i}{2 + 3i}$

38. $\dfrac{34 + 2i}{2 - 4i}$

39. $\dfrac{5 - \sqrt{-16}}{-8 + \sqrt{-4}}$

40. $\dfrac{3 - \sqrt{-9}}{2 - \sqrt{-1}}$

41. $\dfrac{2 + i\sqrt{3}}{3 + i}$

42. $\dfrac{3}{4 - i\sqrt{2}}$

In Exercises 43–50, compute each absolute value.

43. $|2 + 3i|$ **44.** $|-5 - i|$ **45.** $|-7 + 7i|$ **46.** $\left| \dfrac{1}{2} - \dfrac{1}{4}i \right|$

47. $|-6i|$ **48.** $|5i|$ **49.** $\left| \dfrac{-3i}{2 + i} \right|$ **50.** $\left| \dfrac{5i}{i - 2} \right|$

In Exercises 51–54, use a calculator and write each expression in $a + bi$ form.

51. $(2.3 + 4.5i)(8.9 - 3.2i)$ **52.** $(6.73 - 3.25i)^2 + (1.75 + 2.21i)$

53. $\dfrac{4.7 + 11.2i}{5.2 - 3.7i} - (6.5 - 7.4i)$ **54.** $\dfrac{29.8 - 45.3i}{-7.4 + 27.3i}$

In Exercises 55–58, factor each expression over the set of complex numbers.

55. $x^2 + 4$ **56.** $16a^2 + 9$ **57.** $50m^2 + 2n^2$ **58.** $64a^4 + 4b^2$

59. Verify that $1 - i$ is a square root of $-2i$ by showing that $(1 - i)^2 = -2i$.

60. Verify that $2 - i$ is a square root of $3 - 4i$ by showing that $(2 - i)^2 = 3 - 4i$.

61. Show that the quotient $\dfrac{a + bi}{c + di}$ is a complex number.

In Exercises 62–67, let $z = a + bi$. Assume that $\bar{z}$ represents the conjugate of $z : \bar{z} = a - bi$.

62. Show that $|z| = |\bar{z}|$.

63. Show that $|z| + |\bar{z}| = 2|z|$.   **64.** Show that $|z| \, |\bar{z}| = |z|^2$.

65. Show that $|z + \bar{z}| = \sqrt{4a^2}$.   **66.** Show that $\sqrt{z\bar{z}} = |z|$.

67. Show that $z = \bar{z}$ if and only if z is a real number.

68. Show that the addition of two complex numbers is commutative. Do this by adding the complex numbers $a + bi$ and $c + di$ in both orders, and observing that the sums are equal.

69. Show that the multiplication of two complex numbers is commutative. Do this by multiplying the complex numbers $a + bi$ and $c + di$ in both orders, and observing that the products are equal.

70. Show that the addition of complex numbers is associative.

71. Find three examples of complex numbers that are reciprocals of their own conjugates.

72. Show that if at least one of the numbers x and y is not negative, then $\sqrt{xy} = \sqrt{x}\sqrt{y}$. (*Hint*: Note that because $(i\sqrt{x})^2 = -x$, then $\sqrt{-x} = i\sqrt{x}$. Let both x and y be positive numbers and evaluate $\sqrt{x(-y)}$ and $\sqrt{x}\sqrt{-y}$.)

CHAPTER SUMMARY

Key Words

absolute value (1.2)
absolute value of a complex number (1.8)
additive identity element (1.2)
additive inverse (1.2)

associative properties of addition and multiplication (1.2)
base of an exponential expression (1.3)
binomial (1.5)
closed interval (1.2)

Key Ideas

(1.1)   Rational numbers are either terminating or repeating decimals.

        Sets of numbers can be graphed on the number line.

(1.2)   Irrational numbers are nonterminating, nonrepeating decimals.

        The set of real numbers $\mathfrak{R}$ is the set of all decimals.

        If $x \geq 0$, then $|x| = x$. If $x < 0$, then $|x| = -x$.

        $|x| \geq 0$

        The distance between points a and b on a number line is $d = |a - b|$.

(1.3)   For any natural number n, $x^n = \overbrace{x \cdot x \cdot x \cdot \cdots \cdot x}^{n \text{ factors of } x}$.

        Rules of exponents: If no denominators are 0, then

$$x^m x^n = x^{m+n} \qquad (x^m)^n = x^{mn} \qquad (xy)^n = x^n y^n \qquad \left(\frac{x}{y}\right)^n = \frac{x^n}{y^n}$$

$$x^0 = 1 \qquad x^{-n} = \frac{1}{x^n} \qquad \frac{x^m}{x^n} = x^{m-n} \qquad \left(\frac{y}{x}\right)^{-n} = \left(\frac{x}{y}\right)^n$$

(1.4) $a^{1/n} = b$ if and only if $b^n = a$. $\sqrt[n]{a} = a^{1/n}$

$\sqrt[n]{a} = b$ if and only if $b^n = a$. $\sqrt[n]{a}\sqrt[n]{a} = a$

If m and n are positive integers and a and $\sqrt[n]{a}$ are real numbers, then

$$a^{m/n} = (\sqrt[n]{a})^m = \sqrt[n]{a^m}$$

If all radicals are real numbers and there are no divisions by 0, then

$$\sqrt[n]{ab} = \sqrt[n]{a}\sqrt[n]{b} \qquad \sqrt[n]{\frac{a}{b}} = \frac{\sqrt[n]{a}}{\sqrt[n]{b}} \qquad \sqrt[m]{\sqrt[n]{a}} = \sqrt[n]{\sqrt[m]{a}} = \sqrt[mn]{a}$$

(1.5) $(x + y)^2 = x^2 + 2xy + y^2$ $(x - y)^2 = x^2 - 2xy + y^2$ $(x + y)(x - y) = x^2 - y^2$

(1.6) $x^2 - y^2 = (x + y)(x - y)$ $x^2 + 2xy + y^2 = (x + y)^2$

$x^3 + y^3 = (x + y)(x^2 - xy + y^2)$ $x^2 - 2xy + y^2 = (x - y)^2$

$x^3 - y^3 = (x - y)(x^2 + xy + y^2)$

(1.7)   **Properties of fractions:**

The denominator of a fraction can never be 0.

$$\frac{a}{b} = \frac{c}{d} \text{ if and only if } ad = bc. \qquad \frac{a}{b} = \frac{-a}{-b} = -\frac{a}{-b} = -\frac{-a}{b}$$

$$a \cdot 1 = a, \quad \frac{a}{1} = a, \quad \frac{a}{a} = 1 \qquad -\frac{a}{b} = \frac{a}{-b} = \frac{-a}{b} = -\frac{-a}{-b}$$

$$\frac{a}{b} \cdot \frac{c}{d} = \frac{ac}{bd} \qquad \frac{a}{b} \div \frac{c}{d} = \frac{ad}{bc} \qquad \frac{a}{b} + \frac{c}{b} = \frac{a + c}{b} \qquad \frac{a}{b} - \frac{c}{b} = \frac{a - c}{b}$$

$$\frac{a}{b} = \frac{ax}{bx} \quad \text{provided } x \neq 0$$

(1.8)   The complex numbers $a + bi$ and $c + di$ are equal if and only if $a = c$ and $b = d$.

$(a + bi) + (c + di) = (a + c) + (b + d)i$

$(a + bi)(c + di) = (ac - bd) + (ad + bc)i$

$|a + bi| = \sqrt{a^2 + b^2}$

REVIEW EXERCISES

In Review Exercises 1–4, assume that $\mathbf{A} = \{2, 3, 5, 7, 11\}$, $\mathbf{B} = \{2, 4, 6, 8, 10\}$, *and* $\mathbf{C} = \{1, 2, 3, 4, 5\}$. *Find each set.*

1. $\mathbf{A} \cap \mathbf{B}$ **2.** $\mathbf{A} \cup \mathbf{C}$ **3.** $\mathbf{A} \cap \mathbf{B} \cap \mathbf{C}$ **4.** $\mathbf{A} \cup (\mathbf{B} \cap \varnothing)$

5. Change $\frac{25}{27}$ to decimal form.   **6.** Change $0.8\overline{61}$ to a common fraction.

7. Graph the prime numbers between 10 and 20.   **8.** Does the decimal 2.3773 represent an irrational number? Explain.

In Review Exercises 9–12, graph each interval on the number line.

9. $-3 < x \le 5$ **10.** $x \ge 0$ or $x < -1$ **11.** $(-2, 4]$ **12.** $[-2, 4) \cup (0, 5]$

In Review Exercises 13–14, assume that x = −2, y = 3, and z = −1. Find the value of each expression.

13. $|xyz|$

14. $\dfrac{|xz|}{y} - \left|\dfrac{y}{z}\right|$

In Review Exercises 15–20, indicate which field property justifies each given expression.

15. $(a + b) + 2 = a + (b + 2)$

16. $a + 7 = 7 + a$

17. $(6 + a) + 0 = 6 + a$

18. $4(2x) = (4 \cdot 2)x$

19. $(a + b) \cdot \dfrac{1}{a + b} = 1$

20. $1(a - b) = a - b$

In Review Exercises 21–28, simplify each expression. Express all answers with positive exponents only. Assume that all variables represent positive numbers.

21. $\left(\dfrac{x^3}{y^2}\right)^3$

22. $-(x^3 y^{-2})^2$

23. $\left(\dfrac{x^{-2} y^2}{2}\right)^3$

24. $\left(\dfrac{x^2 y}{3x^{-1}}\right)^4$

25. $\left(\dfrac{3x^2 y^{-2}}{x^2 y^2}\right)^{-2}$

26. $\left(\dfrac{a^{-3} b^2}{ab^{-3}}\right)^{-2}$

27. $\left(\dfrac{-3x^3 y}{xy^3}\right)^{-2}$

28. $\left(-\dfrac{2m^{-2} n^0}{4m^2 n^{-1}}\right)^{-3}$

In Review Exercises 29–36, evaluate each expression if x = 0, y = −1, and z = 2.

29. $x^2(3^x)$

30. $y^3(3^y)$

31. $z^{yz}(z^{xy})$

32. $(z^z y^z)^y$

33. zy^x

34. $(y^z y^{yz})^x$

35. y^{xyz}

36. $\left(\dfrac{4z}{y^z}\right)^y$

In Review Exercises 37–48, simplify each expression.

37. $4^{1/2}$

38. $8^{2/3}$

39. $-8^{2/3}$

40. $(-8)^{5/3}$

41. $\left(\dfrac{16}{81}\right)^{3/4}$

42. $\left(\dfrac{32}{243}\right)^{2/5}$

43. $\left(\dfrac{8}{27}\right)^{-2/3}$

44. $\left(\dfrac{16}{625}\right)^{-3/4}$

45. $\sqrt{36}$

46. $-\sqrt{49}$

47. $\sqrt{\dfrac{9}{25}}$

48. $\sqrt[3]{\dfrac{27}{125}}$

In Review Exercises 49–56, assume that all variables are positive numbers. Simplify each expression.

49. $(x^{12} y^2)^{1/2}$

50. $\left(\dfrac{x^{14}}{y^4}\right)^{-1/2}$

51. $\left(\dfrac{-c^{2/3} c^{5/3}}{c^{-2/3}}\right)^{1/3}$

52. $\left(\dfrac{a^{-1/4} a^{3/4}}{a^{9/2}}\right)^{-1/2}$

53. $\sqrt{x^2 y^4}$

54. $\sqrt[3]{x^3}$

55. $\sqrt[4]{\dfrac{m^8 n^4}{p^{12}}}$

56. $\sqrt[5]{\dfrac{a^{15} b^{10}}{c^5}}$

In Review Exercises 57–60, assume that all variables are unrestricted numbers. Simplify each expression.

57. $(x^{16} y^4 c^2)^{1/2}$

58. $\left(\dfrac{a^{14}}{b^6}\right)^{-1/2}$

59. $\sqrt{x^4 y^8}$

60. $\sqrt{a^2 b^2}$

In Review Exercises 61–68, rationalize each denominator and simplify.

61. $\dfrac{2}{\sqrt{5}}$

62. $\dfrac{8}{\sqrt{8}}$

63. $\dfrac{1}{\sqrt[3]{2}}$

64. $\dfrac{2}{\sqrt[3]{25}}$

65. $\dfrac{2}{\sqrt{3} - 1}$

66. $\dfrac{-2}{\sqrt{3} - \sqrt{2}}$

67. $\dfrac{2x}{\sqrt{x} - 2}$

68. $\dfrac{\sqrt{y}}{\sqrt{x} + \sqrt{y}}$

In Review Exercises 69–74, simplify each expression and combine terms.

69. $\sqrt{50} + \sqrt{8}$

70. $\sqrt{12} + \sqrt{3} - \sqrt{27}$

71. $(\sqrt{2} + \sqrt{3})^2$

72. $(2 + \sqrt{3})(\sqrt{3} - 2)$

73. $(\sqrt{2} + 1)(\sqrt{3} + 1)$

74. $(\sqrt[3]{3} - 2)(\sqrt[3]{9} + 2\sqrt[3]{3} + 4)$

In Review Exercises 75–78, indicate whether the given expression is a polynomial, and if so, give its degree and then indicate whether it is a monomial, a binomial, or a trinomial.

75. $x^3 - 8$

76. $8x^2 - 8x - 8$

77. $\sqrt{3}x^2$

78. $3x^2 - \sqrt{x}$

In Review Exercises 79–82, perform the indicated operations and simplify.

79. $3x^2(x - 1) - 2x(x + 3) - x^2(x + 2)$

80. $(3x + y)(2x - 3y)$

81. $(4a + 2b)(2a - 3b)$

82. $(x + 3)(3x^2 + x - 1)$

In Review Exercises 83–84, perform each division. Express all answers without using negative exponents.

83. $\dfrac{8a^2b^3 + 12a^4b^2}{4a^5b^2}$

84. $\dfrac{16x^2y - 8xyz + 4yz^2}{12x^2y^2z^2}$

In Review Exercises 85–88, simplify each fraction by performing a long division. If there is a nonzero remainder, write the answer in $\text{quotient} + \dfrac{\text{remainder}}{\text{divisor}}$ *form.*

85. $\dfrac{2x^3 + 13x^2 + 17x - 14}{2x + 7}$

86. $\dfrac{x^4 + x^3 + 2x^2 + x + 1}{x^2 + 1}$

87. $\dfrac{3x^2 + 1}{x + 1}$

88. $\dfrac{x^3 - x + 5}{x^2 + 3x}$

In Review Exercises 89–102, completely factor each polynomial, if possible.

89. $3x^3 - 3x$

90. $5x^3 - 5$

91. $6x^2 + 7x - 24$

92. $3a^2 + ax - 3a - x$

93. $8x^3 - 125$

94. $6x^2 - 20x - 16$

95. $x^2 + 6x + 9 - 4t^2$

96. $3x^2 - 1 + 5x$

97. $8z^3 + 343$

98. $1 + 14b + 49b^2$

99. $121z^2 + 4 - 44z$

100. $64y^3 - 1000$

101. $2xy - 4zx - wy + 2zw$

102. $x^8 + x^4 + 1$

In Review Exercises 103–118, perform the indicated operations and simplify.

103. $\dfrac{x^2 - 4x + 4}{x + 2} \cdot \dfrac{x^2 + 5x + 6}{x - 2}$

104. $\dfrac{2x^2 - 11x + 15}{x^2 - 6x + 8} \cdot \dfrac{x^2 - 2x - 8}{x^2 - x - 6}$

105. $\dfrac{2x^2 + x - 3}{3x^2 - 7x + 4} \div \dfrac{10x + 15}{3x^2 - x - 4}$

106. $\dfrac{x^2 + 7x + 12}{x^3 + 8x^2 + 4x} \div \dfrac{x^2 - 9}{x^2}$

107. $\dfrac{x^2 + x - 6}{x^2 - x - 6} \cdot \dfrac{x^2 - x - 6}{x^2 + x - 2} \div \dfrac{x^2 - 4}{x^2 - 5x + 6}$

108. $\left(\dfrac{2x + 6}{x + 5} \div \dfrac{2x^2 - 2x - 4}{x^2 - 25}\right) \dfrac{x^2 - x - 2}{x^2 - 2x - 15}$

109. $\dfrac{2}{x - 4} + \dfrac{3x}{x + 5}$

110. $\dfrac{5x}{x - 2} - \dfrac{3x + 1}{x + 3}$

111. $\dfrac{x}{x - 1} + \dfrac{x}{x - 2} + \dfrac{x}{x - 3}$

112. $\dfrac{x}{x + 1} - \dfrac{3x + 7}{x + 2} + \dfrac{2x + 1}{x + 2}$

113. $\dfrac{3(x + 1)}{x} - \dfrac{5(x^2 + 3)}{x^2} + \dfrac{x}{x + 1}$

114. $\dfrac{3x}{x + 1} + \dfrac{x^2 + 4x + 3}{x^2 + 3x + 2} - \dfrac{x^2 + x - 6}{x^2 - 4}$

115. $\dfrac{\dfrac{5x}{2}}{\dfrac{3x^2}{8}}$

116. $\dfrac{\dfrac{3x}{y}}{\dfrac{6x}{y^2}}$

117. $\dfrac{\dfrac{1}{x} + \dfrac{1}{y}}{x - y}$

118. $\dfrac{\dfrac{1}{x} + \dfrac{1}{y}}{\dfrac{1}{y} - \dfrac{1}{x}}$

In Review Exercises 119–130, perform all indicated operations and express all answers in a + bi form.

119. i^{53}

120. i^{103}

121. $(2 + 3i) - (-4 + 2i)$

122. $(3 - \sqrt{-36}) + (\sqrt{-16} + 2)$

123. $(3 + 4i)(1 - 2i)$

124. $(3 + \sqrt{-9})(\sqrt{2} - \sqrt{-25})$

125. $\dfrac{3}{i}$

126. $-\dfrac{2}{i^3}$

127. $\dfrac{3}{1 + i}$

128. $-\dfrac{2i}{2 - i}$

129. $\dfrac{3 + i}{1 - i}$

130. $\dfrac{\sqrt{2} + 3i}{i - \sqrt{3}}$

In Review Exercises 131–134, find each absolute value.

131. $|2 + 7i|$

132. $|3 - i|$

133. $\left|\dfrac{3i}{-2 - i}\right|$

134. $\left|\dfrac{1 + i}{1 - i}\right|$

2 Equations

The topic of this chapter is equations, one of the most fundamental concepts in algebra. Equations are used in almost every academic discipline and vocational area, especially in chemistry, physics, medicine, economics, electronics, and business.

2.1 LINEAR EQUATIONS AND THEIR SOLUTIONS

An **equation** is a statement indicating that two quantities are equal. An equation can be either true or false. For example, the equation $2 + 2 = 4$ is true, and the equation $2 + 3 = 6$ is false. An equation such as $3x - 2 = 10$ is true or false depending on the value of x. If $x = 4$, the equation is true, as the following work shows:

$$3x - 2 = 10$$
$$3(4) - 2 = 10$$
$$12 - 2 = 10$$
$$10 = 10$$

Because 4 makes the equation true, we say it **satisfies** the equation. The equation is false for all other numbers x.

The set of all permissible replacements for a variable such as x is called the **domain of the variable**. In the equation $3x - 2 = 10$, the domain of x consists of all real numbers. Of all the numbers in the domain, only the number 4 makes the equation true. Thus, 4 is called a **solution** or **root** of the equation. The set of all solutions of an equation is called the **solution set** of the equation. The solution set of $3x - 2 = 10$ is $\{4\}$. To **solve** an equation means to find all of its solutions.

Example 1   Find the domain of x in the equation $\sqrt{x} = \dfrac{2}{x - 1}$. Assume that all expressions represent real numbers.

Solution   The domain of x consists of all real numbers that are permissible replacements for x. For $\sqrt{x}$ to be a real number, x must be nonnegative, and for $\dfrac{2}{x - 1}$ to be a real number, x cannot be 1. Thus, the domain of x is the set of all nonnegative real numbers, except 1.   ∎

For some equations, every number in the domain of the variable is a solution of the equation. Such equations are called **identities**. The equation

$$x^2 - 9 = (x + 3)(x - 3)$$

is an identity because every real number x is a solution. For other equations, no number in the domain of the variable is a solution. These equations are called **impossible equations** or **contradictions**. The equation

$$3x + 3 = 3x + 2$$

has no solutions.

An equation whose solution set contains some, but not all, of the members of the domain of the variable is called a **conditional equation**. In this section, we will discuss methods to solve many of these equations.

> **Definition.**   If two equations, each with one variable, have the same solution set, they are called **equivalent equations**.

There are several properties that we can use to transform equations into equivalent, but less complicated, equations. If we use these properties, the resulting equation will be equivalent to the original equation, and the two equations will have the same solution set.

> **The Addition and Subtraction Properties of Equality.**   If a, b, and c are real numbers and $a = b$, then
>
> $$a + c = b + c \quad \text{and} \quad a - c = b - c$$
>
> **The Multiplication and Division Properties of Equality.**   If a, b, and c are real numbers, $a = b$, and $c \neq 0$, then
>
> $$ac = bc \quad \text{and} \quad \frac{a}{c} = \frac{b}{c}$$
>
> **The Substitution Property of Equality.**   In an equation, a quantity may be substituted for its equal without changing the truth of the equation.

The easiest equations to solve are the **first-degree** or **linear equations**.

> **Definition.**   A **linear equation in one variable**, say x, is any equation that can be written in the form
>
> $$ax + c = 0 \quad (a \text{ and } c \text{ are real numbers and } a \neq 0)$$

To solve the linear equation $2x + 3 = 0$, we subtract 3 from both sides of the equation and then divide both sides by 2 to obtain

$$2x + 3 = 0$$
$$2x + 3 - 3 = 0 - 3$$
$$2x = -3$$
$$\frac{2x}{2} = -\frac{3}{2}$$
$$x = -\frac{3}{2}$$

The solution of $2x + 3 = 0$ is $-\frac{3}{2}$. To show that this solution satisfies the equation, we substitute $-\frac{3}{2}$ for x in the original equation and simplify:

$$2x + 3 = 0$$
$$2\left(-\frac{3}{2}\right) + 3 \stackrel{?}{=} 0$$
$$-3 + 3 \stackrel{?}{=} 0$$
$$0 = 0$$

Because both sides of the equation are equal, the solution checks.

As this example suggests, every linear equation has exactly one solution.

Example 2 Find the solution set for $3(x + 2) = 5x + 2$.

Solution Proceed as follows:

$$3(x + 2) = 5x + 2$$

$3x + 6 = 5x + 2$	Use the distributive property and remove parentheses.
$3x - 3x + 6 = 5x - 3x + 2$	Subtract $3x$ from both sides.
$6 = 2x + 2$	Simplify.
$6 - 2 = 2x + 2 - 2$	Subtract 2 from both sides.
$4 = 2x$	Simplify.
$\dfrac{4}{2} = \dfrac{2x}{2}$	Divide both sides by 2.
$2 = x$	Simplify.

Because all of the equations above are equivalent, the solution set of the final equation is the same as the solution set of the original equation. Thus, the solution set of the original equation is $\{2\}$. Verify that 2 satisfies the equation. ■

Example 3   Solve the equation $\frac{3}{2}x - \frac{2}{3} = \frac{1}{5}x$.

Solution   To clear the equation of fractions, multiply both sides by the least common denominator of the three fractions. Then proceed as follows:

$$\frac{3}{2}x - \frac{2}{3} = \frac{1}{5}x$$

$$30\left(\frac{3}{2}x - \frac{2}{3}\right) = 30\left(\frac{1}{5}x\right) \qquad \text{Multiply both sides by 30.}$$

$$45x - 20 = 6x \qquad \text{Remove parentheses and simplify.}$$

$$45x - 20 + 20 = 6x + 20 \qquad \text{Add 20 to both sides.}$$

$$45x = 6x + 20 \qquad \text{Simplify.}$$

$$45x - 6x = 6x - 6x + 20 \qquad \text{Subtract } 6x \text{ from both sides.}$$

$$39x = 20 \qquad \text{Simplify.}$$

$$\frac{39x}{39} = \frac{20}{39} \qquad \text{Divide both sides by 39.}$$

$$x = \frac{20}{39} \qquad \text{Simplify.}$$

The solution of the given equation is $\frac{20}{39}$. Verify that this solution satisifies the equation.   ■

Example 4   Solve the equations   **a.** $3(x + 5) = 3(1 + x)$ and   **b.** $5 + 5(x + 2) - 2x = 3x + 15$.

Solution   Proceed as follows:

a.
$$3(x + 5) = 3(1 + x)$$
$$3x + 15 = 3 + 3x \qquad \text{Remove parentheses.}$$
$$3x - 3x + 15 = 3 + 3x - 3x \qquad \text{Subtract } 3x \text{ from both sides.}$$
$$15 = 3 \qquad \text{Simplify.}$$

Because the final equation is false, it has no solutions. Thus, the original equation has no solutions either. Its solution set is $\varnothing$.

b.
$$5 + 5(x + 2) - 2x = 3x + 15$$
$$5 + 5x + 10 - 2x = 3x + 15 \qquad \text{Remove parentheses.}$$
$$3x + 15 = 3x + 15 \qquad \text{Simplify.}$$

Because both sides of the final equation are the same, they are always equal, and every value of x will make the equation true. Thus, the solution set of the original equation is the set of all real numbers. This equation is an identity.   ■

Equations that contain fractions with polynomials in their denominators are called **rational equations**. When solving these equations, we often multiply both sides of an equation by a quantity that contains a variable. However, when we do this, it is possible to inadvertently multiply both sides of an equation by 0 and obtain a solution that makes the denominator of a fraction 0. In this case, we have found a false solution, called an **extraneous solution**. We must be careful to exclude from the solution set of an equation any value that makes the denominator of a fraction 0 and thus is not in the domain of the variable. The following equation has an extraneous solution.

$$\frac{x+1}{x-2} = \frac{3}{x-2}$$

$$(x-2)\left(\frac{x+1}{x-2}\right) = (x-2)\left(\frac{3}{x-2}\right) \qquad \text{Multiply both sides by } x-2.$$

$$x + 1 = 3 \qquad \text{Simplify.}$$

$$x = 2 \qquad \begin{array}{l}\text{Subtract 1 from both sides and} \\ \text{simplify.}\end{array}$$

If we attempt to check this solution by substituting 2 for x in the original equation, we obtain 0's in the denominator. Thus, 2 is not a solution. The solution set of this equation is $\varnothing$.

Example 5 Solve the equation $\dfrac{x+2}{x+3} + \dfrac{1}{x^2 + 2x - 3} = 1$.

Solution Note that x cannot be -3, because that would cause the denominator of the first fraction to be 0. To find any other restricted values, factor the trinomial in the denominator of the second fraction:

$$x^2 + 2x - 3 = (x+3)(x-1)$$

This denominator will be 0 if $x = -3$ or $x = 1$. Thus, x cannot be either -3 or 1. Solve the equation as follows:

$$\frac{x+2}{x+3} + \frac{1}{(x^2 + 2x - 3)} = 1$$

$$\frac{x+2}{x+3} + \frac{1}{(x+3)(x-1)} = 1 \qquad \begin{array}{l}\text{Factor the second} \\ \text{denominator.}\end{array}$$

$$(x+3)(x-1)\left[\frac{x+2}{x+3} + \frac{1}{(x+3)(x-1)}\right] = (x+3)(x-1)1 \qquad \begin{array}{l}\text{Multiply both sides by} \\ (x+3)(x-1).\end{array}$$

$$(x+3)(x-1)\frac{x+2}{x+3} + (x+3)(x-1)\frac{1}{(x+3)(x-1)} = (x+3)(x-1)1 \qquad \text{Remove brackets.}$$

$$(x-1)(x+2) + 1 = (x+3)(x-1) \qquad \text{Simplify.}$$

$$x^2 + x - 2 + 1 = x^2 + 2x - 3 \qquad \text{Multiply the binomials.}$$

$$x - 1 = 2x - 3 \qquad \begin{array}{l}\text{Add } -x^2 \text{ to both sides} \\ \text{and combine terms.}\end{array}$$

$$2 = x \qquad \text{Add 3 and } -x \text{ to both sides.}$$

Because 2 is in the domain of x, it is a solution. However, it is always a good idea to check the solution. Show that 2 is a root by verifying that it satisfies the given equation by substituting 2 for x in the original equation and simplifying:

$$\frac{x + 2}{x + 3} + \frac{1}{x^2 + 2x - 3} = 1$$

$$\frac{2 + 2}{2 + 3} + \frac{1}{2^2 + 2(2) - 3} \stackrel{?}{=} 1$$

$$\frac{4}{5} + \frac{1}{5} \stackrel{?}{=} 1$$

$$1 = 1$$

Because 2 satisifies the original equation, it is a solution.   ■

EXERCISE 2.1

In Exercises 1–8, each quantity represents a real number. Find the domain of x.

1. $x + 3 = 1$ **2.** $\frac{1}{2}x - 7 = 14$ **3.** $\frac{1}{x} = 12$ **4.** $\frac{3}{x - 2} = 2$

5. $\sqrt{x} = 4$ **6.** $\sqrt[3]{x} = 64$ **7.** $\frac{1}{x - 3} = \frac{5}{x + 2}$ **8.** $\frac{24}{\sqrt{x - 3}} = 13$

In Exercises 9–22, solve each equation, if possible. Classify each one as an identity, a conditional equation, or an equation with no solutions.

9. $2x + 5 = 15$ **10.** $3x + 2 = x + 8$

11. $2(x + 2) = 2x + 5$ **12.** $3(x + 2) - x = 2(x + 3)$

13. $\frac{x + 7}{2} = 7$ **14.** $\frac{x}{2} - 7 = 14$

15. $2(a + 1) = 3(a - 2) - a$ **16.** $x^2 = (x + 4)(x - 4) + 16$

17. $3(x - 3) = \frac{6x - 18}{2}$ **18.** $x(x + 2) = (x + 1)^2$

19. $\frac{3}{b - 3} = 1$ **20.** $x^2 - 8x + 15 = (x - 3)(x + 5)$

21. $2x^2 + 5x - 3 = (2x - 1)(x + 3)$ **22.** $2x^2 + 5x - 3 = 2x\left(x + \frac{19}{2}\right)$

In Exercises 23–64, solve each equation, if possible.

23. $2x + 7 = 10 - x$ **24.** $9x - 3 = 15 + 3x$ **25.** $\frac{5}{3}z - 8 = 7$ **26.** $\frac{4}{3}y + 12 = -4$

27. $\frac{z}{5} + 2 = 4$ **28.** $\frac{3p}{7} - p = -4$ **29.** $\frac{3x - 2}{3} = 2x + \frac{7}{3}$ **30.** $\frac{7}{2}x + 5 = x + \frac{15}{2}$

31. $5(x - 2) = 2x + 8$ **32.** $5(r - 4) = -5(r - 4)$

33. $2(2x + 1) - \frac{3x}{2} = \frac{-3(4 + x)}{2}$ **34.** $(x - 2)(x - 3) = (x + 3)(x + 4)$

35. $7(2x + 5) - 6(x + 8) = 7$

36. $(t + 1)(t - 1) = (t + 2)(t - 3) + 4$

37. $(x - 2)(x + 5) = (x - 3)(x + 2)$

38. $\dfrac{3x + 1}{20} = \dfrac{1}{2}$

39. $\dfrac{3}{2}(3x - 2) - 10x - 4 = 0$

40. $a(a - 3) + 5 = (a - 1)^2$

41. $x(x + 2) = (x + 1)^2 - 1$

42. $\dfrac{2 + x}{3} + \dfrac{x + 7}{2} = 4x + 1$

43. $\dfrac{(y + 2)^2}{3} = y + 2 + \dfrac{y^2}{3}$

44. $2x - \dfrac{7}{6} + \dfrac{x}{6} = \dfrac{4x + 2}{6}$

45. $2(s + 2) + (s + 3)^2 = s(s + 5) + 2\left(\dfrac{17}{2} + s\right)$

46. $\dfrac{3}{x} + \dfrac{1}{2} = \dfrac{4}{x}$

47. $\dfrac{2}{x + 1} + \dfrac{1}{3} = \dfrac{1}{x + 1}$

48. $\dfrac{3}{x - 2} + \dfrac{1}{x} = \dfrac{3}{x - 2}$

49. $\dfrac{9x + 6}{x(x + 3)} = \dfrac{7}{x + 3}$

50. $x + \dfrac{2(-2x + 1)}{3x + 5} = \dfrac{3x^2}{3x + 5}$

51. $\dfrac{2}{(n - 7)(n + 2)} = \dfrac{4}{(n + 3)(n + 2)}$

52. $\dfrac{2}{a - 2} + \dfrac{1}{a + 1} = \dfrac{1}{a^2 - a - 2}$

53. $\dfrac{2x + 3}{x^2 + 5x + 6} + \dfrac{3x - 2}{x^2 + x - 6} = \dfrac{5x - 2}{x^2 - 4}$

54. $\dfrac{3x}{x^2 + x} - \dfrac{2x}{x^2 + 5x} = \dfrac{x + 2}{x^2 + 6x + 5}$

55. $\dfrac{3x + 5}{x^3 + 8} + \dfrac{3}{x^2 - 4} = \dfrac{2(3x - 2)}{(x - 2)(x^2 - 2x + 4)}$

56. $\dfrac{1}{n + 8} - \dfrac{3n - 4}{5n^2 + 42n + 16} = \dfrac{1}{5n + 2}$

57. $\dfrac{1}{11 - n} - \dfrac{2(3n - 1)}{-7n^2 + 74n + 33} = \dfrac{1}{7n + 3}$

58. $\dfrac{4}{a^2 - 13a - 48} - \dfrac{2}{a^2 - 18a + 32} = \dfrac{1}{a^2 + a - 6}$

59. $\dfrac{5}{y + 4} + \dfrac{2}{y + 2} = \dfrac{6}{y + 2} - \dfrac{1}{y^2 + 6y + 8}$

60. $\dfrac{6}{2a - 6} - \dfrac{3}{3 - 3a} = \dfrac{1}{a^2 - 4a + 3}$

61. $\dfrac{3y}{6 - 3y} + \dfrac{2y}{2y + 4} = \dfrac{8}{4 - y^2}$

62. $\dfrac{3 + 2a}{a^2 + 6 + 5a} - \dfrac{2 - 3a}{a^2 - 6 + a} = \dfrac{5a - 2}{a^2 - 4}$

63. $\dfrac{a}{a + 2} - 1 = -\dfrac{3a + 2}{a^2 + 4a + 4}$

64. $\dfrac{x - 1}{x + 3} + \dfrac{x - 2}{x - 3} = \dfrac{1 - 2x}{3 - x}$

2.2 APPLICATIONS OF LINEAR EQUATIONS

In this section, we will apply equation-solving techniques to word problems. The list of steps below provides a strategy to follow.

1. Read the problem several times until you understand the facts that are given. What information is given? What are you asked to find? Draw a sketch or diagram, if possible, to help you visualize the facts of the problem.
2. Pick a variable to represent the quantity that is to be found, and write a sentence telling what that variable represents. Express all of the other quantities mentioned in the problem as expressions involving this single variable.

3. Organize the data and find a way of expressing the same quantity in two different ways. This may involve a formula from geometry, finance, or physics.
4. Set up an equation indicating that the two quantities found in Step 3 are equal.
5. Solve the equation.
6. Answer the questions posed by the problem. Has all the requested information been found?
7. Check the answers in the words of the problem.

This list does not apply to all situations, but it can be applied to a wide range of problems with only slight modifications. The following examples use these steps to solve a variety of word problems.

Example 1   The denominator of a fraction exceeds its numerator by 3. If 5 is added to the numerator and 3 is subtracted from the denominator, the resulting fraction equals 2. Find the original fraction.

Solution   Use n to represent the number in the numerator of the fraction. The denominator exceeds the numerator by 3, so the denominator must be $n + 3$. The fraction can now be expressed as $\frac{n}{n + 3}$. If you add 5 to the numerator and subtract 3 from the denominator, the result is 2. Set up the following equation and solve for n:

$$\frac{n + 5}{n + 3 - 3} = 2$$

$$\frac{n + 5}{n} = 2$$

$$n + 5 = 2n$$

$$5 = n$$

Because the numerator of the original fraction is n, and $n = 5$, the denominator is $5 + 3$ or 8. The desired fraction is $\frac{5}{8}$. To check this result, note that if 5 is added to the numerator of the fraction $\frac{5}{8}$, and 3 is subtracted from the denominator, the result is $\frac{10}{5}$, which is equal to 2.   ■

Example 2   A mathematics student has scores of 74%, 78%, and 70% on three examinations. What score is needed on a fourth examination for the student to earn an average grade of 80%?

Solution   Let x be the required grade on the fourth examination. On the one hand, the average grade is one-fourth of the sum of the four grades; on the other hand, the average is to be 80.

The average of the four grades	=	the required average grade.

$$\frac{74 + 78 + 70 + x}{4} = 80$$

Simplify and solve this equation for x:

$$\frac{222 + x}{4} = 80$$

$$222 + x = 320$$

$$x = 98$$

To score an average of 80% on all four examinations, the student must score 98% on the fourth exam.   ■

Example 3 A woman invests $10,000, part at 9% annual interest and the rest at 14%. In each case, the interest is compounded annually. The total annual income is $1275. How much is invested at each rate?

Solution Let x represent the amount invested at 9% interest. Then $10,000 - x$ (the rest of $10,000) represents the amount invested at 14% interest. The annual income from any investment is the product of the interest rate and the amount invested. The total income from these two investments can be expressed in two different ways: as $1275 (which is given) and as the sum of the incomes of the two investments:

| The income from the 9% investment | + | the income from the 14% investment | = | the total income. |

$$9\% \text{ of} \begin{pmatrix} \text{amount invested} \\ \text{at } 9\% \end{pmatrix} + 14\% \text{ of} \begin{pmatrix} \text{amount invested} \\ \text{at } 14\% \end{pmatrix} = \begin{pmatrix} \text{total} \\ \text{income} \end{pmatrix}$$

$$0.09(x) \qquad + \qquad 0.14(10,000 - x) \qquad = \qquad 1275$$

Multiply both sides of this equation by 100 to clear it of decimal fractions, and solve for x:

$$9x + 14(10,000 - x) = 127,500$$

$$9x + 140,000 - 14x = 127,500$$

$$-5x = -12,500$$

$$x = 2500$$

The investment at 9% was $2500, while $10,000 - $2500, or $7500, was invested at 14%. These amounts are correct because 9% of $2500 is $225, and 14% of $7500 is $1050. The total income from both investments is $225 + $1050, or $1275.   ■

Example 4 If a man can paint a house in six days and his daughter can paint the same house in eight days, how long will it take them to paint the house working together?

Solution Let n represent the number of days it would take them to paint the house if they work together. In one day, the two working together can paint $\frac{1}{n}$ of the house. In one day, the father working alone can paint $\frac{1}{6}$ of the house. In one day, the daughter working alone can paint $\frac{1}{8}$ of the house. The work that they can do together in one day is the sum of what each can do in one day. This gives the equation

The part of the house the father can do in one day		the part of the house the daughter can do in one day		the part of the house they can do together in one day.
	+		=	

$$\frac{1}{6} + \frac{1}{8} = \frac{1}{n}$$

Multiply both sides of this equation by $24n$ to clear the fractions, and solve for n:

$$24n\left(\frac{1}{6} + \frac{1}{8}\right) = 24n\left(\frac{1}{n}\right)$$
$$4n + 3n = 24$$
$$7n = 24$$
$$n = \frac{24}{7}$$

The paint job will take $\frac{24}{7}$, or $3\frac{3}{7}$, days. To check this answer, verify that if the father paints $\frac{1}{6}$ of the house each day and the daughter paints $\frac{1}{8}$ of the house each day, and if they work together for $\frac{24}{7}$ days, then one complete house is painted:

$$\frac{1}{6} \cdot \frac{24}{7} + \frac{1}{8} \cdot \frac{24}{7} \stackrel{?}{=} 1$$
$$\frac{4}{7} + \frac{3}{7} \stackrel{?}{=} 1$$
$$1 = 1 \qquad \blacksquare$$

Example 5   If a bottle holding 3 liters of milk contains $3\frac{1}{2}\%$ butterfat, how much skimmed milk must be added to dilute the milk to 2% butterfat?

Solution   Let L represent the number of liters of skimmed milk that must be added to the milk. Then $3 + L$ represents the number of liters in the final mixture. The butterfat in the final mixture is the sum of the butterfat in the original 3 liters of milk and the butterfat in the skimmed milk added. This leads to the following equation:

The butterfat in the milk		the butterfat in the skimmed milk		the butterfat in the total mixture.
	+		=	

$$3\frac{1}{2}\% \text{ of 3 liters} + 0\% \text{ of } L \text{ liters} = 2\% \text{ of } (3 + L) \text{ liters}$$

$$0.035(3) + 0 = 0.02(3 + L)$$

Multiply both sides by 1000 to remove the decimal fractions, and solve the resulting equation for L:

$$35(3) = 20(3 + L)$$
$$105 = 60 + 20L$$
$$45 = 20L$$
$$\frac{9}{4} = L$$

To dilute the milk to a 2% mixture, $2\frac{1}{4}$ liters of skimmed milk must be added. To check this answer, note that the final mixture contains $0.02(5.25) = 0.105$ liters of pure butterfat, and that this is equal to the amount of pure butterfat, $0.035(3) = 0.105$ liters, in the original solution.   ∎

Example 6   A man leaves home driving at the rate of 50 miles per hour. After discovering that he forgot his wallet, his daughter drives at the rate of 65 miles per hour to catch up to him. How long will it take to overtake him and return the wallet if the man had a 15-minute head start?

Solution   Uniform-motion problems are based on the formula $d = rt$, where d is distance, r is rate, and t is time. You can organize the information of this problem in a chart such as Figure 2-1. Let t represent the number of hours the daughter must drive to overtake her father. Because the father had a 15-minute, or $\frac{1}{4}$ hour, head start, he was on the road for $(t + \frac{1}{4})$ hours.

	d	$=$	r	$\cdot$	t
man	$50\left(t + \frac{1}{4}\right)$		50		$t + \frac{1}{4}$
daughter	$65t$		65		t

Figure 2-1

Set up the following equation and solve for t:

The distance the man drove	=	the distance the daughter drove.

$$50\left(t + \frac{1}{4}\right) = 65t$$

$$50t + \frac{25}{2} = 65t$$

$$\frac{25}{2} = 15t$$

$$\frac{5}{6} = t$$

It will take the daughter $\frac{5}{6}$ hours, or 50 minutes, to overtake her father.   ∎

EXERCISE 2.2

Solve each word problem.

1. If a certain number is added to both the numerator and the denominator of the fraction $\frac{3}{5}$, the resulting fraction is $\frac{5}{6}$. Find the number.

2. The denominator of a fraction exceeds the numerator by 1. If 4 is added to the numerator of the fraction, and 3 is subtracted from the denominator, the result is 7. Find the original fraction.

3. One number is 3 more than twice another. Their sum is 54. Find the numbers.

4. The sum of three consecutive odd integers is 69. Find the integers.

5. John scored 5 points higher on his midterm, and 13 points higher on his final, than on his first exam. What did he score on that first exam if his mean (average) score was 90?

6. Sally took four tests in science class. On each successive test, her score improved by three points. If her mean score is 69.5%, what did she get on the first test?

7. An executive invests some money at 7% and some money at 6% annual interest. If a total of $22,000 is invested and the annual return is $1420, how much is invested at 7%?

8. A student invested some money at 8% and twice as much at 9% annual interest. The total income from these two investments is $2080. How much was invested at each rate?

9. An adult ticket for a college basketball game costs $2.50, and a student ticket costs $1.75. The total receipts from the game were $1217.25, and 585 tickets were sold. How many of these were student tickets?

10. Of the 800 tickets sold to a movie, 480 were adult tickets. The gate receipts totaled $2080. What was the cost of a student ticket if an adult ticket cost $3?

11. If a woman can mow a yard with a lawn tractor in 2 hours, and another woman can mow the same lawn with a push mower in 4 hours, how long will it take them if they work together?

12. A garden hose can fill a swimming pool in 3 days, and a larger hose can fill the pool in 2 days. How long will it take to fill the pool if both hoses are used?

13. An empty swimming pool fills in 10 hours. When full, this same swimming pool drains in 19 hours. How long will it take to fill this pool if the drain is accidentally left open?

14. Sam stuffs shrimp in his part-time job as a seafood chef. Sam can stuff 1000 shrimp in 6 hours. If his sister Sally helps him, they can stuff 1000 shrimp in 4 hours. If Sam is sent home sick, how long will it take Sally to stuff 500 shrimp?

15. A small car radiator has a 6-liter capacity. If the liquid in the radiator is 40% antifreeze, how much liquid must be replaced with pure antifreeze to bring the mixture up to a 50% solution?

16. A nurse has 1 liter of a solution that is 20% alcohol. How much pure alcohol must she add to bring it to a solution that contains 25% alcohol?

17. If there are 400 cubic centimeters of a chemical in 1 liter of solution, how many cubic centimeters of water must be added to dilute it to a 25% solution? (There are 1000 cubic centimeters in 1 liter.)

18. A swimming pool contains 15,000 gallons of water. How many gallons of chlorine must be added to "shock the pool" and bring the water to a $\frac{3}{100}$% solution?

19. Assume that an automobile engine will run on a mixture of gasoline and a substitute fuel. Also assume that gasoline costs $1.50 per gallon and that the substitute fuel costs 40¢ per gallon. What percent of the mixture must be substitute fuel to bring the total cost down to $1.00 per gallon?

20. How many liters of water must evaporate to turn 12 liters of a 24% salt solution into a 36% salt solution?

21. John drove to his uncle's house in a distant city in 5 hours. When he returned home, there was less traffic, and the trip took only 3 hours. If John drove 26 miles per hour faster on the return trip, how fast did he drive each way?

22. Suzi drove home for spring vacation at 60 miles per hour, but her brother Jim, who left at the same time, could drive at only 48 miles per hour. When Suzi got home, Jim still had 60 miles to go. How far did Suzi drive?

23. Two cars leave Pima Community College traveling in opposite directions. One car travels at 60 miles per hour and the other car at 64 miles per hour. In how many hours will they be 310 miles apart?

24. Some bank robbers leave town, speeding at 70 miles per hour. Ten minutes later, the police give chase, traveling at 78 miles per hour. How long will it take the police to overtake the robbers?

25. Two young lovers are standing on a beach 440 yards apart when they recognize each other. They begin to run toward each other, the girl at 8 miles per hour and the boy at 10 miles per hour. In how many seconds will they meet?

26. One morning John drove 5 hours before stopping to eat. After lunch, he increased his speed by 10 miles per hour in order to complete a 430-mile trip in 8 hours of driving time. How fast did he drive in the morning?

27. A motorboat goes 5 miles upstream in the same time that it requires to go 7 miles downstream. The river flows at 2 miles per hour. What is the speed of the boat in still water?

28. A plane can fly 340 miles per hour in still air. If it can fly 200 miles downwind in the same amount of time it can fly 140 miles upwind, what is the velocity of the wind?

29. A child has equal numbers of nickels, dimes, and quarters. The coins are worth $3.20. How many of each type are there?

30. Maria has twice as many quarters as dimes. If all the dimes were quarters and all the quarters were dimes, she would have 60¢ less. How much money does she have?

31. A farmer wishes to mix 2400 pounds of cattle feed that is to be 14% crude protein. Barley, which is 11.7% crude protein, represents 25% of the mixture. The remaining 75% is made up of oats (11.8% crude protein) and soybean oil meal (44.5% crude protein). How many pounds of barley, oats, and soybean oil meal ought to be used?

32. If the farmer in Exercise 31 wants only 20% of the mixture to be barley, how many pounds of barley, oats, and soybean oil meal ought to be used?

2.3   QUADRATIC AND POLYNOMIAL EQUATIONS

An equation such as $2x^2 - 11x - 21 = 0$ or $3x^2 - x - 2 = 0$ is called a **quadratic** or **second-degree equation**. In general, the following definition applies.

> **Definition.**   A **quadratic equation** is an equation that can be written in the form $ax^2 + bx + c = 0$, where a, b, and c are real numbers and $a \neq 0$.

Many quadratic equations can be solved by factoring. To do so, we use the following theorem.

> **The Zero-Factor Theorem.**   If a and b are real numbers, and
>
> if $ab = 0$, then $a = 0$ or $b = 0$.

To prove the zero-factor theorem, we suppose that $ab = 0$. If $a = 0$, we are finished because at least one of a or b is 0.

If $a \neq 0$, then a has a reciprocal $\frac{1}{a}$, and we can multiply both sides of the equation $ab = 0$ by $\frac{1}{a}$ to obtain

$$ab = 0$$

$$\frac{1}{a}(ab) = \frac{1}{a}(0)$$

$$\left(\frac{1}{a} \cdot a\right)b = 0$$

$$1b = 0$$

$$b = 0$$

Thus, if $a \neq 0$, then b must be 0, and the theorem is proved.

Example 1   Solve the quadratic equation $2x^2 - 9x - 35 = 0$.

Solution   The left-hand side of the equation can be factored, and the equation can be written as

$$(2x + 5)(x - 7) = 0$$

If either factor is 0, the product will be 0. So use the zero-factor theorem and set each factor equal to 0. Then solve for x.

$$
\begin{array}{ccc}
2x + 5 = 0 & \text{or} & x - 7 = 0 \\
2x = -5 & & x = 7 \\
x = -\dfrac{5}{2} & &
\end{array}
$$

Because the product $(2x + 5)(x - 7)$ can be 0 only if one of its factors is 0, the numbers $-\frac{5}{2}$ and 7 are the only solutions of the given equation. Verify that each of these solutions satisfies the equation $2x^2 - 9x - 35 = 0$.   ∎

Example 2   Solve the equation $\dfrac{2}{x} + 1 = 3x$.

Solution   When both sides of this equation are multiplied by x, the result is a quadratic equation that can be solved by factoring:

$$\frac{2}{x} + 1 = 3x$$

$$x\left(\frac{2}{x} + 1\right) = x(3x) \qquad \text{Multiply both sides by } x.$$

$$2 + x = 3x^2 \qquad \text{Remove parentheses and simplify.}$$

$$0 = 3x^2 - x - 2 \qquad \text{Add } -2 \text{ and } -x \text{ to both sides.}$$

$$0 = (3x + 2)(x - 1) \qquad \text{Factor the trinomial.}$$

$$3x + 2 = 0 \quad \text{or} \quad x - 1 = 0 \qquad \text{Set each factor equal to 0.}$$

$$3x = -2 \qquad\qquad x = 1$$

$$x = -\frac{2}{3}$$

Verify that both solutions satisfy the equation.   ∎

Many polynomial equations of higher degree can also be solved by factoring.

Example 3   Solve the equations   **a.** $6x^3 - x^2 - 2x = 0$ and   **b.** $x^4 - 5x^2 + 4 = 0$.

Solution   **a.**
$$6x^3 - x^2 - 2x = 0$$

$$x(6x^2 - x - 2) = 0 \qquad \text{Factor out } x.$$

$$x(3x - 2)(2x + 1) = 0 \qquad \text{Factor the trinomial.}$$

$$x = 0 \quad \text{or} \quad 3x - 2 = 0 \quad \text{or} \quad 2x + 1 = 0 \qquad \text{Set each factor}$$
equal to 0.

$$3x = 2 \qquad\qquad 2x = -1$$

$$x = \frac{2}{3} \qquad\qquad x = -\frac{1}{2}$$

Verify that each solution satisfies the given equation.

b.
$$x^4 - 5x^2 + 4 = 0$$

$$(x^2 - 4)(x^2 - 1) = 0 \qquad \text{Factor the trinomial.}$$

$$(x + 2)(x - 2)(x + 1)(x - 1) = 0 \qquad \text{Factor each difference of two squares.}$$

$$x + 2 = 0 \quad \text{or} \quad x - 2 = 0 \quad \text{or} \quad x + 1 = 0 \quad \text{or} \quad x - 1 = 0$$

$$x = -2 \qquad\quad x = 2 \qquad\quad x = -1 \qquad\quad x = 1$$

Verify that each solution satisfies the given equation.   ∎

There are many types of nonpolynomial equations that can be solved by factoring. To solve such equations, we often use a theorem that states that equal powers of equal real numbers are equal.

Theorem.   If a and b are real numbers, n is an integer, and $a = b$, then
$$a^n = b^n$$

When we raise both sides of an equation to the same power, the resulting equation might not be equivalent to the original equation. For example, if we raise both sides of the equation

1. $x = 4$ with a solution set of $\{4\}$

to the second power, we obtain the equation

2. $x^2 = 16$ with a solution set of $\{4, -4\}$

Equations 1 and 2 are not equivalent because they have different solution sets. The solution -4 of Equation 2 does not satisfy Equation 1. Because raising both sides of an equation to the same power can introduce extraneous solutions, we must check all suspected solutions to be certain that they satisfy the original equation. The following equation has an extraneous solution.

$$x - x^{1/2} - 6 = 0$$
$$(x^{1/2} - 3)(x^{1/2} + 2) = 0 \qquad\qquad \text{Factor } x - x^{1/2} - 6 = 0.$$
$$x^{1/2} - 3 = 0 \quad \text{or} \quad x^{1/2} + 2 = 0 \qquad \text{Set each factor equal to 0.}$$
$$x^{1/2} = 3 \quad \Big| \qquad x^{1/2} = -2$$

Because equal powers of equal numbers are equal, we can square both sides of the previous equations to obtain

$$(x^{1/2})^2 = (3)^2 \quad \text{or} \quad (x^{1/2})^2 = (-2)^2$$
$$x = 9 \quad \Big| \qquad x = 4$$

The root $x = 9$ satisfies the original equation and the root $x = 4$ does not, as the following check shows:

$$
\begin{array}{ll}
\text{If } x = 9 & \text{If } x = 4 \\
x - x^{1/2} - 6 = 0 & x - x^{1/2} - 6 = 0 \\
9 - 9^{1/2} - 6 \overset{?}{=} 0 & 4 - 4^{1/2} - 6 \overset{?}{=} 0 \\
9 - 3 - 6 \overset{?}{=} 0 & 4 - 2 - 6 \overset{?}{=} 0 \\
0 = 0 & -4 \neq 0
\end{array}
$$

Thus, 9 is the only solution of the original equation.

Example 4   Solve the equation $2x^{2/5} - 5x^{1/5} - 3 = 0$.

Solution   Proceed as follows:

$$2x^{2/5} - 5x^{1/5} - 3 = 0$$

$$(2x^{1/5} + 1)(x^{1/5} - 3) = 0 \qquad \text{Factor } 2x^{2/5} - 5x^{1/5} - 3.$$

$$2x^{1/5} + 1 = 0 \qquad \text{or} \qquad x^{1/5} - 3 = 0 \qquad \text{Set each factor}$$

$$2x^{1/5} = -1 \qquad\qquad\qquad x^{1/5} = 3 \qquad \text{equal to 0.}$$

$$x^{1/5} = -\frac{1}{2}$$

Because equal powers of equal numbers are equal, you can raise both sides of each of the previous equations to the fifth power to obtain

$$(x^{1/5})^5 = \left(-\frac{1}{2}\right)^5 \qquad \text{or} \qquad (x^{1/5})^5 = (3)^5$$

$$x = -\frac{1}{32} \qquad\qquad\qquad x = 243$$

Verify that each solution satisfies the given equation.   ∎

If a quadratic expression in a quadratic equation factors easily, the factoring method is very convenient. However, quadratic expressions do not always factor easily. For example, it would be difficult to factor the left-hand side of the equation $3x^2 + 25x + 1 = 0$, because it cannot be factored by using integers only. To solve such equations, we need to develop other methods. We begin by considering the equation $x^2 = c$. If c is positive, its two real roots can be found by adding $-c$ to both sides, factoring the binomial $x^2 - c$, setting each factor equal to 0, and solving for x.

$$x^2 = c$$

$$x^2 - c = 0$$

$$x^2 - (\sqrt{c})^2 = 0$$

$$(x - \sqrt{c})(x + \sqrt{c}) = 0$$

$$x - \sqrt{c} = 0 \qquad \text{or} \qquad x + \sqrt{c} = 0$$

$$x = \sqrt{c} \qquad\qquad\qquad x = -\sqrt{c}$$

Hence, the roots of the equation $x^2 = c$ are $x = \sqrt{c}$ and $x = -\sqrt{c}$. This fact is often called the **square root property**.

The Square Root Property.   If $c > 0$, then the equation $x^2 = c$ has two real roots:

$$x = \sqrt{c} \qquad \text{or} \qquad x = -\sqrt{c}$$

Example 5   Solve the equation $x^2 - 8 = 0$.

Solution   Solve the equation for x^2 and apply the square root property:

$$x^2 - 8 = 0$$
$$x^2 = 8$$

$$x = \sqrt{8} \quad \text{or} \quad x = -\sqrt{8}$$
$$x = 2\sqrt{2} \quad | \quad x = -2\sqrt{2} \qquad \sqrt{8} = \sqrt{4}\sqrt{2} = 2\sqrt{2}.$$

Verify that each root satisfies the equation $x^2 - 8 = 0$.   ■

Example 6   Solve the equation $(x + 4)^2 = 1$.

Solution   Apply the square root property to $(x + 4)^2 = 1$ and solve for x:

$$(x + 4)^2 = 1$$
$$x + 4 = \sqrt{1} \quad \text{or} \quad x + 4 = -\sqrt{1}$$
$$x + 4 = 1 \quad | \quad x + 4 = -1$$
$$x = -3 \quad | \quad x = -5$$

Verify that each root satisfies the original equation.   ■

Completing the Square

We now discuss another method, called **completing the square**, that can be used to solve quadratic equations. To use this method to solve the equation $x^2 - 10x + 24 = 0$, for example, we begin by adding -24 to both sides of the equation to get the constant term on the right of the equals sign.

$$x^2 - 10x = -24$$

Next we add 25 to both sides of the equation, to make the left-hand side a perfect trinomial square, and combine terms:

$$x^2 - 10x + 25 = -24 + 25$$
$$x^2 - 10x + 25 = 1$$

When the trinomial on the left-hand side is factored, we obtain

$$(x - 5)^2 = 1$$

which is an equation that we can solve by using the square root property:

$$x - 5 = 1 \quad \text{or} \quad x - 5 = -1$$
$$x = 6 \quad | \quad x = 4$$

Both solutions satisfy the original equation.

Example 7   Use the method of completing the square to solve the equation $x^2 + 4x - 5 = 0$.

Solution   Rewrite this equation in the form

$$x^2 + 4x \quad = 5$$

And add 4 to both sides to get

$$x^2 + 4x + 4 = 5 + 4$$
$$x^2 + 4x + 4 = 9$$

Adding 4 to both sides makes the trinomial on the left-hand side a perfect trinomial square. Now solve for x by using the square root property:

$$x^2 + 4x + 4 = 9$$
$$(x + 2)^2 = 9$$

$$x + 2 = 3 \qquad \text{or} \qquad x + 2 = -3$$
$$x = 1 \qquad | \qquad x = -5$$

Verify that each root satisfies the original equation.   ∎

A question is raised by the previous two examples. What number must be added to $x^2 + bx$ to make a perfect trinomial square? There is a pattern in the following list of perfect trinomial squares. We look at the right-hand side of each equation for a pattern relating the constant term and the coefficient of x.

$$(x + 1)^2 = x^2 + 2x + 1$$
$$(x + 3)^2 = x^2 + 6x + 9$$
$$(x + 5)^2 = x^2 + 10x + 25$$
$$(x - 7)^2 = x^2 - 14x + 49$$
$$(x - 8)^2 = x^2 - 16x + 64$$
$$(x - a)^2 = x^2 - 2ax + a^2$$

In each case, the constant term that completes the square is the square of one-half of the coefficient of x. This fact is applied in the next example.

Example 8   Solve the equation $x(x + 3) = 2$.

Solution   Remove parentheses to obtain

$$x^2 + 3x = 2$$

To solve this equation, begin by calculating the number to be added to both sides of the equation to complete the square. One-half of 3 (the coefficient of x) is $\frac{3}{2}$. The square of $\frac{3}{2}$ is $\frac{9}{4}$, so add $\frac{9}{4}$ to both sides of the above equation and solve for x:

$$x^2 + 3x + \frac{9}{4} = 2 + \frac{9}{4}$$

$$x^2 + 3x + \left(\frac{3}{2}\right)^2 = \frac{8}{4} + \frac{9}{4}$$

$$\left(x + \frac{3}{2}\right)^2 = \frac{17}{4}$$

$$x + \frac{3}{2} = \frac{\sqrt{17}}{2} \qquad \text{or} \qquad x + \frac{3}{2} = -\frac{\sqrt{17}}{2} \qquad \text{Use the square root property.}$$

$$x = \frac{-3 + \sqrt{17}}{2} \qquad\qquad\qquad x = \frac{-3 - \sqrt{17}}{2}$$

Verify that each of these roots satisifies the original equation.　■

In Example 8, the coefficient of the x^2 term is 1. This is a necessary condition before the square can be completed. To solve a quadratic equation involving the variable x by completing the square, follow these steps.

1. Make sure that the coefficient of x^2 is 1. If it is not, make it 1 by dividing both sides of the equation by the coefficient of x^2.
2. If necessary, add a number to both sides of the equation to get the constant on the right-hand side of the equation.
3. Complete the square.
   a. Identify the coefficient of x.
   b. Take half the coefficient of x.
   c. Square half the coefficient of x.
   d. Add that square to both sides of the equation.
4. Factor the trinomial square and combine terms.
5. Solve the resulting quadratic equation by applying the square root property.

Example 9　Solve $6x^2 + 5x - 6 = 0$.

Solution　Begin by dividing both sides by 6 to make the coefficient of x^2 equal to 1. Then proceed as follows.

$$6x^2 + 5x - 6 = 0$$

$$x^2 + \frac{5}{6}x - 1 = 0 \qquad\qquad \text{Divide both sides by 6.}$$

$$x^2 + \frac{5}{6}x = 1 \qquad\qquad \text{Add 1 to both sides.}$$

$$x^2 + \frac{5}{6}x + \frac{25}{144} = \frac{25}{144} + 1 \qquad \text{Take } \tfrac{1}{2} \text{ of } \tfrac{5}{6} \text{ to get } \tfrac{5}{12}. \text{ Then square } \tfrac{5}{12} \text{ to get } \tfrac{25}{144}. \text{ Add this to both sides.}$$

$$\left(x + \frac{5}{12}\right)^2 = \frac{169}{144} \qquad\qquad \text{Factor and combine terms.}$$

Now apply the square root property.

$$x + \frac{5}{12} = \sqrt{\frac{169}{144}} \qquad \text{or} \qquad x + \frac{5}{12} = -\sqrt{\frac{169}{144}}$$

$$x + \frac{5}{12} = \frac{13}{12} \qquad\qquad\qquad x + \frac{5}{12} = -\frac{13}{12}$$

$$x = -\frac{5}{12} + \frac{13}{12} \qquad\qquad x = -\frac{5}{12} - \frac{13}{12}$$

$$x = \frac{8}{12} \qquad\qquad\qquad x = -\frac{18}{12}$$

$$x = \frac{2}{3} \qquad\qquad\qquad x = -\frac{3}{2}$$

Verify that both roots check. ∎

The Quadratic Formula

The method of completing the square can be used to solve the **general quadratic equation** $ax^2 + bx + c = 0$, where $a \neq 0$. The solution is as follows:

$$ax^2 + bx + c = 0$$

$$x^2 + \frac{bx}{a} + \frac{c}{a} = 0 \qquad \text{Divide both sides by } a.$$

$$x^2 + \frac{b}{a}x = -\frac{c}{a} \qquad \text{Add } -\frac{c}{a} \text{ to both sides.}$$

$$x^2 + \frac{b}{a}x + \frac{b^2}{4a^2} = \frac{b^2}{4a^2} - \frac{c}{a} \qquad \text{Add } \frac{b^2}{4a^2} \text{ to both sides to complete the square.}$$

$$x^2 + \frac{b}{a}x + \left(\frac{b}{2a}\right)^2 = \frac{b^2}{4a^2} - \frac{4ac}{4aa} \qquad \text{Rewrite the left-hand side, and get a common denominator on the right-hand side.}$$

$$\left(x + \frac{b}{2a}\right)^2 = \frac{b^2 - 4ac}{4a^2} \qquad \text{Factor the left-hand side and add the fractions on the right-hand side.}$$

We can now apply the square-root property.

$$x + \frac{b}{2a} = \frac{\sqrt{b^2 - 4ac}}{2a} \qquad \text{or} \qquad x + \frac{b}{2a} = -\frac{\sqrt{b^2 - 4ac}}{2a}$$

$$x = \frac{-b}{2a} + \frac{\sqrt{b^2 - 4ac}}{2a} \qquad\qquad x = \frac{-b}{2a} - \frac{\sqrt{b^2 - 4ac}}{2a} \qquad \text{Add } -\frac{b}{2a} \text{ to both sides of each equation.}$$

$$x = \frac{-b + \sqrt{b^2 - 4ac}}{2a} \qquad\qquad x = \frac{-b - \sqrt{b^2 - 4ac}}{2a} \qquad \text{Add the fractions on the right-hand side of each equation.}$$

These two values of x are the roots of the equation $ax^2 + bx + c = 0$. They are usually combined into a single expression, called the **quadratic formula**.

> **The Quadratic Formula.**   The solutions of the general quadratic equation $ax^2 + bx + c = 0$, where $a \neq 0$, are
>
> $$x = \frac{-b \pm \sqrt{b^2 - 4ac}}{2a}$$

The above expression should be read twice, once using the plus sign, and once using the minus sign. The quadratic formula implies that

$$x = \frac{-b + \sqrt{b^2 - 4ac}}{2a} \qquad \text{or} \qquad x = \frac{-b - \sqrt{b^2 - 4ac}}{2a}$$

Example 10   Use the quadratic formula to solve the equation $2x^2 + 8x + 7 = 0$.

Solution   In this equation, $a = 2$, $b = 8$, and $c = 7$. Substitute these numbers into the quadratic formula and simplify:

$$x = \frac{-b \pm \sqrt{b^2 - 4ac}}{2a}$$

$$= \frac{-8 \pm \sqrt{8^2 - 4(2)(7)}}{2(2)}$$

$$= \frac{-8 \pm \sqrt{8}}{4} \qquad\qquad 8^2 - 4(2)(7) = 64 - 56 = 8$$

$$= \frac{-8 \pm 2\sqrt{2}}{4}$$

$$x = -2 + \frac{\sqrt{2}}{2} \qquad \text{or} \qquad x = -2 - \frac{\sqrt{2}}{2}$$

Both values satisfy the original equation.   ∎

Example 11   Use the quadratic formula to solve the equation $x^2 - 4x + 5 = 0$.

Solution   In this equation, $a = 1$, $b = -4$, and $c = 5$. Substitute these numbers into the quadratic formula and simplify:

$$x = \frac{-b + \sqrt{b^2 - 4ac}}{2a}$$

$$= \frac{-(-4) \pm \sqrt{(-4)^2 - 4(1)(5)}}{2(1)}$$

$$= \frac{4 \pm \sqrt{16 - 20}}{2}$$

$$= \frac{4 \pm \sqrt{-4}}{2}$$

$$= \frac{4 \pm 2i}{2}$$

$$= 2 \pm i$$

The two solutions, $x = 2 + i$ and $x = 2 - i$, are complex conjugates.   ∎

EXERCISE 2.3

In Exercises 1–30, solve each equation by factoring. Check all answers.

1. $x^2 - x - 6 = 0$ **2.** $x^2 + 8x + 15 = 0$ **3.** $x^2 - 144 = 0$ **4.** $x^2 + 4x = 0$

5. $2x^2 + x - 10 = 0$ **6.** $3x^2 + 4x - 4 = 0$ **7.** $5x^2 - 13x + 6 = 0$ **8.** $2x^2 + 5x - 12 = 0$

9. $15x^2 + 16x = 15$ **10.** $6x^2 - 25x = -25$ **11.** $12x^2 + 9 = 24x$ **12.** $24x^2 + 6 = 24x$

13. $x^3 + 9x^2 + 20x = 0$ **14.** $x^3 + 4x^2 - 21x = 0$

15. $6a^3 - 5a^2 - 4a = 0$ **16.** $8b^3 - 10b^2 + 3b = 0$

17. $y^4 - 26y^2 + 25 = 0$ **18.** $y^4 - 13y^2 + 36 = 0$

19. $x^4 - 37x^2 + 36 = 0$ **20.** $x^4 - 50x^2 + 49 = 0$

21. $2y^4 - 46y^2 + 180 = 0$ **22.** $2x^4 - 102x^2 + 196 = 0$

23. $z^{3/2} - z^{1/2} = 0$ **24.** $r^{5/2} - r^{3/2} = 0$

25. $2m^{2/3} + 3m^{1/3} - 2 = 0$ **26.** $6t^{2/5} + 11t^{1/5} + 3 = 0$

27. $x - 13x^{1/2} + 12 = 0$ **28.** $p + p^{1/2} - 20 = 0$

29. $2t^{1/3} + 3t^{1/6} - 2 = 0$ **30.** $z^3 - 7z^{3/2} - 8 = 0$

In Exercises 31–38, use the square root property to solve each equation. You may need to factor an expression.

31. $x^2 = 9$ **32.** $x^2 = 20$ **33.** $y^2 - 50 = 0$ **34.** $x^2 - 75 = 0$

35. $(x - 1)^2 = 4$ **36.** $(y + 2)^2 - 49 = 0$ **37.** $a^2 + 2a + 1 = 9$ **38.** $x^2 - 6x + 9 = 25$

In Exercises 39–50, solve each equation by completing the square.

39. $x^2 - 8x + 15 = 0$ **40.** $x^2 + 10x + 21 = 0$ **41.** $x^2 + x - 6 = 0$ **42.** $x^2 - 9x + 20 = 0$

43. $x^2 - 25x = 0$ **44.** $x^2 + x = 0$ **45.** $3x^2 - 4 = -4x$ **46.** $5x = 12 - 2x^2$

47. $x^2 + 5 = -5x$ **48.** $x^2 + 1 = -4x$ **49.** $3x^2 = 1 - 4x$ **50.** $2x^2 = 3x + 1$

In Exercises 51–62, solve each equation by using the quadratic formula.

51. $x^2 - 12 = 0$ **52.** $x^2 - 20 = 0$ **53.** $2x^2 - x - 15 = 0$ **54.** $6x^2 + x - 2 = 0$

55. $5x^2 - 9x - 2 = 0$ **56.** $4x^2 - 4x - 3 = 0$

57. $2x^2 + 2x - 4 = 0$ **58.** $3x^2 + 18x + 15 = 0$

59. $-3x^2 = 5x + 1$ **60.** $2x(x + 3) = -1$ **61.** $5x\left(x + \frac{1}{5}\right) = 3$ **62.** $7x^2 = 2x + 2$

In Exercises 63–72, change each equation to quadratic form and solve by any method.

63. $x + 1 = \dfrac{12}{x}$ **64.** $x - 2 = \dfrac{15}{x}$ **65.** $8x - \dfrac{3}{x} = 10$ **66.** $15x - \dfrac{4}{x} = 4$

67. $\dfrac{5}{x} = \dfrac{4}{x^2} - 6$

68. $\dfrac{6}{x^2} + \dfrac{1}{x} = 12$

69. $x\left(30 - \dfrac{13}{x}\right) = \dfrac{10}{x}$

70. $x\left(20 - \dfrac{17}{x}\right) - \dfrac{10}{x} = 0$

71. $(a - 2)(a + 4) = 2a(a - 3)$

72. $\dfrac{a + 4}{2a} = \dfrac{a - 2}{3}$

In Exercises 73–78, solve each quadratic equation. The roots will be complex numbers.

73. $x^2 + 2x + 2 = 0$

74. $x^2 + 4x + 8 = 0$

75. $x^2 + 4x = -5$

76. $x^2 = -(2x + 5)$

77. $x^2 - \dfrac{2}{3}x = -\dfrac{2}{9}$

78. $x^2 + \dfrac{5}{4} = x$

2.4   MORE ON QUADRATIC EQUATIONS AND RADICAL EQUATIONS

It is possible to discover what types of numbers will be roots of a given quadratic equation without solving that equation. Suppose that the coefficients a ($a \neq 0$), b, and c in the quadratic equation $ax^2 + bx + c = 0$ are real numbers. The two solutions of this equation are given by the quadratic formula

$$x = \frac{-b \pm \sqrt{b^2 - 4ac}}{2a}$$

If $b^2 - 4ac$ is a positive number or 0, then the solutions are real numbers. On the other hand, if $b^2 - 4ac$ is a negative number, then the solutions are nonreal complex numbers. The expression $b^2 - 4ac$ is called the **discriminant**. Its value determines the nature of the roots of any given quadratic equation. The possibilities are summarized in the following table:

If a, b, and c are real numbers and	
If $b^2 - 4ac$ is . . .	**the solutions are . . .**
positive	real numbers and unequal
0	real numbers and equal
negative	nonreal complex numbers and complex conjugates

If a, b, and c are rational numbers and	
If $b^2 - 4ac$ is . . .	**the solutions are . . .**
0	rational numbers and equal
a nonzero perfect square	rational numbers and unequal
positive and not a perfect square	irrational numbers and unequal

Example 1   Determine the nature of the roots of the quadratic equation $3x^2 + 4x + 1 = 0$.

Solution   Calculate the discriminant, $b^2 - 4ac$, as follows:

$$b^2 - 4ac = 4^2 - 4(3)(1)$$
$$= 16 - 12$$
$$= 4$$

Because a, b, and c are rational numbers and the discriminant is a perfect square, the two roots of the given quadratic equation are rational and unequal.   ■

Example 2   If k is a constant, many quadratic equations are represented by the equation

$$(k - 2)x^2 + (k + 1)x + 4 = 0$$

What values of k will give an equation with roots that are real and equal?

Solution   Compute the discriminant and demand that it be 0.

$$b^2 - 4ac = (k + 1)^2 - 4(k - 2)(4)$$
$$0 = k^2 + 2k + 1 - 16k + 32$$
$$0 = k^2 - 14k + 33$$
$$0 = (k - 3)(k - 11)$$

$$k - 3 = 0 \quad \text{or} \quad k - 11 = 0$$
$$k = 3 \quad | \quad k = 11$$

If $k = 3$, then the equation $(k - 2)x^2 + (k + 1)x + 4 = 0$ becomes

$$(3 - 2)x^2 + (3 + 1)x + 4 = 0$$

or

$$x^2 + 4x + 4 = 0$$

Solve this equation to show that its roots are real and equal:

$$x^2 + 4x + 4 = 0$$
$$(x + 2)(x + 2) = 0$$
$$x + 2 = 0 \quad \text{or} \quad x + 2 = 0$$
$$x = -2 \quad | \quad x = -2$$

Similarly, $k = 11$ produces an equation with roots that are real and equal.   ■

There are many types of equations that, while not quadratic equations, can be put in quadratic form. These equations can then be solved by using techniques for solving quadratic equations.

Example 3   Solve the equation $\dfrac{1}{x-1} + \dfrac{3}{x+1} = 2$.

Solution   Because the denominator of a fraction cannot be 0, x cannot be 1 or -1. If either 1 or -1 appears as a suspected root, it must be discarded. Solve the equation as follows:

$$\frac{1}{x-1} + \frac{3}{x+1} = 2$$

$$(x+1)(x-1)\left[\frac{1}{x-1} + \frac{3}{x+1}\right] = 2(x+1)(x-1) \qquad \text{Multiply both sides by } (x+1)(x-1).$$

$$(x+1) + 3(x-1) = 2(x^2-1) \qquad \text{Remove brackets and simplify}$$

$$4x - 2 = 2x^2 - 2 \qquad \text{Remove parentheses and simplify}$$

$$0 = 2x^2 - 4x \qquad \text{Add } 2 - 4x \text{ to both sides.}$$

The resulting equation is a quadratic equation. Factor the binomial $2x^2 - 4x$, set each factor equal to 0, and solve for x:

$$0 = 2x^2 - 4x$$

$$0 = 2x(x-2)$$

$$2x = 0 \qquad \text{or} \qquad x - 2 = 0$$

$$x = 0 \qquad | \qquad x = 2$$

Because the numbers 0 and 2 are in the domain of x, each number is a root of the original equation. Verify this by checking each root.   ∎

Radical Equations

Equations that contain radicals with variables in the radicand are called **radical equations**. To solve such equations, we will use the rule that states that equal numbers raised to equal powers are equal.

Example 4   Solve the equation $\sqrt{x+3} - 4 = 7$.

Solution

$$\sqrt{x+3} - 4 = 7$$

$$\sqrt{x+3} = 11 \qquad \text{Add 4 to both sides.}$$

$$(\sqrt{x+3})^2 = (11)^2 \qquad \text{Square both sides.}$$

$$x + 3 = 121 \qquad \text{Simplify.}$$

$$x = 118 \qquad \text{Add } -3 \text{ to both sides.}$$

Squaring both sides of an equation does not guarantee a result that is equivalent to the original equation. Thus, you *must* check the suspected root in the original equation:

$$\sqrt{x + 3} - 4 = 7$$
$$\sqrt{118 + 3} - 4 \stackrel{?}{=} 7$$
$$\sqrt{121} - 4 \stackrel{?}{=} 7$$
$$11 - 4 \stackrel{?}{=} 7$$
$$7 = 7$$

Because it checks, 118 is a solution of $\sqrt{x + 3} - 4 = 7$.   ∎

Example 5   Solve the equation $\sqrt{x + 3} = 3x - 1$.

Solution   Rid this equation of the radical by squaring both sides of the equation and then simplify:

$$\sqrt{x + 3} = 3x - 1$$
$$x + 3 = (3x - 1)^2 \qquad \text{Square both sides.}$$
$$x + 3 = 9x^2 - 6x + 1 \qquad \text{Remove parentheses.}$$
$$0 = 9x^2 - 7x - 2 \qquad \text{Add } -x - 3 \text{ to both sides.}$$

The right side of this equation factors as

$$0 = (9x + 2)(x - 1)$$

and the possible solutions can be found by solving two linear equations:

$$9x + 2 = 0 \qquad \text{or} \qquad x - 1 = 0$$
$$x = -\frac{2}{9} \qquad \qquad \qquad x = 1$$

You *must* check each proposed root in the given equation.

$$\sqrt{x + 3} = 3x - 1 \qquad \text{or} \qquad \sqrt{x + 3} = 3x - 1$$
$$\sqrt{-\frac{2}{9} + 3} \stackrel{?}{=} 3\left(-\frac{2}{9}\right) - 1 \qquad \qquad \sqrt{1 + 3} \stackrel{?}{=} 3(1) - 1$$
$$\sqrt{\frac{25}{9}} \stackrel{?}{=} -\frac{2}{3} - 1 \qquad \qquad \sqrt{4} \stackrel{?}{=} 2$$
$$\frac{5}{3} \neq -\frac{5}{3} \qquad \qquad \qquad 2 = 2$$

Thus, $-\frac{2}{9}$ is not a solution. | Thus, 1 is a solution.

The only solution of the equation $\sqrt{x + 3} = 3x - 1$ is the number 1.   ∎

Example 6   Solve the equation $\sqrt[3]{x^3 + 56} = x + 2$.

Solution   To eliminate the radical, cube both sides of the equation. Then proceed as follows:

$$\sqrt[3]{x^3 + 56} = x + 2$$
$$(\sqrt[3]{x^3 + 56})^3 = (x + 2)^3$$
$$x^3 + 56 = x^3 + 6x^2 + 12x + 8$$
$$0 = 6x^2 + 12x - 48 \qquad \text{Add } -x^3 \text{ and } -56 \text{ to both sides.}$$
$$0 = x^2 + 2x - 8 \qquad \text{Divide both sides by 6.}$$
$$0 = (x + 4)(x - 2)$$

$$x + 4 = 0 \qquad \text{or} \qquad x - 2 = 0$$
$$x = -4 \qquad | \qquad x = 2$$

Check each apparent solution to determine if either is extraneous.

For $x = -4$:	For $x = 2$:
$\sqrt[3]{x^3 + 56} = x + 2$	$\sqrt[3]{x^3 + 56} = x + 2$
$\sqrt[3]{(-4)^3 + 56} \overset{?}{=} -4 + 2$	$\sqrt[3]{2^3 + 56} \overset{?}{=} 2 + 2$
$\sqrt[3]{-64 + 56} \overset{?}{=} -2$	$\sqrt[3]{8 + 56} \overset{?}{=} 4$
$\sqrt[3]{-8} \overset{?}{=} -2$	$\sqrt[3]{64} \overset{?}{=} 4$
$-2 = -2$	$4 = 4$

Because both values satisfy the given equation, the solution set is $\{-4, 2\}$.   ■

Example 7   Solve the equation $\sqrt{2x + 3} + \sqrt{x - 2} = 4$.

Solution   To simplify the work, rewrite the equation in the form

$$\sqrt{2x + 3} = 4 - \sqrt{x - 2}$$

so that its left-hand side contains only one radical. Square both sides to get

$$(\sqrt{2x + 3})^2 = (4 - \sqrt{x - 2})^2$$
$$2x + 3 = 16 - 8\sqrt{x - 2} + (x - 2)$$

Combine terms and add $-x - 14$ to both sides, leaving the single radical by itself on the right-hand side of the equation:

$$2x + 3 = 14 - 8\sqrt{x - 2} + x$$
$$x - 11 = -8\sqrt{x - 2}$$

Square both sides again, simplify, and solve the resulting equation for x:

$$x^2 - 22x + 121 = 64(x - 2)$$
$$x^2 - 86x + 249 = 0$$
$$(x - 3)(x - 83) = 0$$

$$x - 3 = 0 \qquad \text{or} \qquad x - 83 = 0$$
$$x = 3 \qquad | \qquad x = 83$$

Substituting these proposed roots into the given equation shows that 83 does not check. Thus, 83 is an extraneous solution and must be discarded. However, 3 does satisfy the given equation. Hence, 3 is a root.   ∎

EXERCISE 2.4

In Exercises 1–6, use the discriminant to determine the nature of the roots of each equation. **Do not solve the equations***.*

1. $x^2 + 6x + 9 = 0$

2. $x^2 - 5x + 2 = 0$

3. $3x^2 - 2x + 5 = 0$

4. $9x^2 + 42x + 49 = 0$

5. $10x^2 + 29x = 21$

6. $10x^2 + x = 21$

7. Find two values of k such that the equation $x^2 + kx + 3k - 5 = 0$ will have a double root; that is, the roots will be equal.

8. For what value of b will the solutions of equation $x^2 - 2bx + b^2 = 0$ be equal?

9. Does the equation $1492x^2 + 1984x - 1776 = 0$ have any roots that are real numbers?

10. Does the equation $2004x^2 + 10x + 1985 = 0$ have any roots that are real numbers?

In Exercises 11–52, find all real solutions of each equation, if possible.

11. $\dfrac{1}{x} + \dfrac{3}{x+2} = 2$

12. $\dfrac{1}{x-1} + \dfrac{1}{x-4} = \dfrac{5}{4}$

13. $\dfrac{1}{x+1} + \dfrac{5}{2x-4} = 1$

14. $\dfrac{x(2x+1)}{x-2} = \dfrac{10}{x-2}$

15. $x + 1 + \dfrac{x+2}{x-1} = \dfrac{3}{x-1}$

16. $\dfrac{1}{4-y} = \dfrac{1}{4} + \dfrac{1}{y+2}$

17. $\dfrac{24}{a} - 11 = \dfrac{-12}{a+1}$

18. $\dfrac{36}{b} - 17 = \dfrac{-24}{b+1}$

19. $x^{-4} - 13x^{-2} + 36 = 0$

20. $2x^{-4} - 11x^{-2} + 14 = 0$

21. $\sqrt{x-2} = 5$

22. $3\sqrt{x+1} = \sqrt{6}$

23. $\sqrt{x+3} = 2\sqrt{x}$

24. $\sqrt{16x+4} = \sqrt{x+4}$

25. $\sqrt{6}\sqrt{x^2-2} = \sqrt{-13x-7}$

26. $\sqrt{x^2+1} = \dfrac{\sqrt{-7x+11}}{\sqrt{6}}$

27. $\sqrt[3]{7x+1} = 4$

28. $\sqrt[3]{11a-40} = 5$

29. $\sqrt[4]{30t+25} = 5$

30. $\sqrt[4]{3z+1} = 2$

31. $\sqrt{x^2+21} = x+3$

32. $\sqrt{5-x^2} = -(x+1)$

33. $\sqrt{y+2} = 4 - y$

34. $\sqrt{3z+1} = z - 1$

35. $x - \sqrt{7x-12} = 0$

36. $x - \sqrt{4x-4} = 0$

37. $x + 4 = \sqrt{\dfrac{6x+6}{5}} + 3$

38. $\sqrt{\dfrac{8x+43}{3}} - 1 = x$

39. $\sqrt{\dfrac{x^2-1}{x-2}} = 2\sqrt{2}$

40. $\dfrac{\sqrt{x^2-1}}{\sqrt{3x-5}} = \sqrt{2}$

41. $\sqrt[3]{x^3+7} = x + 1$

42. $\sqrt[3]{x^3-7} + 1 = x$

43. $\sqrt[3]{8x^3+61} = 2x + 1$

44. $\sqrt[3]{8x^3-37} = 2x - 1$

45. $\sqrt{x+3} = \sqrt{2x+8} - 1$

46. $3 + \sqrt{x+7} = \sqrt{x-4}$

47. $\sqrt{y+8} - \sqrt{y-4} = -2$

48. $\sqrt{r} + \sqrt{r+2} = 2$

49. $\sqrt{2b + 3} - \sqrt{b + 1} = \sqrt{b - 2}$

50. $\sqrt{a + 1} + \sqrt{3a} = \sqrt{5a + 1}$

51. $\sqrt{\sqrt{b} + \sqrt{b + 8}} = 2$

52. $\sqrt{\sqrt{x + 19} - \sqrt{x - 2}} = \sqrt{3}$

53. If r_1 and r_2 are the two solutions of the equation $ax^2 + bx + c = 0$, find the values of $r_1 + r_2$ and $r_1 r_2$ in terms of a, b, and c. These results can be used to check the solutions of a quadratic equation.

54. Find five consecutive integers a, b, c, d, and e such that $a^2 + b^2 + c^2 = d^2 + e^2$.

2.5   APPLICATIONS OF QUADRATIC EQUATIONS

The solutions of many word problems involve quadratic equations.

Example 1   The length of a rectangle exceeds its width by 3 feet. If the area of the rectangle is 40 square feet, find its dimensions.

Solution   Let w equal the width of the rectangle. Then, $w + 3$ represents its length. See Figure 2-2. Use the formula for the area of a rectangle ($area = length \cdot width$) to express the area as $(w + 3)w$, which is equal to 40. This leads to the equation

The length of the rectangle	$\cdot$	the width of the rectangle	$=$	the area of the rectangle.

$$(w + 3) \cdot w = 40$$

Solve this equation for w as follows:

$$w^2 + 3w = 40$$
$$w^2 + 3w - 40 = 0$$
$$(w - 5)(w + 8) = 0$$
$$w - 5 = 0 \quad \text{or} \quad w + 8 = 0$$
$$w = 5 \quad | \quad w = -8$$

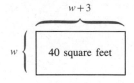

Figure 2-2

You are asked for the dimensions of the given rectangle. Its width, w, is 5, and its length is $w + 3$, or 8. The solution $w = -8$ must be discarded because a rectangle's width cannot be negative. Verify that this solution is correct by observing that a rectangle with dimensions of 5 feet by 8 feet does have an area of 40 square feet.   ∎

Example 2   A man drives 600 miles to a business convention. On the return trip, he increases his speed by 10 miles per hour and saves 2 hours of driving time. How fast did he drive in each direction?

Solution   Let s represent the car's speed (in miles per hour) driving to the convention. On the return trip, his speed was $s + 10$ miles per hour. Recall that the distance traveled by an object moving at a constant rate for a certain time is given by the formula $d = rt$. Thus,

$$t = \frac{d}{r}$$

Organize the information given in this problem in a chart such as Figure 2-3.

	d	$=$	r	$\cdot$	t
outbound trip	600		s		$\dfrac{600}{s}$
return trip	600		$s + 10$		$\dfrac{600}{s + 10}$

Figure 2-3

Although neither the outbound nor the return travel time is given, you know the difference of those two times. This fact can be used to form the equation

The longer time of the outbound trip	$-$	the shorter time of the return trip	$=$	the difference in travel times.

$$\frac{600}{s} - \frac{600}{s + 10} = 2$$

Multiply both sides of this equation by the common denominator $s(s + 10)$ to clear the equation of fractions, and solve for s.

$$s(s + 10)\left(\frac{600}{s} - \frac{600}{s + 10}\right) = 2s(s + 10)$$
$$600(s + 10) - 600s = 2s(s + 10)$$
$$600s + 6000 - 600s = 2s^2 + 20s$$
$$6000 = 2s^2 + 20s$$
$$0 = 2s^2 + 20s - 6000$$
$$0 = s^2 + 10s - 3000$$
$$0 = (s - 50)(s + 60)$$
$$s - 50 = 0 \quad \text{or} \quad s + 60 = 0$$
$$s = 50 \quad | \quad s = -60$$

The solution $s = -60$ must be discarded because it is negative. The man drove 50 miles per hour to the convention, and $50 + 10$, or 60 miles per hour on the return trip. These answers are correct because a 600-mile trip at 50 miles per hour would take $\frac{600}{50}$ or 12 hours; at 60 miles per hour, that same trip would take only 10 hours—2 hours less. ∎

Example 3   If an object is thrown straight up into the air with an initial velocity of 144 feet per second, its height is given by the formula $h = 144t - 16t^2$, where h represents its height (in feet) and t represents the elapsed time (in seconds) since it was thrown. When does the object return to the point from which it was thrown?

Solution   When this object returns to its starting point, its height is again 0. Hence, set h equal to 0, and solve for t:

$$h = 144t - 16t^2$$
$$0 = 144t - 16t^2$$
$$0 = 16t(9 - t)$$

$$16t = 0 \quad \text{or} \quad 9 - t = 0$$
$$t = 0 \quad | \quad t = 9$$

At $t = 0$, the object's height is 0 because it is just being released. When $t = 9$, the height is again 0, and the object has returned to its starting point.   ∎

EXERCISE 2.5

Solve each word problem.

1. The product of two consecutive even natural numbers is 48. Find the numbers.

2. The product of the first and last of three consecutive odd integers is 45. Find the sum of the three integers.

3. The length of a rectangle is 4 feet longer than its width. Its area is 32 square feet. What are its dimensions?

4. The length of a rectangle is 5 times longer than its width. Its area is 125 square feet. What is its perimeter?

5. The side of a square is 4 centimeters shorter than the side of a second square. The sum of the areas of the squares is 106 square centimeters. Find the length of the side of the larger square.

6. The base of a triangle is one-third as long as its height. How long is the base if the triangle's area is 24 square meters?

7. A cyclist rides from DeKalb to Rockford, a distance of 40 miles. His return trip takes 2 hours longer because his speed decreases by 10 miles per hour. How fast does he ride each way?

8. A farmer drives a tractor from one town to another, a distance of 120 kilometers. He drives 10 kilometers per hour faster on the return trip, cutting 1 hour off the time. How fast does he drive each way?

9. If the speed were increased by 10 miles per hour, a 420-mile trip would take 1 hour less time. How long does the trip take at the slower speed?

10. By increasing her usual speed by 25 kilometers per hour, a bus driver decreases the time on a 25-kilometer trip by 10 minutes. What is the usual speed?

11. The height of a projectile fired upward with an initial velocity of 400 feet per second is given by the formula $h = -16t^2 + 400t$, where h is the height in feet and t is the time in seconds. Find the time required for the projectile to return to earth.

12. The height of an object tossed upward with an initial velocity of 104 feet per second is given by the formula $h = 104t - 16t^2$, where h is the height in feet and t is the time in seconds. In how many seconds will the object return to its point of departure?

13. An object will fall s feet in t seconds, where $s = 16t^2$. How long will it take a penny to hit the ground if it is dropped from the top of the Sears Tower in Chicago? (*Hint*: The tower is 1454 feet tall.)

14. The height of an object thrown upward with an initial velocity of 32 feet per second is given by the formula $h = 32t - 16t^2$, where t is the time in seconds. How long will it take the object to reach a height of 16 feet?

15. Two pipes are used to fill a water storage tank. The first pipe can fill the tank in 4 hours. The two pipes together can fill the tank in 2 hours less time than the second pipe alone. How long would it take for the second pipe to fill the tank?

16. A hose can fill a swimming pool in 6 hours. Another hose requires 3 more hours to fill the pool than the two hoses together. How long would it take the second hose to fill the pool?

17. Kristy can mow a lawn in 1 hour less time than her brother Steven. Together they can finish in 5 hours. How long would it take Kristy if she worked alone?

18. Working together, Sarah and Heidi can milk the cows in 2 hours. If they work alone, it takes Heidi 3 hours longer than it takes Sarah. How long does it take Heidi to milk the cows alone?

19. Is it possible for a rectangle to have a width that is 3 units shorter than its diagonal and a length that is 4 units longer than its diagonal?

20. If two opposite sides of a square are increased by 10 meters and the other sides are decreased by 8 meters, the area of the rectangle that is formed is 63 square meters. Find the area of the original square.

21. Maude and Matilda have each invested some money for their retirement. Maude invested $1000 more than Matilda, but at an interest rate that was 1% less. Last year Maude received interest of $280 on her investment and Matilda received $240. At what rates were their investments made?

22. Scott and Laura have both invested some money. Scott invested $3000 more than Laura and at a 2% higher interest rate. Scott received $800 interest and Laura received $400. How much did Scott invest?

23. Some mathematics professors at a college would like to purchase a $150 microwave oven for their department workroom. If four of the professors do not contribute, everyone's share will increase by $10. How many mathematics professors teach at the college?

24. A farmer intends to construct a windscreen by planting pine trees in a quarter-mile row. His daughter points out that 44 fewer trees will be needed if they are planted 1 foot farther apart. If the farmer takes her advice, how many trees will he plant? (*Hint*: 1 mile = 5280 feet.) A row starts and ends with a tree.

25. If a wagon wheel had 10 more spokes, the angle between spokes would decrease by 6°. How many spokes does the wheel have?

26. A merchant could sell all of his calculators at list price for a total of $180. If he had 3 more calculators, he could sell each one for $10 less and still receive $180. Find the list price of each calculator.

27. If one leg of a right triangle is 14 meters shorter than the other leg and the hypotenuse is 26 meters, find the length of the two legs.

28. One leg of a right triangle is 2 centimeters longer than the other and its length is the average length of the shorter leg and the hypotenuse. Find the length of the leg.

2.6   LITERAL EQUATIONS

Many equations, called **literal equations**, contain several variables. Often these equations are formulas such as $F = \frac{9}{5}C + 32$, the formula that converts degrees Celsius to degrees Fahrenheit. If we need to change a large number of Fahrenheit readings to degrees Celsius, it is tedious to substitute each value of F into the above formula and then repeatedly solve the equation for C. A more efficient way, especially if we have a calculator, is to solve the formula for C, substitute each value of F (the Fahrenheit readings) into the rearranged formula, and evaluate C directly.

Example 1   Solve the formula $F = \frac{9}{5}C + 32$ for C.

Solution   Solve for C by using the techniques for solving linear equations:

$$F = \frac{9}{5}C + 32$$

$$F - 32 = \frac{9}{5}C \qquad \text{Subtract 32 from both sides.}$$

$$\frac{5}{9}(F - 32) = C \qquad \text{Multiply both sides by } \tfrac{5}{9}.$$

This result can be written in the alternate form

$$C = \frac{5F - 160}{9}$$

■

Example 2   The formula $A = p + prt$ is used to determine the amount of money in a savings account at the end of a specific period of time. A represents the amount, p represents the original deposit (the principal), r represents the rate of simple interest per unit of time, and t represents the number of units of time. Solve this formula for p.

Solution   Factor p from both terms on the right-hand side of the equation, and divide both sides by $(1 + rt)$ to solve for p:

$$A = p + prt$$

$$A = p(1 + rt) \qquad \text{Factor out } p.$$

$$\frac{A}{1 + rt} = p \qquad \text{Divide both sides by } 1 + rt.$$

■

Example 3   Solve the equation $y(xy + 3x) + 7 = 0$ for x.

Solution
$$y(xy + 3x) + 7 = 0$$

$$xy^2 + 3xy + 7 = 0 \qquad \text{Remove parentheses.}$$

$$xy^2 + 3xy = -7 \qquad \text{Add } -7 \text{ to both sides.}$$

$$x(y^2 + 3y) = -7 \qquad \text{Factor out } x.$$

$$x = \frac{-7}{y^2 + 3y} \qquad \text{Divide both sides by } y^2 + 3y.$$

■

Example 4   Solve the equation $y(xy + 3x) + 7 = 0$ for y.

Solution   Use the distributive property and remove parentheses first:

$$y(xy + 3x) + 7 = 0$$

$$xy^2 + 3xy + 7 = 0$$

This equation is of the form $ay^2 + by + c = 0$, with $a = x$, $b = 3x$, and $c = 7$.
Solve for y by using the quadratic formula:

$$y = \frac{-b \pm \sqrt{b^2 - 4ac}}{2a}$$

$$y = \frac{-3x \pm \sqrt{9x^2 - 28x}}{2x}$$

■

EXERCISE 2.6

In Exercises 1–36, solve for the specified variable.

1. $k = 2.2p$ for p

2. $p = 2l + 2w$ for w

3. $A = \frac{1}{2}h(b_1 + b_2)$ for b_2

4. $V = \frac{1}{3}\pi r^2 h$ for h

5. $V = \frac{1}{3}\pi r^2 h$ for r

6. $z = \frac{x - \mu}{\sigma}$ for μ

7. $P_n = L + \frac{si}{f}$ for s

8. $P_n = L + \frac{si}{f}$ for f

9. $\frac{1}{r} = \frac{1}{r_1} + \frac{1}{r_2}$ for r

10. $\frac{1}{r} = \frac{1}{r_1} + \frac{1}{r_2}$ for r_1

11. $l = a + (n - 1)d$ for n

12. $l = a + (n - 1)d$ for d

13. $C = \frac{5}{9}(F - 32)$ for F

14. $S = \pi h(r + h)$ for r

15. $x + y = \frac{7y + 1}{3x}$ for y

16. $x - y = \frac{9y - 2}{4x}$ for y

17. $S = \frac{a - lr}{1 - r}$ for a

18. $S = \frac{a - lr}{1 - r}$ for r

19. $R = \dfrac{1}{\dfrac{1}{r_1} + \dfrac{1}{r_2} + \dfrac{1}{r_3}}$ for r_1

20. $R = \dfrac{1}{\dfrac{1}{r_1} + \dfrac{1}{r_2} + \dfrac{1}{r_3}}$ for r_3

21. $xy = ax - y^2$ for x

22. $xy = ax - y^2$ for y

23. $a = (n - 2)\dfrac{180}{n}$ for n

24. $V = \pi h^2\left(r - \dfrac{h}{3}\right)$ for r

25. $V = (B_1 + B_2 + \sqrt{B_1 B_2})\, h$ for h

26. $V = \frac{1}{6}h(B + B' + 4M)$ for h

27. $A = \frac{1}{2}r^2(\theta - \phi)$ for θ

28. $r = \dfrac{x + y}{1 - xy}$ for x

29. $r = \dfrac{x + y}{1 - xy}$ for y

30. $A = \frac{1}{2}r^2(\theta - \phi)$ for ϕ

31. $y - y_1 = \dfrac{y_2 - y_1}{x_2 - x_1}(x - x_1)$ for y

32. $y - y_1 = \dfrac{y_2 - y_1}{x_2 - x_1}(x - x_1)$ for x

33. $x^2y - 3xy + 2xy^2 = 0$ for x **34.** $x^2y - 3xy + 2xy^2 = 0$ for y

35. $\dfrac{y^2}{x} - \dfrac{y}{x-1} = 1$ for y **36.** $\dfrac{y^2}{x} - \dfrac{y}{x-1} = 1$ for x

37. In a certain city, the cost of electricity is given by the formula

$$C = 0.10n + 8$$

where C is the cost and n is the number of kilowatt hours used. Solve the formula for n and find the number of kilowatt hours used for costs of $28, $58, and $78.

38. A woman invests $5000. Solve the formula $A = p + prt$ for t and find the length of time it will take for her to double her money at rates of 5% and 12%.

39. For delivering newspapers, Juan is paid $5 per day plus 4 cents for each newspaper delivered. How many newspapers must he deliver to earn $11?

40. While waiting for her car to be repaired, Sue rents a car for $15 per day plus 22 cents per mile. If she keeps the rental car for three days, how many miles can she drive for a total cost of $155?

CHAPTER SUMMARY

Key Words

completing the square   (2.3)   literal equations   (2.6)
conditional equations   (2.1)   quadratic equations   (2.3)
discriminant   (2.4)   quadratic formula   (2.3)
domain of a variable   (2.1)   radical equations   (2.4)
equation   (2.1)   rational equations   (2.1)
equivalent equations   (2.1)   root of an equation   (2.1)
extraneous root   (2.1)   second-degree equations   (2.3)
general quadratic equation   (2.3)   solution of an equation   (2.1)
identity   (2.1)   solution set   (2.1)
linear equations   (2.1)   zero-factor theorem   (2.3)

Key Ideas

(2.1)   Any algebraic expression that represents a number may be added to or subtracted from both sides of an equation: If $a = b$ and c is a number, then

$$a + c = b + c \quad \text{and} \quad a - c = b - c$$

Both sides of an equation may be multiplied or divided by any algebraic expression that represents a nonzero number: If $a = b$ and $c \neq 0$, then

$$ac = bc \quad \text{and} \quad a/c = b/c$$

Either side of an equation can be simplified, provided the value represented by that side remains the same. That is, a quantity may be substituted for its equal.

(2.2)   Use the following steps to solve a word problem:

1. Read the problem.
2. Pick a variable to represent the quantity to be found.
3. Form an equation.

4. Solve the equation.

5. Check the answer in the words of the problem.

(2.3) If a and b are two numbers and $ab = 0$, then either $a = 0$ or $b = 0$.

If $a = b$, then $a^2 = b^2$.

The equation $x^2 = c$, with $c > 0$, has two real roots:

$$x = \sqrt{c} \qquad \text{or} \qquad x = -\sqrt{c}$$

Quadratic formula: $\quad x = \dfrac{-b \pm \sqrt{b^2 - 4ac}}{2a}$

(2.4) If $b^2 - 4ac > 0$, then the roots of $ax^2 + bx + c = 0$, where $a \neq 0$, are unequal real numbers.

If $b^2 - 4ac = 0$, then the roots of $ax^2 + bx + c = 0$, where $a \neq 0$, are equal real numbers.

If $b^2 - 4ac < 0$, then the roots of $ax^2 + bx + c = 0$, where $a \neq 0$, are nonreal complex numbers that are complex conjugates.

(2.6) Solve literal equations, or formulas, by using the same techniques as for solving linear equations.

REVIEW EXERCISES

In Review Exercises 1–6, find the domain of x.

1. $3x + 7 = 4$

2. $x + \dfrac{1}{x} = 2$

3. $x^2 = x(x - 5)$

4. $3(x + 2) = 2(x - 3)$

5. $\sqrt{x} = 4$

6. $\dfrac{1}{x - 2} = \dfrac{2}{x - 3}$

In Review Exercises 7–14, solve each equation and classify each one as an identity, a conditional equation, or an equation with no solution.

7. $3(x + 4) = 28$

8. $\dfrac{3}{2}x = 7(x + 11)$

9. $8(3x - 5) - 4(2x + 3) = 12$

10. $\dfrac{x + 3}{x + 4} + \dfrac{x + 3}{x + 2} = 2$

11. $\dfrac{3}{x - 1} = \dfrac{1}{2}$

12. $\dfrac{8x^2 + 72x}{9 + x} = 8x$

13. $\dfrac{3x}{x - 1} - \dfrac{5}{x + 3} = 3$

14. $x + \dfrac{1}{2x - 3} = \dfrac{2x^2}{2x - 3}$

In Review Exercises 15–18, solve each equation by factoring.

15. $2x^2 - x - 6 = 0$ 16. $12x^2 + 13x = 4$ 17. $5x^2 - 8x = 0$ 18. $27x^2 = 30x - 8$

In Review Exercises 19–22, solve each equation by completing the square.

19. $x^2 - 8x + 15 = 0$ 20. $3x^2 + 18x = -24$ 21. $5x^2 - x - 1 = 0$ 22. $5x^2 - x = 0$

In Review Exercises 23–26, solve each equation by using the quadratic formula.

23. $x^2 + 5x - 14 = 0$ **24.** $3x^2 - 25x = 18$ **25.** $5x^2 = 1 - x$ **26.** $-5 = a^2 + 2a$

In Review Exercises 27–34, solve by any method.

27. $\dfrac{3x}{2} - \dfrac{2x}{x-1} = x - 3$ **28.** $\dfrac{12}{x} - \dfrac{x}{2} = x - 3$ **29.** $x^4 - 2x^2 + 1 = 0$ **30.** $x^4 + 36 = 37x^2$

31. $\sqrt{x-1} + x = 7$ **32.** $\sqrt{5-x} + \sqrt{5+x} = 4$

33. $\sqrt{a+9} - \sqrt{a} = 3$ **34.** $\sqrt{y+5} + \sqrt{y} = 1$

35. Find the value of k that will cause the roots of $kx^2 + 4x + 12 = 0$ to be equal.

36. Find the values of k that will cause the roots of $4y^2 + (k+2)y = 1 - k$ to be equal.

In Review Exercises 37–44, solve each word problem.

37. A liter of fluid is 50% alcohol. How much water must be added to dilute it to a 20% solution?

38. Scott can wash 37 windows in 3 hours, while Bill can wash 27 windows in 2 hours. How long will it take the two of them to wash 100 windows?

39. A tank can be filled in 9 hours by one pipe, and in 12 hours by another. How long will it take both pipes to fill the empty tank?

40. How many ounces of pure zinc must be alloyed with 20 ounces of brass that is 30% zinc and 70% copper, to produce brass that is 40% zinc?

41. A bank lends $10,000, part of it at 11% annual interest and the rest at 14%. If the annual income is $1265, how much was lent at each rate?

42. George can paint his summer cottage in 9 hours less time than his son can. Working together, they can paint the house in 20 hours. How long would it take each, if they worked alone?

43. A farmer wishes to enclose a rectangular garden with 300 yards of fencing. A river runs along one side of the garden, so no fencing is needed there. What will be the dimensions of the rectangle if the area is 10,450 square yards?

44. A jet plane, flying 120 miles per hour faster than a propeller-driven plane, travels 3520 miles in 3 hours less time than the propeller plane requires to fly the same distance. How fast does each plane fly?

In Review Exercises 45–50, solve each equation for the indicated variable.

45. $\dfrac{1}{f} = \dfrac{1}{f_1} + \dfrac{1}{f_2}$ for f_1 **46.** $\dfrac{1}{C} = \dfrac{1}{C_1} + \dfrac{1}{C_2}$ for C

47. $S = \dfrac{a - lr}{1 - r}$ for l **48.** $s(s-a)(s-b)(s-c) = A^2$ for b

49. $\dfrac{y}{x} = \dfrac{x+y}{2}$ for y **50.** $\dfrac{y}{x} = \dfrac{x+y}{2}$ for x

3 Functions and Their Graphs

Mathematical expressions often indicate relationships between several quantities. For example, the formula $d = rt$ indicates that distance traveled depends on the rate of speed and elapsed time. Because d depends on the values r and t, we say that d is a function of r and t. The concept of function is the topic of this chapter.

3.1 THE CARTESIAN COORDINATE SYSTEM

Any point on a number line is associated with a single real number, called its **coordinate**. René Descartes (1596–1650) is credited with the idea of associating each point in the *plane* with a *pair* of real numbers.

Descartes' idea is based on two perpendicular number lines, usually called the **x-axis** and **y-axis**, which divide the plane into four **quadrants**, numbered as shown in Figure 3-1. These axes intersect at a point called the **origin**, which is the zero point on each number line. The positive direction on the x-axis is to the right, the positive direction on the y-axis is upward, and the same unit distance is used on both axes. These axes determine a **rectangular coordinate system**, or a **Cartesian coordinate system** in honor of its inventor.

To **plot** the point associated with the pair of real numbers $(-3, 2)$, for example, we start at the origin, count 3 units to the left, and then 2 units up. See Figure 3-2. Point P, called the **graph** of the pair $(-3, 2)$, lies in the second quadrant.

René Descartes(1596–1650) Descartes merged algebra and geometry into a single subject called analytical geometry.

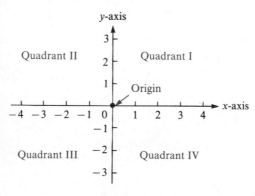

Figure 3-1

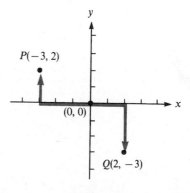

Figure 3-2

The pair $(-3, 2)$ gives the **coordinates** of the point P. Point Q with coordinates $(2, -3)$ lies in the fourth quadrant. The coordinates of the origin are $(0, 0)$.

Note that the pairs $(-3, 2)$ and $(2, -3)$ represent different points. Because the order of the real numbers of the pair (a, b) is important, such pairs are called **ordered pairs**. The first coordinate, a, of the ordered pair (a, b) is called the **x-coordinate** or the **abscissa**. The second coordinate, b, is called the **y-coordinate**, or the **ordinate**. It is proper to say "the point P with coordinates (a, b)," but it is acceptable to say simply, "the point $P(a, b)$." The set of all points determined by ordered pairs (x, y) is called the **xy-plane** or the **Cartesian plane**.

Example 1 Graph the points $P(2, 3)$, $Q(-4, -2)$, $R(-5, 0)$, $S(0, 4)$, and $T(3, -2)$.

Solution

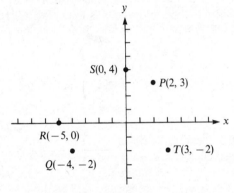

$P(2, 3)$ lies in Quadrant I.

$Q(-4, -2)$ lies in Quadrant III.

$R(-5, 0)$ lies on the x-axis. It is not in any quadrant.

$S(0, 4)$ lies on the y-axis. It is not in any quadrant.

$T(3, -2)$ lies in Quadrant IV.

Figure 3-3   ∎

The **graph of an equation** in the variables x and y is the set of all points on the xy-plane with coordinates (x, y) that satisfy the equation. To graph an equation by plotting points, we choose a value for one of the variables, say x, and calculate the corresponding value for y. We then plot the resulting point (x, y). We do this for several points and join them with a line or curve. If the graph is a curve, the more points we plot, the more accurate the final graph will be.

Example 2 Graph the equation $x + 2y = 5$.

Solution Pick some values for either x or y, substitute those values in the equation, and solve for the other variable. For example, if $y = -1$, you can find x as follows:

$$x + 2y = 5$$
$$x + 2(-1) = 5 \qquad \text{Substitute } -1 \text{ for } y.$$
$$x - 2 = 5 \qquad \text{Simplify.}$$
$$x = 7 \qquad \text{Add 2 to both sides.}$$

Thus, one ordered pair that satisfies the equation is $(7, -1)$.

If $x = 0$, you have

$$x + 2y = 5$$
$$\mathbf{0} + 2y = 5 \qquad \text{Substitute 0 for } x.$$
$$2y = 5 \qquad \text{Simplify.}$$
$$y = \frac{5}{2} \qquad \text{Divide both sides by 2.}$$

Thus, another ordered pair that satisfies the equation is $(0, \frac{5}{2})$.

The ordered pairs $(7, -1)$ and $(0, \frac{5}{2})$ and others that satisfy the equation $x + 2y = 5$ are shown in the table of values in Figure 3-4. Plot each of these ordered pairs on a rectangular coordinate system, as in the figure. Later, we will show that all points (x, y) that satisfy an equation such as $x + 2y = 5$ lie on a straight line. The line that joins the five points is the graph of the equation.

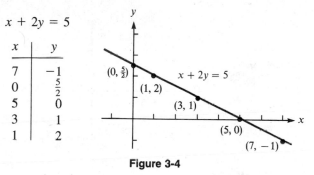

$x + 2y = 5$

x	y
7	-1
0	$\frac{5}{2}$
5	0
3	1
1	2

Figure 3-4

Because the line in Example 2 intersects the y-axis at the point $(0, \frac{5}{2})$, the number $\frac{5}{2}$ is called the **y-intercept** of the line. Similarly, 5 is the **x-intercept** of the line.

Example 3   The graph of the equation $4x - 3y = 12$ is a line. Use the x- and y-intercepts to graph it.

Solution   To find the y-intercept, substitute 0 for x and solve for y:

$$4x - 3y = 12$$
$$4(\mathbf{0}) - 3y = 12 \qquad \text{Substitute 0 for } x.$$
$$-3y = 12 \qquad \text{Simplify.}$$
$$y = -4 \qquad \text{Divide both sides by } -3.$$

Because the y-intercept is -4, the line intersects the y-axis at the point $(0, -4)$.

To find the x-intercept, substitute 0 for y and solve for x:

$$4x - 3y = 12$$
$$4x - 3(\mathbf{0}) = 12 \qquad \text{Substitute 0 for } y.$$
$$4x = 12 \qquad \text{Simplify.}$$
$$x = 3 \qquad \text{Divide both sides by 4.}$$

Because the x-intercept is 3, the line intersects the x-axis at the point $(3, 0)$.

Although these two points are sufficient to draw the line, find and plot a third point as a check. If $x = 6$, for example, then $y = 4$. Thus, the line passes through $(6, 4)$. The graph appears in Figure 3-5.

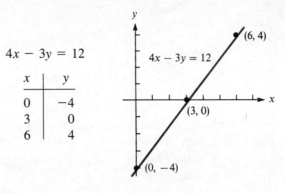

$$4x - 3y = 12$$

x	y
0	−4
3	0
6	4

Figure 3-5

Example 4 Graph the equation $3(y + 2) = 2x - 3$.

Solution First solve the equation for y. This will make it easier to find pairs (x, y) that satisfy the equation:

$$3(y + 2) = 2x - 3$$

$$3y + 6 = 2x - 3 \qquad \text{Remove parentheses.}$$

$$3y = 2x - 9 \qquad \text{Add } -6 \text{ to both sides.}$$

$$y = \frac{2}{3}x - 3 \qquad \text{Divide both sides by 3.}$$

Substitute numbers for x and calculate the corresponding values of y. For example, let $x = 3$:

$$y = \frac{2}{3}x - 3$$

$$y = \frac{2}{3}(3) - 3 \qquad \text{Substitute 3 for } x.$$

$$y = 2 - 3 \qquad \text{Simplify.}$$

$$y = -1$$

Thus, the point $(3, -1)$ lies on the graph. To find another point, let $x = 0$ and calculate y:

$$y = \frac{2}{3}x - 3$$

$$y = \frac{2}{3}(0) - 3 \qquad \text{Substitute 0 for } x.$$

$$y = -3 \qquad \text{Simplify.}$$

Thus, the point $(0, -3)$ also lies on the graph. These two solutions and several others are plotted in Figure 3-6. The graph of the equation is the line that passes through all of the points.

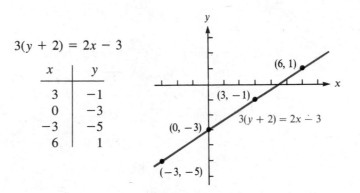

$3(y + 2) = 2x - 3$

x	y
3	-1
0	-3
-3	-5
6	1

Figure 3-6

The Distance Formula

To distinguish between two points on the Cartesian plane, subscript notation is often used. One point might be denoted as $P(x_1, y_1)$, read as "point P with co-ordinates of x sub 1 and y sub 1," and the second as $Q(x_2, y_2)$. We can derive a formula to calculate the distance between two points $P(x_1, y_1)$ and $Q(x_2, y_2)$.

If $P(x_1, y_1)$ and $Q(x_2, y_2)$ are two points in Figure 3-7, then the right triangle PQR can be formed, with R having coordinates (x_2, y_1). Because the line segment RQ is vertical, the square of its length is $(y_2 - y_1)^2$. Because PR is horizontal, the square of its length is $(x_2 - x_1)^2$. By the Pythagorean theorem, we know that the square of the hypotenuse of right triangle PQR is equal to the sum of the squares of the two legs. Thus, we have

$$d^2 = (x_2 - x_1)^2 + (y_2 - y_1)^2$$

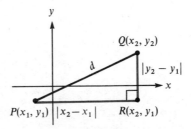

Figure 3-7

Because equal positive numbers have equal positive square roots, we can take the positive square root of both sides of this equation to obtain the **distance formula**.

> **The Distance Formula.**   The distance d between points (x_1, y_1) and (x_2, y_2) is given by
>
> $$d = \sqrt{(x_2 - x_1)^2 + (y_2 - y_1)^2}$$

Example 5   Find the distance $d(PQ)$ between the points $P(-1, -2)$ and $Q(-7, 8)$.

Solution   Let (x_1, y_1) be $P(-1, -2)$ and (x_2, y_2) be $Q(-7, 8)$. Substitute these coordinates into the distance formula and simplify:

$$d(PQ) = \sqrt{(x_2 - x_1)^2 + (y_2 - y_1)^2}$$
$$d(PQ) = \sqrt{[-7 - (-1)]^2 + [8 - (-2)]^2}$$
$$= \sqrt{(-6)^2 + (10)^2}$$
$$= \sqrt{36 + 100}$$
$$= \sqrt{136}$$
$$= \sqrt{4 \cdot 34}$$
$$= 2\sqrt{34}$$

∎

The Midpoint Formula

The **midpoint** of a line segment is the point on the segment that is midway between the endpoints. The x-coordinate of the midpoint is found by averaging the x-coordinates of the segment's endpoints. Similarly, the y-coordinate of the midpoint is the average of the y-coordinates of the segment's endpoints. For example, the midpoint of the segment joining $P(3, 4)$ and $Q(5, 7)$ is

$$M\left(\frac{3 + 5}{2}, \frac{4 + 7}{2}\right) \qquad \text{or} \qquad M\left(4, \frac{11}{2}\right)$$

> **The Midpoint Formula.**   The midpoint of the line segment joining points $P(x_1, y_1)$ and $Q(x_2, y_2)$ is the point M with coordinates of
>
> $$\left(\frac{x_1 + x_2}{2}, \frac{y_1 + y_2}{2}\right)$$

Proof   To prove that M is the midpoint of segment PQ, we must show that $d(PM) = d(MQ)$, and that points P, Q, and M all lie on the same line. See Figure 3-8. To show that $d(PM) = d(MQ)$, we use the distance formula and observe that the distances are equal:

$$d(PM) = \sqrt{\left(\frac{x_1 + x_2}{2} - x_1\right)^2 + \left(\frac{y_1 + y_2}{2} - y_1\right)^2}$$

$$= \sqrt{\left(\frac{x_2 - x_1}{2}\right)^2 + \left(\frac{y_2 - y_1}{2}\right)^2}$$

$$= \sqrt{\frac{(x_2 - x_1)^2}{4} + \frac{(y_2 - y_1)^2}{4}}$$

$$= \frac{1}{2}\sqrt{(x_2 - x_1)^2 + (y_2 - y_1)^2}$$

$$d(MQ) = \sqrt{\left(x_2 - \frac{x_1 + x_2}{2}\right)^2 + \left(y_2 - \frac{y_1 + y_2}{2}\right)^2}$$

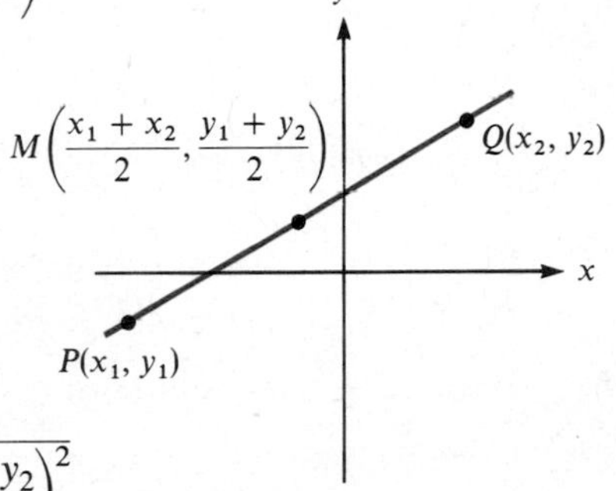

$$= \sqrt{\left(\frac{x_2 - x_1}{2}\right)^2 + \left(\frac{y_2 - y_1}{2}\right)^2}$$

$$= \sqrt{\frac{(x_2 - x_1)^2}{4} + \frac{(y_2 - y_1)^2}{4}}$$

$$= \frac{1}{2}\sqrt{(x_2 - x_1)^2 + (y_2 - y_1)^2}$$

Figure 3-8

Thus, $d(PM) = d(MQ)$. The distance between P and Q is

$$d(PQ) = \sqrt{(x_2 - x_1)^2 + (y_2 - y_1)^2}$$

Because $d(PM) + d(MQ) = d(PQ)$, points P, Q, and M must lie on the same line. The theorem is proved. ☐

Example 6 The midpoint of the line segment joining $P(-3, 2)$ and point $Q(x_2, y_2)$ is $M(1, 4)$. Find the coordinates of Q.

Solution Let $P(x_1, y_1)$ be $P(-3, 2)$, and let $M(x_M, y_M)$ be $M(1, 4)$. Find the coordinates x_2 and y_2 of $Q(x_2, y_2)$ as follows:

$$x_M = \frac{x_1 + x_2}{2} \qquad \text{and} \qquad y_M = \frac{y_1 + y_2}{2}$$

$$1 = \frac{-3 + x_2}{2} \qquad\qquad 4 = \frac{2 + y_2}{2}$$

$$2 = -3 + x_2 \qquad\qquad 8 = 2 + y_2 \qquad \text{Multiply both sides by 2.}$$

$$5 = x_2 \qquad\qquad 6 = y_2$$

The coordinates of point Q are $(5, 6)$. ∎

EXERCISE 3.1

In Exercises 1–12, graph each point on the Cartesian plane. Indicate the quadrant in which the point lies or the axis on which it lies.

1. $(2, 5)$ **2.** $(-3, 4)$ **3.** $(-4, -5)$ **4.** $(6, -2)$

5. $(5, 2)$ **6.** $(3, -4)$ **7.** $(4, 0)$ **8.** $(-2, 2)$

9. $(0, 2)$ **10.** $(3, 4)$ **11.** $(-7, 0)$ **12.** $(0, -5)$

In Exercises 13–20, use the x- and y-intercepts to graph each equation.

13. $x + y = 5$ **14.** $x - y = 3$ **15.** $2x - y = 4$ **16.** $3x + y = 9$

17. $3x + 2y = 6$ **18.** $4x - 5y = 20$ **19.** $2x + 7y = 14$ **20.** $3x - 5y = 15$

In Exercises 21–28, first solve each equation for y, and then graph each equation.

21. $y - 2x = 7$ **22.** $y + 3 = -4x$ **23.** $6x - 3y = 10$ **24.** $4x + 8y - 1 = 0$

25. $2(x - y) = 3x + 2$ **26.** $5(x + 2) = 3y - x$

27. $3x + y = 3(x - 1)$ **28.** $2(y - x) = 3(x + 2)$

In Exercises 29–34, find the distance between P and the point $(0, 0)$. Assume that a and b are positive numbers.

29. $P(4, -3)$ **30.** $P(-5, 12)$ **31.** $P(-3, 2)$ **32.** $P(5, 0)$

33. $P(a, b)$ **34.** $P(a, -a)$

In Exercises 35–46, find the distance between P and Q. Assume that all variables represent positive numbers.

35. $P(3, 7); Q(6, 3)$ **36.** $P(4, -6); Q(-1, 6)$

37. $P(0, 5); Q(6, -3)$ **38.** $P(-2, -15); Q(-9, -39)$

39. $P(3, 3); Q(5, 5)$ **40.** $P(6, -3); Q(-3, 2)$

41. $P(3, 7); Q(5, 7)$ **42.** $P(4, -6); Q(4, -8)$

43. $P(\pi, 2); Q(\pi, 5)$ **44.** $P(\sqrt{3}, \pi); Q(\pi, \sqrt{3})$

45. $P(x, 0); Q(0, y)$ **46.** $P(a, b); Q(b, a)$

In Exercises 47–54, find the midpoint of the line segment PQ.

47. $P(2, 4); Q(6, 8)$ **48.** $P(3, -6); Q(-1, -6)$

49. $P(2, -5); Q(-2, 7)$ **50.** $P(0, 3); Q(-10, -13)$

51. $P(0, 0); Q(\sqrt{5}, \sqrt{5})$ **52.** $P(\sqrt{3}, 0); Q(0, \sqrt{3})$

53. $P(a, 2); Q(2, a)$ **54.** $P(a, b); Q(0, 0)$

In Exercises 55–58, one endpoint P and the midpoint M of line segment PQ are given. Find the coordinates of the other endpoint, Q.

55. $P(1, 4); M(3, 5)$ **56.** $P(2, -7); M(-5, 6)$ **57.** $P(5, -5); M(5, 5)$ **58.** $P(-7, 3); M(0, 0)$

59. Show that a triangle with vertices at $(13, -2)$, $(9, -8)$, and $(5, -2)$ is isosceles (has two equal sides).

60. Show that a triangle with vertices at $(-1, 2)$, $(3, 1)$, and $(4, 5)$ is isosceles.

61. Show that the points $(-3, -1)$, $(3, 1)$, $(1, 7)$, and $(-5, 5)$ are the vertices of a rhombus by showing that its four sides are equal.

62. Use the distance formula to find the radius of a circle with center at $(4, -2)$ and passing through $(6, -9)$.

63. Find the center of a circle if the endpoints of a diameter are $(-17, 6)$ and $(9, -12)$.

64. Find the coordinates of the two points on the y-axis that are $\sqrt{74}$ units from the point $(5, -3)$.

65. The diagonals of a square with area 50 square units lie on the x- and y-axes. Find the coordinates of the vertices.

66. Show that the distance between the points (a, b) and (c, d) is equal to the distance between the point $(a - c, b - d)$ and the origin.

67. In Illustration 1, points M and N are midpoints of AC and BC, respectively. Find the length of MN.

68. In Illustration 2, points M and N are midpoints of AC and BC, respectively. Show that $d(MN) = \frac{1}{2}[d(AB)]$.

Illustration 1   **Illustration 2**

3.2   THE SLOPE OF A NONVERTICAL LINE

The **slope of a nonvertical line** drawn in the Cartesian plane is a measure of its tilt or inclination. Consider the line l in Figure 3-9**a**, which passes through $P(x_1, y_1)$ and $Q(x_2, y_2)$. If line RQ is perpendicular to the x-axis, and PR is perpendicular to the y-axis, then triangle PRQ is a right triangle, and point R has coordinates of (x_2, y_1). The distance from R to Q, called the **rise**, is the change in the y-coordinates of R and Q: $y_2 - y_1$. The horizontal distance from P to R, called the **run**, is the change in the x-coordinates of points P and R: $x_2 - x_1$. The slope of the line is the *rise* divided by the *run*. The letter m often represents the slope of a line:

$$\text{slope of line } PQ = m = \frac{\text{rise}}{\text{run}} = \frac{y_2 - y_1}{x_2 - x_1}$$

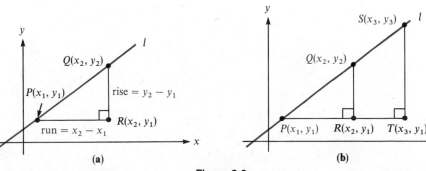

Figure 3-9

In Figure 3-9**b**, point S represents an arbitrary third point on line l. Because triangles PRQ and PTS are similar, their corresponding sides are in proportion. Thus, the ratios of the rise to the run in the two triangles are equal, and we have

$$m = \frac{y_2 - y_1}{x_2 - x_1} = \frac{y_3 - y_1}{x_3 - x_1}$$

This implies that the slope of a nonvertical line is constant and can be calculated using *any* two points on the line. Furthermore, if point P is on a line with slope m, and the ratio of rise to run of a segment PQ is also m, then point Q is on the line.

Definition.   If $P(x_1, y_1)$ and $Q(x_2, y_2)$ are two points on a nonvertical line l in the Cartesian plane, then the slope of l is given by

$$m = \frac{y_2 - y_1}{x_2 - x_1} \quad \text{provided } x_1 \neq x_2$$

If $x_1 = x_2$, then line l is a vertical line and has no defined slope.

Example 1   Find the slope of the line passing through the points $P(-1, -2)$ and $Q(7, 8)$.

Solution   Let $x_1 = -1$, $y_1 = -2$, $x_2 = 7$, and $y_2 = 8$ in the formula that defines slope:

$$m = \frac{y_2 - y_1}{x_2 - x_1}$$

$$= \frac{8 - (-2)}{7 - (-1)}$$

$$= \frac{10}{8}$$

$$= \frac{5}{4}$$

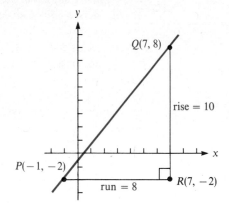

Figure 3-10

Thus, the slope of the line is $\frac{5}{4}$. See Figure 3-10.   ■

Example 2   The graph of the equation $5x + 2y = 10$ is a line. Find its slope.

Solution   Determine the coordinates of two points on the line by finding two ordered pairs (x, y) that satisfy the equation $5x + 2y = 10$. If $x_1 = 0$, for example, then $y_1 = 5$. Thus, the point $(0, 5)$ lies on the line. If $x_2 = -2$, then $y_2 = 10$. Thus, the point $(-2, 10)$ also lies on the line. Substitute $x_1 = 0$, $y_1 = 5$, $x_2 = -2$, and $y_2 = 10$ into the formula for the slope of a line, and simplify.

$$m = \frac{y_2 - y_1}{x_2 - x_1} = \frac{10 - 5}{-2 - 0} = \frac{5}{-2} = -\frac{5}{2}$$

The slope of the line is $-\frac{5}{2}$.   ∎

As a point moves along a nonvertical line l from $P(x_1, y_1)$ to $Q(x_2, y_2)$, its y-coordinate changes by an amount $y_2 - y_1$, often denoted by Δy (read as "the change in y" or "delta y"). The difference of the x-coordinates, $x_2 - x_1$, is often denoted by Δx. See Figure 3-11. Thus,

$$m = \frac{y_2 - y_1}{x_2 - x_1} = \frac{\Delta y}{\Delta x}$$

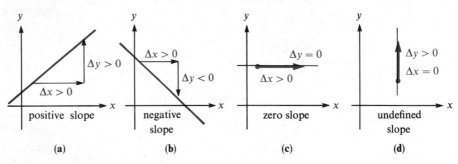

Figure 3-11

If a line rises as it moves to the right—that is, if increasing x-values result in increasing y-values—then the slope of the line is positive. See Figure 3-12**a**. If a line drops as it moves to the right—that is, if increasing x-values result in decreasing y-values—then the slope of the line is negative. See Figure 3-12**b**. If a line is horizontal, then the slope of the line is 0 because the difference between the y-coordinates of any two points on the line is 0. See Figure 3-12**c**. If a line is vertical, the slope is undefined because the denominator of the formula that defines slope—the difference of the x-coordinates—is 0. See Figure 3-12**d**.

A theorem relates parallel lines to their slopes.

Theorem.   Nonvertical parallel lines have the same slope, and distinct lines having the same slope are parallel.

Proof   Suppose that the nonvertical lines l_1 and l_2 of Figure 3-13 are parallel and have slopes of m_1 and m_2, respectively. Then the right triangles ABC and DEF are similar, and it follows that

$$m_1 = \frac{\text{rise of } l_1}{\text{run of } l_1}$$

$$= \frac{\text{rise of } l_2}{\text{run of } l_2}$$

$$= m_2$$

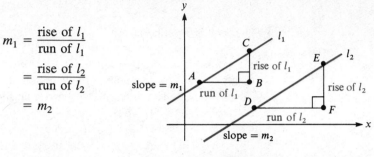

Figure 3-13

Thus, two nonvertical parallel lines have the same slope.

In the exercises, you will be asked to prove that distinct lines having the same slope are parallel.   □

It is also true that two lines with no defined slope are vertical lines, and vertical lines are parallel.

Example 3   The line passing through $P(-2, 3)$ and $Q(3, -1)$ is parallel to the line passing through $R(4, -5)$ and $S(x, 7)$. Find x.

Solution   Because the lines PQ and RS are parallel, the slopes of PQ and RS are equal. Find the slope of each line, set them equal to each other, and solve the resulting equation:

slope of PQ = slope of RS

$$\frac{-1 - 3}{3 - (-2)} = \frac{7 - (-5)}{x - 4}$$

$$\frac{-4}{5} = \frac{12}{x - 4} \qquad \text{Simplify.}$$

$$-4(x - 4) = 5 \cdot 12 \qquad \text{Multiply both sides by } 5(x - 4).$$

$$-4x + 16 = 60 \qquad \text{Remove parentheses and simplify.}$$

$$-4x = 44 \qquad \text{Add } -16 \text{ to both sides.}$$

$$x = -11 \qquad \text{Divide both sides by } -4.$$

Thus, x is -11. The line passing through $P(-2, 3)$ and $Q(3, -1)$ is parallel to the line passing through $R(4, -5)$ and $S(-11, 7)$.   ■

If the product of two numbers is equal to -1, the numbers are called **negative reciprocals**. The following theorem describes the relation between the slopes of perpendicular lines.

> **Theorem.** If two nonvertical lines are perpendicular, their slopes are negative reciprocals of each other.
>
> If the slopes of two lines are negative reciprocals, the lines are perpendicular.

Proof Suppose that l_1 and l_2 are lines with slopes m_1 and m_2 that intersect at the origin. Let $P(a, b)$ be a point on l_1, and let $Q(c, d)$ be a point on l_2. Neither point P nor point Q can be the origin. See Figure 3-14.

First, suppose lines l_1 and l_2 are perpendicular. Then triangle POQ is a right triangle with right angle at O, and by the Pythagorean theorem, $d(OP)^2 + d(OQ)^2 = d(PQ)^2$. Proceed as follows:

$$d(OP)^2 + d(OQ)^2 = d(PQ)^2$$
$$(a - 0)^2 + (b - 0)^2 + (c - 0)^2 + (d - 0)^2 = (a - c)^2 + (b - d)^2$$
$$a^2 + b^2 + c^2 + d^2 = a^2 - 2ac + c^2 + b^2 - 2bd + d^2$$
$$0 = -2ac - 2bd$$
$$bd = -ac$$

1.
$$\frac{b}{a} \cdot \frac{d}{c} = -1$$

The coordinates of P are (a, b) and the coordinates of O are $(0, 0)$. Using the definition of slope, we have

$$m_1 = \frac{b - 0}{a - 0} = \frac{b}{a}$$

Similarly, we have

$$m_2 = \frac{d}{c}$$

Substitute m_1 for $\frac{b}{a}$ and m_2 for $\frac{d}{c}$ in Equation 1 to obtain

$$m_1 m_2 = -1$$

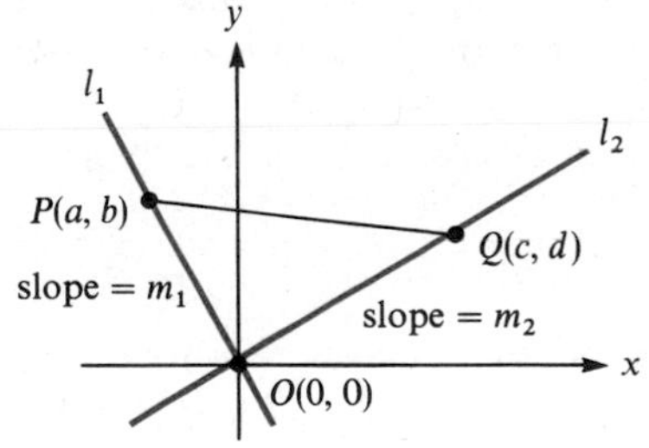

Figure 3-14

Hence, if lines l_1 and l_2 are perpendicular, then they have slopes that are negative reciprocals of each other.

Conversely, suppose that the slopes of lines l_1 and l_2 are negative reciprocals of each other. Because the steps of the proof are reversible, we have that $d(OP)^2 + d(OQ)^2 = d(PQ)^2$. By the Pythagorean theorem, triangle POQ is a right triangle. Thus, l_1 and l_2 are perpendicular. □

It is also true that a line with slope of 0 is horizontal and thus is perpendicular to a vertical line that has no defined slope.

Example 4   Two lines intersect at the point $P(-5, 3)$. One passes through the point $Q(-1, -3)$ and the other passes through $R(1, 7)$. Are the lines perpendicular?

Solution   First find the slopes of lines PQ and PR:

$$\text{slope of } PQ = \frac{\Delta y}{\Delta x} = \frac{-3 - 3}{-1 - (-5)} = -\frac{6}{4} = -\frac{3}{2}$$

$$\text{slope of } PR = \frac{\Delta y}{\Delta x} = \frac{7 - 3}{1 - (-5)} = \frac{4}{6} = \frac{2}{3}$$

Because the slopes of these lines are negative reciprocals of each other, the lines are perpendicular.   ∎

EXERCISE 3.2

In Exercises 1–10, find the slope of the line passing through each pair of points, if possible.

1. $P(2, 5)$; $Q(3, 10)$ **2.** $P(3, -1)$; $Q(5, 3)$ **3.** $P(3, -2)$; $Q(-1, 5)$ **4.** $P(3, 7)$; $Q(6, 16)$

5. $P(8, -7)$; $Q(4, 1)$ **6.** $P(5, 17)$; $Q(17, 17)$

7. $P(-4, 3)$; $Q(-4, -3)$ **8.** $P(2, \sqrt{7})$; $Q(\sqrt{7}, 2)$

9. $P(a + b, c)$; $Q(b + c, a)$ assume $a \neq c$.   **10.** $P(b, 0)$; $Q(a + b, a)$ assume $a \neq 0$.

In Exercises 11–18, find two points on the line and determine the slope of the line.

11. $y = 3x + 2$ **12.** $y = 5x - 8$ **13.** $5x - 10y = 3$ **14.** $8y + 2x = 5$

15. $3(y + 2) = 2x - 3$ **16.** $4(x - 2) = 3y + 2$

17. $3(y + x) = 3(x - 2)$ **18.** $2x + 5 = 2(y + x)$

In Exercises 19–24, determine whether the lines with the given slopes are parallel, perpendicular, or neither.

19. $m_1 = 3$, $m_2 = -\dfrac{1}{3}$ **20.** $m_1 = \dfrac{2}{3}$, $m_2 = \dfrac{3}{2}$

21. $m_1 = \sqrt{8}$, $m_2 = 2\sqrt{2}$ **22.** $m_1 = 1$, $m_2 = -1$

23. $m_1 = -\sqrt{2}$, $m_2 = \dfrac{\sqrt{2}}{2}$ **24.** $m_1 = 2\sqrt{7}$, $m_2 = \sqrt{28}$

In Exercises 25–30, determine if the line through the given points and the line through $R(-3, 5)$ and $S(2, 7)$ are parallel, perpendicular, or neither.

25. $P(2, 4)$; $Q(7, 6)$ **26.** $P(-3, 8)$; $Q(-13, 4)$

27. $P(-4, 6)$; $Q(-2, 1)$ **28.** $P(0, -9)$; $Q(4, 1)$

29. $P(a, a)$; $Q(3a, 6a)$ and $a \neq 0$ **30.** $P(b, b)$; $Q(-b, 6b)$ and $b \neq 0$

In Exercises 31–32, find the slopes of lines PQ and PR, and determine if points P, Q, and R lie on the same line.

31. $P(-2, 8)$; $Q(-6, 9)$; $R(2, 5)$ **32.** $P(1, -1)$; $Q(3, -2)$; $R(-3, 0)$

In Exercises 33–36, determine which, if any, of the three lines PQ, PR, and QR are perpendicular.

33. $P(5, 4)$; $Q(2, -5)$; $R(8, -3)$ **34.** $P(8, -2)$; $Q(4, 6)$; $R(6, 7)$

35. $P(0, 0)$; $Q(a, b)$; $R(-b, a)$ **36.** $P(1, 3)$; $Q(1, 9)$; $R(7, 3)$

37. Show that points $A(-1, -1)$, $B(-3, 4)$, and $C(4, 1)$ are the vertices of a right triangle.

38. Show that points $D(0, 1)$, $E(-1, 3)$, and $F(3, 5)$ are the vertices of a right triangle.

39. Show that points $A(1, -1)$, $B(3, 0)$, $C(2, 2)$, and $D(0, 1)$ are the vertices of a square.

40. Show that points $E(-1, -1)$, $F(3, 0)$, $G(2, 4)$, and $H(-2, 3)$ are the vertices of a square.

41. Show that points $A(-2, -2)$, $B(3, 3)$, $C(2, 6)$, and $D(-3, 1)$ are the vertices of a parallelogram. (Show that both pairs of opposite sides are parallel.)

42. Show that points $E(1, -2)$, $F(5, 1)$, $G(3, 4)$, and $H(-3, 4)$ are the vertices of a trapezoid. (Show that only one pair of opposite sides are parallel.)

43. In Illustration 1, points M and N are midpoints of CB and BA, respectively. Show that MN is parallel to AC.

44. In Illustration 2, $d(AB) = d(AC)$. Show that AD is perpendicular to BC.

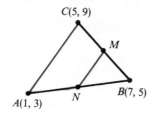

Illustration 1

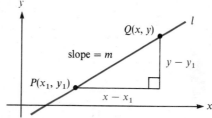

Illustration 2

45. Prove that if two lines have the same slope, they are parallel.

3.3 EQUATIONS OF LINES

Suppose that the nonvertical line l of Figure 3-15 has a slope of m and passes through the point $P(x_1, y_1)$. If $Q(x, y)$ is another point on that line, then by the definition of *slope*, we have

$$m = \frac{y - y_1}{x - x_1}$$

or

$$y - y_1 = m(x - x_1)$$

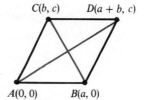

Figure 3-15

This suggests the following theorem.

Theorem. The equation of the line passing through $P(x_1, y_1)$ with slope m is

$$y - y_1 = m(x - x_1)$$

Proof   The proof that the equation $y - y_1 = m(x - x_1)$ is the equation of the line l has two parts. We must show

1.   that every point on the line has coordinates that satisfy the equation, and
2.   that every point whose coordinates satisfy the equation lies on the line.

Part 1.   Let $Q(a, b)$ be any point on the line except point $P(x_1, y_1)$. Substitute a for x and b for y in the equation $y - y_1 = m(x - x_1)$ and solve for m:

$$y - y_1 = m(x - x_1)$$

$$b - y_1 = m(a - x_1) \qquad \text{Substitute } a \text{ for } x \text{ and } b \text{ for } y.$$

$$m = \frac{b - y_1}{a - x_1} \qquad \text{Divide both sides by } a - x_1.$$

Because the right side of this equation represents the slope, m, of the line, we have the identity

$$m = m$$

Thus, the coordinates of the point Q satisfy the equation of the line. The coordinates of point $P(x_1, y_1)$ itself satisfy the equation $y - y_1 = m(x - x_1)$, because $y_1 - y_1 = m(x_1 - x_1)$ reduces to $0 = 0$. Thus, *any point on line l has coordinates that satisfy the equation.*

Part 2.   Suppose that the coordinates of the point $R(a, b)$ satisfy the equation $y - y_1 = m(x - x_1)$, and that point R is not point P. Then $b - y_1 = m(a - x_1)$, and

$$m = \frac{b - y_1}{a - x_1}$$

Thus, the slope of the line RP is m. Because there is only *one* line with slope m passing through P, and that is line l, point R must lie on line l. Thus, *any point with coordinates that satisfy the equation $y - y_1 = m(x - x_1)$ lies on the line l.*
The theorem is proved.   □

Point-Slope Form

Because the equation $y - y_1 = m(x - x_1)$ displays the coordinates of a fixed point on the line and the line's slope, it is called the **point-slope form** of the equation of a line.

The Point-Slope Form of the Equation of a Line.   The equation of the line passing through the point $P(x_1, y_1)$ and having a slope of m is

$$y - y_1 = m(x - x_1)$$

Example 1   Find the equation of the line passing through the point $P(3, -1)$ with a slope of $-\frac{5}{3}$. Then solve that equation for y.

Solution   Substitute 3 for x_1, -1 for y_1, and $-\frac{5}{3}$ for m in the point-slope form of the equation of a line:

$$y - y_1 = m(x - x_1)$$

$$y - (-1) = -\frac{5}{3}(x - 3) \qquad \text{Substitute 3 for } x_1, -1 \text{ for } y_1, \text{ and } -\frac{5}{3} \text{ for } m.$$

$$y + 1 = -\frac{5}{3}x + 5 \qquad \text{Remove parentheses.}$$

$$y = -\frac{5}{3}x + 4 \qquad \text{Add } -1 \text{ to both sides.} \qquad \blacksquare$$

We can use the concept of slope to help graph the equation obtained in Example 1. We first plot the point $P(3, -1)$ as in Figure 3-16. Because the slope is $-\frac{5}{3}$, $\Delta y = -5$ provided that $\Delta x = 3$. Thus, we can locate another point Q on the line by moving 3 units to the right of P and then 5 units down. The graph of the equation is the line PQ.

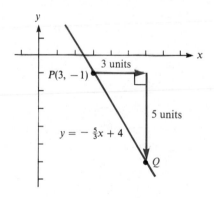

Figure 3-16

Example 2   Find the equation of the line passing through the points $P(3, 7)$ and $Q(-5, 3)$. Solve the equation for y.

Solution   Let $P(x_1, y_1)$ be the point $P(3, 7)$ and $Q(x_2, y_2)$ be the point $Q(-5, 3)$. Then find the slope of the line:

$$m = \frac{y_2 - y_1}{x_2 - x_1} = \frac{3 - 7}{-5 - 3} = \frac{-4}{-8} = \frac{1}{2}$$

Finally, substitute the coordinates of $P(3, 7)$ and $\frac{1}{2}$ for m in the point-slope form of the equation of a line and solve for y:

$$y - y_1 = m(x - x_1)$$

$$y - 7 = \frac{1}{2}(x - 3)$$

$$y = \frac{1}{2}x - \frac{3}{2} + 7 \qquad \text{Remove parentheses and add 7 to both sides.}$$

$$y = \frac{1}{2}x + \frac{11}{2}$$

The same result would be obtained if you had used the point $P(-5, 3)$.   ■

Slope-Intercept Form

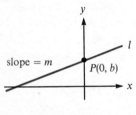

Figure 3-17

If the y-intercept of the line with slope m shown in Figure 3-17 is b, the line intersects the y-axis at $P(0, b)$.

We can write the equation of this line by substituting 0 for x_1, and b for y_1 into the point-slope form of the equation of the line and simplifying:

$$y - y_1 = m(x - x_1)$$

$$y - b = m(x - 0) \qquad \text{Substitute 0 for } x_1 \text{ and } b \text{ for } y_1.$$

$$y = mx + b \qquad \text{Simplify and add } b \text{ to both sides.}$$

Because the equation $y = mx + b$ displays both the slope and the y-intercept of a line, it is called the **slope-intercept form** of the equation of a line.

The Slope-Intercept Form of the Equation of a Line. The equation of the line with slope m and y-intercept b is

$$y = mx + b$$

Example 3   Use the slope-intercept form to find the equation of the line with slope of $\frac{7}{3}$ and y-intercept of -6.

Solution   Substitute $\frac{7}{3}$ for m and -6 for b into the slope-intercept form of the equation of the line and simplify:

$$y = mx + b$$

$$y = \frac{7}{3}x + (-6) \qquad \text{Substitute } \frac{7}{3} \text{ for } m \text{ and } -6 \text{ for } b.$$

$$y = \frac{7}{3}x - 6 \qquad \text{Simplify.}$$

■

Example 4   Find the slope and the y-intercept of the line $3(y + 2) = 6x - 1$.

Solution   Write the equation in the form $y = mx + b$ to determine the slope m and the y-intercept b.

$$3(y + 2) = 6x - 1$$

$$3y + 6 = 6x - 1 \qquad \text{Remove parentheses.}$$

$$3y = 6x - 7 \qquad \text{Add } -6 \text{ to both sides.}$$

$$y = 2x - \frac{7}{3} \qquad \text{Divide both sides by 3.}$$

The slope of the line is 2, and the y-intercept is $-\frac{7}{3}$.   ∎

Example 5   Find the y-intercept of the line having a slope of 2 and passing through the point $P(3, -5)$.

Solution   Because the line passes through the point $P(3, -5)$, the coordinates $x = 3$ and $y = -5$ must satisfy the equation. To find the y-intercept, substitute 3 for x, -5 for y, and 2 for m in the slope-intercept form and solve for b.

$$y = mx + b$$

$$-5 = 2(3) + b$$

$$-5 = 6 + b$$

$$-11 = b$$

The y-intercept is -11.   ∎

Example 6   Find the equation of a line l passing through the point $P(-2, 9)$ and parallel to the line with equation $y = 4x - 73$.

Solution   The slope of the line with equation $y = 4x - 73$ is **4**. Because line l is parallel to the given line, it also has a slope of 4. Because line l passes through the point $P(-2, 9)$ and has a slope of 4, substitute -2 for x, 9 for y, and 4 for m into the slope-intercept form of the equation of the line and simplify:

$$y = mx + b$$

$$9 = 4(-2) + b$$

$$9 = -8 + b$$

$$17 = b$$

The equation of line l is $y = 4x + 17$.   ∎

Example 7   Use the point-slope form to find the equation of the line perpendicular to the line $y = \frac{1}{3}x + 7$ and passing through the point $(2, 3)$. Write the result in slope-intercept form.

Solution   The slope of the given line $y = \frac{1}{3}x + 7$ is $\frac{1}{3}$. The slope of the required line must be -3, the negative reciprocal of $\frac{1}{3}$. Use the point-slope form to find the equation of a line with slope of -3 and passing through the point $(2, 3)$ as follows:

$$y - y_1 = m(x - x_1)$$
$$y - 3 = -3(x - 2)$$
$$y - 3 = -3x + 6$$
$$y = -3x + 9 \qquad \blacksquare$$

If a line is horizontal, its slope m is 0. We can find its equation by substituting 0 for m in the equation $y = mx + b$ and simplifying:

$$y = mx + b$$
$$y = 0x + b$$
$$y = b$$

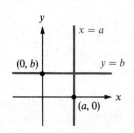

Figure 3-18

The equation of a horizontal line is $y = b$, the y-intercept is b, and the y-coordinate of *every* point on the line is b. See Figure 3-18.

If a line is vertical, it has no defined slope. Thus, it cannot be the graph of any equation of the form $y = mx + b$. Such a line does have an equation, however. If a vertical line passes through a point with an x-coordinate of a, then the x-coordinate of *every* point on the line is a. The equation of that line is $x = a$. See Figure 3-18.

Example 8   Find the equation of the line passing through the points $P(-3, 4)$ and $Q(-3, -2)$.

Solution   Because the x-coordinates of points P and Q are equal, the line has no defined slope. It is a vertical line, with equation $x = -3$. See Figure 3-19. $\blacksquare$

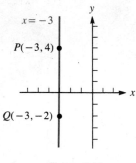

Figure 3-19

General Form

We have shown that the graph of the equation

$$y - y_1 = m(x - x_1)$$

is a line. In this equation, x_1 and y_1 are the coordinates of some fixed point, m is the slope, and x and y are variables. We can rewrite this equation as follows:

$$y - y_1 = m(x - x_1)$$
$$y - y_1 = mx - mx_1 \qquad \text{Remove parentheses.}$$
$$-mx + y = y_1 - mx_1 \qquad \text{Add } y_1 - mx \text{ to both sides.}$$

This final equation is of the form $Ax + By = C$, where A, B, and C are constants: $A = -m$, $B = 1$, and $C = y_1 - mx_1$. In general, for any real numbers A, B, and C (where A and B are not *both* 0), the equation $Ax + By = C$ represents a line and is called the **general form** of the equation of a line. Whenever possible, we will write the general form $Ax + By = C$ so that A, B, and C are integers and $A \geq 0$.

Any equation that can be written in general form is called a **linear equation in
x and y**.

The General Form of the Equation of a Line. If A, B, and C are real
numbers and B is not 0, then the graph of the equation

$$Ax + By = C$$

in a nonvertical line with slope of $-\dfrac{A}{B}$ and y-intercept of $\dfrac{C}{B}$.

If $B = 0$, then the equation $Ax + By = C$ represents a vertical line with x-intercept of $\dfrac{C}{A}$. In the exercises, you will be asked to prove this fact and others related to the general form of the equation of a line.

Example 9   Show that the two lines $3x - 2y = 5$ and $-6x + 4y = 7$ are parallel.

Solution   To show that two lines are parallel, show that they have the same slope. The first equation, $3x - 2y = 5$, is in general form, with $A = 3$, $B = -2$, and $C = 5$. By the previous theorem, the slope of the line is

$$m_1 = -\frac{A}{B} = -\frac{3}{-2} = \frac{3}{2}$$

Similarly, the second equation, $-6x + 4y = 7$, is in general form, with $A = -6$, $B = 4$, and $C = 7$. The slope of this line is

$$m_2 = -\frac{A}{B} = -\frac{-6}{4} = \frac{3}{2}$$

Because the slopes of the two lines are equal, the lines are parallel.   ∎

EXERCISE 3.3

In Exercises 1–8, use the point-slope form to write the equation of the line passing through the given point and having the given slope. Express the answer in slope-intercept form.

1. $P(2, 4)$; $m = 2$ **2.** $P(3, 5)$; $m = -3$

3. $P\left(-\dfrac{3}{2}, \dfrac{1}{2}\right)$; $m = 2$ **4.** $P\left(\dfrac{1}{4}, -2\right)$; $m = -6$

5. $P(5, -5)$; $m = -5$ **6.** $P(0, -8)$; $m = -3$ **7.** $P(\pi, 0)$; $m = 0$ **8.** $P(0, \pi)$; $m = \pi$

In Exercises 9–16, use the slope-intercept form to write the equation of the line with the given slope and y-intercept.

9. $m = 3$, $b = -2$ **10.** $m = -\dfrac{1}{3}$, $b = \dfrac{2}{3}$ **11.** $m = \sqrt{2}$, $b = \sqrt{2}$ **12.** $m = \pi$, $b = \dfrac{1}{\pi}$

13. $m = a$, $b = \dfrac{1}{a}$ **14.** $m = a$, $b = 2a$ **15.** $m = a$, $b = a$ **16.** $m = \dfrac{1}{a}$; $b = a$

In Exercises 17–24, use the slope-intercept form to write the equation of the line passing through the given point and having the given slope. Express the answer in general form.

17. $P(0, 0); m = \dfrac{3}{2}$

18. $P(-3, -7); m = -\dfrac{2}{3}$

19. $P(-3, 5); m = -3$

20. $P(-5, 9); m = 1$

21. $P(0, \sqrt{2}); m = \sqrt{2}$

22. $P(-\sqrt{3}, 0); m = 2\sqrt{3}$

23. $P(r, s); m = \dfrac{s}{r}$

24. $P\left(a, \dfrac{1}{a}\right); m = \dfrac{1}{a}$

In Exercises 25–30, write the equation of the line passing through the two given points. Express the answer in general form.

25. $P(3, 2); Q(2, 3)$

26. $P(0, 5); Q(-5, 0)$

27. $P(3, -2); Q(4, -8)$

28. $P(-7, 5); Q(-5, -7)$

29. $P(0, 0); Q(-9, 13)$

30. $P(-9, -5); Q(5, -9)$

In Exercises 31–34, find, if possible, the slope and the y-intercept of each line.

31. $3(14x + 12) = 5(y + 3)$

32. $-2(3x + 6) = -2y + 15$

33. $2(y - 3) + x = 2y - 7$

34. $3(y + 2) + 2x = 2(y + x)$

In Exercises 35–40, find the slope and the y-intercept of the line with the given properties.

35. passes through $P(-3, 5)$ and $Q(7, -5)$

36. passes through $P(15, 3)$ and $Q(-3, -6)$

37. is parallel to line $y = 3(x - 7)$ and passes through $P(-2, 5)$

38. is parallel to line $3y + 1 = 6(x - 2)$ and passes through $P(5, 0)$

39. is perpendicular to line $y + 3x = 8$ and passes through $P(2, 3)$

40. is perpendicular to line $x = 2y - 7$ and passes through $P(0, 8)$

In Exercises 41–58, write the equation of the line with the given properties. Write the answer in general form.

41. passes through $(3, 4)$ and slope of $\frac{1}{2}$

42. passes through $(-2, 6)$ and slope of $\frac{3}{5}$

43. passes through $(-2, 4)$ and $(5, -4)$

44. passes through $(3, 5)$ and $(-5, 3)$

45. passes through the origin and $(-2, 11)$

46. passes through the origin and $(5, -9)$

47. y-intercept of 7 and passes through $(-3, 4)$

48. y-intercept of -3 and x-intercept of 4

49. y-intercept of 3 and x-intercept of -4

50. y-intercept of -2 and passes through $(4, -3)$

51. slope of $-\frac{2}{3}$ and y-intercept of 10

52. slope of $-\frac{1}{2}$ and x-intercept of 10

53. passes through $(3, -5)$ and is parallel to the x-axis

54. has an x-intercept of -4 and is parallel to the y-axis

55. is perpendicular to the line with equation $y = 3x - 17$ and passes through $(0, -5)$

56. is perpendicular to the line with equation $y = -\frac{1}{3}x + 5$ and has the same y-intercept as the line $3x - 2y = 6$

57. is parallel to the line $3y - 5x = 0$ and has the same x-intercept as the line $5x - 2y = 15$

58. is parallel to the line $5x + 7y = 8$ and has the same y-intercept as the line $x = 7y - 5$

In Exercises 59–64, write the equation of the line with the given properties. Write the answer in slope-intercept form.

59. passes through the origin and is parallel to the line $3x + 2y = 6$

60. passes through the point $P(2, 8)$ and is parallel to the line $2x - 3y = 12$

61. passes through the point $P(-2, 3)$ and is perpendicular to the line $3x + 5y = 25$

62. passes through the origin and is perpendicular to the line $4x - y = 12$

63. passes through the origin and the point $P(r, s)$

64. passes through the points $P(r, s)$ and $Q(s, r)$

65. Water freezes at 32° Fahrenheit, or 0° Celsius. Water boils at 212°F, or 100°C. Find formulas for converting a temperature from one system to the other. (*Hint:* Find the equation of the line passing through points $P(32, 0)$ and $Q(212, 100)$.

66. A temperature of 0° Celsius is equal to a temperature of 273° Kelvin, and 100°C is 373°K. Find formulas for converting a temperature from one system to the other.

67. Assume that the temperature T is related to the altitude x by an equation of the form $T = mx + b$. If $T = 30°$ at an altitude of 12,000 feet, and $T = -30°$ at 30,000 feet, determine the temperature at sea level.

68. The value of the stock of ABC Corporation has been increasing by the same fixed dollar amount each year. The pattern is expected to continue. Let 1980 be the base year corresponding to $x = 0$, with $x = 1, 2, 3, \ldots$ corresponding to later years. ABC stock was selling at $\$37\frac{1}{2}$ in 1985 and at \$45 in 1988. If y represents the price of ABC stock, find the equation $y = mx + b$ that relates x and y, and predict the price in the year 2000.

69. Prove that an equation of the line with x-intercept of a and y-intercept of b can be written in the form

$$\frac{x}{a} + \frac{y}{b} = 1$$

70. Prove that, if $B \neq 0$, the graph of $Ax + By = C$ has a slope of $-A/B$ and a y-intercept of C/B.

71. Prove that, if $B = 0$, the graph of $Ax + By = C$ is a vertical line with x-intercept of C/A.

72. Show that the lines $Ax + By = C$ and $Bx - Ay = C$ are perpendicular.

73. Show that the graphs of $Ax + By = C$ and $kAx + kBy = D$, where $k \neq 0$, are parallel.

74. Use the distance formula to find the equation of the perpendicular bisector of the line segment joining $A(3, 5)$ and $B(5, -3)$. (*Hint:* Let $P(x, y)$ be a point on the perpendicular bisector. Then $d(PA) = d(PB)$.)

75. Use the point-slope form to find the equation of the perpendicular bisector of the line segment joining $A(3, 5)$ and $B(5, -3)$. (*Hint:* First find the midpoint of AB.)

76. Find the x-intercept of the line $y = mx + b$.

77. What is the equation of the y-axis?

78. What is the equation of the x-axis?

Computer Exercises. *In Exercises 79–80, use a computer and the program **Algebra Pak** to perform each computer experiment.* **Write a brief paragraph describing your findings.** *To avoid re-typing an equation, use the* $\boxed{\text{F2}}$ *key to recall the expression previously typed and then make necessary alterations. When the computer asks if you want automatic scaling of the y-axis, answer "N."*

79. Investigate the effect of changing the y-intercept of the graph of a line by graphing $y = 2x + b$ for several values of b. If you choose $b = 1$, for example, enter the function as $2X + 1$. Does the slope of the line change? How does the y-intercept change?

80. Investigate the effect of changing the slope of the graph of a line by graphing $y = mx + 1$ for several values of m. How does the slope of the line change? Does the y-intercept change?

3.4 FUNCTIONS AND FUNCTION NOTATION

Many equations describe a correspondence between two variables. The equation $y = 5x$, for example, describes a correspondence in which each number x determines one value y. Such a correspondence is called a **function**.

> **Definition.**   A **function** is a correspondence that assigns to *each* element x of some set **X** a *single* value y of some set **Y**.
>
> The set **X** is called the **domain** of the function. The value y that corresponds to a particular x is called the **image** of x under the function. The collection of all such images is called the **range** of the function.

Because the value of y depends on the number x that is chosen, y is called the **dependent variable**. The variable x is called the **independent variable**. Unless indicated otherwise, the domain of a function is its implied domain—the set of all real numbers x for which the function is defined.

To decide whether the equation $y = 5 - 7x$, for example, defines y to be a function of x, we must decide if each number x determines exactly one value y. The equation indicates that the product of the numbers x and 7 is subtracted from 5. Because this arithmetic gives a single result for each number x, each x does determine a single value y. Thus, the equation $y = 5 - 7x$ does define y to be a function of x. Because x can represent any real number, the domain of the function is $\mathcal{R}$, the set of real numbers. Because the resulting value of y could be any real number, the range of the function is also $\mathcal{R}$.

The equation $y = \dfrac{3}{x - 2}$ also defines y to be a function of x, because the calculations defined by the equation's right-hand side determine a single value y for each number x. To find the domain of this function, we note that x cannot be 2, because division by zero is not defined. Thus, the domain of the function is the set of all real numbers except 2. Because the fraction has a nonzero numerator, y cannot be zero. Thus, the range of the function is the set of all real numbers except 0.

To indicate that y is a function of x, we use **function notation** and write

$$y = f(x)$$

Strictly speaking, y is the image of x under the function f, and the correspondence itself is the function. Yet it is common to read $y = f(x)$ as "y is a function of x." Functions are often denoted by letters other than f.

Function notation provides a way of indicating the image of a particular number x. The symbol $f(2)$ indicates the image of 2 under the function f:

$$f(x) = 5 - 7x$$
$$f(2) = 5 - 7(2) \qquad \text{Substitute 2 for } x.$$
$$= -9$$

Thus, if $x = 2$, then $y = f(2) = -9$. Similarly, the image of -5 is $f(-5)$, or 40, because

$$f(x) = 5 - 7x$$
$$f(-5) = 5 - 7(-5) \qquad \text{Substitute } -5 \text{ for } x.$$
$$= 40$$

Example 1   Let $g(x) = 3x^2 + x - 4$. Find **a.** $g(-3)$, **b.** $g(k)$, **c.** $g(-t^3)$, and **d.** $g(k + 1)$.

Solution   **a.** $\qquad g(x) = 3x^2 + x - 4$

$$g(-3) = 3(-3)^2 + (-3) - 4 \qquad \text{Substitute } -3 \text{ for } x.$$
$$= 3(9) - 3 - 4$$
$$= 20$$

b. $\quad g(x) = 3x^2 + x - 4$

$$g(k) = 3k^2 + k - 4 \qquad \text{Substitute } k \text{ for } x.$$

c. $\qquad g(x) = 3x^2 + x - 4$

$$g(-t^3) = 3(-t^3)^2 + (-t^3) - 4 \qquad \text{Substitute } -t^3 \text{ for } x.$$
$$= 3t^6 - t^3 - 4$$

d. $\qquad g(x) = 3x^2 + x - 4$

$$g(k + 1) = 3(k + 1)^2 + (k + 1) - 4 \qquad \text{Substitute } k + 1 \text{ for } x.$$
$$= 3(k^2 + 2k + 1) + k + 1 - 4$$
$$= 3k^2 + 6k + 3 + k + 1 - 4$$
$$= 3k^2 + 7k \qquad \text{Combine terms.} \qquad ■$$

By definition, a function is a correspondence that assigns to each element of some set **X**, a single element of a set **Y**. We can visualize this correspondence with the diagram of Figure 3-20a. The function f that assigns the element $y = f(x)$ to the element x is represented by an arrow leaving x and pointing to y. The set of all those elements in **X** from which arrows originate is the domain of the function.

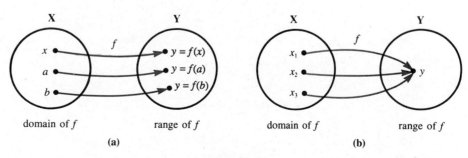

Figure 3-20

The set of all images in **Y** to which arrows point is the range of the function. To be a function, each element of the domain must have a *single* image in the range. However, the same value y may be the image of several values x. In the function shown in Figure 3-20**b**, the single image y corresponds to the three elements x_1, x_2, and x_3 in the domain.

For most of the functions that we will study in this course, the domains and ranges will be sets of real numbers. If f is such a function, then the graph of the function is the set of all points $(x, f(x))$ in the Cartesian plane, where x is any element of the domain of f. In other words, the graph of f is the graph of the equation $y = f(x)$. The graph of the function $f(x) = 5 - 7x$, for example, is a straight line with slope -7 and y-intercept 5. See Figure 3-21.

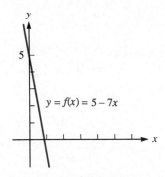

Figure 3-21

The graph of a function f illustrates the domain and range of the function, as well as the correspondence. The interval indicated on the x-axis in Figure 3-22**a** represents the domain of f, because it contains all x for which the function is defined. To the particular number x indicated in Figure 3-22**b** there corresponds its image, the value $f(x)$ on the y-axis. The set of all possible values y that correspond to elements x in the domain is the range of the function; it is the interval on the y-axis indicated in Figure 3-22**c**.

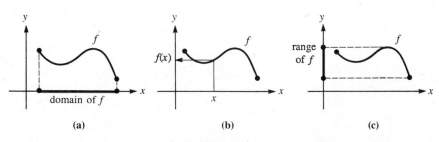

Figure 3-22

Example 2 Graph the function f defined by the equation $f(x) = |x|$, and determine both the domain of f and its range.

Solution The graph of f is the graph of the equation $y = f(x)$, or $y = |x|$. If $x \geq 0$, then $|x|$ is just x itself, and the graph of the function is the graph of $y = x$. When $x = 1$, for example, y is also 1, and the point $(1, 1)$ lies on the graph of the function. So does the point $(2, 2)$, and others in the table beside Figure 3-23**a**. For $x < 0$, $|x| = -x$, and the graph of the function is the graph of $y = -x$. When $x = -1$, y is 1, and the point $(-1, 1)$ lies on the graph. The complete graph appears in Figure 3-23**a**. The domain of f is the set of all real numbers; the range is the set of nonnegative real numbers. See Figure 3-23**b**.

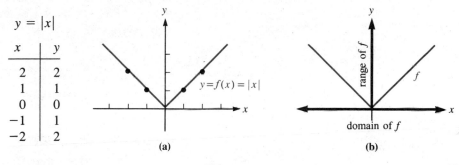

$$y = |x|$$

x	y
2	2
1	1
0	0
−1	1
−2	2

(a)   (b)

Figure 3-23

Linear Functions

The equation of a nonvertical line defines a linear function, one of several functions important in mathematics and its applications.

> **Definition.** A **linear function** is a function determined by an equation of the form
>
> $$f(x) = mx + b$$

Example 3 The cost of electricity is a linear function of x, the number of kilowatt-hours used. If the cost of 100 kilowatt-hours is \$17 and the cost of 500 kilowatt-hours is \$57, what formula expresses the function?

Solution Because c (the cost of electricity) is a linear function of x, there are constants m and b such that

$$c = mx + b$$

Because $c = 17$ when $x = 100$, the point $(x_1, c_1) = (100, 17)$ lies on the straight-line graph of this function. Similarly, $c = 57$ when $x = 500$, and the point $(x_2, c_2) = (500, 57)$ also lies on that line. The slope of the line is

$$m = \frac{c_2 - c_1}{x_2 - x_1}$$

$$= \frac{57 - 17}{500 - 100}$$

$$= \frac{40}{400}$$

$$= 0.10$$

Thus, $m = 0.10$. To determine b, substitute the coordinates of one of the points for x and c. If you choose the point $(100, 17)$, for example, substitute 100 for x, 17 for c, and 0.10 for m into the equation $c = mx + b$, and solve for b:

$$c = mx + b$$
$$17 = 0.10\,(100) + b$$
$$17 = 10 + b$$
$$7 = b$$

Thus, $c = 0.10x + 7$. The electric company charges \$7.00, plus 10¢ per kilowatt-hour used.   ■

Relations and Other Functions

To be a function, a correspondence must assign a single element of the range to each element in the domain. A **relation** is a correspondence in which some elements of the domain might have more than one image. The relation represented in Figure 3-24 does not represent a function, because two values of y correspond to the number x.

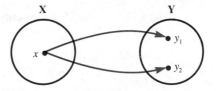

Figure 3-24

> **Definition.** A **relation** is a correspondence that assigns to each element x in a set **X** one or more values y in a set **Y**.
> The set **X** is called the **domain** of the relation. The **range** of the relation is the set of all y values corresponding to numbers x in the domain.

From the definitions of function and relation, it follows that *a function is a relation, but a relation is not necessarily a function.*

If the domain and range of a relation are sets of real numbers, then the relation defines a set of ordered pairs (x, y), where x is an element of the domain and y is a corresponding value in the range. The **graph of the relation** is the graph of all these ordered pairs in the Cartesian plane.

A test, called the **vertical line test**, can be used to determine whether the graph of a relation also represents the graph of a function. If each vertical line that intersects the graph does so exactly once, then each number x determines exactly one y, and the graph represents a function. See Figure 3-25a. If any vertical line intersects the graph more than once, then to some numbers x there correspond more than one value y, and the relation is *not* a function. See Figure 3-25b.

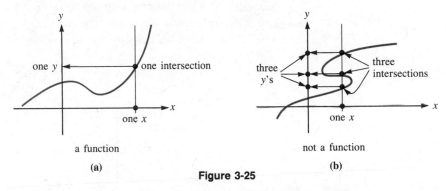

Figure 3-25

Example 4   Graph the points (x, y) for which   **a.** $x = y^2$ and   **b.** $y = \sqrt{x}$. From the graphs, decide whether these relations are functions.

Solution   **a.** Several ordered pairs (x, y) that satisfy the equation are listed in the table of values in Figure 3-26a. Plot those points and draw the graph as shown in the figure. Because some vertical lines that could be drawn would intersect the curve twice, there are some numbers x that determine two values of y. Thus, the graph does not pass the vertical line test, and the relation is *not* a function.

b. Because the symbol $\sqrt{x}$ denotes a nonnegative number, the values of y defined by the equation $y = \sqrt{x}$ are never negative. Several ordered pairs (x, y) that satisfy this equation are plotted to determine the graph in Figure 3-26b. Because every vertical line that could be drawn would intersect the curve exactly once, each number x in the domain determines only one value of y. Thus, the graph passes the vertical line test, and the relation *is* a function.

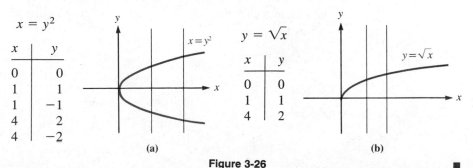

Figure 3-26

Some functions must be defined by using different equations for different intervals of their domains. Such functions are called **piecewise-defined functions**. For example, we will graph the piecewise-defined function f given by

$$y = f(x) = \begin{cases} 0 & \text{if } x < 0 \\ x^2 & \text{if } 0 \le x \le 2 \\ 4 - 2x & \text{if } x > 2 \end{cases}$$

For each number x, we decide which of three schemes will be used to produce the corresponding value of y. For numbers x that are less than zero, the corresponding value of y is the constant 0. For numbers x between 0 and 2 inclusive, the corresponding value of y is determined by the equation $y = x^2$. For numbers x greater than 2, y is determined by the linear function $y = 4 - 2x$. The graph appears in Figure 3-27. The use of solid and open circles indicates that $f(2) = 4$.

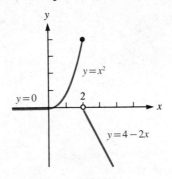

Figure 3-27

The next example illustrates how the concept of function can be used in geometry.

Example 5   Express the area of a square as a function of its perimeter.

Solution   The area of a square is a function of the length s of a side. This can be expressed as

1. $A = f(s) = s^2$

The perimeter P is also a function of the length s of a side, given by

2. $P = g(s) = 4s$

To express the area A as a function of the perimeter P, you must determine a function h such that $A = h(P)$. To do so, solve Equation 2 for s and substitute into Equation 1.

$$P = 4s$$

$$s = \frac{P}{4} \qquad \text{Divide both sides by 4.}$$

$$A = \left(\frac{P}{4}\right)^2 \qquad \text{Substitute } \tfrac{P}{4} \text{ for } s \text{ in Equation 1.}$$

$$= \frac{P^2}{16}$$

Thus,

$$A = h(P) = \frac{P^2}{16}$$

In the context of this example, the domain of h is not its implied domain of the set of real numbers. Because the perimeter of a square is a positive number, the domain of h is the set of positive real numbers. ∎

EXERCISE 3.4

In Exercises 1–12, indicate whether the equation determines y as a function of x. Assume that all variables represent real numbers.

1. $y = x$ **2.** $y - 2x = 0$ **3.** $y^2 = x$ **4.** $|y| = x$

5. $y = x^2 + 4$ **6.** $y - 7 = 7$ **7.** $y^2 - 4x = 1$ **8.** $|x - 2| = y$

9. $|x| = |y|$ **10.** $x = 7$ **11.** $y = 7$ **12.** $|x + y| = 7$

In Exercises 13–20, let the function f be defined by the equation y = f(x), where x and f(x) are real numbers. Find the domain and the range of each function.

13. $f(x) = 3x + 5$ **14.** $f(x) = -5x + 2$ **15.** $f(x) = x^2$ **16.** $f(x) = x^2 + 3$

17. $f(x) = \dfrac{3}{x + 1}$ **18.** $f(x) = \dfrac{-7}{x + 3}$ **19.** $f(x) = \sqrt{x}$ **20.** $f(x) = \sqrt{x^2}$

In Exercises 21–32, let function f be defined by the equation y = f(x), where x and f(x) are real numbers. Find f(2), f(-3), f(k), and f(k² - 1).

21. $f(x) = 3x - 2$ **22.** $f(x) = 5x + 7$ **23.** $f(x) = \dfrac{1}{2}x + 3$ **24.** $f(x) = \dfrac{2}{3}x + 5$

25. $f(x) = x^2$ **26.** $f(x) = 3 - x^2$ **27.** $f(x) = \dfrac{2}{x + 4}$ **28.** $f(x) = \dfrac{3}{x - 5}$

29. $f(x) = \dfrac{1}{x^2 - 1}$ **30.** $f(x) = \dfrac{3}{x^2 + 3}$ **31.** $f(x) = \sqrt{x^2 + 1}$ **32.** $f(x) = \sqrt{x^2 - 1}$

In Exercises 33–38, some correspondences are illustrated. If the correspondence could represent y as a function of x, so indicate. If the illustration could not represent such a function, explain why.

33.
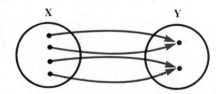

34.
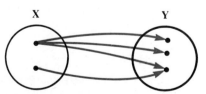

35.
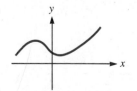

36.
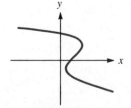

37.

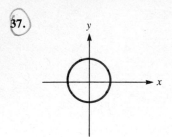

38.

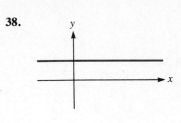

In Exercises 39–40, indicate the domain and range of the function as intervals on the x- and y-axes. Then, on the y-axis, indicate f(x), the image of the indicated number x.

39.

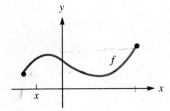

40.

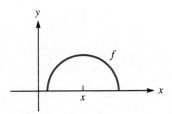

In Exercises 41–60, graph each equation. Use the vertical line test to decide if the equation defines y as a function of x.

41. $y = 2x + 3$ **42.** $y = 3x + 2$ **43.** $2x = 3y - 3$ **44.** $3x = 2(y + 1)$

45. $y = -x^2$ **46.** $y = 1 - x^2$ **47.** $y^2 = x^2$ **48.** $|y| = |x|$

49. $y = x^3$ **50.** $x = y^2$ **51.** $y = |x| + x$ **52.** $|y| = x + 1$

53. $y = \dfrac{1}{x}$ **54.** $y = \dfrac{1}{x^2}$ **55.** $|x| = 3y + 2$ **56.** $|y| = 3x + 2$

57. $|y| = x^2$ **58.** $|x| = y^2$

59. The number y is the greatest integer less than or equal to x.

60. The number y is the least integer greater than or equal to x.

In Exercises 61–64, graph each piecewise-defined function.

61. $y = f(x) = \begin{cases} x + 2 & \text{if } x < 0 \\ 2 & \text{if } x \geq 0 \end{cases}$

62. $y = f(x) = \begin{cases} 1 & \text{if } x < 0 \\ 0 & \text{if } x = 0 \\ -1 & \text{if } x > 0 \end{cases}$

63. $y = f(x) = \begin{cases} 2 & \text{if } x < 0 \\ 2 - x & \text{if } 0 \leq x < 2 \\ x & \text{if } x \geq 2 \end{cases}$

64. $y = f(x) = \begin{cases} 1 & \text{if } x < 0 \\ x^2 & \text{if } 0 \leq x \leq 1 \\ 0 & \text{if } x > 1 \end{cases}$

65. Express the radius of a circle as a function of the circumference.

66. Express the radius of a circle as a function of the area.

67. Express the perimeter of a square as a function of the area.

68. Express the area of a circle as a function of the circumference.

69. Express the volume of a cube as a function of the area of one face.

70. Express the length of the diagonal of a square as a function of the length of an edge.

71. Express the area of a square as a function of the length of a diagonal.

72. Express the circumference of a circle as a function of the area.

73. The Fahrenheit temperature reading F is a linear function of the Celsius reading C. If $C = 0$ when $F = 32$, and the readings are the same at -40, express F as a function of C.

74. The velocity v of a falling object is a linear function of the time t it has been falling. If $v = 15$ when $t = 0$ and $v = 79$ when $t = 2$, express v as a function of t.

75. The cost c of water is a linear function of n, the number of gallons used. If 1000 gallons cost \$4.70, and 9000 gallons cost \$14.30, what equation describes this function?

76. The amount A of money on deposit for t years in an account earning simple interest is a linear function of t. Express that function as an equation if $A = \$272$ when $t = 3$ and $A = \$320$ when $t = 5$.

Computer Exercises. In Exercises 77–80, use a computer and Algebra Pak to graph each function. **Write a brief paragraph describing your findings.**

77. Graph the function given by $y = \sqrt{x + 3}$. Find the domain and the range of the function. What happens to the graph at values of x not in the domain? Select nonautomatic scaling and enter the function by typing SQR(X + 3).

78. Graph the function given by $y = \sqrt{16 - x^2}$. Find the domain and the range of the function. Select nonautomatic scaling and enter the function by typing SQR(16 - X^2).

79. Graph $y = |kx|$ for $k = 2$ and for $k = -2$. What do you observe about the graphs? Select automatic scaling and enter the functions by typing ABS(2X) and ABS(-2X).

80. Graph $y = k|x|$ for $k = 2$ and for $k = -2$. What do you observe about the graphs? Select automatic scaling.

3.5 POLYNOMIAL FUNCTIONS

The linear function defined by the equation $y = mx + b$ is a first-degree polynomial function because its right-hand side is a first-degree polynomial in the variable x. In this section, we will discuss functions involving polynomials of higher degree.

> **Definition.** A **polynomial function in one independent variable**, say x, is defined by an equation of the form $y = P(x)$, where $P(x)$ is a polynomial in the variable x.
>   The **degree** of the polynomial function $y = P(x)$ is the degree of $P(x)$.

A polynomial function of second-degree is called a **quadratic function**.

> **Definition.** A **quadratic function** is a second-degree polynomial function in one variable. It is defined by an equation of the form $y = ax^2 + bx + c$, where a, b, and c are constants and $a \neq 0$.

Example 1   Graph the quadratic functions defined by the equations   **a.** $y = x^2 - 2x - 3$
and   **b.** $y = -\frac{1}{3}x^2 + 3$.

Solution   Plot several points whose coordinates satisfy these equations. Connect them with
smooth curves to obtain the graphs shown in Figure 3-28.

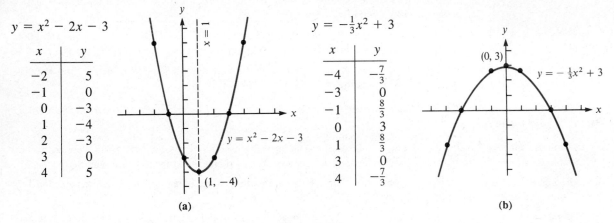

$y = x^2 - 2x - 3$

x	y
-2	5
-1	0
0	-3
1	-4
2	-3
3	0
4	5

$y = -\frac{1}{3}x^2 + 3$

x	y
-4	$-\frac{7}{3}$
-3	0
-1	$\frac{8}{3}$
0	3
1	$\frac{8}{3}$
3	0
4	$-\frac{7}{3}$

(a)   (b)

Figure 3-28   ■

The graphs of quadratic functions are called **parabolas**. They open upward when
$a > 0$ (as in Figure 3-28**a**) and downward when $a < 0$ (as in Figure 3-28**b**). The
bottom point of a parabola that opens upward (or the top point of a parabola that
opens downward) is called the **vertex** of the parabola. The vertex of the parabola
in Figure 3-28**a** is the point $(1, -4)$; the vertex of the parabola in Figure 3-28**b** is
the point $(0, 3)$. The vertical line that passes through the vertex is called the **axis
of symmetry** because it divides the parabola into two congruent halves. The axis
of symmetry in Figure 3-28**a** is the line $x = 1$, and in Figure 3-28**b**, the line $x = 0$ (which is the y-axis).

If a, h, and k are constants and $a \ne 0$, then the graph of the equation

$$y - k = a(x - h)^2$$

is a parabola, because the equation takes the form of a quadratic function $y = ax^2 + bx + c$ when the right-hand side is expanded and the equation solved for y.
The equation $y - k = a(x - h)^2$ displays the coordinates (h, k) of the vertex
of its parabolic graph, as the following discussion will show.

Suppose that $a > 0$, so that the graph of $y - k = a(x - h)^2$ is a parabola
opening upward. The vertex of this parabola is that point on the graph that has the
least possible y-coordinate. Because a is positive and $(x - h)^2$ is nonnegative, the
smallest value attainable by the right-hand side is 0; this occurs when $x = h$. When
$x = h$, the left-hand side of the equation must be 0 also, so the smallest possible
value of y is $y = k$. The parabola's vertex, therefore, is the point (h, k). A similar
argument holds if $a < 0$. Thus, we have the following theorem.

> **Theorem.** The graph of the equation
>
> $$y - k = a(x - h)^2 \qquad (a \neq 0)$$
>
> is a parabola with its vertex at the point (h, k). The parabola opens upward if $a > 0$ and downward if $a < 0$.

We can use the process of completing the square to transform an equation from the form $y = ax^2 + bx + c$ to the form $y - k = a(x - h)^2$, from which we can read the coordinates of the vertex directly.

Example 2   Find the vertex of the parabola determined by $y = 2x^2 - 5x - 3$.

Solution   Complete the square on the right-hand side of the equation:

$$y = 2x^2 - 5x - 3$$

$$y + 3 = 2x^2 - 5x \qquad \text{Add 3 to both sides.}$$

$$y + 3 = 2\left(x^2 - \frac{5}{2}x \qquad\right) \qquad \begin{array}{l}\text{Factor a 2 out of the terms}\\\text{on the right-hand side.}\end{array}$$

Complete the square within the parentheses by adding the square of one-half of the coefficient of x. This number,

$$\left[\frac{1}{2}\left(-\frac{5}{2}\right)\right]^2 = \frac{25}{16}$$

is added within the parentheses. Because of the factor of 2, however, you are really adding twice $\frac{25}{16}$, or $\frac{25}{8}$, to the right-hand side of the equation. You must also add $\frac{25}{8}$ to the left-hand side.

$$y + 3 = 2\left(x^2 - \frac{5}{2}x \qquad\right)$$

$$y + 3 + \frac{25}{8} = 2\left(x^2 - \frac{5}{2}x + \frac{25}{16}\right) \qquad \text{Add } \tfrac{25}{8} \text{ to both sides.}$$

$$y + \frac{49}{8} = 2\left(x - \frac{5}{4}\right)^2 \qquad \text{Combine terms and factor.}$$

$$y - \left(-\frac{49}{8}\right) = 2\left(x - \frac{5}{4}\right)^2$$

The equation is now in the form

$$y - k = a(x - h)^2$$

with $h = \frac{5}{4}$ and $k = -\frac{49}{8}$. The vertex is the point $(h, k) = \left(\frac{5}{4}, -\frac{49}{8}\right)$. ∎

To find the vertex of any parabola defined by the quadratic function $y = ax^2 + bx + c$, we can write its equation in the form $y - k = a(x - h)^2$ by completing the square.

$$y = ax^2 + bx + c$$

$$y = a\left(x^2 + \frac{b}{a}x\right) + c \qquad \text{Factor out } a.$$

$$y + \frac{b^2}{4a} = a\left(x^2 + \frac{b}{a}x + \frac{b^2}{4a^2}\right) + c \qquad \text{Add } \frac{b^2}{4a} \text{ to both sides.}$$

$$y - c + \frac{b^2}{4a} = a\left(x^2 + \frac{b}{a}x + \frac{b^2}{4a^2}\right) \qquad \text{Add } -c \text{ to both sides.}$$

$$y - \left(c - \frac{b^2}{4a}\right) = a\left(x - \frac{-b}{2a}\right)^2 \qquad \text{Factor.}$$

Thus, the vertex is the point with coordinates of $\left(-\dfrac{b}{2a}, c - \dfrac{b^2}{4a}\right)$, and we have the following theorem.

Theorem.   The vertex of the parabola determined by the quadratic function

$$y = ax^2 + bx + c \qquad (a \neq 0)$$

is the point with coordinates $\left(-\dfrac{b}{2a}, c - \dfrac{b^2}{4a}\right)$.

Example 3   A farmer has 400 feet of fencing to enclose a rectangular corral. To save money and fencing, he intends to use one bank of a river as one boundary of the corral. See Figure 3-29. What dimensions would enclose the largest area?

Solution   Let x represent the width of the fenced area. Then $400 - 2x$ represents the length. Because the area of a rectangle is the product of the length and width, you have

$$A = (400 - 2x)x$$

or

$$A = -2x^2 + 400x$$

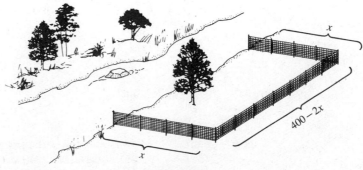

Figure 3-29

Because the coefficient of x^2 is negative, the parabola determined by this function opens downward, and its vertex is its highest point. The x-coordinate of the vertex is the width of the corral of maximum area, and the A-coordinate gives that area. The equation

$$A = -2x^2 + 400x$$

defines a quadratic function with $a = -2$, $b = 400$, and $c = 0$. The vertex of the parabola is the point

$$\left(-\frac{b}{2a}, c - \frac{b^2}{4a}\right) = \left(-\frac{400}{2(-2)}, 0 - \frac{400^2}{4(-2)}\right) = (100, 20{,}000)$$

If the farmer's fence runs 100 feet out from the river, 200 feet parallel to the river, and 100 feet back to the river, it will enclose the maximum area of 20,000 square feet. ∎

Symmetries of Graphs

If the point $(-x, y)$ lies on a graph whenever the point (x, y) does, the graph is **symmetric about the y-axis**. See Figure 3-30a. If the point $(-x, -y)$ lies on a graph whenever the point (x, y) does, the graph is **symmetric about the origin**. See Figure 3-30b. If the point $(x, -y)$ lies on a graph whenever the point (x, y) does, the graph is **symmetric about the x-axis**. See Figure 3-30c.

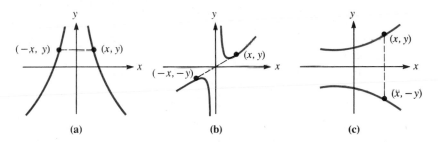

Figure 3-30

To test the graph of a function $y = f(x)$ for y-axis symmetry, we replace x with $-x$. If the equation $y = f(-x)$ is equivalent to the equation $y = f(x)$, then the graph is symmetric to the y-axis. A function with a graph that is symmetric about the y-axis is called an **even function**.

Definition. A function f with the property that $f(-x) = f(x)$ for all x is called an **even function**.

To test the graph of a function $y = f(x)$ for symmetry about the origin, we replace x with $-x$ and y with $-y$. If the equation $-y = f(-x)$ is equivalent to the equation $y = f(x)$, then the graph is symmetric about the origin. A function whose graph is symmetric about the origin is called an **odd function**.

Definition.   A function f with the property that $f(-x) = -f(x)$ for all x is called an **odd function**.

To test a graph for x-axis symmetry, we replace y in its equation with $-y$. If the resulting equation is equivalent to the original equation, then the graph is symmetric to the x-axis. Except for the graph of the function $y = f(x) = 0$, such a graph does not pass the vertical line test and does not represent a function.

In summary, we have the following tests for symmetry.

Tests for Symmetry.   If the point $(-x, y)$ lies on a graph whenever (x, y) does, the graph is **symmetric about the y-axis**: If $y = f(x)$ and $y = f(-x)$ are equivalent, the graph has y-axis symmetry.

If the point $(-x, -y)$ lies on a graph whenever (x, y) does, the graph is **symmetric about the origin**: If $y = f(x)$ and $-y = f(-x)$ are equivalent, the graph is symmetric about the origin.

If the point $(x, -y)$ lies on a graph whenever (x, y) does, the graph is **symmetric about the x-axis**. Except for $f(x) = 0$, no function has a graph with x-axis symmetry.

Example 4   Determine the symmetries of the graph of the polynomial function $y = f(x) = x^3 - x$, and graph the function.

Solution   To check for symmetries, calculate both $f(x)$ and $f(-x)$, and compare the results.

$$y = f(x) = x^3 - x \qquad y = f(-x) = (-x)^3 - (-x)$$
$$= -x^3 + x$$

Because $y = f(x)$ is not equivalent to $y = f(-x)$, there is no symmetry about the y-axis. Because $y = f(x)$ is a function, there is no symmetry about the x-axis. However, if x and y are replaced with $-x$ and $-y$, respectively, the resulting equation is equivalent to the original equation:

$$y = x^3 - x$$
$$-y = (-x)^3 - (-x) \qquad \text{Replace } x \text{ with } -x, \text{ and } y \text{ with } -y.$$
$$-y = -x^3 + x \qquad \text{Simplify.}$$
$$y = x^3 - x \qquad \text{Multiply both sides by } -1.$$

Because the final result is identical to the original equation, the graph is symmetric about the origin.

To graph the polynomial function $y = f(x) = x^3 - x$, it is also useful to know the graph's x-intercepts—the points at which the graph intersects the x-axis. The x-intercepts are the numbers x for which $f(x) = 0$. To find them, solve the equation $x^3 - x = 0$.

$$x^3 - x = 0$$
$$x(x^2 - 1) = 0 \qquad \text{Factor out } x.$$
$$x(x + 1)(x - 1) = 0 \qquad \text{Factor } x^2 - 1.$$
$$x = 0 \quad \text{or} \quad x + 1 = 0 \qquad \text{or} \quad x - 1 = 0 \qquad \text{Set each factor equal to 0.}$$
$$x = -1 \qquad\qquad x = 1$$

Thus, the x-intercepts are are 0, -1, and 1.

Finally, plot a few points for positive x and use the graph's symmetry to draw the rest of the graph. See Figure 3-31.

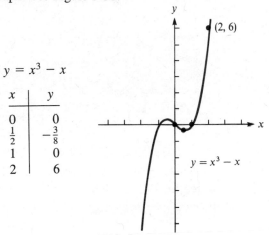

$y = x^3 - x$

x	y
0	0
$\frac{1}{2}$	$-\frac{3}{8}$
1	0
2	6

Figure 3-31

Example 5 Graph the function defined by $y = x^4 - 5x^2 + 4$.

Solution Because the variable x appears only with even exponents, $y = f(x)$ is equivalent to $y = f(-x)$. That is, the value of y is the same whether you evaluate $y = f(x)$ or $y = f(-x)$.

$$y = f(x) = x^4 - 5x^2 + 4$$
$$y = f(-x) = (-x)^4 - 5(-x)^2 + 4$$
$$= x^4 - 5x^2 + 4$$

Hence, this graph is symmetric with respect to the y-axis. It is not symmetric to either the x-axis or the origin, however.

The x-intercepts of the graph are those numbers x for which $y = 0$. Set y equal to 0 and solve the equation for x by factoring:

$$y = x^4 - 5x^2 + 4$$
$$0 = x^4 - 5x^2 + 4$$
$$0 = (x^2 - 4)(x^2 - 1)$$
$$0 = (x + 2)(x - 2)(x + 1)(x - 1)$$
$$x + 2 = 0 \quad \text{or} \quad x - 2 = 0 \quad \text{or} \quad x + 1 = 0 \quad \text{or} \quad x - 1 = 0$$
$$x = -2 \qquad x = 2 \qquad x = -1 \qquad x = 1$$

The x-intercepts are $x = -2$, $x = 2$, $x = -1$, and $x = 1$. To graph this function, plot the intercepts and several other points whose coordinates satisfy the equation, and make use of the symmetry with respect to the y-axis. The graph appears in Figure 3-32.

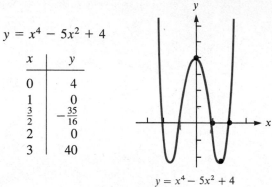

$$y = x^4 - 5x^2 + 4$$

x	y
0	4
1	0
$\frac{3}{2}$	$-\frac{35}{16}$
2	0
3	40

$$y = x^4 - 5x^2 + 4$$

Figure 3-32

Increasing and Decreasing Functions

If the values $f(x)$ increase as x increases on an interval, we say that the function is **increasing** on that interval. See Figure 3-33a. If the values $f(x)$ decrease as x increases on an interval, we say that the function is **decreasing** on that interval. See Figure 3-33b. If the values $f(x)$ remain unchanged as x increases on an interval, we say that the function is **constant** on that interval. See Figure 3-33c.

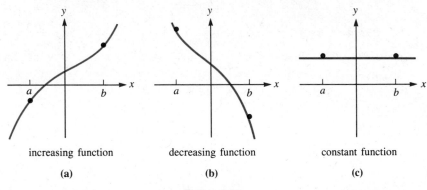

increasing function	decreasing function	constant function
(a)	(b)	(c)

Figure 3-33

Definition.   A function is **increasing** on the interval (a, b) if $f(x_1) < f(x_2)$ whenever $a < x_1 < x_2 < b$.

   A function is **decreasing** on the interval (a, b) if $f(x_1) > f(x_2)$ whenever $a < x_1 < x_2 < b$.

   A function is **constant** on the interval (a, b) if $f(x_1) = f(x_2)$ for all x_1 and x_2 on (a, b).

Example 6   Graph the piecewise-defined function

$$f(x) = \begin{cases} -x & \text{if} \quad x < 0 \\ x^2 & \text{if } 0 \leq x \leq 1 \\ 1 & \text{if} \quad x > 1 \end{cases}$$

and determine the intervals on which it is increasing, decreasing, and constant.

Solution   The graph of the given function appears in Figure 3-34. The function is decreasing on the interval $(-\infty, 0)$, it is increasing on the interal $(0, 1)$, and it is constant on the interval $(1, \infty)$.

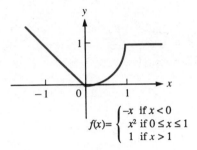

$$f(x) = \begin{cases} -x & \text{if } x < 0 \\ x^2 & \text{if } 0 \leq x \leq 1 \\ 1 & \text{if } x > 1 \end{cases}$$

Figure 3-34   ■

Translating and Stretching Graphs

Figure 3-35a shows the graph of the function $y = x^2 + k$ for three different values of k. The graph of $y = x^2 + 2$ is identical to the graph of $y = x^2$, except that it is shifted 2 units upward. The graph of $y = x^2 - 3$ is identical to the graph of $y = x^2$, except that it is shifted 3 units downward. Such shifts are called **vertical translations**.

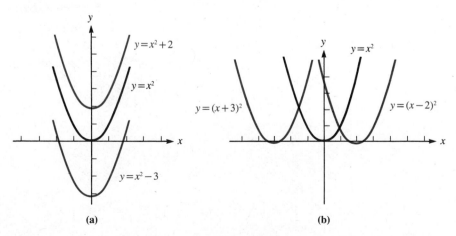

(a)   (b)

Figure 3-35

The graph of $y = (x - h)^2$ is a parabola with vertex at $(h, 0)$. Figure 3-35**b** shows the graph of this equation for three different values of h. The graph of $y = (x - 2)^2$ is identical to the graph of $y = x^2$, except that it is shifted 2 units to the right. The graph of $y = (x + 3)^2$ is identical to the graph of $y = x^2$, except that it is shifted 3 units to the left. Such shifts are called **horizontal translations**.

In general, we can make the following statements.

Let k be a positive number.

Vertical Translations.   The graph of $y = f(x) + k$ is identical to the graph of $y = f(x)$, except that it is translated k units upward.

The graph of $y = f(x) - k$ is identical to the graph of $y = f(x)$, except that it is translated k units downward.

Horizontal Translations.   The graph of $y = f(x - k)$ is identical to the graph of $y = f(x)$, except that it is translated k units to the right.

The graph of $y = f(x + k)$ is identical to the graph of $y = f(x)$, except that it is translated k units to the left.

Example 7   Graph the function $y = (x - 5)^3 + 4$.

Solution   The graph of $y = (x - 5)^3 + 4$ is identical to the graph of $y = (x - 5)^3$, except that it is translated 4 units upward. The graph of $y = (x - 5)^3$ is identical to the graph of $y = x^3$, except that it is translated 5 units to the right. Thus, to graph $y = (x - 5)^3 + 4$, graph $y = x^3$ and translate it 4 units upward and 5 units to the right. See Figure 3-36.

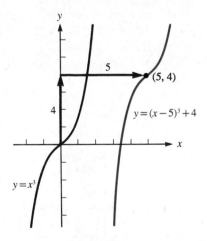

Figure 3-36   ∎

In Figure 3-37**a**, the graph of $y = f(x) = x^2 - 1$ is the black parabola with x-intercepts of 1 and -1. The graph of $y = 3f(x) = 3(x^2 - 1)$ appears in color. For every number x, the value of y determined by $y = 3(x^2 - 1)$ is 3 times the value determined by $y = x^2 - 1$. The graph of $y = x^2 - 1$ has been stretched vertically by a factor of 3 to become the graph of $y = 3(x^2 - 1)$. This stretching does not change the x-intercepts of the graph.

Figure 3-37**b** shows the graph of $y = f(x) = x^2 - 1$ and the graph of $y = f(3x) = (3x)^2 - 1$. If the point (a, b) lies on the graph of $y = x^2 - 1$, the point $(\frac{1}{3}a, b)$ lies on the graph of $y = (3x)^2 - 1$. The graph of $y = x^2 - 1$ has been stretched horizontally by a factor of $\frac{1}{3}$ to become the graph of $y = (3x)^2 - 1$. This stretching does not change the y-intercept of the graph.

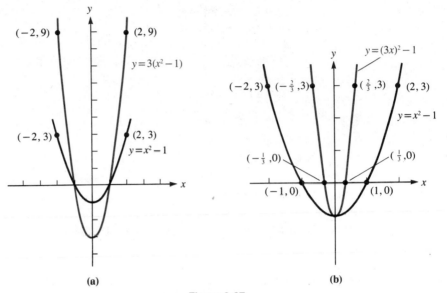

(a)   (b)

Figure 3-37

The effect of vertically stretching a graph by a factor of -1 is to reflect the graph in the x-axis. In Figure 3-38**a**, the colored graph $y = -f(x) = -x^2$ can be obtained by reflecting the graph of $y = x^2$ in the x-axis. The effect of horizontally stretching a graph by a factor of -1 is to reflect the graph in the y-axis. In Figure 3-38**b**, the colored graph $y = f(-x) = (-x)^3$ can be obtained by reflecting the graph of $y = f(x) = x^3$ in the y-axis. Because $(-x)^3 = -x^3$, the reflection of the graph of $y = x^3$ in the x-axis is also the reflection in the y-axis.

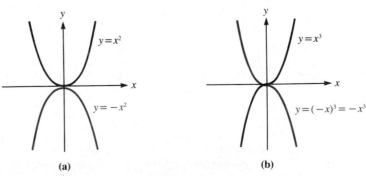

(a)   (b)

Figure 3-38

In general, we can make the following statements.

Vertical and Horizontal Stretchings.   If $k > 0$, the graph of $y = kf(x)$ can be obtained by stretching the graph of $y = f(x)$ vertically by a factor of k. If $k < 0$, the graph of $y = kf(x)$ can be obtained by stretching the reflection of $y = f(x)$ in the x-axis by a factor of $|k|$.

If $k > 0$, the graph of $y = f(kx)$ can be obtained by stretching the graph of $y = f(x)$ horizontally by a factor of $\frac{1}{k}$. If $k < 0$, the graph of $y = f(kx)$ can be obtained by stretching the reflection of $y = f(x)$ in the y-axis by a factor of $|\frac{1}{k}|$.

EXERCISE 3.5

In Exercises 1–6, graph each quadratic function.

1. $y = x^2 - x$ **2.** $y = x^2 + 2x$ **3.** $y = -3x^2 + 2$ **4.** $y = 3x^2 - 4$

5. $y = x^2 - 4x + 1$ **6.** $y = -x^2 - 4x + 1$

In Exercises 7–12, find the vertex of each parabola.

7. $y = x^2 - 4x + 4$ **8.** $y = x^2 - 10x + 25$ **9.** $y = x^2 + 6x - 3$ **10.** $y = -x^2 + 9x - 2$

11. $-2x^2 + 12x - 17 - y = 0$ **12.** $2x^2 + 16x - y + 33 = 0$

In Exercises 13–20, find the symmetries, if any, of the graph of each equation. If the equation represents a function, indicate whether the function is even, odd, or neither.

13. $y = 3x$ **14.** $y = x^2 + 2$ **15.** $y^2 = x$ **16.** $x^2 + y^2 = 25$

17. $y = 3x + 7$ **18.** $|y| = |x|$ **19.** $y = |x|$ **20.** $x = |y|$

In Exercises 21–28, graph each polynomial function.

21. $y = x^3$ **22.** $y = x^4$ **23.** $y = x^3 + x^2$ **24.** $y = x^3 - x$

25. $y = -x^3$ **26.** $y = -x^3 + 1$ **27.** $y = x^4 - 2x^2 + 1$ **28.** $y = x^3 + x^2 - 6x$

In Exercises 29–32, tell where each function is increasing, decreasing, and constant.

29. $y = x^2 - 2x$ **30.** $y = x^3$ **31.** $y = |x - 1| - 1$ **32.** $y = \begin{cases} x & \text{if } x \le 2 \\ 2 & \text{if } x > 2 \end{cases}$

In Exercises 33–38, the graph of each function is a translation of the graph of $y = f(x) = x^2$. Graph each equation.

33. $y = x^2 - 2$ **34.** $y = (x - 2)^2$ **35.** $y = (x + 3)^2$ **36.** $y = x^2 + 5$

37. $y = (x + 1)^2 + 2$ **38.** $y = (x - 3)^2 - 1$

In Exercises 39–46, the graph of each equation is a translation of the graph of $y = f(x) = x^3$. Graph each equation.

39. $y = x^3 + 1$ **40.** $y = x^3 - 3$ **41.** $y = (x - 2)^3$ **42.** $y = (x + 3)^3$

43. $y = (x - 2)^3 - 3$ **44.** $y = (x + 1)^3 + 4$ **45.** $y + 2 = x^3$ **46.** $y - 7 = (x - 5)^3$

In Exercises 47–50, the graph of each equation is a stretching of the graph of $y = f(x) = x^2$. Graph each equation.

47. $y = 2x^2$
48. $y = \frac{1}{2}x^2$
49. $y = -3x^2$
50. $y = -\frac{1}{3}x^2$

In Exercises 51–54, the graph of each equation is a stretching of the graph of $y = f(x) = x^3$. Graph each equation.

51. $y = \left(\frac{1}{2}x\right)^3$
52. $y = \frac{1}{8}x^3$
53. $y = -8x^3$
54. $y = (-2x)^3$

In Exercises 55–64, graph each equation.

55. $y = |x - 2| + 1$
56. $y = |x + 5| - 2$
57. $y = \sqrt{x - 2} + 1$ $(x \geq 2)$
58. $y = \sqrt{x + 5} - 2$ $(x \geq -5)$
59. $y = 3|x|$
60. $y = 2\sqrt{x}$ $(x \geq 0)$
61. $y = \sqrt{2x} + 1$ $(x \geq 0)$
62. $y = |3x| - 2$
63. $y = -(x - \pi)^2 + \pi$
64. $y = \sqrt{\pi - x} + \pi$ $(x \leq \pi)$

65. A parabolic arch has an equation of $x^2 + 20y - 400 = 0$. Find the maximum height of the arch.

66. An object is thrown from the origin of a coordinate system with x-axis along the ground and y-axis vertical. Its path, or **trajectory**, is given by the equation $y = 400x - 16x^2$. What is the object's maximum height?

67. A child throws a ball up a hill that makes an angle of 45° with the horizontal. The ball lands 100 feet up the hill. Its trajectory is a parabola with equation $y = -x^2 + ax$ for some number a. Find a. See Illustration 1.

68. The rectangle in Illustration 2 has a width of x and a perimeter of 100 feet. Find x so that the area of the rectangle is maximum.

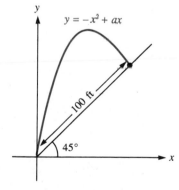

Illustration 1

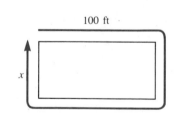

Illustration 2

69. A farmer wants to partition off a rectangular stall in a corner of his barn. The barn walls form two sides of the stall, and the farmer has 50 feet of fencing for the remaining two sides. What dimensions will maximize the area? See Illustration 3.

70. A rancher wishes to enclose a rectangular partitioned corral with 1800 feet of fencing. See Illustration 4. What dimensions of the corral would enclose the largest possible area? What is the maximum area?

71. A 24-inch wide sheet of metal is bent into a rectangular trough; a cross section is shown in Illustration 5. What dimensions will maximize the amount of water carried? That is, what dimensions will maximize the cross-sectional area?

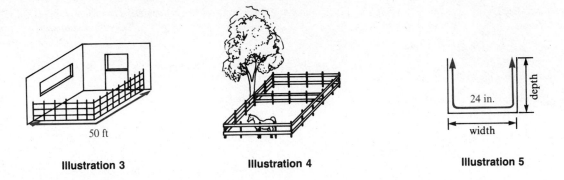

Illustration 3   **Illustration 4**   **Illustration 5**

72. A farmer will use D feet of fencing to enclose a rectangular plot of ground. Show that he will enclose the maximum area if the rectangle is a square.

73. The sum of two numbers is 6, and the sum of the squares of those two numbers is minimum. What are the two numbers?

74. What number most exceeds its square?

75. Find the dimensions of the largest rectangle that can be inscribed in the right triangle ABC shown in Illustration 6.

76. Point P lies in the first quadrant and on the line $x + y = 1$ in such a position that the area of triangle OPA is maximum (see Illustration 7). Find the coordinates of P.

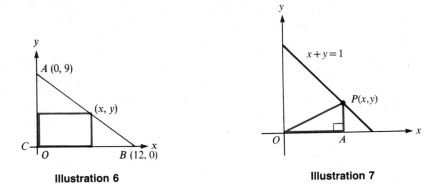

Illustration 6   **Illustration 7**

In Exercises 77–80, refer to this information: At a time t seconds after an object is tossed vertically upward, it reaches a height s in feet given by the equation $s = 80t - 16t^2$.

77. In how many seconds does the object reach its maximum height?

78. In how many seconds does the object return to the point from which it was thrown?

79. What is the maximum height reached by the object?

80. Show that it takes the same amount of time for the object to reach its maximum height as it does for the object to fall back to the point from which it was thrown.

Computer Exercises. *In Exercises 81–84, use a computer and* **Algebra Pak** *to perform each computer experiment.* **Write a brief paragraph describing your findings.** *Enter the square-root function as SQR( ). Select nonautomatic scaling of the y-axis.*

81. Investigate the horizontal translation of the graph of a function by graphing $y = \sqrt{x - k}$ for several values of k. What do you observe?

82. Investigate the vertical translation of the graph of a function by graphing $y = \sqrt{x} + k$ for several values of k. What do you observe?

83. Investigate the horizontal stretching of the graph of a function by graphing $y = \sqrt{kx}$ for several values of k. What do you observe?

84. Investigate the vertical stretching of the graph of a function by graphing $y = k\sqrt{x}$ for several values of k. What do you observe?

3.6   RATIONAL FUNCTIONS

Certain functions, called **rational functions,** are defined by equations of the form

$$y = \frac{P(x)}{Q(x)}$$

where $P(x)$ and $Q(x)$ are polynomials. Because $Q(x)$ appears in the denominator of a fraction, and a denominator cannot be zero, the domain of the rational function must exclude all values of x for which $Q(x) = 0$.

An example of a rational function is given by

$$y = \frac{3x - 2}{2x^2 - 5x - 3}$$

The domain of this function includes all numbers x, *except* those that cause the denominator of the fraction to be zero. To find these excluded values, we set the denominator equal to 0 and solve for x.

$$2x^2 - 5x - 3 = 0$$
$$(2x + 1)(x - 3) = 0$$
$$2x + 1 = 0 \quad \text{or} \quad x - 3 = 0$$
$$x = -\frac{1}{2} \quad \bigg| \quad x = 3$$

The domain of the rational function is the set of all real numbers except $-\frac{1}{2}$ and 3.

Example 1   Graph the rational function defined by $y = \dfrac{x^2 - 4}{x^2 - 1}$.

Solution   To determine the numbers *excluded* from the domain of this function, set its denominator equal to zero and solve for x. You will find the domain to be the set of all real numbers x except $x = 1$ and $x = -1$. To discover what happens to the graph of the rational function when x is near 1, calculate values of y for numbers x close to 1.

$x < 1$			$x > 1$	
x	y		x	y
0.5	5		1.5	-1.4
0.9	16.8		1.1	-13.3
0.99	151.8		1.01	-148.3
0.999	1501.8		1.001	-1498.3

Plotting these points suggests a curve that approaches the vertical line $x = 1$ but never touches it. See Figure 3-39a. Because the variable x in this function has only even exponents, $f(x) = f(-x)$ and the curve is symmetric to the y-axis. Thus, similar behavior occurs near the line $x = -1$.

The y-intercept of the curve is determined by setting x equal to 0 and computing y; the y-intercept is 4.

The numbers $x = 2$ and $x = -2$ reduce the numerator of the rational expression to 0. The corresponding values of y, therefore, are both 0, and the x-intercepts of the curve are 2 and -2. You can sketch most of the curve, as shown in Figure 3-39b.

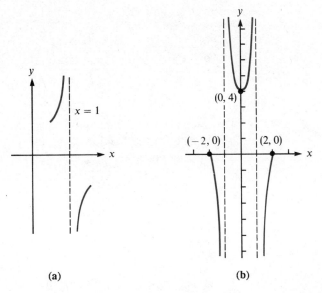

(a)   (b)

Figure 3-39

To discover the shape of the curve to the right of $x = 2$ and to the left of $x = -2$, examine the behavior of

$$y = \frac{x^2 - 4}{x^2 - 1}$$

as x gets very large. Do this by performing a long division and writing the answer in *quotient* + $\dfrac{remainder}{divisor}$ form:

$$
\begin{array}{r}
1 \\
x^2 - 1 \overline{)\, x^2 - 4 } \\
\underline{x^2 - 1 } \\
- 3
\end{array}
$$

Hence,

$$y = \frac{x^2 - 4}{x^2 - 1} = 1 + \frac{-3}{x^2 - 1}$$

If x becomes very large, the denominator of the fraction $-3/(x^2 - 1)$ becomes very large also, and the magnitude of the fraction itself becomes very small. Because the term $-3/(x^2 - 1)$ is negative and approaches 0 as x becomes very large, it becomes negligible, and the value of y (which is less than 1) approaches the value 1. As x increases without bound, the value of the function defined by $y = (x^2 - 4)/(x^2 - 1)$ approaches 1 from below. Hence, as x increases, the curve approaches the line $y = 1$ from below. Because of symmetry, the curve also approaches the line $y = 1$ as x decreases without bound. The graph of the given rational function is shown in Figure 3-40.

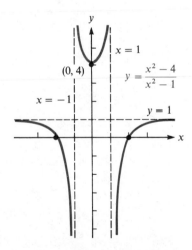

Figure 3-40

The broken lines in Figure 3-40 are called **asymptotes**. The function $y = (x^2 - 4)/(x^2 - 1)$ has two vertical asymptotes, the lines $x = 1$ and $x = -1$, and one horizontal asymptote, the line $y = 1$. Although a curve might intersect a horizontal asymptote when $|x|$ is small, it will approach, but never touch, a horizontal asymptote when $|x|$ grows large. ∎

Let's summarize the work done in the previous example. First you found the vertical asymptotes $x = 1$ and $x = -1$ by setting the denominator of the rational expression equal to 0 and solving that equation for x. Then you found the y-intercept 4 by letting x equal 0 in the rational expression and evaluating y. Next you found the x-intercepts by finding the numbers x (2 and -2) that make the numerator of the rational expression equal to 0, and you found the horizontal asymptote $y = 1$ by performing a long division and ignoring the remainder. Finally, you made use of symmetry to graph the part of the curve to the left of the y-axis.

Example 2   Graph the rational function defined by $y = \dfrac{3x}{x - 2}$.

Solution   First look for symmetry about the y-axis. Because x appears to an odd power, the function is not symmetric about the y-axis. The y-intercept is found by setting $x = 0$ and solving for y; the y-intercept is 0. Because $y = 0$ when $x = 0$, the curve passes through the origin.

The x-intercepts are found by setting the numerator of the rational expression equal to 0 and solving for x:

$$3x = 0$$
$$x = 0$$

Thus, the only x-intercept is at 0.

The vertical asymptotes are found by determining which real numbers lead to zeros in the denominator of the rational expression. If x is replaced by 2 in the denominator $x - 2$, that denominator is 0. Hence, the line $x = 2$ is a vertical asymptote.

The horizontal asymptotes, if any, are found by dividing $3x$ by $x - 2$, and expressing the answer in *quotient* $+$ $\dfrac{remainder}{divisor}$ form.

$$y = \frac{3x}{x - 2} = 3 + \frac{6}{x - 2}$$

As $|x|$ increases without bound, the fraction $\frac{6}{x-2}$ approaches 0, so the curve approaches the line $y = 3$. The line $y = 3$ is a horizontal asymptote.

To discover what happens to the graph when x is greater than 2, pick a value of x that is greater than 2 and find the corresponding value of y. If $x = 3$, for example, the value of y is 9. After plotting the point $(3, 9)$, you can use the intercepts and asymptotes to sketch the graph. This curve, shown in Figure 3-41, is called a **hyperbola**.

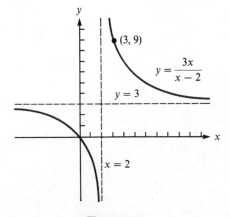

Figure 3-41

Example 3 Graph the rational function defined by $y = \dfrac{1}{x(x-1)^2}$.

Solution The expansion of the denominator gives a third-degree polynomial in which x appears to an odd power. Thus, there is no symmetry about the y-axis. Because $y = f(x)$ and $-y = f(-x)$ are not equivalent, there is no symmetry about the origin either. Because 0 and 1 cause the denominator to be 0, the vertical asymptotes of the graph are the lines $x = 0$ and $x = 1$.

Because x cannot be 0, the graph has no y-intercept. Because y cannot be 0, the graph has no x-intercept.

As $|x|$ grows large, the denominator of the fraction becomes large, and the corresponding values of y approach 0. Thus, the line $y = 0$ is a horizontal asymptote.

The table in Figure 3-42 lists the coordinates of several points lying in regions between the asymptotes. The value of y changes sign at $x = 0$: To the left of the asymptote $x = 0$, y is negative, and to the right of the asymptote, y is positive. This happens because x appears to an odd power in the denominator. The value of y does *not* change sign at $x = 1$: To the left and right of the asymptote $x = 1$, y is positive. The function behaves this way because $(x - 1)$ appears to an even power in the denominator. The graph of the rational function appears in Figure 3-42.

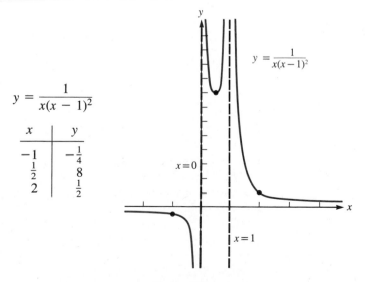

$$y = \frac{1}{x(x-1)^2}$$

x	y
-1	$-\dfrac{1}{4}$
$\dfrac{1}{2}$	8
2	$\dfrac{1}{2}$

Figure 3-42

Example 4   Graph the rational function $y = f(x) = \dfrac{1}{x^2 + 1}$.

Solution   Because x appears only to an even power, $f(-x) = f(x)$. Thus, the graph of the function is symmetric about the y-axis. Because $f(0) = 1$, the graph's y-intercept is 1. Because no numbers x will make the denominator 0, the graph has no vertical asymptotes. The graph does have a horizontal asymptote of $y = 0$, however, because as $|x|$ grows large, $f(x)$ approaches 0. Because the denominator is always positive, the graph lies entirely above the x-axis. See Figure 3-43.

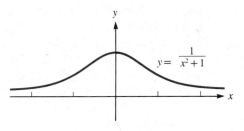

Figure 3-43   ■

Example 5   Graph the rational function defined by $y = \dfrac{x^2 + x - 2}{x - 3}$.

Solution   First factor the numerator of the rational expression:

$$y = \frac{(x - 1)(x + 2)}{x - 3}$$

Most of the information is straightforward:

- The function is not symmetric about the y-axis or origin.
- The y-intercept is $\frac{2}{3}$.
- The x-intercepts are 1 and -2.
- A vertical asymptote is $x = 3$.

Perform the long division and write the rational expression as

$$y = \frac{x^2 + x - 2}{x - 3} = x + 4 + \frac{10}{x - 3}$$

As before, the fraction $10/(x - 3)$ approaches 0 as $|x|$ grows large. This time, however, the graph does not approach a constant. Instead, it approaches the line $y = x + 4$. Because this line is not horizontal, the graph has no horizontal asymptotes. However, it does have a **slant asymptote** or an **oblique asymptote**: the line $y = x + 4$. Put all of this information together and graph the rational function. See Figure 3-44.

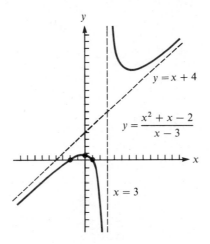

Figure 3-44

Example 6 Discuss the nature of the asymptotes of the following rational functions:

a. $y = \dfrac{x + 2}{x^2 - 1}$ **b.** $y = \dfrac{x^2 + x + 2}{x^2 - 1}$

c. $y = \dfrac{3x^3 + 2x^2 + 2}{x^2 - 1}$ **d.** $y = \dfrac{x^4 + x + 2}{x^2 - 1}$

Solution All four functions have vertical asymptotes at $x = 1$ and $x = -1$ because at these values the denominators are 0. The nature of the remaining asymptotes must be considered.

a. $y = \dfrac{x + 2}{x^2 - 1}$

Because the degree of the numerator is less than the degree of the denominator, long division is not practical. Try a different approach, dividing both numerator and denominator by x^2, which is the largest power of x in the denominator.

$$y = \frac{x + 2}{x^2 - 1} = \frac{\dfrac{x}{x^2} + \dfrac{2}{x^2}}{\dfrac{x^2}{x^2} - \dfrac{1}{x^2}} = \frac{\dfrac{1}{x} + \dfrac{2}{x^2}}{1 - \dfrac{1}{x^2}}$$

The three fractions in the final result, $\dfrac{1}{x}$, $\dfrac{2}{x^2}$, and $\dfrac{1}{x^2}$, all approach 0 as $|x|$ increases without bound. Hence, y approaches

$$\frac{0 + 0}{1 - 0} = 0$$

The graph of this function has a horizontal asymptote of $y = 0$.

In each of the three remaining functions, a long division can be performed.

b. $y = \dfrac{x^2 + x + 2}{x^2 - 1} = 1 + \dfrac{x + 3}{x^2 - 1}$

As $|x|$ increases without bound, the fraction $(x + 3)/(x^2 - 1)$ approaches 0 (for reasons discussed in part **a**), and y approaches 1. This curve has a horizontal asymptote of $y = 1$.

c. $y = \dfrac{3x^3 + 2x^2 + 2}{x^2 - 1} = 3x + 2 + \dfrac{3x + 4}{x^2 - 1}$

Again, the last fraction approaches 0 as $|x|$ increases without bound, and the curve approaches the slant asymptote $y = 3x + 2$.

d. $y = \dfrac{x^4 + x + 2}{x^2 - 1} = x^2 + 1 + \dfrac{x + 3}{x^2 - 1}$

As $|x|$ increases, the fractional part again approaches 0, and the curve approaches $y = x^2 + 1$. However, $y = x^2 + 1$ does not represent a line. This curve has neither horizontal nor slant asymptotes.   ∎

We generalize the results of Example 6 and summarize the techniques discussed in this section as follows:

Perform the following steps when you graph the rational function $y = \dfrac{P(x)}{Q(x)}$, where $\dfrac{P(x)}{Q(x)}$ is in simplified form.

Check for symmetry.   If the polynomials $P(x)$ and $Q(x)$ involve only even powers of x, the graph is symmetric about the y-axis. Otherwise, y-axis symmetry does not exist.

Look for y-intercepts.   Set x equal to 0. The resulting value of y (if any) is the y-intercept of the graph.

Look for x-intercepts.   Set $P(x)$ equal to 0. The real solutions of the equation $P(x) = 0$ (if any) are the x-intercepts of the graph.

Look for vertical asymptotes.   Set $Q(x)$ equal to 0. The real solutions of the equation $Q(x) = 0$ (if any) determine the vertical asymptotes of the graph.

Look for horizontal asymptotes.   If the degree of $P(x)$ is less than the degree of $Q(x)$, then the line $y = 0$ is a horizontal asymptote.

(continued)

(continued)

If the degrees of $P(x)$ and $Q(x)$ are equal, then the line $y = p/q$, where p and q are the lead coefficients of $P(x)$ and $Q(x)$, is a horizontal asymptote. (Be sure that $P(x)$ and $Q(x)$ are written in descending powers of x before applying this rule.)

If the degree of $P(x)$ is greater than the degree of $Q(x)$, then there is no horizontal asymptote.

A graph might intersect a horizontal asymptote when $|x|$ is small. However, it will approach but never touch a horizontal asymptote as $|x|$ grows large.

Look for a slant asymptote

If the degree of $P(x)$ is exactly one greater than the degree of $Q(x)$, there is a slant asymptote. To find it, perform the long division $Q(x)\overline{)P(x)}$ and ignore the remainder.

We have considered only rational functions for which the fraction is in simplified form. We now consider a rational function $y = \frac{P(x)}{Q(x)}$ in which $P(x)$ and $Q(x)$ share a common factor. Graphs of such functions often have gaps, as the next example illustrates.

Example 7   Find the domain of the rational function defined by $y = \dfrac{x^2 - x - 12}{x - 4}$ and graph the function.

Solution   Because the denominator of a fraction cannot be zero, x cannot be 4. Thus, the domain of the function is the set of all real numbers except 4. Now factor the numerator of the rational expression to see that its numerator and denominator share a common factor of $x - 4$.

$$y = \frac{x^2 - x - 12}{x - 4}$$

$$= \frac{(x + 3)(x - 4)}{x - 4}$$

For numbers x *other than* 4, the common factor of $x - 4$ can be divided out. The result is equivalent to the original function *only if you keep the restriction that* $x \neq 4$. Thus,

$$y = \frac{(x + 3)(x - 4)}{(x - 4)} = x + 3 \qquad \text{provided that } x \neq 4$$

When $x = 4$, the function is not defined. The graph of the given rational function appears in Figure 3-45. It is a line with one point missing, the point with x-coordinate of 4.

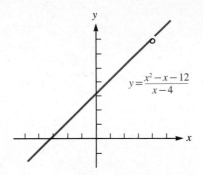

$$y = \frac{x^2 - x - 12}{x - 4}$$

Figure 3-45

EXERCISE 3.6

In Exercises 1–6, find the domain of each rational function. **Do not graph the function.**

1. $y = \dfrac{x^2}{x - 2}$

2. $y = \dfrac{x^3 - 3x^2 + 1}{x + 3}$

3. $y = \dfrac{2x^2 + 7x - 2}{x^2 - 25}$

4. $y = \dfrac{5x^2 + 1}{x^2 + 5}$

5. $y = \dfrac{x - 1}{x^3 - x}$

6. $y = \dfrac{1}{2x^2 - 9x + 9}$

In Exercises 7–34, find all vertical, horizontal, and slant asymptotes, x- and y-intercepts, and symmetries, and then graph each rational function.

7. $y = \dfrac{1}{x - 2}$

8. $y = \dfrac{3}{x + 3}$

9. $y = \dfrac{x}{x - 1}$

10. $y = \dfrac{x}{x + 2}$

11. $y = \dfrac{x + 1}{x + 2}$

12. $y = \dfrac{x - 1}{x - 2}$

13. $y = \dfrac{2x - 1}{x - 1}$

14. $y = \dfrac{3x + 2}{x^2 - 4}$

15. $y = \dfrac{x^2 - 9}{x^2 - 4}$

16. $y = \dfrac{x^2 - 4}{x^2 - 9}$

17. $y = \dfrac{x^2 - x - 2}{x^2 - 4x + 3}$

18. $y = \dfrac{x^2 + 7x + 12}{x^2 - 7x + 12}$

19. $y = \dfrac{x^2 + 2x - 3}{x^3 - 4x}$

20. $y = \dfrac{3x^2 - 4x + 1}{2x^3 + 3x^2 + x}$

21. $y = \dfrac{x^2 - 9}{x^2}$

22. $y = \dfrac{3x^2 - 12}{x^2}$

23. $y = \dfrac{x}{(x + 3)^2}$

24. $y = \dfrac{x}{(x - 1)^2}$

25. $y = \dfrac{x + 1}{x^2(x - 2)}$

26. $y = \dfrac{x - 1}{x^2(x + 2)^2}$

27. $y = \dfrac{x}{x^2 + 1}$

28. $y = \dfrac{x - 1}{x^2 + 2}$

29. $y = \dfrac{3x^2}{x^2 + 1}$

30. $y = \dfrac{x^2 - 9}{2x^2 + 1}$

31. $y = \dfrac{x^2 - 2x - 8}{x - 1}$

32. $y = \dfrac{x^2 + x - 6}{x + 2}$

33. $y = \dfrac{x^3 + x^2 + 6x}{x^2 - 1}$

34. $y = \dfrac{x^3 - 2x^2 + x}{x^2 - 4}$

In Exercises 35–40, the numerator and denominator of the fraction share a common factor. Graph each rational function.

35. $y = \dfrac{x^2}{x}$

36. $y = \dfrac{x^2 - 1}{x - 1}$

37. $y = \dfrac{x^3 + x}{x}$

38. $y = \dfrac{x^3 - x^2}{x - 1}$

39. $y = \dfrac{x^2 - 2x + 1}{x - 1}$

40. $y = \dfrac{2x^2 + 3x - 2}{x + 2}$

41. Can a rational function have more than one horizontal asymptote? Explain.

42. Can a rational function have more than one slant asymptote? Explain.

43. Show that the graph of $y = \dfrac{ax + b}{cx^2 + d}$, where a, b, c, and d are nonzero constants, has the horizontal aysmptote $y = 0$.

44. Show that the graph of $y = \dfrac{ax^2 + b}{cx^2 + d}$, where a, b, c, and d are nonzero constants, has the horizontal asymptote $y = \dfrac{a}{c}$.

45. Show that the graph of $y = \dfrac{ax^3 + b}{cx^2 + d}$, where a, b, c, and d are nonzero constants, has the slant asymptote $y = \dfrac{a}{c}x$.

46. Graph the rational function $y = \dfrac{x^3 + 1}{x}$, and explain why the curve is said to have a *parabolic asymptote*.

Computer Exercises. *In Exercises 47–50, use a computer and* **Algebra Pak** *to perform each computer experiment.* **Write a brief paragraph describing your findings.** *Select nonautomatic scaling of the y-axis.*

47. Investigate the positioning of the vertical asymptotes of a rational function by graphing $y = \dfrac{x}{x - k}$ for several values of k. If you choose $k = 2$, for example, enter the function by typing X/(X − 2). What do you observe?

48. Investigate the positioning of the vertical asymptotes of a rational function by graphing $y = \dfrac{x}{x^2 - k}$ for $k = 4$, for $k = 1$, for $k = -1$, and for $k = 0$. If you choose $k = 4$, for example, enter the function by typing X/(X^2 − 4). What do you observe?

49. Determine the range of the rational function $y = \dfrac{kx^2}{x^2 + 1}$ for several values of k. What do you observe?

50. Investigate the positioning of the x-intercepts of a rational function by graphing $y = \dfrac{x^2 - k}{x}$ for $k = 4$, for $k = 1$, for $k = -1$, and for $k = 0$. What do you observe?

3.7 ALGEBRA AND COMPOSITION OF FUNCTIONS

It is possible to perform some arithmetic with functions.

Definition.   If the ranges of functions f and g are subsets of the real numbers, then four new functions can be defined.

The sum of f and g, denoted as $f + g$, is defined by

$$(f + g)(x) = f(x) + g(x)$$

The difference of f and g, denoted as $f - g$, is defined by

$$(f - g)(x) = f(x) - g(x)$$

The product of f and g, denoted as $f \cdot g$, is defined by

$$(f \cdot g)(x) = f(x) \cdot g(x)$$

The quotient of f and g, denoted as f/g, is defined by

$$(f/g)(x) = \frac{f(x)}{g(x)} \qquad \text{provided } g(x) \neq 0$$

The domain of each of these functions (unless otherwise restricted) is the set of all real numbers x that are in the domain of *both* f and g. In the case of the quotient f/g, there is the further restriction that $g(x) \neq 0$.

Example 1   Let $f(x) = 3x + 1$ and $g(x) = 2x - 3$. Find   **a.** $f + g$ and   **b.** $f \cdot g$, and give the domain of each.

Solution   **a.** $(f + g)(x) = f(x) + g(x)$

$\qquad\qquad\qquad = (3x + 1) + (2x - 3)$ Substitute $3x + 1$ for $f(x)$ and $2x - 3$ for $g(x)$.

$\qquad\qquad\qquad = 5x - 2$ Combine terms.

Thus, the function $f + g$ is defined by $(f + g)(x) = 5x - 2$.

The domain of $f + g$ is the set of all real numbers that are in the domain of both f and g. Because the domain of both f and g is the set of real numbers, the domain of $f + g$ is also the set of real numbers.

b. $(f \cdot g)(x) = f(x) \cdot g(x)$

$\qquad\qquad\qquad = (3x + 1)(2x - 3)$ Substitute $3x + 1$ for $f(x)$ and $2x - 3$ for $g(x)$.

$\qquad\qquad\qquad = 6x^2 - 7x - 3$

Thus, the function $f \cdot g$ is defined by $(f \cdot g)(x) = 6x^2 - 7x - 3$.

The domain of $f \cdot g$ is the set of real numbers.   ∎

Example 2   Let $f(x) = x^2 - 4$ and $g(x) = \sqrt{x}$. If all calculations involve real numbers only, find the functions   **a.** $f + g$,   **b.** $f \cdot g$,   **c.** f/g, and   **d.** g/f, and give the domain of each.

Solution   Because any real number can be squared, the domain of f is the set of real numbers. Because all calculations are to involve real numbers only, the domain of g is the set of nonnegative real numbers: $\{x : x \geq 0\}$.

a. $(f + g)(x)$ is defined as $f(x) + g(x)$. Hence,

$$(f + g)(x) = x^2 - 4 + \sqrt{x}$$

The domain consists of the numbers x that are common to the domains of f and g. The domain of $f + g$ is $\{x : x \geq 0\}$.

b. $(f \cdot g)(x)$ is defined as $f(x) \cdot g(x)$. Hence,

$$(f \cdot g)(x) = (x^2 - 4) \cdot \sqrt{x}$$
$$= x^2\sqrt{x} - 4\sqrt{x}$$

The domain consists of those numbers that are common to the domains of f and g. The domain of $f \cdot g$ is $\{x : x \geq 0\}$.

c. $(f/g)(x)$ is defined as $\dfrac{f(x)}{g(x)}$. Hence,

$$(f/g)(x) = \frac{x^2 - 4}{\sqrt{x}}$$

Its domain consists of those numbers common to the domains of f and g except 0, because division by 0 is not defined. The domain of f/g is $\{x : x > 0\}$.

d. $(g/f)(x)$ is defined as $\dfrac{g(x)}{f(x)}$. Hence,

$$(g/f)(x) = \frac{\sqrt{x}}{x^2 - 4}$$

The domain consists of those numbers common to the domains of f and g except the number 2, because division by 0 is not defined. Thus, the domain of g/f is $\{x : x \geq 0 \quad \text{and} \quad x \neq 2\}$.   ■

Example 3   Find $(f + g)(3)$ if $f(x) = x^2 + 1$ and $g(x) = 2x + 1$.

Solution   First find $(f + g)(x)$.

$$(f + g)(x) = f(x) + g(x)$$
$$= x^2 + 1 + 2x + 1$$
$$= x^2 + 2x + 2$$

Then find $(f + g)(3)$.

$$(f + g)(3) = 3^2 + 2(3) + 2$$
$$= 9 + 6 + 2$$
$$= 17$$ ∎

Example 4   Let $h(x) = x^2 + 3x + 2$. Find functions f and g such that   **a.** $h = f + g$ and **b.** $h = f \cdot g$.

Solution   **a.** Several answers are possible. One answer is $f(x) = x^2$ and $g(x) = 3x + 2$. Then

$$h(x) = x^2 + 3x + 2$$
$$= (x^2) + (3x + 2)$$
$$= f(x) + g(x)$$
$$= (f + g)(x)$$

Another answer is $f(x) = x^2 + 2x$ and $g(x) = x + 2$.

b. Several answers are possible. One answer is suggested by factoring the polynomial $x^2 + 3x + 2$:

$$h(x) = x^2 + 3x + 2 = (x + 1)(x + 2)$$

Let $f(x) = x + 1$ and $g(x) = x + 2$. Then

$$h(x) = x^2 + 3x + 2$$
$$= (x + 1)(x + 2)$$
$$= f(x) \cdot g(x)$$
$$= (f \cdot g)(x)$$

Another answer is $f(x) = 3$ and $g(x) = \dfrac{x^2}{3} + x + \dfrac{2}{3}$.   ∎

Composition of Functions

Often one quantity is a function of a second quantity that depends, in turn, on a third quantity. A farmer's income, for example, is a function of the number of bushels of grain harvested. But the harvest, in turn, is a function of the number of inches of rainfall received (among other things). Such chains of dependence can be analyzed mathematically by considering the **composition of functions**.

We suppose that $y = f(x)$ and $y = g(x)$ define two functions. Any number x in the domain of g will produce the corresponding value $g(x)$ in the range of g. If this value $g(x)$ is in the domain of the function f, then $g(x)$ is an acceptable input into the function f, and a corresponding value $f(g(x))$ is determined. This two-step process defines a new function, called a **composite function**, denoted by $f \circ g$.

To visualize the domain of the composite function $f \circ g$, refer to Figure 3-46. Note that the value of the independent variable x must be in the domain of the function g, because x is the number used as input into the function g. Also note that $g(x)$ must be in the domain of function f, because $g(x)$ is the number used as input to the function f. Hence, the domain of $f \circ g$ consists of all those elements x that are permissible inputs into g, and for which each $g(x)$ is a permissible input into f.

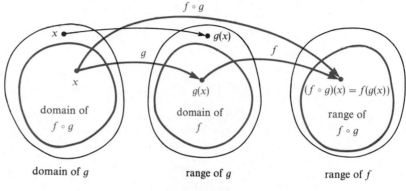

Figure 3-46

Definition.   The **composite function** $f \circ g$ is defined by

$$(f \circ g)(x) = f(g(x))$$

The **domain of the composite function** $f \circ g$ consists of all those elements in the domain of g for which $g(x)$ is an element in the domain of f.

To illustrate composition of functions, we suppose that $f(x) = x + 1$ and $g(x) = x^2$, and calculate $(f \circ g)(2)$. We do so by calculating $f(g(2))$. Because $g(x) = x^2$, $g(2) = 2^2 = 4$. Because 4 is in the domain of f, calculate $f(4)$, which is $4 + 1 = 5$. Hence,

$$(f \circ g)(2) = f(g(2))$$
$$= f(4)$$
$$= 5$$

Example 5   If $f(x) = 2x + 7$ and $g(x) = 4x + 1$, find   **a.** $(f \circ g)(x)$ and   **b.** $(g \circ f)(x)$.

Solution   **a.** $(f \circ g)(x) = f(g(x)) = f(4x + 1)$
$$= 2(4x + 1) + 7$$
$$= 8x + 9$$

b. $(g \circ f)(x) = g(f(x)) = g(2x + 7)$
$$= 4(2x + 7) + 1$$
$$= 8x + 29$$

∎

Example 5 illustrates that composition of functions is *not* commutative. In general, $(f \circ g)(x)$ is not equal to $(g \circ f)(x)$.

Example 6   Find two functions f and g such that $(f \circ g)(x) = \sqrt{x^2 + 2}$.

Solution   Let $g(x) = x^2 + 2$ and $f(x) = \sqrt{x}$. Then

$$(f \circ g)(x) = f(g(x)) = f(x^2 + 2)$$
$$= \sqrt{x^2 + 2}$$

Thus, $g(x) = x^2 + 2$ and $f(x) = \sqrt{x}$ are two functions such that $(f \circ g)(x) = \sqrt{x^2 + 2}$. There are many other answers. For example, if $g(x) = x^2$ and $f(x) = \sqrt{x + 2}$, then

$$(f \circ g)(x) = f(g(x)) = f(x^2) = \sqrt{x^2 + 2} \qquad \blacksquare$$

Example 7   Let $f(x) = \sqrt{x}$ and $g(x) = x - 3$. Find the domain of functions **a.** f, **b.** g, **c.** $f \circ g$, and **d.** $g \circ f$.

Solution   **a.** If the range is to be a subset of the real numbers, the domain of f is the set of nonnegative real numbers: $\{x : x \geq 0\}$.

b. The domain of g is the set of real numbers.

c. Because $g(x)$ must be in the domain of f, the value $g(x)$ must be nonnegative. Thus,

$$g(x) \geq 0$$
$$x - 3 \geq 0 \qquad \text{Because } g(x) = x - 3.$$
$$x \geq 3$$

Hence, the domain of $f \circ g$ is $\{x : x \geq 3\}$.

d. The domain of $g \circ f$ contains those numbers x for which $f(x)$ is also in the domain of g. Because the domain of g includes all real numbers, the domain of $g \circ f$ is just the domain of f itself: $\{x : x \geq 0\}$. $\blacksquare$

Example 8   Let $f(x) = \sqrt{1 - x}$ and $g(x) = x^2$. If all expressions represent real numbers, find **a.** $(f \circ g)(x)$ and **b.** $(g \circ f)(x)$, and give the domain of each function.

Solution   Begin by finding the domain of f. Because all expressions represent real numbers, the radicand $1 - x$ cannot be negative, so x must be less than or equal to 1. The domain of f is the set $\{x : x \leq 1\}$.

Now find the domain of g. Because the square of any real number is still a real number, the domain of g is the set of real numbers.

a. $(f \circ g)(x) = f(g(x))$
$$= f(x^2)$$
$$= \sqrt{1 - x^2}$$

The domain of $f \circ g$ consists of those numbers x for which $g(x)$ is in the domain of f. Thus, the value $g(x)$ must be less than or equal to 1.

$$g(x) \le 1$$
$$x^2 \le 1 \qquad \text{Because } g(x) = x^2.$$

Hence, the domain of $f \circ g$ is $\{x : -1 \le x \le 1\}$.

b. $(g \circ f)(x) = g(f(x))$
$$= g(\sqrt{1 - x})$$
$$= (\sqrt{1 - x})^2$$
$$= 1 - x$$

The domain of $g \circ f$ consists of those numbers x for which $f(x)$ is in the domain of g. Because any real number is acceptable to g, the domain of $g \circ f$ is just the domain of f: $\{x : x \le 1\}$. ∎

EXERCISE 3.7

In Exercises 1–4, let $f(x) = 2x + 1$ and $g(x) = 3x - 2$. Find each function, and determine its domain.

1. $f + g$ **2.** $f - g$ **3.** $f \cdot g$ **4.** f/g

In Exercises 5–8, let $f(x) = x^2 + x$ and $g(x) = x^2 - 1$. Find each function, and determine its domain.

5. $f - g$ **6.** $f + g$ **7.** f/g **8.** $f \cdot g$

In Exercises 9–16, let $f(x) = x^2 - 1$ and $g(x) = 3x - 2$. Find each value, if possible.

9. $(f + g)(2)$ **10.** $(f + g)(-3)$ **11.** $(f - g)(0)$ **12.** $(f - g)(-5)$

13. $(f \cdot g)(2)$ **14.** $(f \cdot g)(-1)$ **15.** $(f/g)(\frac{2}{3})$ **16.** $(f/g)(1)$

In Exercises 17–24, find two functions f and g such that the given correspondence can be expressed as the function indicated. Several answers are possible.

17. $y = 3x^2 + 2x; f + g$ **18.** $y = 3x^2; f \cdot g$

19. $y = \dfrac{3x^2}{x^2 - 1}; f/g$ **20.** $y = 5x + x^4; f - g$

21. $y = x(3x^2 + 1); f - g$ **22.** $y = (3x - 2)(3x + 2); f + g$

23. $y = x^2 + 7x - 18; f \cdot g$ **24.** $y = 5x^5; f/g$

In Exercises 25–28, let $f(x) = 2x - 5$ and $g(x) = 5x - 2$. Find each value.

25. $(f \circ g)(2)$ **26.** $(g \circ f)(-3)$ **27.** $(f \circ f)(-\frac{1}{2})$ **28.** $(g \circ g)(\frac{3}{5})$

In Exercises 29–32, let $f(x) = 3x^2 - 2$ and $g(x) = 4(x + 1)$. Find each value.

29. $(f \circ g)(-3)$ **30.** $(g \circ f)(3)$ **31.** $(f \circ f)(\sqrt{3})$ **32.** $(g \circ g)(-4)$

In Exercises 33–36, let $f(x) = 3x$ and $g(x) = x + 1$. Find each composite function, and determine its domain.

33. $f \circ g$ **34.** $g \circ f$ **35.** $f \circ f$ **36.** $g \circ g$

In Exercises 37–40, let f(x) = x² and g(x) = 2x. Find each composite function, and determine its domain.

37. $g \circ f$ **38.** $f \circ g$ **39.** $f \circ f$ **40.** $g \circ g$

In Exercises 41–44, let f(x) = √x and g(x) = x² + 1. Find each composite function, and determine its domain.

41. $f \circ g$ **42.** $g \circ f$ **43.** $f \circ f$ **44.** $g \circ g$

In Exercises 45–48, let f(x) = √(x + 1) and g(x) = x² − 1. Find each composite function, and determine its domain.

45. $g \circ f$ **46.** $f \circ g$ **47.** $g \circ g$ **48.** $f \circ f$

In Exercises 49–60, find two functions f and g such that the composition f ∘ g expresses the given correspondence. Several answers are possible.

49. $y = 3x - 2$ **50.** $y = 7x - 5$ **51.** $y = x^2 - 2$ **52.** $y = x^3 - 3$

53. $y = (x - 2)^2$ **54.** $y = (x - 3)^3$ **55.** $y = \sqrt{x + 2}$ **56.** $y = \dfrac{1}{x - 5}$

57. $y = \sqrt{x} + 2$ **58.** $y = \dfrac{1}{x} - 5$ **59.** $y = x$ **60.** $y = 3$

61. Let $f(x) = 3x$. Show that $(f + f)(x) = f(x + x)$.   **62.** Let $g(x) = x^2$. Show that $(g + g)(x) \neq g(x + x)$.

63. Let $f(x) = \dfrac{x - 1}{x + 1}$. Find $(f \circ f)(x)$.   **64.** Let $g(x) = \dfrac{x}{x - 1}$. Find $(g \circ g)(x)$.

3.8 INVERSE FUNCTIONS

By definition, each element x in the domain of a function has a *single* image y. For some functions, distinct numbers x in the domain have the *same* image. See Figure 3-47a. For other functions, called **one-to-one functions**, distinct numbers x have distinct images. See Figure 3-47b.

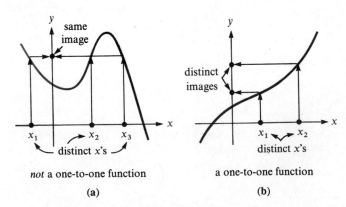

not a one-to-one function
(a)

a one-to-one function
(b)

Figure 3-47

> **Definition.**   A function f from a set **X** to a set **Y** is called **one-to-one** if and only if different elements in the domain of f have different images in the range of f: if x_1 and x_2 are two elements in the domain of f and $x_1 \neq x_2$, then $f(x_1) \neq f(x_2)$.

Example 1   Determine if the functions   **a.** $f(x) = x^2$ and   **b.** $f(x) = x^3$ are one-to-one.

Solution   **a.** The function $f(x) = x^2$ is *not* one-to-one because different elements in the domain (2 and -2, for example) have the same image ($f(2) = f(-2) = 4$).

b. The function $f(x) = x^3$ *is* one-to-one because two distinct numbers x produce different images $f(x)$—two different numbers have different cubes.   ∎

A **horizontal line test** can be used to determine whether the graph of a function represents a one-to-one function. If any horizontal line intersects the graph of a function more than once, the function is *not* one-to-one. See Figure 3-48**a**. If every horizontal line that intersects the graph does so only once, the function *is* one-to-one. See Figure 3-48**b**.

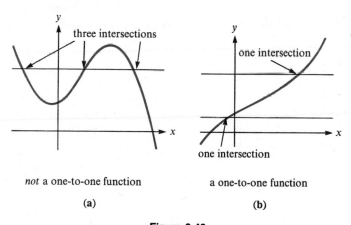

not a one-to-one function   a one-to-one function

(a)   (b)

Figure 3-48

Figure 3-49**a** illustrates a function f from a set **X** to a set **Y**. Because several arrows point to a single y, the function f is not one-to-one. If the arrows of Figure 3-49**a** were reversed, the diagram would not represent a function, because to some y of set **Y** would correspond several values x in set **X**. If the arrows of the one-to-one function f in Figure 3-49**b** were reversed, the diagram would still represent a function.

This "backwards function" is called the **inverse** of the original function f and is denoted by the symbol f^{-1}. The -1 indicates the inverse of function f and is *not* an exponent.

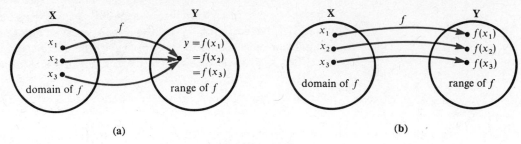

(a)   (b)

Figure 3-49

Refer to the one-to-one function f and its inverse f^{-1} shown in Figure 3-50. To the element x in the domain of f there corresponds its image $f(x)$ in the range of f. This element $f(x)$, however, is also in the domain of f^{-1}. The image of $f(x)$ under the function f^{-1} is $f^{-1}(f(x))$, which is the original number x. Thus, $(f^{-1} \circ f)(x) = f^{-1}(f(x)) = x$. Similarly, if y is an element of the domain of f^{-1}, then $(f \circ f^{-1})(y) = f(f^{-1}(y)) = y$.

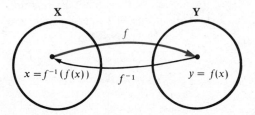

Figure 3-50

Theorem.   If f is a one-to-one function with domain $\mathbf{X}$ and range $\mathbf{Y}$, then there is a one-to-one function f^{-1} with domain $\mathbf{Y}$ and range $\mathbf{X}$ such that

$$(f^{-1} \circ f)(x) = x$$

and

$$(f \circ f^{-1})(y) = y$$

To show that one function is the inverse of another, we must show that their composition is the **identity function**—the function that assigns x itself as the image of each real number x.

Example 2   Show that the function $f(x) = x^3$ is the inverse of $g(x) = \sqrt[3]{x}$.

Solution   Show that the composition of f and g (in both directions) is the identity function.

$$(f \circ g)(x) = f(g(x)) = f(\sqrt[3]{x}) = (\sqrt[3]{x})^3 = x$$
$$(g \circ f)(x) = g(f(x)) = g(x^3) = \sqrt[3]{x^3} = x$$

■

If f is a one-to-one function defined by $y = f(x)$, then f^{-1} reverses the correspondence of f. That is, if $f(a) = b$, then $f^{-1}(b) = a$. To determine f^{-1}, we interchange the variables x and y in the equation $y = f(x)$. The resulting equation, $x = f(y)$, defines the inverse function, f^{-1}. If we can solve $x = f(y)$ for y, we will have expressed the inverse as the equation $y = f^{-1}(x)$.

Example 3   Find the inverse of the function defined by $y = \dfrac{3}{2}x + 2$, and verify the result.

Solution   To find the inverse of $y = \dfrac{3}{2}x + 2$, interchange the x and y to obtain

$$x = \frac{3}{2}y + 2$$

To express this equation in the form $y = f^{-1}(x)$, solve it for y:

$$x = \frac{3}{2}y + 2$$
$$2x = 3y + 4$$
$$2x - 4 = 3y$$
$$y = \frac{2x - 4}{3}$$

Thus, the inverse of the function defined by $y = \dfrac{3}{2}x + 2$ is $y = \dfrac{2x - 4}{3}$.

To verify that

$$f(x) = \frac{3}{2}x + 2 \quad \text{and} \quad f^{-1}(x) = \frac{2x - 4}{3}$$

are inverses of each other, show that their composition is the identity function:

$$(f \circ f^{-1})(x) = f(f^{-1}(x)) = f\left(\frac{2x - 4}{3}\right) = \frac{3}{2}\left(\frac{2x - 4}{3}\right) + 2 = x - 2 + 2 = x$$

Similarly,

$$(f^{-1} \circ f)(x) = f^{-1}(f(x)) = f^{-1}\left(\frac{3}{2}x + 2\right) = \frac{2\left(\dfrac{3}{2}x + 2\right) - 4}{3} = \frac{3x + 4 - 4}{3} = x$$

■

Because x and y are interchanged when finding the inverse, the point (b, a) lies on the graph of $y = f^{-1}(x)$ whenever the point (a, b) lies on the graph of $y = f(x)$. Thus, the graph of a function and its inverse are reflections of each other in the line $y = x$.

Example 4   Find the inverse of $y = f(x) = x^3 + 3$, and graph both the function and its inverse on a single set of coordinate axes.

Solution   The inverse of the function $y = x^3 + 3$ is found by interchanging x and y. Hence, the inverse is

$$x = y^3 + 3$$

Solve this equation for y to obtain

$$x - 3 = y^3$$
$$y = \sqrt[3]{x - 3}$$

Thus, $f^{-1}(x) = \sqrt[3]{x - 3}$.

The graphs appear in Figure 3-51. Note that $y = x$ is the axis of symmetry.

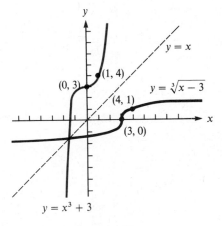

Figure 3-51

Example 5   The function defined by the equation $f(x) = x^2 + 3$ is not one-to-one. However, the function becomes one-to-one if the domain is restricted to a carefully chosen subset of the real numbers, such as the set $\{x : x \leq 0\}$. **a.** Find the range of f,   **b.** find the inverse of f along with its domain and range, and   **c.** graph each function.

Solution   **a.** The function f is defined by $f(x) = y = x^2 + 3$, with domain $\{x : x \leq 0\}$. If x is replaced with numbers from this domain, y ranges over the values 3 and above. Thus, the range of f is the set $\{y : y \geq 3\}$.

b. To find the inverse of f, interchange x and y in the equation that defines f, and solve for y:

$$y = x^2 + 3 \qquad \text{where } x \leq 0$$
$$x = y^2 + 3 \qquad \text{where } y \leq 0 \qquad \text{Interchange } x \text{ and } y.$$
$$x - 3 = y^2 \qquad \text{where } y \leq 0 \qquad \text{Add } -3 \text{ to both sides.}$$

To solve this equation for y, take the square root of both sides. Because $y \leq 0$, you have

$$-\sqrt{x - 3} = y \qquad \text{where } y \leq 0$$

Thus, the inverse of f is defined by the equation

$$y = f^{-1}(x) = -\sqrt{x - 3}$$

It has domain $\{x : x \geq 3\}$ and range $\{y : y \leq 0\}$.

c. The graphs of these two functions appear in Figure 3-52. Note that the line of symmetry is $y = x$.

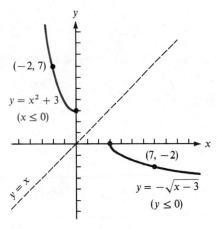

Figure 3-52

If a function is defined by the equation $y = f(x)$, we can often find the domain of f by inspection. Finding the range can be more difficult. One way to find the range of f is to find the domain of f^{-1}.

Example 6   Find the domain and the range of the function defined by

$$y = f(x) = \frac{2}{x} + 3$$

Solution   Because x cannot be 0, the domain of f is $\{x : x \text{ is a real number and } x \neq 0\}$. To find the range of f, find the domain of f^{-1}. Find f^{-1} as follows:

$$y = \frac{2}{x} + 3$$

$$x = \frac{2}{y} + 3 \qquad \text{Interchange } x \text{ and } y.$$

$$yx = 2 + 3y \qquad \text{Multiply both sides by } y.$$

$$yx - 3y = 2 \qquad \text{Add } -3y \text{ to both sides.}$$

$$y(x - 3) = 2 \qquad \text{Factor out } y.$$

$$y = \frac{2}{x - 3} \qquad \text{Divide both sides by } x - 3.$$

This final equation defines f^{-1} whose domain is $\{x : x \text{ is a real number and } x \neq 3\}$. Because the range of f is equal to the domain of f^{-1}, the range of f is

$$\{y : y \text{ is a real number and } y \neq 3\}$$

■

EXERCISE 3.8

In Exercises 1–10, determine if each function is one-to-one.

1. $y = 3x$ **2.** $y = \dfrac{1}{2}x$ **3.** $y = x^2 + 3$ **4.** $y = x^4 - x^2$

5. $y = x^3 - x$ **6.** $y = x^2 - x$ **7.** $y = |x|$ **8.** $y = |x - 3|$

9. $y = 5$ **10.** $x = \sqrt{y - 5}$ and $y \geq 5$

In Exercises 11–14, use the horizontal line test to determine if the graph represents a one-to-one function.

11.

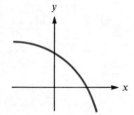

12.

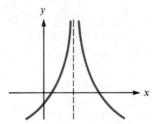

13.

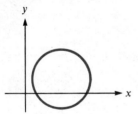

14.

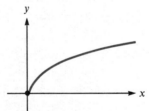

In Exercises 15–18, verify that the functions are inverses by determining that $f \circ g$ and $g \circ f$ are the identity function.

15. $f(x) = 5x;\ g(x) = \dfrac{1}{5}x$ **16.** $f(x) = 3x + 2;\ g(x) = \dfrac{x - 2}{3}$

17. $f(x) = \dfrac{x + 1}{x};\ g(x) = \dfrac{1}{x - 1}$ **18.** $f(x) = \dfrac{x + 1}{x - 1};\ g(x) = \dfrac{x + 1}{x - 1}$

In Exercises 19–26, each equation defines a one-to-one function f. Determine f^{-1} and verify that $f \circ f^{-1}$ and $f^{-1} \circ f$ are the identity function.

19. $y = 3x$ **20.** $y = \dfrac{1}{3}x$ **21.** $y = 3x + 2$ **22.** $y = 2x - 5$

23. $y = \dfrac{1}{x + 3}$ **24.** $y = \dfrac{1}{x - 2}$ **25.** $y = \dfrac{1}{2x}$ **26.** $y = \dfrac{1}{x^3}$

In Exercises 27–36, find the inverse of each one-to-one function, and graph both the given function and its inverse on one set of coordinate axes.

27. $y = 5x$

28. $y = \dfrac{3}{2}x$

29. $y = 2x - 4$

30. $y = \dfrac{3}{2}x - 2$

31. $y = \dfrac{1}{2x}$

32. $y = \dfrac{1}{x - 3}$

33. $2x + y = 4$

34. $3x + 2y = 6$

35. $x - y = 2$

36. $x + y = 0$

In Exercises 37–42, the function f defined by the given equation is one-to-one on the given domain. Find $f^{-1}(x)$.

37. $f(x) = x^2 - 3 \quad \{x : x \le 0\}$

38. $f(x) = \dfrac{1}{x^2} \qquad \{x : x > 0\}$

39. $f(x) = x^4 - 8 \quad \{x : x \ge 0\}$

40. $f(x) = \dfrac{-1}{x^4} \qquad \{x : x < 0\}$

41. $f(x) = \sqrt{4 - x^2} \quad \{x : 0 \le x \le 2\}$

42. $f(x) = \sqrt{x^2 - 1} \quad \{x : x \le -1\}$

In Exercises 43–46, find the domain and the range of f. Find the range by finding the domain of f^{-1}.

43. $f(x) = \dfrac{x}{x - 1}$

44. $f(x) = \dfrac{x - 2}{x + 3}$

45. $f(x) = \dfrac{1}{x} - 2$

46. $f(x) = \dfrac{3}{x} - \dfrac{1}{2}$

3.9 PROPORTION AND VARIATION

Quantities may be compared by using the concept of **ratio**. We might say, for example, that the *ratio* of men to women is 3 to 2, or that gasoline and oil are to be mixed in a *ratio* of 50 to 1.

> **Definition.** A **ratio** is the comparison of two numbers by their indicated quotient.

This definition implies that a ratio is a fraction. Thus, the denominator of a ratio cannot be 0. Some examples of ratios are

$$\frac{3}{5} \qquad \frac{x + 1}{9} \qquad \frac{a}{b} \qquad \frac{x^2 - 4}{x + 5}$$

> **Definition.** A **proportion** is a statement indicating that two ratios are equal.

Some examples of proportions are

$$\frac{2}{3} = \frac{4}{6} \qquad \frac{x}{y} = \frac{3}{5} \qquad \frac{x^2 + 8}{2(x + 3)} = \frac{17(x + 3)}{2}$$

In the proportion $\frac{a}{b} = \frac{c}{d}$, the numbers b and c are called the **means** of the proportion, and the numbers a and d are called the **extremes**. Because the equality $\frac{a}{b} = \frac{c}{d}$ is equivalent to $ad = bc$, we have the following theorem.

Theorem.   In any proportion, the product of the means is equal to the product of the extremes.

Example 1   Solve the proportion $\dfrac{x}{5} = \dfrac{2}{x + 3}$ for the variable x.

Solution
$$\frac{x}{5} = \frac{2}{x + 3}$$

$x(x + 3) = 5 \cdot 2$ The product of the means equals the product of the extremes.

$x^2 + 3x = 10$ Remove parentheses and simplify.

$x^2 + 3x - 10 = 0$ Add -10 to both sides.

$(x - 2)(x + 5) = 0$ Factor.

$x - 2 = 0$ or $x + 5 = 0$ Set each factor equal to 0.

$\qquad x = 2$ | $\qquad x = -5$

Thus $x = 2$ or $x = -5$.   ■

Example 2   Gasoline and oil for a lawn mower are to be mixed in a 50 to 1 ratio. How many ounces of oil should be mixed with 6 gallons of gasoline?

Solution   First, express 6 gallons as $6 \cdot 128$, or 768 ounces. Let x represent the number of ounces of oil needed. Set up and solve the proportion.

$$\frac{50}{1} = \frac{768}{x}$$

$50x = 768$ The product of the means equals the product of the extremes.

$x = \dfrac{768}{50}$ Divide both sides by 50.

$x \approx 15$ Read $\approx$ as "is approximately equal to."

Thus, approximately 15 ounces of oil should be added to 6 gallons of gasoline.   ■

Two variables are said to be **directly proportional** if their ratio is a constant. The variables x and y are directly proportional if

$$\frac{y}{x} = k \qquad \text{or} \qquad y = kx$$

where k is a constant.

Definition.   The words "*y* **varies directly** as *x*," or "*y* is **directly proportional** to *x*," mean that $y = kx$ for some real constant k. The number k is called the **constant of proportionality**.

Example 3   Distance traveled in a given time is directly proportional to the speed. If a car travels 70 miles at 30 miles per hour, how far will it travel in the same time at 45 miles per hour?

Solution   Let *d* represent distance traveled, and let *s* represent the speed. "Distance is directly proportional to the speed" translates into the formula $d = ks$. The constant of proportionality *k* can be evaluated by using the fact that $d = 70$ when $s = 30$:

$$d = ks$$

$$70 = k \cdot 30$$

$$\frac{7}{3} = k$$

Substitute $\frac{7}{3}$ for *k* in the formula $d = ks$. To evaluate the distance traveled at 45 miles per hour, let $s = 45$ in the formula $d = \frac{7}{3}s$ and simplify:

$$d = \frac{7}{3}s$$

$$= \frac{7}{3}(45)$$

$$= 105$$

In the time it took to go 70 miles at 30 miles per hour, the car could travel 105 miles at 45 miles per hour.   ∎

Example 4   The area of a circle varies directly as the square of its radius. Find the constant of proportionality.

Solution   Let *A* represent the area and *r* represent the radius. "The area varies directly as the square of the radius" translates into the formula $A = kr^2$. You know that $A = \pi r^2$ is the formula for the area of any circle. The constant of proportionality *k* is the number π.   ∎

Definition.   The words "*y* **varies inversely** with *x*," or "*y* is **inversely proportional** to *x*," mean that $y = \dfrac{k}{x}$ for some real constant k.

Example 5   The intensity of illumination from a light source varies inversely with the square of the distance from the source. If the intensity is 100 lumens at a distance of 20 feet, what will be the intensity at 30 feet?

Solution   If I is the intensity and d is the distance from the light source, then the phrase "intensity varies inversely as the square of the distance" translates into the formula

$$I = \frac{k}{d^2}$$

Evaluate k by using the fact that $I = 100$ when $d = 20$. Substitute these values into the formula $I = \frac{k}{d^2}$ and solve for k:

$$I = \frac{k}{d^2}$$

$$100 = \frac{k}{20^2}$$

$$40,000 = k$$

Use $k = 40,000$ in the formula to find the intensity at 30 feet:

$$I = \frac{40,000}{d^2}$$

$$= \frac{40,000}{30^2}$$

$$= \frac{40,000}{900}$$

$$= \frac{400}{9}$$

At 30 feet, the intensity would be $\frac{400}{9}$ lumens per square centimeter.   ■

Definition.   The words "y **varies jointly** as w and x" mean that $y = kwx$ for some real constant k.

Example 6   The energy an object has because of its motion is called *kinetic energy*. The kinetic energy of an object varies jointly as its mass and the square of its velocity. A 25-gram mass moving at a rate of 30 centimeters per second has a kinetic energy of 11,250 dyne-centimeters. What would be the kinetic energy of a 10-gram mass that is moving at 40 centimeters per second?

Solution   Let E, m, and v represent the kinetic energy, mass, and velocity, respectively. The phrase "energy varies jointly as its mass and the square of its velocity" translates into the formula

$$E = kmv^2$$

The constant k can be evaluated by using the given information: $E = 11{,}250$, $m = 25$, and $v = 30$. Substitute these values into the formula $E = kmv^2$, and solve for k:

$$E = kmv^2$$
$$11{,}250 = k \cdot 25 \cdot 30^2$$
$$11{,}250 = k \cdot 22{,}500$$
$$\frac{1}{2} = k$$

Substitute $\frac{1}{2}$ for k in the formula to evaluate E when $m = 10$ and $v = 40$:

$$E = kmv^2$$
$$= \frac{1}{2}mv^2$$
$$= \frac{1}{2} \cdot 10 \cdot 40^2$$
$$= 8000$$

A 10-gram mass that is moving at 40 centimeters per second has a kinetic energy of 8000 dyne-centimeters. ■

The preceding terminology can be used in various combinations. In each of the following statements, the wording on the left translates into the formula on the right:

y varies directly as x and inversely as z:	$y = k\dfrac{x}{z}$
y varies jointly as the square of x and the cube root of z:	$y = kx^2\sqrt[3]{z}$
y varies jointly as x and the square root of z and inversely as the cube root of t:	$t = \dfrac{kx\sqrt{z}}{\sqrt[3]{t}}$
y varies inversely as the product of x and z:	$y = \dfrac{k}{xz}$

EXERCISE 3.9

In Exercises 1–4, solve each proportion.

1. $\dfrac{4}{x} = \dfrac{2}{7}$ **2.** $\dfrac{5}{2} = \dfrac{x}{6}$ **3.** $\dfrac{x}{2} = \dfrac{3}{x+1}$ **4.** $\dfrac{x+5}{6} = \dfrac{7}{8-x}$

In Exercises 5–6, set up and solve a proportion to answer each question.

5. The ratio of girls to boys in class is 3 to 5. If there are 30 boys in the class, how many girls are there?

6. The ratio of lime to sand in mortar is 3 to 7. To make mortar, how much lime must be mixed with 21 bags of sand?

In Exercises 7–12, determine the constant of proportionality.

7. y is directly proportional to x. If $x = 30$, then $y = 15$.

8. z is directly proportional to t. If $t = 7$, then $z = 21$.

9. I is inversely proportional to R. If $R = 20$, then $I = 50$.

10. R is inversely proportional to the square of I. If $I = 25$, then $R = 100$.

11. E varies jointly as I and R. If $R = 25$ and $I = 5$, then $E = 125$.

12. z is directly proportional to the sum of x and y. If $x = 2$ and $y = 5$, then $z = 28$.

13. y is directly proportional to x. If $y = 15$ when $x = 4$, find y when $x = \frac{7}{5}$.

14. w is directly proportional to z. If $w = -6$ when $z = 2$, find w when $z = -3$.

15. P varies jointly with r and s. If $P = 16$ when $r = 5$ and $s = -8$, find P when $r = 2$ and $s = 10$.

16. m varies jointly as the square of n and the square root of q. If $m = 24$ when $n = 2$ and $q = 4$, find m when $n = 5$ and $q = 9$.

17. The volume of a gas varies directly as the temperature and inversely as the pressure. When the temperature of a certain gas is $330°$, the pressure is 40 pounds per square inch and the volume is 20 cubic feet. Find the volume if the pressure increases 10 pounds per square inch and the temperature decreases to $300°$.

18. The force required to stretch a spring a distance d is directly proportional to d. A force of 5 newtons stretches a spring 0.2 meter. What force will stretch the spring 0.35 meter?

19. The distance that an object falls in t seconds varies directly as the square of t. An object falls 16 feet in 1 second. How long will it take to fall 144 feet?

20. The power, in watts, dissipated as heat in a resistor varies jointly as the resistance, in ohms, and the square of the current, in amperes. A 10-ohm resistor carrying a current of 1 ampere dissipates 10 watts. How much power is dissipated in a 5-ohm resistor carrying a current of 3 amperes?

21. The power, in watts, dissipated as heat in a resistor varies directly as the square of the voltage and inversely as the resistance. If 20 volts are placed across a 20-ohm resistor, it will dissipate 20 watts. What voltage across a 10-ohm resistor will dissipate 40 watts?

22. The time required for one complete swing of a pendulum is called the **period** of the pendulum. The period varies directly as the square root of its length. A 1-meter pendulum has a period of 1 second. What will be the length of a pendulum with a period of 2 seconds?

23. The pitch, or frequency, of a vibrating string varies directly as the square root of the tension. If a string vibrates at a frequency of 144 hertz due to a tension of 2 pounds, what would be its frequency if the tension was 18 pounds?

24. The gravitational attraction between two massive objects varies jointly as their masses and inversely as the square of the distance between them. What happens to this force if each mass is tripled, and the distance between them is doubled?

25. The area of an equilateral triangle varies directly as the square of a side. What is the constant of proportionality?

26. The diagonal of a cube varies directly as a side. What is the constant of proportionality?

CHAPTER SUMMARY

Key Words

asymptote (3.6)

axis of symmetry (3.5)

linear function (3.4)

means of a proportion (3.9)

composition of functions (3.7)

constant function (3.5)

degree of a polynomial function (3.5)

dependent variable (3.4)

direct variation (3.9)

domain of a function (3.4)

extremes of a proportion (3.9)

function (3.4)

horizontal line test (3.8)

hyperbola (3.6)

identity function (3.8)

image (3.4)

independent variable (3.4)

inverse function (3.8)

inverse variation (3.9)

joint variation (3.9)

one-to-one function (3.8)

parabola (3.5)

polynomial function (3.5)

proportion (3.9)

quadratic function (3.5)

range of a function (3.4)

ratio (3.9)

rational function (3.6)

relation (3.4)

slope of a nonvertical line (3.2)

symmetry (3.5)

vertex (3.5)

vertical line test (3.4)

x-intercept (3.1)

y-intercept (3.1)

Key Ideas

(3.1)   **The distance formula:** $d = \sqrt{(x_2 - x_1)^2 + (y_2 - y_1)^2}$

The midpoint formula: The midpoint of the line segment joining (x_1, y_1) and (x_2, y_2) is the point M with coordinates

$$\left(\frac{x_1 + x_2}{2}, \frac{y_1 + y_2}{2}\right)$$

(3.2)   The slope, m, of a line passing through (x_1, y_1) and (x_2, y_2) is given by

$$m = \frac{y_2 - y_1}{x_2 - x_1} \qquad (x_2 \neq x_1)$$

Nonvertical parallel lines have the same slope.

Perpendicular lines have slopes that are negative reciprocals of each other, provided neither line is vertical.

(3.3)   **Point-slope form of the equation of a line:** $y - y_1 = m(x - x_1)$

Slope-intercept form of the equation of a line: $y = mx + b$

General form of the equation of a line: $Ax + By = C$

(3.4)   A **function** is a correspondence that assigns to each element x in the domain a single value y in the range.

A **relation** is a correspondence that assigns to each element x in the domain one or more values y in the range.

The **graph of a function** f is the graph of all points $(x, f(x))$, where x is in the domain of the function.

To decide if a graph represents a function, use the vertical line test.

(3.5)   The graph of a quadratic function given by $y = ax^2 + bx + c$ with $a \neq 0$ is a parabola with vertex at $\left(-\dfrac{b}{2a}, c - \dfrac{b^2}{4a}\right)$.

Tests for Symmetry:

If $(-x, y)$ lies on a graph whenever (x, y) does, the graph is symmetric about the y-axis.

If $(x, -y)$ lies on a graph whenever (x, y) does, the graph is symmetric about the x-axis.

If $(-x, -y)$ lies on a graph whenever (x, y) does, the graph is symmetric about the origin.

If $y = f(x)$ and $y = f(-x)$ are equivalent, the graph of the function f is symmetric about the y-axis.

If $y = f(x)$ and $-y = f(-x)$ are equivalent, the graph of the function f is symmetric about the origin.

Vertical translations: If $k > 0$, the graph of $\begin{cases} y = f(x) + k \\ y = f(x) - k \end{cases}$ is identical to the graph of

$y = f(x)$, except that it is translated k units $\begin{cases} \text{upward} \\ \text{downward.} \end{cases}$

Horizontal translations: If $h > 0$, the graph of $\begin{cases} y = f(x - h) \\ y = f(x + h) \end{cases}$ is identical to the graph

of $y = f(x)$, except that it is a translated h units to the $\begin{cases} \text{right} \\ \text{left.} \end{cases}$

(3.6)   To graph the rational function $\dfrac{P(x)}{Q(x)}$:

1. Check symmetries.   4. Look for vertical asymptotes.
2. Look for the y-intercept.   5. Look for a horizontal asymptote.
3. Look for the x-intercepts.   6. Look for a slant asymptote.

(3.7)   If the ranges of functions f and g are subsets of the real numbers, then

$$(f + g)(x) = f(x) + g(x) \qquad (f \cdot g)(x) = f(x) \cdot g(x)$$
$$(f - g)(x) = f(x) - g(x) \qquad (f/g)(x) = f(x)/g(x) \quad \text{provided } g(x) \neq 0$$

The domain of these functions (unless otherwise restricted) is the set of all real numbers x that are in the domain of *both* f and g, and, in the case of the quotient f/g, there is the further restriction that $g(x) \neq 0$.

Composition of functions:

$$(f \circ g)(x) = f(g(x))$$

The domain of $f \circ g$ is the set of all x for which $g(x)$ is in the domain of f.

(3.8)   The composition of a one-to-one function f and its inverse f^{-1} is the identity function:

$$(f \circ f^{-1})(x) = (f^{-1} \circ f)(x) = x$$

The graph of a function is symmetric to the graph of its inverse. The axis of symmetry is the line $y = x$.

To find the inverse of f, exchange x and y in the equation $y = f(x)$ and solve for y.

(3.9)   **Types of variation:**

1. y varies directly as x means that $y = kx$.

2. y varies inversely as x means that $y = \dfrac{k}{x}$.

3. y varies jointly with x and z means that $y = kxz$.

REVIEW EXERCISES

In Review Exercises 1–4, find **a.** *the length* *and* **b.** *the midpoint of the line segment PQ.*

1. $P(-3, 7); Q(3, -1)$

2. $P(0, 5); Q(-12, 10)$

3. $P(-\sqrt{3}, 9); Q(-\sqrt{3}, -7)$

4. $P(a, -a); Q(-a, a)$ $(a > 0)$

In Review Exercises 5–8, find the slope of the line PQ, if possible.

5. $P(3, -5); Q(1, 7)$

6. $P(2, 7); Q(-5, -7)$

7. $P(b, a); Q(a, b)$

8. $P(a + b, b); Q(b, b - a)$

In Review Exercises 9–18, write the equation of the line with the given properties. Express the answer in general form.

9. The line passes through the origin and the point $(-5, 7)$.

10. The line passes through $(-2, 1)$ and has a slope of -4.

11. The line passes through $(7, -5)$ and $(4, 1)$.

12. The line has a slope of $\frac{2}{3}$ and a y-intercept of 3.

13. The line has a slope of 0 and passes through $(-5, 17)$.

14. The line has no defined slope and passes through $(-5, 17)$.

15. The line passes through $(8, -2)$ and is parallel to the line segment joining $(2, 4)$ and $(4, -10)$.

16. The line passes through $(8, -2)$ and is perpendicular to the line segment joining $(2, 4)$ and $(4, -10)$.

17. The line is parallel to $3x - 4y = 7$ and passes through $(2, 0)$.

18. The line passes through $(7, 0)$ and is perpendicular to the line $3y + x - 4 = 0$.

In Review Exercises 19–24, determine if the given equation determines y as a function of x. If it does, give the domain and range, and graph the function.

19. $y = 5x - 2$

20. $y = |x - 2|$

21. $y = \sqrt{x - 1}$

22. $y^2 = 2x + 7$

23. $y = x^2 - x + 1$

24. $y = \dfrac{x}{|x|}$

In Review Exercises 25–32, graph each polynomial function. If the graph is a parabola, find the coordinates of its vertex.

25. $y - 3 = (x + 2)^2$

26. $y - 1 = (x + 3)^2$

27. $y = x^2 - 3x - 2$

28. $x^2 + y - 2x + 4 = 0$

29. $y = x^4 + 2x^2 + 1$

30. $y = x^3 - x$

31. $y = x^3 + 2$

32. $y = (x + 2)^3$

In Review Exercises 33–38, graph each rational function.

33. $y = \dfrac{2}{x - 4}$

34. $y = \dfrac{x^2 - 9}{x^2 - 4}$

35. $y = \dfrac{4x}{x - 3}$

36. $y = \dfrac{x^2 + 5x + 6}{x - 1}$

37. $y = \dfrac{x - 2}{x^2 - 3x - 4}$

38. $y = \dfrac{x^2 + 4}{x^2 - 3x - 4}$

In Review Exercises 39–46, let $f(x) = 2x^2 - x$ and $g(x) = x + 3$. Find each value.

39. $f(2)$

40. $g(-3)$

41. $(f + g)(2)$

42. $(f - g)(-1)$

43. $(f \cdot g)(0)$

44. $(f/g)(2)$

45. $(f \circ g)(-2)$

46. $(g \circ f)(4)$

In Review Exercises 47–52, let $f(x) = \sqrt{x + 1}$ and $g(x) = x^2 - 1$.

47. Find $(f + g)(x)$.   **48.** Find $(f \cdot g)(x)$.   **49.** Find $(f \circ g)(x)$.   **50.** Find $(g \circ f)(x)$.

51. Determine the domain of f/g.   **52.** Determine the domain of $f \circ g$.

In Review Exercises 53–58, determine if the given function is one-to-one. If so, determine its inverse function.

53. $y = 7x$ **54.** $y = 7x^2$ **55.** $y = \dfrac{1}{x - 1}$ **56.** $y = \dfrac{3}{x^3}$

57. $y = \dfrac{x + 2}{x - 3}$ **58.** $y = |x| + 3$

In Review Exercises 59–60, find the domain and the range of each function. Find the range by finding the domain of the function's inverse.

59. $y = f(x) = \dfrac{2x + 1}{2x - 1}$ **60.** $y = f(x) = \dfrac{2}{x} + \dfrac{1}{2}$

61. Hooke's law states that the force required to stretch a spring is proportional to the amount of stretch. If a 3-pound force stretches a spring 5 inches, what force would stretch the spring 3 inches?

62. The volume of gas in a balloon varies directly as its temperature and inversely as the pressure. If the volume is 400 cubic centimeters when the temperature is 300°K and the pressure is 25 dynes per square centimeter, find the volume when the temperature is 200°K and the pressure is 20 dynes per square centimeter.

63. A moving body has a kinetic energy proportional to the square of its velocity. By what factor does the kinetic energy of an automobile increase if its speed increases from 30 miles per hour to 50 miles per hour?

64. The electrical resistance of a wire varies directly as the length of the wire and inversely as the square of its diameter. A 1000-foot length of wire, 0.05 inch in diameter, has a resistance of 200 ohms. What would be the resistance of a 1500-foot length of wire that is 0.08 inch in diameter?

4

Inequalities and Absolute Value

The topic of this chapter is inequalities—mathematical statements that indicate that two quantities are *not* equal. The inequality relations $\neq$, $<$, $\leq$, $>$, and $\geq$ were introduced in Chapter 1. We now examine their properties more closely.

4.1 LINEAR INEQUALITIES

In this section we will develop techniques for solving inequalities. We begin by defining some terms.

> **Definition.** If a and b are real numbers, then a is greater than b (written $a > b$) if and only if $a - b$ is positive.
>   Similarly, a is less than b (written $a < b$) if and only if $b - a$ is positive.

Note that if $a < b$, then $b > a$. If a is greater than or equal to b, then we write $a \geq b$. Similarly, if a is less than or equal to b, then we write $a \leq b$.

Example 1   **a.** $5 > 3$ because $5 - 3$ is a positive real number, which is 2.

b. $2 < 7$ because $7 - 2$ is a positive real number, which is 5.

c. $3 \leq 3$ because 3 is equal to 3.

d. $2 \leq 3$ because 2 is less than 3.

e. $x > x - 1$ because $x - (x - 1) = x - x + 1 = 1$, and 1 is positive.   ■

We now state without proof several theorems involving $<$. Similar theorems exist for the other relations.

The first theorem states that any real number can be added to (or subtracted from) both sides of an inequality to obtain another inequality with the same order (direction).

Theorem.   Let a, b, and c represent real numbers.

If $a < b$,   then $a + c < b + c$.

If $a < b$,   then $a - c < b - c$.

The second theorem states that both sides of an inequality can be multiplied (or divided) by a positive number to obtain another inequality with the same order.

Theorem.   Let a, b, and c represent real numbers.

If $a < b$ and $c > 0$,   then $ca < cb$.

If $a < b$ and $c > 0$,   then $\dfrac{a}{c} < \dfrac{b}{c}$.

The third theorem states that both sides of an inequality can be multiplied (or divided) by a negative number to obtain another inequality in the *opposite* order.

Theorem.   Let a, b, and c represent real numbers.

If $a < b$ and $c < 0$,   then $ca > cb$.

If $a < b$ and $c < 0$,   then $\dfrac{a}{c} > \dfrac{b}{c}$.

The previous theorems indicate that inequalities are solved like equations. However, *we must always remember to change the order of an inequality when multiplying or dividing both sides of the inequality by a negative number.*

The $<$ relation is neither reflexive (a number cannot be less than itself) nor symmetric (if a is less than b, then b cannot be less than a). However, the relation *is* transitive.

Theorem.   If a, b, and c are real numbers with $a < b$ and $b < c$, then $a < c$.

Similarly, the relations $>$, $\leq$, and $\geq$ are transitive.

Inequalities such as $ax + c < 0$ or $ax + c \geq 0$, where $a \neq 0$, are called **linear inequalities**. Numbers that make an inequality true when substituted for the variable are called **solutions** of the inequality. Two inequalities with exactly the same solutions are called **equivalent inequalities**.

Example 2 Solve the inequality $3(x + 2) < 8$.

Solution Proceed as with equations:

$$3(x + 2) < 8$$
$$3x + 6 < 8 \qquad \text{Remove parentheses.}$$
$$3x < 2 \qquad \text{Add } -6 \text{ to both sides.}$$
$$x < \frac{2}{3} \qquad \text{Divide both sides by 3.}$$

Any value of x that is less than $\frac{2}{3}$ is in the solution set of the given inequality. This solution set can be expressed in interval notation as $(-\infty, \frac{2}{3})$. The symbol $-\infty$, read as "negative infinity," indicates the interval is unbounded on the left. The graph of the solution set is shown in Figure 4-1.

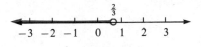

Figure 4-1   ▪

Example 3 Solve the inequality $-5(x - 2) < 20 + x$.

Solution Proceed as with equations, removing the parentheses and solving for x:

$$-5(x - 2) < 20 + x$$
$$-5x + 10 < 20 + x$$
$$-6x + 10 < 20$$
$$-6x < 10$$

Now divide both sides of the inequality by -6, which changes the order (direction) of the inequality:

$$x > -\frac{10}{6}$$

or

$$x > -\frac{5}{3}$$

The graph of the solution set of the inequality $-5(x - 2) < 20 + x$ is shown in Figure 4-2. The interval represented by the graph can be expressed in interval notation as $(-\frac{5}{3}, \infty)$.

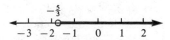

Figure 4-2   ▪

The statement that x is between two numbers, such as 2 and 5, implies two inequalities: $x > 2$ and $x < 5$. It is customary to write both inequalities as a single expression, $2 < x < 5$, which means that "x is greater than 2, and x is less than 5." The word *and* indicates that the two inequalities hold true simultaneously.

To express that x is *not* between 2 and 5, we must convey the idea that "either x is greater than or equal to 5, or x is less than or equal to 2." Note the either/or in this statement. This either/or statement is written symbolically as "$x \leq 2$ or $x \geq 5$." It is incorrect to string the two together as $2 \geq x \geq 5$, because that would mean that $2 \geq 5$, which is false.

Example 4 Solve the inequality $5 < 3x - 7 \leq 8$.

Solution There are two inequalities involved in this expression. Because a constant can be added to both sides of each inequality, add 7 to each part of the *double inequality* to get

$$5 + 7 < 3x - 7 + 7 \leq 8 + 7$$

or

$$12 < 3x \leq 15$$

Then, divide each part by 3 to obtain

$$4 < x \leq 5$$

This solution is represented graphically on the number line in Figure 4-3. The interval represented by the graph can be expressed in interval notation as (4, 5].

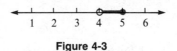

Figure 4-3 ∎

Example 5 Solve the inequality $3 + x \leq 3x + 1 < 7x - 2$.

Solution Because it is impossible to isolate the variable x between the inequality signs, you must solve each of the two indicated inequalities separately.

$$
\begin{array}{ccc}
3 + x \leq 3x + 1 & \quad \text{and} \quad & 3x + 1 < 7x - 2 \\
3 \leq 2x + 1 & & 1 < 4x - 2 \\
2 \leq 2x & & 3 < 4x \\
1 \leq x & & \dfrac{3}{4} < x
\end{array}
$$

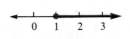

Figure 4-4

You are interested only in those values of x that are greater than or equal to 1 *and also* greater than $\frac{3}{4}$. Because any values greater than or equal to 1 will also be greater than $\frac{3}{4}$, the solution is $x \geq 1$, which can be represented graphically as shown in Figure 4-4. The interval represented by the graph can be expressed in interval notation as $[1, \infty)$. ∎

It is possible for an inequality to be true for all values of its variable. For example, $x < x + 1$ is true for all values of x. It is also possible for an inequality to have no solutions. The inequality $x > x + 1$ is true for no values of x.

EXERCISE 4.1

In Exercises 1–20, solve each inequality, graph the solution set, and write the answer in interval notation.

1. $3x + 2 < 5$ **2.** $-2x + 4 < 6$ **3.** $3x + 2 \geq 5$ **4.** $-2x + 4 \geq 6$

5. $-5x + 3 > -2$ **6.** $4x - 3 > -4$ **7.** $-5x + 3 \leq -2$ **8.** $4x - 3 \leq -4$

9. $2(x - 3) \leq -2(x - 3)$ **10.** $3(x + 2) \leq 2(x + 5)$

11. $\frac{3}{5}x + 4 > 2$ **12.** $\frac{1}{4}x - 3 > 5$

13. $\frac{x + 3}{4} < \frac{2x - 4}{3}$ **14.** $\frac{x + 2}{5} > \frac{x - 1}{2}$

15. $\frac{6(x - 4)}{5} \geq \frac{3(x + 2)}{4}$ **16.** $\frac{3(x + 3)}{2} < \frac{2(x + 7)}{3}$

17. $\frac{5}{9}(a + 3) - a \geq \frac{4}{3}(a - 3) - 1$ **18.** $\frac{2}{3}y - y \leq -\frac{3}{2}(y - 5)$

19. $\frac{2}{3}a - \frac{3}{4}a < \frac{3}{5}(a + \frac{2}{3}) + \frac{1}{3}$ **20.** $\frac{1}{4}b + \frac{2}{3}b - \frac{1}{2} > \frac{1}{2}(b + 1) + b$

In Exercises 21–40, solve each double inequality, graph the solution set, and write the answer in interval notation.

21. $4 < 2x - 8 \leq 10$ **22.** $3 \leq 2x + 2 < 6$ **23.** $9 \geq \frac{x - 4}{2} > 2$ **24.** $5 < \frac{x - 2}{6} < 20$

25. $0 \leq \frac{4 - x}{3} \leq 5$ **26.** $0 \geq \frac{5 - x}{2} \geq -10$

27. $-2 \geq \frac{1 - x}{2} \geq -10$ **28.** $-2 \leq \frac{1 - x}{2} < 10$

29. $-3x > -2x > -x$ **30.** $-3x < -2x < -x$ **31.** $x < 2x < 3x$ **32.** $x > 2x > 3x$

33. $2x + 1 < 3x - 2 < 12$ **34.** $2 - x < 3x + 5 < 18$

35. $2 + x < 3x - 2 < 5x + 2$ **36.** $x > 2x + 3 > 4x - 7$

37. $3 + x > 7x - 2 > 5x - 10$ **38.** $2 - x < 3x + 1 < 10x$

39. $x \leq x + 1 \leq 2x + 3$ **40.** $-x \geq -2x + 1 \geq -3x + 1$

In Exercises 41–44, find which values of x make each expression positive.

41. $2(x - 1)$ **42.** $-3(x - 1)$ **43.** $-4(x + 4)$ **44.** $5(x + 4)$

In Exercises 45–48, find which values of x make each expression negative.

45. $-2(x + 4)$ **46.** $3(x - 1)$ **47.** $5(x - 3)$ **48.** $-4(x + 1)$

In Exercises 49–56, solve the word problem.

49. An airline pilot flies at altitudes from 27,720 feet to 33,000 feet. Express this interval in miles. (*Hint*: 1 mile = 5280 feet.)

50. The perimeter of a rectangle is to be between 180 inches and 200 inches. What is the range of values for its length if its width is to be 40 inches?

51. The perimeter of an equilateral triangle is to be between 50 centimeters and 60 centimeters. What is the range of values for the length of one side?

52. The perimeter of a square is to be from 25 meters to 60 meters. What is the range of values for its area?

53. Express the relationship $20 < l < 30$ in terms of P, where $P = 2l + 2w$.

54. Express the relationship $10 < C < 20$ in terms of F, where $F = \frac{9}{5}C + 32$.

55. Andy can spend up to $275 on a compact disc player and some compact discs. If he can buy a disc player for $150 and discs for $9.75 each, what is the greatest number of discs that he can buy?

56. Mary wants to spend less than $600 for a VCR and some videotapes. If the VCR of her choice costs $425 and video tapes cost $7.50 each, how many videotapes can she buy?

In Exercises 57–60, find the domain of the variable x.

57. $y = \sqrt{3x - 4}$ 58. $y = \sqrt{6 - 4x}$ 59. $y = \dfrac{1}{\sqrt{3x + 2}}$ 60. $y = \dfrac{5}{\sqrt{5 - 7x}}$

4.2 ABSOLUTE VALUE

We now review the definition of the absolute value of a number and examine its consequences in greater detail.

Definition.   The absolute value of the real number x, denoted as $|x|$, is defined as

If $x \geq 0$, then $|x| = x$.

If $x < 0$, then $|x| = -x$.

Whether x is positive, 0, or negative, the previous definition enables us to find the absolute value of x. If x is positive or 0, the absolute value of x is x. If x is negative, the absolute value of x is the positive number $-x$. Thus, for all real numbers,

$$|x| \geq 0$$

Example 1   Find   **a.** $|0|$,   **b.** $|7|$,   **c.** $|-3|$,   **d.** $-|-7|$,   and **e.** $|x - 2|$.

Solution   **a.** By the first part of the definition, $|0| = 0$.

b. Because 7 is positive, by the first part of the definition, $|7| = 7$.

c. Because -3 is negative, by the second part of the definition,

$$|-3| = -(-3) = 3$$

d. The expression $-|-7|$ means "the negative of the absolute value of -7." Thus,

$$-|-7| = -7$$

e. To denote the absolute value of a variable quantity, you must give a conditional answer. Either

$$|x - 2| = x - 2 \quad \text{when } x - 2 \geq 0 \quad \text{(or when } x \geq 2\text{)}$$

or

$$|x - 2| = -(x - 2) = 2 - x \quad \text{when } x - 2 < 0 \quad \text{(or when } x < 2\text{)} \quad \blacksquare$$

To develop a strategy for solving equations that contain absolute symbols, we consider the equation $|x| = a$ where a is positive. If x is positive or 0, then $|x| = x$ and

$$x = a$$

On the other hand, if x is negative, then $|x| = -x$ and $-x = a$ or

$$x = -a$$

Thus, we have the following theorem.

Theorem.   The equation $|x| = a$ (for $a \geq 0$) is equivalent to the statement

$$x = a \qquad \text{or} \qquad x = -a$$

The absolute value of x can be interpreted as the distance on a number line that a point is from the origin. For example, the solutions of the equation $|x| = a$ are represented by the two points that lie exactly a units from the origin. See Figure 4-5a shown on page 202.

To solve an inequality that contains absolute values and an "is less than" symbol, we consider the inequality $|x| < a$. If x is positive or 0, then $|x| = x$ and

$$x < a$$

If x is negative, then $|x| = -x$ and $-x < a$ or

$$x > -a$$

Thus, if x is positive or 0, then x is less than a, and if x is negative, then x is greater than $-a$. This implies that x must be between $-a$ and a. The two inequalities $x < a$ and $x > -a$ are equivalent to the double inequality $-a < x < a$. This proves the following theorem.

Theorem.   The inequality $|x| < a$ (for $a > 0$) is equivalent to the double inequality

$$-a < x < a$$

The symbol $<$ can be replaced by $\leq$.

The solutions to the inequality $|x| < a$ are the coordinates of the points on a number line that are less than a units from the origin. See Figure 4-5**b**.

To solve an inequality that contains absolute values and an "is greater than" symbol, we consider the inequality $|x| > a$ where a is positive. If x is positive or 0, then $|x| = x$, and

$$x > a$$

If x is negative, then $|x| = -x$ and $-x > a$ or

$$x < -a$$

Thus, if x is positive or 0, then x is greater than a, and if x is negative, then x is less than $-a$. This implies that we can write the solutions of the inequality $|x| > a$ as

$$x < -a \qquad \text{or} \qquad x > a$$

This proves the following theorem.

Theorem.   The inequality $|x| > a$ (for $a > 0$) is equivalent to the statement

$$x < -a \qquad \text{or} \qquad x > a$$

Likewise, $|x| \geq a$ $(a > 0)$ is equivalent to $x \leq -a$ or $x \geq a$.

The solutions of the inequality $|x| > a$ are the coordinates of the points on a number line that are more than a units from the origin. See Figure 4-5**c**.

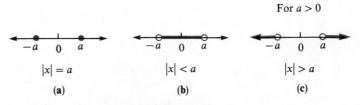

Figure 4-5

Example 2   Solve the equation $|3x - 5| = 7$.

Solution   The equation $|3x - 5| = 7$ is equivalent to the two equations

$$3x - 5 = 7 \quad \text{or} \quad 3x - 5 = -7$$

which can be solved separately as follows:

$$3x - 5 = 7 \qquad \text{or} \qquad 3x - 5 = -7$$
$$3x = 12 \qquad\qquad\qquad 3x = -2$$
$$x = 4 \qquad\qquad\qquad x = -\frac{2}{3}$$

Figure 4-6   The graph of the solution set consists of the two points shown in Figure 4-6.   ■

Example 3   Solve the equation $|2x| = |x - 3|$.

Solution   The equation $|2x| = |x - 3|$ will be satisfied if the quantities $2x$ and $x - 3$ are either equal to each other or negatives of each other. This idea determines two equations, which can be solved separately:

$$2x = x - 3 \qquad \text{or} \qquad 2x = -(x - 3)$$
$$x = -3 \qquad\qquad\qquad 2x = -x + 3$$
$$3x = 3$$
$$x = 1$$

Both -3 and 1 satisfy the given equation.   ■

Example 4   Find the solution set of the inequality $|x - 2| < 7$, and graph it on a number line.

Solution   The inequality $|x - 2| < 7$ is equivalent to the double inequality

$$-7 < x - 2 < 7$$

Add 2 to each part of the double inequality to obtain

$$-5 < x < 9$$

Figure 4-7   The graph of this solution is shown in Figure 4-7. The solution set can be expressed in interval notation as $(-5, 9)$.   ■

Example 5   Solve the inequality $\left|\dfrac{2x + 3}{2}\right| + 7 \geq 12$, and graph its solution set.

Solution   Begin by adding -7 to both sides of the given inequality to obtain

$$\left|\frac{2x + 3}{2}\right| \geq 5$$

This result is equivalent to the following two inequalities, which can be solved separately.

$$\frac{2x + 3}{2} \geq 5 \qquad \text{or} \qquad \frac{2x + 3}{2} \leq -5$$

$$2x + 3 \geq 10 \qquad \qquad 2x + 3 \leq -10$$

$$2x \geq 7 \qquad \qquad \qquad 2x \leq -13$$

$$x \geq \frac{7}{2} \qquad \qquad \qquad x \leq -\frac{13}{2}$$

Figure 4-8

The graph of the solution set is shown in Figure 4-8. The solution set can be expressed as the union of two intervals: $(-\infty, -\frac{13}{2}] \cup [\frac{7}{2}, \infty)$.   ∎

Example 6   Solve the inequality $0 < |x - 5| \leq 3$, and graph its solution set.

Solution   The double inequality $0 < |x - 5| \leq 3$ consists of two inequalities, which may be solved separately. The first inequality, $0 < |x - 5|$, is true for all values of x *except* 5. The second inequality, $|x - 5| \leq 3$, is equivalent to the double inequality $-3 \leq x - 5 \leq 3$, or $2 \leq x \leq 8$. Hence, the solution set consists of all numbers x such that $x \neq 5$ and $2 \leq x \leq 8$. Note that the point $x = 5$ is missing in the graph of the solution set. See Figure 4-9.

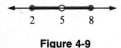

Figure 4-9

In interval notation, the solution set can be expressed as $[2, 5) \cup (5, 8]$.   ∎

Example 7   Express the double inequality $-3 \leq x \leq 7$ as an absolute value inequality.

Solution   Halfway between -3 and 7 is their average.

$$\frac{-3 + 7}{2} = 2$$

Subtracting 2 from each part of the inequality gives

$$-3 \leq x \leq 7$$
$$-3 - 2 \leq x - 2 \leq 7 - 2$$
$$-5 \leq x - 2 \leq 5$$

This statement is equivalent to the absolute value inequality

$$|x - 2| \leq 5$$ ∎

Recall that if a is a positive number or zero, then

$$\sqrt{a^2} = a$$

However, if a is unrestricted, we must use absolute value symbols to guarantee that $\sqrt{a^2}$ is nonnegative. In such a case we write

$$\sqrt{a^2} = |a| \qquad \text{or} \qquad |a| = \sqrt{a^2}$$

This fact can be used to solve certain inequalities.

Example 8   Solve the inequality $|x + 2| > |x + 1|$, and write the solution set using interval notation.

Solution

$$|x + 2| > |x + 1|$$
$$\sqrt{(x + 2)^2} > \sqrt{(x + 1)^2} \qquad \text{Use } |a| = \sqrt{a^2}.$$
$$(x + 2)^2 > (x + 1)^2 \qquad \text{Square both sides.}$$
$$x^2 + 4x + 4 > x^2 + 2x + 1$$
$$2x > -3$$
$$x > -\frac{3}{2}$$

In interval notation, the solution set is $(-\frac{3}{2}, \infty)$. Check several solutions to verify that this interval does not contain any extraneous solutions. ∎

There are three other properties of absolute values that are often useful.

Properties of Absolute Value.   For all real numbers a and b,

1. $|ab| = |a|\,|b|$

2. $\left|\dfrac{a}{b}\right| = \dfrac{|a|}{|b|}$ $(b \neq 0)$

3. $|a + b| \leq |a| + |b|$ (the triangle inequality)

Amalie Noether (1882–1935) Emmy Noether worked in abstract algebra. Einstein called her the most creative woman mathematical genius since higher education of women began.

Property 1 states that the absolute value of a product is the product of the absolute values. Property 2 states that the absolute value of a quotient is the quotient of the absolute values. Property 3 states that the absolute value of a sum is either equal to or less than the sum of the absolute values.

EXERCISE 4.2

In Exercises 1–12, write each expression without using absolute value symbols.

1. $|7|$

2. $|-9|$

3. $|0|$

4. $|3 - 5|$

5. $|5| - |-3|$

6. $|-3| + |5|$

7. $|\pi - 2|$

8. $|\pi - 4|$

9. $|x|$

10. $|x^2|$

11. $|x - 3|$ and $x > 3$

12. $|x - 3|$ and $x < 3$

In Exercises 13–36, solve each equation for x, if possible.

13. $|x + 2| = 2$

14. $|2x + 5| = 3$

15. $|3x - 1| = 5$

16. $|7x - 5| = 3$

17. $\left|\dfrac{3x - 4}{2}\right| = 5$

18. $\left|\dfrac{10x + 1}{2}\right| = \dfrac{9}{2}$

19. $\left|\dfrac{2x - 4}{5}\right| = 2$

20. $\left|\dfrac{3x + 11}{7}\right| = 1$

21. $\left|\dfrac{x - 3}{4}\right| = -2$

22. $\left|\dfrac{x - 5}{3}\right| = 0$

23. $\left|\dfrac{4x - 2}{x}\right| - 3 = 0$

24. $\left|\dfrac{2(x - 3)}{3x}\right| - 8 = -2$

25. $|x| = x$ **26.** $|x| + x = 2$ **27.** $|x + 3| = |x|$ **28.** $|x + 5| = |5 - x|$

29. $|x - 3| = |2x + 3|$ **30.** $|x - 2| = |3x + 8|$

31. $|x + 2| = |x - 2|$ **32.** $|2x - 3| = |3x - 5|$

33. $\left|\dfrac{x + 3}{2}\right| = |2x - 3|$ **34.** $\left|\dfrac{x - 2}{3}\right| = |6 - x|$

35. $\left|\dfrac{3x - 1}{2}\right| = \left|\dfrac{2x + 3}{3}\right|$ **36.** $\left|\dfrac{5x + 2}{3}\right| = \left|\dfrac{x - 1}{4}\right|$

In Exercises 37–52, solve each inequality, express the solution set in interval notation, and graph it.

37. $|x - 3| < 6$ **38.** $|x - 2| \geq 4$ **39.** $|x + 3| > 6$ **40.** $|x + 2| \leq 4$

41. $|2x + 4| \geq 10$ **42.** $|5x - 2| < 7$ **43.** $|3x + 5| + 1 \leq 9$ **44.** $|2x - 7| - 3 > 2$

45. $2|x + 3| > 0$ **46.** $5|x - 3| \leq 0$ **47.** $\left|\dfrac{5x + 2}{3}\right| < 1$ **48.** $\left|\dfrac{3x + 2}{4}\right| > 2$

49. $3\left|\dfrac{3x - 1}{2}\right| > 5$ **50.** $2\left|\dfrac{8x + 2}{5}\right| \leq 1$ **51.** $\dfrac{|x - 1|}{-2} > -3$ **52.** $\dfrac{|2x - 3|}{-3} < -1$

In Exercises 53–62, solve each double inequality, express the solution set in interval notation, and graph it.

53. $0 < |2x + 1| < 3$ **54.** $0 < |2x - 3| < 1$ **55.** $8 > |3x - 1| > 3$ **56.** $8 \geq |4x - 1| > 5$

57. $2 \leq \left|\dfrac{x - 5}{3}\right| < 4$ **58.** $3 < \left|\dfrac{x - 3}{2}\right| < 5$ **59.** $10 > \left|\dfrac{x - 2}{2}\right| > 4$ **60.** $5 \geq \left|\dfrac{x + 2}{3}\right| > 1$

61. $3 \leq \left|\dfrac{x + 1}{3}\right| + 1 < 4$ **62.** $7 > \left|\dfrac{3x + 1}{2}\right| - 1 \geq 1$

In Exercises 63–70, express each inequality using absolute value symbols.

63. $3 \leq x \leq 7$ **64.** $-3 < x < 9$ **65.** $\dfrac{9}{2} \geq x \geq -\dfrac{1}{2}$ **66.** $\dfrac{11}{2} > x > \dfrac{3}{2}$

67. $x \neq 2$ and $0 < x < 4$ **68.** $x \neq 3$ and $-1 < x < 7$

69. $x \neq 0$ and $-5 < x < 5$ **70.** $x \neq 3$ and $-6 < x < 0$

In Exercises 71–78, solve each inequality and express the solution set using interval notation.

71. $|x + 1| \geq |x|$ **72.** $|x + 1| < |x + 2|$

73. $|2x + 1| < |2x - 1|$ **74.** $|3x - 2| \geq |3x + 1|$

75. $|x + 1| < |x|$ **76.** $|x + 1| \geq |x + 2|$

77. $|2x + 1| \geq |2x - 1|$ **78.** $|3x - 2| < |3x + 1|$

Computer Exercises. *In Exercises 79–80, use a computer and **Algebra Pak** to solve each inequality.*

79. Solve the inequality $|x - 3| < 2$ by first writing it as $|x - 3| - 2 < 0$. Then graph the function $y = |x - 3| - 2$, and find those numbers x for which the values of y are negative: that is, find those numbers x for which the graph of the function lies below the x-axis. Enter the function as ABS $(X - 3) - 2$.

80. Use the method of Exercise 79 to solve the inequality $|3x - 2| > 5$.

4.3   INEQUALITIES IN TWO VARIABLES

It is possible to use graphing techniques to find solution sets of **inequalities in two variables.**

Example 1   Graph the inequality $3x + 2y < 6$.

Solution   Because of the trichotomy property of real numbers, one and only one of the following expressions is true for each ordered pair (x, y):

$$3x + 2y < 6 \quad \text{or} \quad 3x + 2y = 6 \quad \text{or} \quad 3x + 2y > 6$$

The graph of the function defined by the equation $3x + 2y = 6$ is a straight line. The graph of each of the two inequalities is a half-plane, one on each side of that line. For this reason, think of the graph of the equation as a boundary separating the two half-planes. The graph associated with the equation $3x + 2y = 6$ is shown in Figure 4-10.

To determine which half-plane is the graph of the inequality $3x + 2y < 6$, substitute into the inequality the coordinates of some point that does not lie on the boundary. In this case, the origin is a convenient choice:

$$3x + 2y < 6$$
$$3(0) + 2(0) < 6$$
$$0 < 6$$

Because the coordinates of the origin satisfy the inequality, the origin is a member of the half-plane represented by $3x + 2y < 6$. The graph of the inequality $3x + 2y < 6$ is shown in Figure 4-11. Note that the boundary, $3x + 2y = 6$, is drawn with a *broken* line because it is *not* included in the graph of $3x + 2y < 6$.

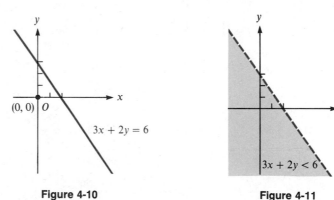

Figure 4-10   Figure 4-11

Example 2   Graph the inequality $3x + 2y \geq 6$.

Solution   The boundary of this inequality is the same line as in the previous example. If you substitute the coordinates of the origin into the given inequality $3x + 2y \geq 6$, you find that they do *not* satisfy that inequality: $0 \geq 6$ is a false statement. Therefore, the origin is *not* in the half-plane represented by the inequality $3x + 2y \geq 6$. The graph of $3x + 2y \geq 6$ is shown in Figure 4-12. Because the boundary *is* included in the graph of this solution set, use a solid line, rather than a broken line, to indicate this boundary.

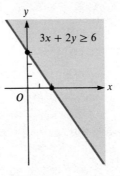

$3x + 2y \geq 6$

Figure 4-12   ■

In a similar fashion, we can graph inequalities in two unknowns with terms of higher degree than 1.

Example 3   Graph the inequality $y < x^2 - 2x - 3$.

Solution   First graph the parabola $y = x^2 - 2x - 3$ with a broken curve because it is not part of the solution set of the inequality $y < x^2 - 2x - 3$. See Figure 4-13. Substitute the coordinates of the origin $(0, 0)$ into the inequality:

$$y < x^2 - 2x - 3$$
$$0 < (0)^2 - 2(0) - 3$$
$$0 < -3$$

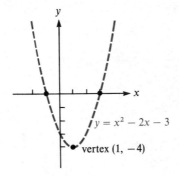

$y = x^2 - 2x - 3$

vertex $(1, -4)$

Figure 4-13

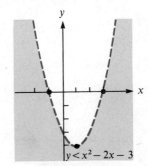

$y < x^2 - 2x - 3$

Figure 4-14

Because the statement $0 < -3$ is false, the origin is *not* in the graph of $y < x^2 - 2x - 3$, and the graph of the inequality is *below* the broken curve, as shown in Figure 4-14.   ■

Example 4   Graph the inequality $y \geq \dfrac{1}{x^2}$.

Solution   The graph of the rational function $y = \frac{1}{x^2}$ is shown in Figure 4-15. It is drawn as a solid curve because it *is* included in the solution set of $y \geq \frac{1}{x^2}$. To determine which regions are part of the solution set, substitute into the inequality the coordinates of some points, such as points P and Q in Figure 4-15, that do not lie on the curve. Point $P(2, 2)$ is a part of the solution set because its coordinates satisfy the inequality: $2 \geq \frac{1}{4}$ is a true statement. Similarly, point $Q(-2, 2)$ is also in the solution set. The graph of the inequality $y \geq \frac{1}{x^2}$ is shown in Figure 4-16.

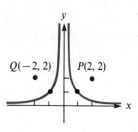

Figure 4-15

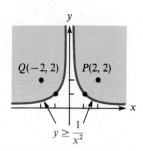

Figure 4-16

EXERCISE 4.3

In Exercises 1–40, graph each inequality.

1. $y < 3x + 7$

2. $y > x + 2$

3. $y > 2x - 9$

4. $y < 3 - x$

5. $y \geq \frac{1}{2}x + 1$

6. $y \leq \frac{1}{3}x + 2$

7. $2x - 3y \leq 12$

8. $3x + 4y \geq 12$

9. $2x + 5y > 10$

10. $3x - 5y < 15$

11. $y > 2x^2$

12. $y < 5x^2$

13. $x(x + 5) \leq x^2 + y$

14. $3x(x + 1) \geq 3x^2 + y$

15. $y > 0$

16. $x < 1$

17. $x \leq 2$

18. $y \leq 0$

19. $y \leq x^2 - 2$

20. $y \geq x^2 + 2$

21. $y > x^2 - 1$

22. $y < 2 - x^2$

23. $y > x^3$

24. $y \leq x^3$

25. $x^2 - x + y > 0$

26. $x^2 - 3x - y < 0$

27. $y < x^3 - x$

28. $y > x^3 - x^2$

29. $y < |x|$

30. $y > |x|$

31. $y \geq |x - 1|$

32. $y < |x - 1|$

33. $|y| < |x|$

34. $|y| > |x|$

35. $y < \dfrac{1}{x^2}$

36. $y < \dfrac{1}{x}$

37. $y \leq \dfrac{1}{x}$

38. $y \geq \dfrac{1}{x + 1}$

39. $xy < 0$

40. $xy > 0$

4.4 QUADRATIC AND RATIONAL INEQUALITIES

If $a \neq 0$, expressions such as $ax^2 + bx + c < 0$ and $ax^2 + bx + c > 0$ are called **quadratic inequalities**. We will discuss two methods for solving these inequalities.

Example 1   Solve the quadratic inequality $x^2 - x - 6 > 0$.

Solution 1   First solve the quadratic equation $x^2 - x - 6 = 0$:

$$x^2 - x - 6 = 0$$
$$(x + 2)(x - 3) = 0$$
$$x + 2 = 0 \quad \text{or} \quad x - 3 = 0$$
$$x = -2 \quad | \quad x = 3$$

The graphs of these solutions on the number line establish the three intervals shown in Figure 4-17.

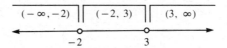

Figure 4-17

The solutions of the inequality $x^2 - x - 6 > 0$ will be the numbers in one or more of these intervals. To decide which intervals are solutions, test a number in each interval and see if it satisfies the inequality.

$-6 \in (-\infty, -2)$	$0 \in (-2, 3)$	$5 \in (3, \infty)$
$x^2 - x - 6 > 0$	$x^2 - x - 6 > 0$	$x^2 - x - 6 > 0$
$(-6)^2 - (-6) - 6 > 0$	$0^2 - 0 - 6 > 0$	$5^2 - 5 - 6 > 0$
$36 + 6 - 6 > 0$	$0 - 0 - 6 > 0$	$25 - 5 - 6 > 0$
$36 > 0$	$-6 > 0$	$14 > 0$
This is true.	This is not true.	This is true.

Thus, the solution is $(-\infty, -2) \cup (3, \infty)$. The graph of the solution is shown in Figure 4-18.

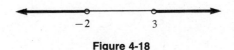

Figure 4-18

Solution 2   Another method relies exclusively on the number line and a notation that keeps track of the signs of the factors of the quadratic inequality $x^2 - x - 6 > 0$.

At $x = 3$, the factor $x - 3$ is 0. For values of x that are less than 3, the factor $x - 3$ is negative; for values of x that are greater than 3, the factor $x - 3$ is positive. Similarly, the value of the factor $x + 2$ is 0 when $x = -2$. For values of x that are less than -2, the factor $x + 2$ is negative; for values of x that are greater than -2, the factor $x + 2$ is positive.

Because the positiveness of the product $(x - 3)(x + 2)$ depends on the signs of $x - 3$ and $x + 2$, keep track of the positiveness or negativeness of these factors by using plus and minus signs in the **sign graph** shown in Figure 4-19. Only to the left of -2 and to the right of $+3$ do the signs of both factors agree; only there is the product positive.

The graph of the solution set is shown in Figure 4-19.

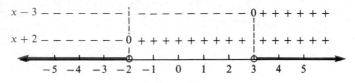

Figure 4-19   ■

Example 2   Solve the inequality $x(x + 3) < -2$.

First remove parentheses and add 2 to both sides to make the right-hand side 0. Then solve the inequality $x^2 + 3x + 2 < 0$.

Solution 1   First solve the quadratic equation $x^2 + 3x + 2 = 0$:

$$x^2 + 3x + 2 = 0$$
$$(x + 2)(x + 1) = 0$$
$$x + 2 = 0 \quad \text{or} \quad x + 1 = 0$$
$$x = -2 \quad | \quad x = -1$$

The graphs of these solutions on the number line establish the three intervals shown in Figure 4-20.

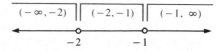

Figure 4-20

The solutions of the inequality $x^2 + 3x + 2 < 0$ will be the numbers in one or more of these intervals. To decide which intervals are solutions, test a number in each interval and see if it satisfies the inequality.

$-7 \in (-\infty, -2)$	$-\dfrac{3}{2} \in (-2, -1)$	$0 \in (-1, \infty)$
$x^2 + 3x + 2 < 0$	$x^2 + 3x + 2 < 0$	$x^2 + 3x + 2 < 0$
$(-7)^2 + 3(-7) + 2 < 0$	$\left(-\dfrac{3}{2}\right)^2 + 3\left(-\dfrac{3}{2}\right) + 2 < 0$	$0^2 + 3(0) + 2 < 0$
$49 - 21 + 2 < 0$		$0 + 0 + 2 < 0$
$30 < 0$	$\dfrac{9}{4} + \left(-\dfrac{9}{2}\right) + 2 < 0$	$2 < 0$
	$-\dfrac{1}{4} < 0$	
This is not true.	This is true.	This is not true.

Thus, the solution is the interval $(-2, -1)$. The graph of this solution is shown in Figure 4-21.

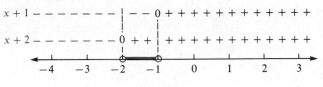

$$\begin{array}{cc} -2 & -1 \end{array}$$

Figure 4-21

Solution 2   Keep track of the positiveness or negativeness of the factors $x + 2$ and $x + 1$ by constructing the sign graph in Figure 4-22. Only between -2 and -1 do the factors have opposite signs. Here, the product is negative. The graph of the solution set is shown in Figure 4-22.

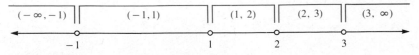

Figure 4-22   ∎

An inequality that contains a fraction with a polynomial denominator is called a **rational inequality**. To solve these inequalities, we can use the same techniques used to solve quadratic inequalities.

Example 3   Solve the rational inequality $\dfrac{x^2 - x - 2}{x^2 - 4x + 3} \le 0$.

Solution 1   The intervals are found by solving the two quadratic equations $x^2 - x - 2 = 0$ and $x^2 - 4x + 3 = 0$. The solutions of the first equation are -1 and 2, and the solutions of the second equation are 1 and 3. The graphs of these solutions establish the five intervals shown in Figure 4-23.

$(-\infty, -1)$		$(-1, 1)$		$(1, 2)$		$(2, 3)$		$(3, \infty)$

$$\begin{array}{ccccc} & -1 & \quad & 1 & \quad 2 \quad\quad 3 \end{array}$$

Figure 4-23

To decide which of these intervals are solutions, test a number in each interval and see if it satisfies the inequality. Numbers in the intervals $(-1, 1)$ and $(2, 3)$ satisfy the rational inequality, but numbers in the intervals $(-\infty, -1)$, $(1, 2)$, and $(3, \infty)$ do not. The graph of the solution set is shown in Figure 4-24. Because the numbers $x = -1$ and $x = 2$ make the numerator 0, they satisfy the inequality

$$\frac{x^2 - x - 2}{x^2 - 4x + 3} \le 0$$

Thus, their graphs are drawn as solid circles to show that -1 and 2 are included in the solution set. Because the numbers 1 and 3 give 0s in the denominator of the fraction, the circles at $x = 1$ and $x = 3$ are open circles to show that 1 and 3 are not in the solution set.

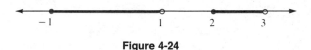

Figure 4-24

Solution 2   Factor each of the trinomials, and write the given inequality in the form

$$\frac{(x - 2)(x + 1)}{(x - 3)(x - 1)} \le 0$$

Construct a sign graph to keep track of the signs of each of the four factors. See Figure 4-25. The value of the given rational expression must be negative or 0. The only way this rational expression can be negative is to have an odd number of negative factors; the only way that the expression can be 0 is for its numerator to be 0. From the sign graph, it is apparent that only between -1 and 1 and between 2 and 3 is there an odd number of negative factors. Only there is the rational expression negative. If $x = 2$ or $x = -1$, the numerator is 0, and therefore the fraction is 0 also.

The graph of the solution set is shown in Figure 4-25. The circles at $x = -1$ and $x = 2$ are solid to show that these two values are included in the solution set. The circles at $x = 1$ and $x = 3$ are left open to indicate that these two values are not included in the solution set. If they were included, the denominator of the fraction would be 0.

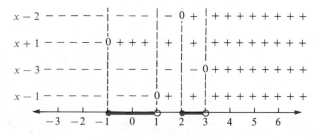

Figure 4-25

Example 4   Solve the inequality $\dfrac{6}{x} > 2$.

Solution   To get a 0 on the right-hand side, add -2 to both sides of the inequality. Then combine terms on the left-hand side.

$$\frac{6}{x} > 2$$

$$\frac{6}{x} - 2 > 0$$

$$\frac{6}{x} - \frac{2x}{x} > 0$$

$$\frac{6 - 2x}{x} > 0$$

The given inequality now has the form of a rational inequality. The intervals are found by solving the equations $6 - 2x = 0$ and $x = 0$. The solution of the first equation is 3, and the solution of the second equation is 0. This determines the intervals $(-\infty, 0)$, $(0, 3)$, and $(3, \infty)$. Because only the numbers in the interval $(0, 3)$ satisfy the original inequality, the solution set is $(0, 3)$. The graph is shown in Figure 4-26.

Figure 4-26

To use the sign-graph method, construct a sign graph as in Figure 4-27 to determine that the solution set is $(0, 3)$.

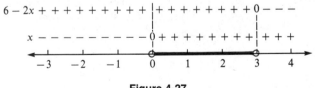

Figure 4-27   ■

It is tempting to solve Example 4 by multiplying both sides of the inequality $\dfrac{6}{x} > 2$ by x and solving the inequality $6 > 2x$. However, this is not correct. Multiplying both sides of $\dfrac{6}{x} > 2$ by x gives $6 > 2x$ only when x is positive. If x is negative, multiplying both sides of $\dfrac{6}{x} > 2$ by x will reverse the direction of the $>$ symbol, and the inequality $\dfrac{6}{x} > 2$ will be equivalent to $6 < 2x$. If we fail to consider both cases, we will obtain an incorrect answer.

EXERCISE 4.4

In Exercises 1–42, find the solution set for each inequality.

1. $x^2 + 7x + 12 < 0$

2. $x^2 - 13x + 12 \leq 0$

3. $x^2 - 5x + 6 \geq 0$

4. $6x^2 + 5x - 6 > 0$

5. $x^2 + 5x + 6 < 0$

6. $x^2 + 9x + 20 \geq 0$

7. $6x^2 + 5x + 1 \geq 0$

8. $x^2 + 9x + 20 < 0$

9. $6x^2 - 5x < -1$

10. $9x^2 + 24x > -16$

11. $2x^2 \geq 3 - x$

12. $9x^2 \leq 24x - 16$

13. $\dfrac{x + 3}{x - 2} < 0$

14. $\dfrac{x + 3}{x - 2} > 0$

15. $\dfrac{x(x + 1)}{x^2 - 1} > 0$

16. $\dfrac{x^2 - 4}{x^2 - 9} < 0$

17. $\dfrac{x^2 + 5x + 6}{x^2 + x - 6} \geq 0$

18. $\dfrac{x^2 + 10x + 25}{x^2 - x - 12} \leq 0$

19. $\dfrac{x^2 + 7x + 6}{x^2 + 8x + 15} \leq 0$

20. $\dfrac{x^2 - 8x + 15}{x^2 - 12x + 35} \geq 0$

21. $\dfrac{x^2 + 13x + 36}{x^2 + 4x + 4} > 0$

22. $\dfrac{6x^2 - 3x - 3}{x^2 - 2x - 8} < 0$

23. $\dfrac{3}{x} > 2$

24. $\dfrac{3}{x} < 2$

25. $\dfrac{6}{x} < 4$

26. $\dfrac{6}{x} > 4$

27. $\dfrac{3}{x - 2} \leq 5$

28. $\dfrac{3}{x + 2} \leq 4$

29. $\dfrac{3}{x + 2} \geq 4$

30. $\dfrac{-1}{x - 1} \geq 3$

31. $\dfrac{6}{x^2 - 1} < 1$

32. $\dfrac{6}{x^2 - 1} > 1$

33. $\dfrac{3}{x} - x > 2$

34. $\dfrac{1}{x} + x < 2$

35. $\dfrac{8}{x} + x < \dfrac{1}{x}$

36. $\dfrac{8}{x} + x > \dfrac{1}{x}$

37. $x(x - 1)(x - 2) > 0$

38. $(x^2 + 1)(x^2 - x - 6) \geq 0$

39. $x^3 \geq x$

40. $(x^2 - 4)(x^2 + 5x + 6) < 0$

41. $\left| \dfrac{x + 2}{x - 3} \right| \leq 4$

42. $\left| \dfrac{x}{2x + 1} \right| > 5$

In Exercises 43–46, find the domain of the variable x.

43. $y = \sqrt{x^2 - 9}$

44. $y = \sqrt{25 - x^2}$

45. $y = \sqrt{x^2 + 4x + 3}$

46. $y = \dfrac{12}{\sqrt{x - \dfrac{1}{x}}}$

Computer Exercises. *In Exercises 47–50, use a computer and **Algebra Pak** to solve each inequality. Select nonautomatic scaling.*

47. Solve the inequality $x^2 - 3x + 2 > 0$ by graphing the function $y = x^2 - 3x + 2$ and finding those numbers x for which the values of y are positive: that is, find those numbers x for which the graph of the function lies above the x-axis.

48. Use the method of Exercise 47 to solve the inequality $2x^2 - x \leq 6$.

49. Use the method of Exercise 47 to solve the inequality $\dfrac{x^2 + x}{x - 1} \geq 0$. Enter the function by typing

$(X\text{\^{}}2 + X)/(X - 1)$.

50. Use the method of Exercise 47 to solve the inequality $\dfrac{x - 1}{x^2 - 4} \leq 0$.

CHAPTER SUMMARY

Key Words

absolute value   (4.2)   *quadratic inequalities*   (4.4)
equivalent inequalities   (4.1)   *rational inequalities*   (4.4)
inequalities in two variables   (4.3)   *sign graph*   (4.4)
linear inequalities   (4.1)   *solution to an inequality*   (4.1)

Key Ideas

(4.1)   Let a, b, and c be real numbers:

If $a < b$, then $a + c < b + c$ and $a - c < b - c$.

If $a < b$ and $c > 0$, then $ac < bc$ and $\dfrac{a}{c} < \dfrac{b}{c}$.

If $a < b$ and $c < 0$, then $ac > bc$ and $\dfrac{a}{c} > \dfrac{b}{c}$.

If $a < b$ and $b < c$, then $a < c$.

Similar relationships hold for $>$, $\leq$, and $\geq$.

(4.2)   If $x \geq 0$, then $|x| = x$. If $x < 0$, then $|x| = -x$.

If $a \geq 0$, then $|x| = a$ is equivalent to $x = a$ or $x = -a$.

If $a > 0$, then $|x| < a$ is equivalent to $-a < x < a$.

If $a > 0$, then $|x| > a$ is equivalent to $x > a$ or $x < -a$.

Properties of absolute value:

1. $|ab| = |a|\,|b|$

2. $\left|\dfrac{a}{b}\right| = \dfrac{|a|}{|b|}$ $(b \neq 0)$

3. $|a + b| \leq |a| + |b|$

(4.3)   Graphs of inequalities in two variables are regions of the Cartesian plane.

(4.4)   Solve quadratic and rational inequalities by the test-number method or by making a sign graph.

REVIEW EXERCISES

In Review Exercises 1–10, find the solution set for each inequality. Graph each solution set on a number line.

1. $2x - 9 < 5$

2. $5x + 3 \geq 2$

3. $\dfrac{3}{7}x + 5 > 10$

4. $\dfrac{2}{3}x - 5 > x$

5. $\dfrac{5(x - 1)}{2} < x$

6. $3(x - 1) > x + 1$

7. $3(x - 5) < x - 1$

8. $0 \leq \dfrac{3 + x}{2} < 4$

9. $-2 \leq \dfrac{2x + 5}{2} < 10$

10. $-x \leq x \leq 2x$

In Review Exercises 11–14, evaluate each expression.

11. $|-8|$

12. $-|3 - 5|$

13. $-|3|\,|-3|$

14. $-\dfrac{|3|}{|5 - 2|}$

In Review Exercises 15–20, solve for x.

15. $|x + 3| < 3$

16. $|3x - 7| \geq 1$

17. $|x + 1| = 6$

18. $|2x - 1| = |2x + 1|$

19. $\left|\dfrac{x + 2}{3}\right| < 1$

20. $1 < |2x + 3| < 4$

In Review Exercises 21–26, graph each inequality.

21. $y \leq 3x + 7$

22. $2x - y \geq 3y - 5$

23. $y > 0.5x^2$

24. $y < \dfrac{|x|}{x}$

25. $x^2 + 3x + y > 0$

26. $|y| < x$

In Review Exercises 27–32, solve each quadratic inequality.

27. $x^2 - 2x - 3 < 0$

28. $2x^2 + x - 3 > 0$

29. $5x^2 + 32x \geq 21$

30. $7x^2 + 8x + 1 \leq 0$

31. $6x^2 + 19x < -10$

32. $2x^2 - x > 15$

In Review Exercises 33–38, solve each rational inequality.

33. $\dfrac{x^2 + x - 2}{x - 3} \geq 0$

34. $\dfrac{x^2 + 6x - 7}{x + 1} < 0$

35. $\dfrac{2x^2 - 5x - 3}{x + 1} \geq 0$

36. $\dfrac{3x^2 - 5x - 2}{x - 3} \leq 0$

37. $x < \dfrac{2}{x + 1}$

38. $x \leq \dfrac{3}{x - 2}$

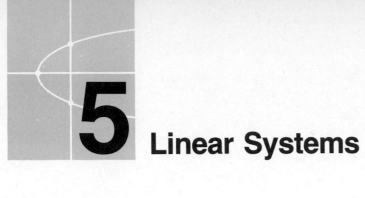

Linear Systems

Any collection of m equations in n variables is called a **system of equations**. For example, the equations

$$\begin{cases} 3x + y = 1 \\ -x + 2y = 9 \end{cases}$$

form a system of two linear equations in two variables. In this chapter, we discuss several methods for solving such systems.

5.1 SYSTEMS OF LINEAR EQUATIONS

Consider the system of linear equations

$$\begin{cases} 3x + y = 1 \\ -x + 2y = 9 \end{cases}$$

There are infinitely many ordered pairs of real numbers (x, y) that satisfy the first equation, and there are infinitely many ordered pairs of real numbers (x, y) that satisfy the second equation. However, there is only one ordered pair of real numbers (x, y) that satisfies both equations simultaneously. This ordered pair is called the **simultaneous solution** or just the **solution** of the system of equations. The process of finding a solution to a system of equations is called **solving the system**. We begin by discussing three methods for solving the previous system of equations.

The Graphing Method

Use the following steps to solve a system of two equations in two variables by graphing.

---

1. On a single coordinate grid, graph each equation.
2. Read the coordinates of the point or points where *all* of the graphs intersect. These coordinates give (often approximately) the solutions of the system.
3. If no point is common to all of the graphs, the system has no solution.

The results of Example 1 illustrate the various possibilities that can occur when graphing two linear equations in two variables.

Example 1 Use the graphing method to solve each system:

a. $\begin{cases} 3x + y = 1 \\ -x + 2y = 9 \end{cases}$ **b.** $\begin{cases} 2x - 3y = 4 \\ 4x = -4 + 6y \end{cases}$ **c.** $\begin{cases} y = 4 - x \\ 2x + 2y = 8 \end{cases}$

Solution **a.** The graph of each equation is a straight line. See Figure 5-1a. The solution of this system is given by the coordinates of the point where the lines intersect, $(-1, 4)$. Thus, the solution of the system is $x = -1$ and $y = 4$. Verify that these values satisfy each equation.

b. The graph of each equation is a straight line, and the lines are parallel. See Figure 5-1b. Thus, this system has no solutions.

c. The graph of each equation is a straight line, and they coincide. See Figure 5-1c. Thus, this system has infinitely many solutions. All ordered pairs whose coordinates satisfy one of the equations satisfy the other also.

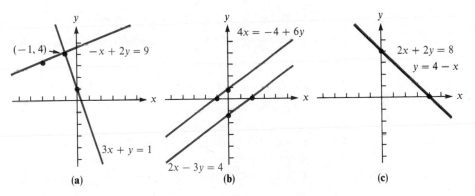

Figure 5-1

If a system of equations has at least one solution, as in part **a** of Example 1, the system is called **consistent**. If it has no solutions, as in part **b**, it is called **inconsistent**.

If a system of n linear equations in n variables has exactly one solution, as in part **a**, or no solutions, as in part **b**, the equations in the system are called **independent**. If a system of linear equations has infinitely many solutions, as in part **c**, the equations of the system are called **dependent**.

The Substitution Method

Use the following steps to solve a system of two equations in two variables by substitution.

1. Solve one equation for a variable, say y.
2. Substitute the expression obtained for y for all occurrences of y in the other equation.
3. Solve the resulting equation.
4. Substitute the solution found in Step 3 in the equation found in Step 1 and solve for y.

Example 2   Use the substitution method to solve the system

$$\begin{cases} 3x + y = 1 \\ -x + 2y = 9 \end{cases}$$

Solution   Solve the first equation for y to obtain $y = 1 - 3x$. Then, substitute $1 - 3x$ for y in the second equation.

$$\begin{cases} 3x + y = 1 \rightarrow y = \boxed{1 - 3x} \\ -x + 2y = 9 \end{cases}$$

This substitution gives one linear equation in one variable, which can be solved for x:

$$\begin{aligned} -x + 2(1 - 3x) &= 9 \\ -x + 2 - 6x &= 9 \\ -7x &= 7 \\ x &= -1 \end{aligned}$$

To find y, substitute -1 for x in the equation $y = 1 - 3x$ and simplify:

$$\begin{aligned} y &= 1 - 3x \\ &= 1 - 3(-1) \\ &= 1 + 3 \\ &= 4 \end{aligned}$$

The solution to the system is $(-1, 4)$.   ■

The Addition Method

The addition method is based on three algebraic manipulations that will transform a system of equations into an **equivalent system**, which has the same solutions.

1. The positions of any two equations of the system can be interchanged.
2. Both sides of any equation of the system can be multiplied or divided by a nonzero constant.
3. Any equation of the system can be altered by adding to its sides a constant multiple of the corresponding sides of another equation of the system.

Example 3 Use the addition method to solve the system

$$\begin{cases} 3x + y = 1 \\ -x + 2y = 9 \end{cases}$$

Solution Multiply both sides of the second equation by 3, and add the result to the first equation to eliminate the variable x:

$$
\begin{array}{r}
3x + y = 1 \\
-3x + 6y = 27 \\
\hline
7y = 28 \\
y = 4 \qquad \text{Divide both sides by 7.}
\end{array}
$$

Substitute 4 for y into either of the original equations and solve for x. For example, use the first equation:

$$
\begin{array}{r}
3x + y = 1 \\
3x + 4 = 1 \\
3x = -3 \\
x = -1
\end{array}
$$

Thus, the solution of the given system is $(-1, 4)$. ∎

Example 4 Use the addition method to solve the system

$$\begin{cases} x + 2y = 3 \\ 2x + 4y = 6 \end{cases}$$

Solution Multiply both sides of the first equation by -2, and add the results to the second equation:

$$
\begin{cases}
-2x - 4y = -6 \\
\underline{2x + 4y = 6} \\
\,0 = 0
\end{cases}
$$

Although the result $0 = 0$ is true, it does not lead to a solution for y. Because the second equation is twice the first, the two equations in this system are essentially the same; they are equivalent. If each equation were graphed, the resulting lines would coincide. The pairs of coordinates of all points on the *one* line described by this system make up the infinite set of solutions to the given system. The system is consistent, and the equations are dependent. ∎

Example 5 Solve the system

$$\begin{cases} x + y = 3 \\ x + y = 2 \end{cases}$$

Solution   Multiply both sides of the second equation by -1, and add the results to the first equation:

$$\left\{ \begin{array}{r} x + y = 3 \\ -x - y = -2 \end{array} \right.$$
$$\overline{ 0 = 1}$$

Because this result is impossible, the system has no solutions. If each equation in this system were graphed, the graphs would be parallel lines. This system is inconsistent, and the equations are independent.   ∎

Example 6 shows how to use the substitution and addition methods to solve a system of three equations in three variables.

Example 6   Solve the system

$$\begin{array}{ll} \textbf{1.} & \\ \textbf{2.} & \left\{ \begin{array}{r} x + 2y + z = 8 \\ 2x + y - z = 1 \\ x + y - 2z = -3 \end{array} \right. \\ \textbf{3.} & \end{array}$$

Solution   Use Equations 1 and 2 to eliminate the variable z. Then use Equations 1 and 3 to eliminate the variable z. You can then solve the resulting system of two equations in the two variables x and y.

Add Equations 1 and 2 to eliminate the variable z:

$$\begin{array}{ll} \textbf{1.} & \\ \textbf{2.} & \left\{ \begin{array}{r} x + 2y + z = 8 \\ 2x + y - z = 1 \end{array} \right. \end{array}$$
$$\overline{3x + 3y = 9}$$

Divide both sides of the resulting equation by 3 to obtain

$$\textbf{4.} \quad x + y = 3$$

The variable z can be eliminated again by using Equations 1 and 3. Multiply both sides of Equation 1 by the constant 2, and add the result to Equation 3:

$$\begin{array}{ll} & \\ \textbf{3.} & \left\{ \begin{array}{r} 2x + 4y + 2z = 16 \\ x + y - 2z = -3 \end{array} \right. \\ \textbf{5.} & \overline{3x + 5y = 13} \end{array}$$

Equations 4 and 5 form a system of two equations in two variables. Solve this system by substitution as follows:

$$\begin{array}{ll} \textbf{4.} & \\ \textbf{5.} & \left\{ \begin{array}{r} x + y = 3 \rightarrow y = \boxed{3 - x} \\ 3x + 5y = 13 \end{array} \right. \end{array}$$

$$\begin{array}{r} 3x + 5(3 - x) = 13 \\ 3x + 15 - 5x = 13 \\ -2x = -2 \\ x = 1 \end{array}$$

Substitute 1 for x in the equation $y = 3 - x$ to find y:

$$y = 3 - 1$$
$$y = 2$$

To find z, substitute 1 for x, and 2 for y in any of the original equations. You will find that $z = 3$. The solution to the given system is the triple

(1, 2, 3)

Because this solution is unique, the given system is consistent, and the equations in the system are independent. Verify that $x = 1$, $y = 2$, and $z = 3$ satisfy each equation in the original system.   ■

Example 7   An airplane flies 600 miles with the wind for 2 hours. The return trip against the wind takes 3 hours. Find the speed of the wind and the airspeed of the plane.

Solution   Let a represent the airspeed of the plane, and w represent the speed of the wind. The groundspeed of the plane on the outbound trip is the combined speed $a + w$. On the return trip, against a headwind, the groundspeed is $a - w$. The information of this problem, organized in the chart in Figure 5-2, gives a system of two equations in the two variables a and w.

Because $d = rt$, you have

$$\begin{cases} 600 = 2(a + w) \\ 600 = 3(a - w) \end{cases}$$

or

$$\begin{cases} 300 = a + w \\ 200 = a - w \end{cases}$$

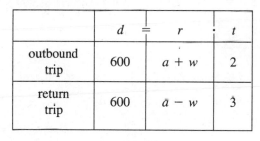

	d	$=$	r	$\cdot$	t
outbound trip	600		$a + w$		2
return trip	600		$a - w$		3

Figure 5-2

Add the equations together to get

$$500 = 2a$$

or

$$a = 250$$

To calculate w, substitute 250 for a in a previous equation such as $300 = a + w$ and solve for w:

$$300 = a + w$$
$$300 = 250 + w$$
$$w = 50$$

The plane could do 250 miles per hour in still air. With a 50-mile-per-hour tailwind, the groundspeed would be 300 miles per hour, and the 600-mile trip would take 2 hours. With a 50-mile-per-hour headwind, the groundspeed would be 200 miles per hour, and the 600-mile trip would take 3 hours. The answers check.   ■

EXERCISE 5.1

In Exercises 1–4, solve each system of equations by the graphing method.

1. $\begin{cases} y = -3x + 5 \\ x - 2y = -3 \end{cases}$

2. $\begin{cases} x - 2y = -3 \\ 3x + y = -9 \end{cases}$

3. $\begin{cases} 3x + 2y = 2 \\ -2x + 3y = 16 \end{cases}$

4. $\begin{cases} x + y = 0 \\ 7x + 8y = 1 \end{cases}$

In Exercises 5–12, solve each system of equations by the substitution method, if possible.

5. $\begin{cases} y = x \\ y = 2x \end{cases}$

6. $\begin{cases} y = 2x - 1 \\ x + y = 5 \end{cases}$

7. $\begin{cases} 2x + 3y = 0 \\ y = 3x - 11 \end{cases}$

8. $\begin{cases} 2x + y = 3 \\ y = 5x - 11 \end{cases}$

9. $\begin{cases} 4x + 3y = 3 \\ 2x - 6y = -1 \end{cases}$

10. $\begin{cases} 4x + 5y = 4 \\ 8x - 15y = 3 \end{cases}$

11. $\begin{cases} x + 3y = 1 \\ 2x + 6y = 3 \end{cases}$

12. $\begin{cases} x - 3y = 14 \\ 3(x - 12) = 9y \end{cases}$

In Exercises 13–24, solve each system of equations by the addition method, if possible.

13. $\begin{cases} 5x - 3y = 12 \\ 2x - 3y = 3 \end{cases}$

14. $\begin{cases} 2x + 3y = 8 \\ -5x + y = -3 \end{cases}$

15. $\begin{cases} x - 7y = -11 \\ 8x + 2y = 28 \end{cases}$

16. $\begin{cases} 3x + 9y = 9 \\ -x + 5y = -3 \end{cases}$

17. $\begin{cases} 3(x - y) = y - 9 \\ 5(x + y) = -15 \end{cases}$

18. $\begin{cases} 2(x + y) = y + 1 \\ 3(x + 1) = y - 3 \end{cases}$

19. $\begin{cases} 2 = \dfrac{1}{x + y} \\[2mm] 2 = \dfrac{3}{x - y} \end{cases}$

20. $\begin{cases} \dfrac{1}{x + y} = 12 \\[2mm] \dfrac{3x}{y} = -4 \end{cases}$

21. $\begin{cases} 5x = 10y \\ -x + 2y = 0 \end{cases}$

22. $\begin{cases} -0.3x + 0.1y = -0.1 \\ 6x - 2y = 2 \end{cases}$

23. $\begin{cases} \dfrac{x + y}{2} + \dfrac{x - y}{5} = 2 \\[2mm] x = \dfrac{y}{2} + 1 \end{cases}$

24. $\begin{cases} \dfrac{3}{2}x + \dfrac{1}{3}y = 2 \\[2mm] \dfrac{2}{3}x + \dfrac{1}{9}y = 1 \end{cases}$

In Exercises 25–40, solve each system of equations, if possible.

25. $\begin{cases} x + y + z = 3 \\ 2x + y + z = 4 \\ 3x + y - z = 5 \end{cases}$

26. $\begin{cases} x - y - z = 0 \\ x + y - z = 0 \\ x - y + z = 2 \end{cases}$

27. $\begin{cases} x - y + z = 0 \\ x + y + 2z = -1 \\ -x - y + z = 0 \end{cases}$

28. $\begin{cases} 2x + y - z = 7 \\ x - y + z = 2 \\ x + y - 3z = 2 \end{cases}$

29. $\begin{cases} 2x + y = 4 \\ x - z = 2 \\ y + z = 1 \end{cases}$

30. $\begin{cases} 3x + y + z = 0 \\ 2x - y + z = 0 \\ 2x + y + z = 0 \end{cases}$

31. $\begin{cases} x + y + z = 6 \\ 2x + y + 3z = 17 \\ x + y + 2z = 11 \end{cases}$

32. $\begin{cases} x + y + z = 3 \\ 2x + y + z = 6 \\ x + 2y + 3z = 2 \end{cases}$

33. $\begin{cases} x + y + z = 3 \\ x + z = 2 \\ 2x + 2y + 2z = 3 \end{cases}$

34. $\begin{cases} x + y + z = 3 \\ x + z = 2 \\ 2x + y + 2z = 5 \end{cases}$

35. $\begin{cases} x + y & = 2 \\ y + z = 2 \\ x & - z = 0 \end{cases}$

36. $\begin{cases} x + y & = 2 \\ y + z = 2 \\ 3x + 3y & = 2 \end{cases}$

37. $\begin{cases} 3x + 4y + 2z = 4 \\ 6x - 2y + z = 4 \\ 3x - 8y - 6z = -3 \end{cases}$

38. $\begin{cases} x + 2y - z = 2 \\ 2x - y = -1 \\ 3x + y + z = 1 \end{cases}$

39. $\begin{cases} (x + y) + (y + z) + (z + x) = 6 \\ (x - y) + (y - z) + (z - x) = 0 \\ x + y + 2z = 4 \end{cases}$

40. $\begin{cases} (x + y) + (y + z) = 1 \\ (x + z) + (x + y) = 3 \\ (x - y) - (x - z) = -1 \end{cases}$

In Exercises 41–52, use two or three variables to solve each word problem.

41. Five tables and three chairs cost $695, and three tables and five chairs cost $561. How much does one table and one chair cost?

42. A farmer raises corn and soybeans on 350 acres of land. He plants 100 more acres of corn than of soybeans. How many acres of each does he plant?

43. A rectangle with a perimeter of 190 centimeters has a width that is 25 centimeters less than its length. Find the area.

44. A boat can travel 30 kilometers downstream in three hours and can make the return trip in five hours. Find the speed of the boat in still water.

45. The sum of three numbers is 19. If the first number is doubled, the sum is 22; if the second number is doubled, the sum is 25. Find the numbers.

46. A woman invested $22,000 in three bank accounts paying 5%, 6%, and 7% annual interest. She invested $2000 more at 6% than at 5%. The total annual interest she earned was $1370. How much did she invest at each rate?

47. A college student earns $99.25 per week working three part-time jobs. Half of the student's 30-hour work week is spent cooking hamburgers at a fast-food chain, earning $2.85 per hour. In addition, the student earns $3.15 per hour working at a gas station and $5 per hour doing janitorial work. How many hours per week does the student work at each job?

48. Flutter Hi-Fi, Wow Stereo, and Rumble Electronics buy a total of 175 cassette tape decks from High Hiss Distributors each month. Because Rumble Electronics buys 25 more tape decks than the other two stores combined, Rumble's cost is only $160 per unit. The decks cost Wow Stereo $165 each and Flutter Hi-Fi $170 each. How many decks does each retailer buy each month if High Hiss receives $28,500 each month from the sale of tape decks to these companies?

49. A collection of nickels, dimes, and quarters has a value of $3.40. There are twice as many dimes as quarters, and there are 32 coins in all. How many of each kind are there?

50. A factory manufactures widgets, gidgets, and gadgets at a monthly cost of $6850 for 2150 units. It costs $2 to make a widget, $3 to make a gidget, and $4 to make a gadget. A widget sells for $3, a gidget for $4.50, and a gadget for $5.50. The monthly profit is $2975. How many of each item are manufactured?

51. Suppose that a retailer obtains a product from three sources, A, B, and C. The retailer buys as many units from source A as from the other two combined. She must pay source A at the rate of $4 per unit and sources B and C at the rate of $5 per unit. If she requires 100 units per month to satisfy customer demand, and if her cost for one month's supply is $450, how many units does she buy from each source?

52. The sum of the angles of a triangle is 180°. The largest angle is 20° greater than the sum of the other two, and is 10° greater than 3 times the smallest. How large is each angle?

5.2   GAUSSIAN ELIMINATION AND MATRIX METHODS

There is a method, called **Gaussian elimination**, that can be used to solve systems of linear equations. In this method, a system of equations is transformed into an equivalent system that can be solved by a process called **back substitution**.

In Gaussian elimination, it is convenient to solve a system of equations by working with only the coefficients of the variables. All of the information needed to find a solution of

$$\begin{cases} x + 2y + z = 8 \\ 2x + y - z = 1 \\ x + y - 2z = -3 \end{cases}$$

for example, is contained in the following rectangular array of numbers, called a **matrix**:

$$\begin{bmatrix} 1 & 2 & 1 & \vdots & 8 \\ 2 & 1 & -1 & \vdots & 1 \\ 1 & 1 & -2 & \vdots & -3 \end{bmatrix}$$

Each row in this matrix represents one of the equations of the system. The first row, for example, represents the equation $x + 2y + z = 8$.

Because the matrix above has three rows and four columns, it is called a 3×4 (read as "3 by 4") matrix. The 3×3 matrix to the left of the broken line is called the **coefficient matrix**. The entire matrix is called the **augmented matrix**.

Example 1   Use Gaussian elimination to solve the system

$$\begin{array}{l} \textbf{1.} \\ \textbf{2.} \\ \textbf{3.} \end{array} \quad \begin{cases} x + 2y + z = 8 \\ 2x + y - z = 1 \\ x + y - 2z = -3 \end{cases}$$

Solution   First multiply each term in Equation 1 by -2, and add the result to Equation 2 to obtain Equation 4. Then multiply each term of Equation 1 by -1, and add the result to Equation 3 to obtain Equation 5. This gives an equivalent system with the same solution as the original system. The system is shown both in equation form and in matrix form. Note that the variable x does not appear in Equation 4 or 5:

$$\begin{array}{l} \textbf{1.} \\ \textbf{4.} \\ \textbf{5.} \end{array} \quad \begin{cases} x + 2y + z = 8 \\ - 3y - 3z = -15 \\ - y - 3z = -11 \end{cases} \qquad \begin{bmatrix} 1 & 2 & 1 & \vdots & 8 \\ 0 & -3 & -3 & \vdots & -15 \\ 0 & -1 & -3 & \vdots & -11 \end{bmatrix}$$

Now divide both sides of Equation 4 by -3 to obtain Equation 6,

$$\begin{array}{l} \textbf{1.} \\ \textbf{6.} \\ \textbf{5.} \end{array} \quad \begin{cases} x + 2y + z = 8 \\ y + z = 5 \\ - y - 3z = -11 \end{cases} \qquad \begin{bmatrix} 1 & 2 & 1 & \vdots & 8 \\ 0 & 1 & 1 & \vdots & 5 \\ 0 & -1 & -3 & \vdots & -11 \end{bmatrix}$$

and add Equation 6 to Equation 5 to obtain Equation 7.

$$\begin{matrix} \textbf{1.} \\ \textbf{6.} \\ \textbf{7.} \end{matrix} \quad \begin{cases} x + 2y + z = 8 \\ y + z = 5 \\ -2z = -6 \end{cases} \qquad \left[\begin{array}{ccc|c} 1 & 2 & 1 & 8 \\ 0 & 1 & 1 & 5 \\ 0 & 0 & -2 & -6 \end{array} \right]$$

Finally, divide both sides of Equation 7 by -2 to obtain the system

$$\begin{matrix} \textbf{1.} \\ \textbf{6.} \\ \end{matrix} \quad \begin{cases} x + 2y + z = 8 \\ y + z = 5 \\ z = 3 \end{cases} \qquad \left[\begin{array}{ccc|c} 1 & 2 & 1 & 8 \\ 0 & 1 & 1 & 5 \\ 0 & 0 & 1 & 3 \end{array} \right]$$

The system can now be solved by back substitution. Because $z = 3$, you can substitute 3 for z in Equation 6, and solve for y:

6. $y + z = 5$
$\ \ \ y + 3 = 5$
$\ \ \ \ \ \ \ \ \ \ y = 2$

You can now substitute 2 for y and 3 for z in Equation 1, and solve for x:

1. $x + 2y + z = 8$
$\ \ x + 2(2) + 3 = 8$
$\ \ \ \ \ \ \ \ \ \ \ \ \ \ x = 1$

The solution of the given system of equations is the ordered triple

$\quad$ (1, 2, 3)

Verify that $x = 1$, $y = 2$, and $z = 3$ satisfy each of the original three equations.

$\blacksquare$

Note the triangular formation of 0's in the final matrix in Example 1. A matrix such as

$$\left[\begin{array}{ccc|c} 1 & 2 & 1 & 8 \\ 0 & 1 & 1 & 5 \\ 0 & 0 & 1 & 3 \end{array} \right]$$

is said to be in **triangular form** if all entries below the diagonal running from the upper left to the lower right are 0.

Example 2   Use matrices to solve the system

$$\begin{cases} x + 2y + 3z = 4 \\ 2x - y - 2z = 0 \\ x - 3y - 3z = -2 \end{cases}$$

Solution   This system is represented by the augmented matrix

$$\left[\begin{array}{ccc|c} 1 & 2 & 3 & 4 \\ 2 & -1 & -2 & 0 \\ 1 & -3 & -3 & -2 \end{array}\right]$$

Change the individual rows of this matrix by adding multiples of one row to another. When you combine rows in this way, you are adding multiples of the coefficients of one equation to the corresponding coefficients of another, always obtaining coefficients that form an equivalent set of equations. The goal is to produce a matrix in triangular form that represents an equivalent system of equations.

To begin, use the 1 in the upper left-hand corner of the augmented matrix to "zero out" the rest of the first column. The notation "$(-2)R1 + R2 \rightarrow R2$" means "multiply row one by -2 and add the result to row two to get a new row two."

$$(-2)R1 + R2 \rightarrow R2$$

$$\left[\begin{array}{ccc|c} 1 & 2 & 3 & 4 \\ 2 & -1 & -2 & 0 \\ 1 & -3 & -3 & -2 \end{array}\right] \Leftrightarrow \left[\begin{array}{ccc|c} 1 & 2 & 3 & 4 \\ 0 & -5 & -8 & -8 \\ 1 & -3 & -3 & -2 \end{array}\right]$$

Read $\Leftrightarrow$ as "is equivalent to."

Next, multiply row one by -1, and add the result to row three to get a new row three:

$$(-1)R1 + R3 \rightarrow R3$$

$$\left[\begin{array}{ccc|c} 1 & 2 & 3 & 4 \\ 0 & -5 & -8 & -8 \\ 1 & -3 & -3 & -2 \end{array}\right] \Leftrightarrow \left[\begin{array}{ccc|c} 1 & 2 & 3 & 4 \\ 0 & -5 & -8 & -8 \\ 0 & -5 & -6 & -6 \end{array}\right]$$

Because each row of a matrix represents an equation, multiplying an entire row by a constant is equivalent to multiplying both sides of an equation by a constant. Hence, multiply row two by -1 to eliminate some minus signs and add the result to row three to get

$$(-1)R2 \rightarrow R2 \qquad\qquad R2 + R3 \rightarrow R3$$

$$\left[\begin{array}{ccc|c} 1 & 2 & 3 & 4 \\ 0 & -5 & -8 & -8 \\ 0 & -5 & -6 & -6 \end{array}\right] \Leftrightarrow \left[\begin{array}{ccc|c} 1 & 2 & 3 & 4 \\ 0 & 5 & 8 & 8 \\ 0 & -5 & -6 & -6 \end{array}\right] \Leftrightarrow \left[\begin{array}{ccc|c} 1 & 2 & 3 & 4 \\ 0 & 5 & 8 & 8 \\ 0 & 0 & 2 & 2 \end{array}\right]$$

Finally, multiply the third row by $\frac{1}{2}$:

$$\left(\tfrac{1}{2}\right)R3 \rightarrow R3$$

$$\left[\begin{array}{ccc|c} 1 & 2 & 3 & 4 \\ 0 & 5 & 8 & 8 \\ 0 & 0 & 2 & 2 \end{array}\right] \Leftrightarrow \left[\begin{array}{ccc|c} 1 & 2 & 3 & 4 \\ 0 & 5 & 8 & 8 \\ 0 & 0 & 1 & 1 \end{array}\right]$$

This final matrix is in triangular form and represents the system

1. $\begin{cases} x + 2y + 3z = 4 \\ 5y + 8z = 8 \\ z = 1 \end{cases}$
2.
3.

To solve this system by back substitution, substitute 1 for z in Equation 2, and solve for y:

2. $5y + 8z = 8$

$5y + 8(1) = 8$

$y = 0$

Then, substitute 0 for y and 1 for z in Equation 1, and solve for x:

1. $x + 2y + 3z = 4$

$x + 2(0) + 3(1) = 4$

$x = 1$

The solution of the original system is the ordered triple

$(1, 0, 1)$

Verify that $x = 1$, $y = 0$, and $z = 1$ satisfy each of the original equations. ∎

In the preceding example, one matrix was transformed into another by certain manipulations called **elementary row operations**.

Elementary Row Operations.

Type 1 row operation: two rows of a matrix can be interchanged.

Type 2 row operation: the elements of a row of a matrix can be multiplied by a nonzero constant.

Type 3 row operation: a row of a matrix can be altered by adding to it a multiple of any other row.

Each type of elementary row operation yields a new matrix representing a system of equations with the same solution as the original system. A type 1 row operation is equivalent to writing the equations of a system in a different order. A type 2 row operation is equivalent to multiplying both sides of an equation by a nonzero constant, and a type 3 row operation is equivalent to adding a multiple of one equation to another.

In a type 3 row operation, it is important to remember which row is being changed and which row is causing that change. For example, if twice row three is added to row one, it is row one that is being altered. Row three, which was used to accomplish the change, stays the same.

Any matrix that can be obtained from another matrix by a sequence of elementary row operations is called **row equivalent** to the original matrix. Thus, the symbol ⇔ can be read as "is row equivalent to." We will use row operations to write a matrix in a special form called **echelon form**.

Echelon Form of a Matrix.   A matrix is said to be in **echelon form** if it has the following properties:

1.   The lead entry (the first nonzero entry) of each row is 1.
2.   Lead entries appear farther to the right as you move down the rows of the matrix.
3.   Rows containing only 0's are at the bottom of the matrix.

Example 3   Use matrix methods to solve the following system of four equations in three variables:

$$\begin{cases} x + 2y - z = 6 \\ x - y + z = -2 \\ 2x + z = 1 \\ x + y + z = 2 \end{cases}$$

Solution   Note that this system has more equations than it has variables. To solve it, form the augmented matrix and use elementary row operations to write the augmented matrix in echelon form.

$$(-1)R1 + R2 \rightarrow R2$$

$$\begin{bmatrix} 1 & 2 & -1 & | & 6 \\ 1 & -1 & 1 & | & -2 \\ 2 & 0 & 1 & | & 1 \\ 1 & 1 & 1 & | & 2 \end{bmatrix} \Leftrightarrow \begin{bmatrix} 1 & 2 & -1 & | & 6 \\ 0 & -3 & 2 & | & -8 \\ 2 & 0 & 1 & | & 1 \\ 1 & 1 & 1 & | & 2 \end{bmatrix}$$

$$(-2)R1 + R3 \rightarrow R3$$
$$(-1)R1 + R4 \rightarrow R4$$

$$\Leftrightarrow \begin{bmatrix} 1 & 2 & -1 & | & 6 \\ 0 & -3 & 2 & | & -8 \\ 0 & -4 & 3 & | & -11 \\ 0 & -1 & 2 & | & -4 \end{bmatrix}$$

$$(-1)R4 \rightarrow R4$$

$$\Leftrightarrow \begin{bmatrix} 1 & 2 & -1 & | & 6 \\ 0 & -3 & 2 & | & -8 \\ 0 & -4 & 3 & | & -11 \\ 0 & 1 & -2 & | & 4 \end{bmatrix}$$

$(4)R4 + R3 \rightarrow R3$

$(3)R4 + R2 \rightarrow R2$

$$\Leftrightarrow \begin{bmatrix} 1 & 2 & -1 & | & 6 \\ 0 & 0 & -4 & | & 4 \\ 0 & 0 & -5 & | & 5 \\ 0 & 1 & -2 & | & 4 \end{bmatrix}$$

$\left(-\frac{1}{4}\right)R2 \rightarrow R2$

$\left(-\frac{1}{5}\right)R3 \rightarrow R3$

$$\Leftrightarrow \begin{bmatrix} 1 & 2 & -1 & | & 6 \\ 0 & 0 & 1 & | & -1 \\ 0 & 0 & 1 & | & -1 \\ 0 & 1 & -2 & | & 4 \end{bmatrix}$$

$R2 \leftrightarrow R4$ **(exchange $R2$ and $R4$)**

$$\Leftrightarrow \begin{bmatrix} 1 & 2 & -1 & | & 6 \\ 0 & 1 & -2 & | & 4 \\ 0 & 0 & 1 & | & -1 \\ 0 & 0 & 1 & | & -1 \end{bmatrix}$$

$(-1)R3 + R4 \rightarrow R4$

$$\Leftrightarrow \begin{bmatrix} 1 & 2 & -1 & | & 6 \\ 0 & 1 & -2 & | & 4 \\ 0 & 0 & 1 & | & -1 \\ 0 & 0 & 0 & | & 0 \end{bmatrix}$$

This final matrix is in echelon form. All lead entries are 1, the lead entries appear farther to the right as you move down the rows of the matrix, and the row of 0's is last. This matrix represents the following system of equations, which can be solved by back substitution:

1. $\qquad x + 2y - z = 6$
2. $\qquad\quad\; y - 2z = 4$
3. $\qquad\qquad\quad\; z = -1$
4. $\qquad 0x + 0y - 0z = 0$

Because Equation 4 is satisfied by *all* numbers x, y, and z, it is unnecessary and can be ignored. From Equation 3, you know that $z = -1$. So substitute -1 for z in Equation 2 and solve for y:

2. $\qquad y - 2z = 4$

$\qquad y - 2(-1) = 4$

$\qquad\quad y + 2 = 4$

$\qquad\qquad y = 2$

Then substitute -1 for z and 2 for y in Equation 1 and solve for x:

1. $x + 2y - z = 6$

$x + 2(2) - (-1) = 6$

$x + 4 + 1 = 6$

$x = 1$

Thus, the solution of the original system of equations is the ordered triple

$(x, y, z) = (1, 2, -1)$

The system is consistent because it has a solution. Check this solution to verify that it satisfies each of the four original equations.   ∎

Example 4   Use matrices to solve the system

$$\begin{cases} x + 2y + z = 8 \\ 2x + y - z = 1 \\ x - y - 2z = -7 \end{cases}$$

Solution   Set up the augmented matrix for the system, and use row operations to reduce it to echelon form:

$$(-2)R1 + R2 \rightarrow R2$$
$$(-1)R1 + R3 \rightarrow R3$$

$$\begin{bmatrix} 1 & 2 & 1 & \vdots & 8 \\ 2 & 1 & -1 & \vdots & 1 \\ 1 & -1 & -2 & \vdots & -7 \end{bmatrix} \Leftrightarrow \begin{bmatrix} 1 & 2 & 1 & \vdots & 8 \\ 0 & -3 & -3 & \vdots & -15 \\ 0 & -3 & -3 & \vdots & -15 \end{bmatrix}$$

$$\left(-\tfrac{1}{3}\right)R2 \rightarrow R2$$
$$\left(-\tfrac{1}{3}\right)R3 \rightarrow R3$$

$$\Leftrightarrow \begin{bmatrix} 1 & 2 & 1 & \vdots & 8 \\ 0 & 1 & 1 & \vdots & 5 \\ 0 & 1 & 1 & \vdots & 5 \end{bmatrix}$$

$$(-1)R2 + R3 \rightarrow R3$$

$$\Leftrightarrow \begin{bmatrix} 1 & 2 & 1 & \vdots & 8 \\ 0 & 1 & 1 & \vdots & 5 \\ 0 & 0 & 0 & \vdots & 0 \end{bmatrix}$$

This matrix is in echelon form and represents the system

1. $\begin{cases} x + 2y + z = 8 \\ y + z = 5 \\ 0x + 0y + 0z = 0 \end{cases}$
2.
3.

As before, the bottom equation can be ignored. Solve this system by back substitution. First, solve Equation 2 for the variable y:

$$y = 5 - z$$

Then, substitute $5 - z$ for y in Equation 1 and solve for x:

1.
$$x + 2y + z = 8$$
$$x + 2(5 - z) + z = 8$$
$$x + 10 - 2z + z = 8$$
$$x + 10 - z = 8$$
$$x = -2 + z$$

The solution of this system is

$$(x, y, z) = (-2 + z, 5 - z, z)$$

There is no unique solution for this system. The variable z can be any real number, but once you pick a number z, the numbers x and y are determined. For example, if z were equal to 3, then x would equal $-2 + 3$, or 1, and y would equal $5 - 3$, or 2. Thus, one possible solution to this system is $x = 1$, $y = 2$, and $z = 3$. Picking $z = 2$ gives another solution: $(0, 3, 2)$. Because this system has infinitely many solutions, it is consistent, but the equations are dependent. ∎

Example 5 Use matrices to solve the system

$$\begin{cases} w + x + y + z = 3 \\ x + 2y - z = -2 \\ w \quad - y - z = 2 \\ 2w + x \qquad = 2 \end{cases}$$

Solution Row reduce the augmented matrix as follows:

$$(-1)R1 + R3 \rightarrow R3$$

$$\left[\begin{array}{cccc|c} 1 & 1 & 1 & 1 & 3 \\ 0 & 1 & 2 & -1 & -2 \\ 1 & 0 & -1 & -1 & 2 \\ 2 & 1 & 0 & 0 & 2 \end{array}\right] \Leftrightarrow \left[\begin{array}{cccc|c} 1 & 1 & 1 & 1 & 3 \\ 0 & 1 & 2 & -1 & -2 \\ 0 & -1 & -2 & -2 & -1 \\ 2 & 1 & 0 & 0 & 2 \end{array}\right]$$

$$(-1)R3 \rightarrow R3$$
$$(-2)R1 + R4 \rightarrow R4$$

$$\Leftrightarrow \left[\begin{array}{cccc|c} 1 & 1 & 1 & 1 & 3 \\ 0 & 1 & 2 & -1 & -2 \\ 0 & 1 & 2 & 2 & 1 \\ 0 & -1 & -2 & -2 & -4 \end{array}\right]$$

$$R3 + R4 \rightarrow R4$$

$$\Leftrightarrow \begin{bmatrix} 1 & 1 & 1 & 1 & | & 3 \\ 0 & 1 & 2 & -1 & | & -2 \\ 0 & 1 & 2 & 2 & | & 1 \\ 0 & 0 & 0 & 0 & | & -3 \end{bmatrix}$$

The last row of the final matrix represents the equation

$$0w + 0x + 0y + 0z = -3$$

Obviously, *no* values of w, x, y, and z could make $0 = -3$. Because a solution must satisfy *each* equation, the given system has no solution. Hence, it is inconsistent.   ∎

EXERCISE 5.2

In Exercises 1–28, write the matrix represented by each system of equations in echelon form. Then solve the system by back substitution. If you have a computer and **Algebra Pak**, *you may use them to help row reduce each matrix. Select manual mode.*

1. $\begin{cases} 2x + y = 3 \\ x - 3y = 5 \end{cases}$

2. $\begin{cases} x + 2y = -1 \\ 3x - 5y = 19 \end{cases}$

3. $\begin{cases} x - 7y = -2 \\ 5x - 2y = -10 \end{cases}$

4. $\begin{cases} 3x - y = 3 \\ 2x + y = -3 \end{cases}$

5. $\begin{cases} 2x - y = 5 \\ x + 3y = 6 \end{cases}$

6. $\begin{cases} 3x - 5y = -25 \\ 2x + y = 5 \end{cases}$

7. $\begin{cases} x - 2y = 3 \\ -2x + 4y = 6 \end{cases}$

8. $\begin{cases} 3x - y = 7 \\ -x + \dfrac{1}{3}y = -\dfrac{7}{3} \end{cases}$

9. $\begin{cases} x - y + z = 3 \\ 2x - y + z = 4 \\ x + 2y - z = -1 \end{cases}$

10. $\begin{cases} 2x + y - z = 1 \\ x + y - z = 0 \\ 3x + y + 2z = 2 \end{cases}$

11. $\begin{cases} x + y - z = -1 \\ 3x + y = 4 \\ y - 2z = -4 \end{cases}$

12. $\begin{cases} 3x + y = 7 \\ x - z = 0 \\ y - 2z = -8 \end{cases}$

13. $\begin{cases} x - y + z = 2 \\ 2x + y + z = 5 \\ 3x - 4z = -5 \end{cases}$

14. $\begin{cases} x + z = -1 \\ 3x + y = 2 \\ 2x + y + 5z = 3 \end{cases}$

15. $\begin{cases} x + y + 2z = 4 \\ -x - y - 3z = -5 \\ 2x + y + z = 2 \end{cases}$

16. $\begin{cases} 2x - y + z = 6 \\ 3x + y - z = 2 \\ -x + 3y - 3z = 8 \end{cases}$

17. $\begin{cases} x + y = -2 \\ 3x - y = 6 \\ 2x + 2y = -4 \\ x - y = 4 \end{cases}$

18. $\begin{cases} x - y = -3 \\ 2x + y = -3 \\ 3x - y = -7 \\ 4x + y = -7 \end{cases}$

19. $\begin{cases} x + 2y + z = 4 \\ 3x - y - z = 2 \end{cases}$

20. $\begin{cases} x + 2y - 3z = -5 \\ 5x + y - z = -11 \end{cases}$

21. $\begin{cases} w + x - y + z = 2 \\ 2w - x - 2y + z = 0 \\ w - 2x - y + z = -1 \end{cases}$

22. $\begin{cases} w + x = 1 \\ w + y = 0 \\ x + z = 0 \end{cases}$

23. $\begin{cases} x + 2y + z = 4 \\ x - y + z = 1 \\ 2x + y + 2z = 2 \end{cases}$

24. $\begin{cases} x + y = 3 \\ 2x + y = 1 \\ 3x + 2y = 2 \end{cases}$

25. $\begin{cases} 2x - 2y + 3z + t = 2 \\ x + y + z + t = 5 \\ -x + 2y - 3z + 2t = 2 \\ x + y + 2z - t = 4 \end{cases}$

26. $\begin{cases} x + y + 2z + t = 1 \\ x + 2y + z + t = 2 \\ 2x + y + z + t = 4 \\ x + y + z + 2t = 3 \end{cases}$

27. $\begin{cases} x + y + t = 4 \\ x + z + t = 2 \\ 2x + 2y + z + 2t = 8 \\ x - y + z - t = -2 \end{cases}$

28. $\begin{cases} x - y + 2z + t = 3 \\ 3x - 2y - z - t = 4 \\ 2x + y + 2z - t = 10 \\ x + 2y + z - 3t = 8 \end{cases}$

In Exercises 29–32, use matrix methods to solve each system of equations.

29. $\begin{cases} x^2 + y^2 + z^2 = 14 \\ 2x^2 + 3y^2 - 2z^2 = -7 \\ x^2 - 5y^2 + z^2 = 8 \end{cases}$

    (*Hint*: Solve first as a linear system in x^2, y^2, and z^2.)

30. $\begin{cases} \dfrac{3}{x} + \dfrac{1}{y} + \dfrac{1}{z} = 4 \\[2mm] \dfrac{1}{x} - \dfrac{3}{y} - \dfrac{2}{z} = -3 \\[2mm] \dfrac{7}{x} - \dfrac{9}{y} + \dfrac{3}{z} = -14 \end{cases}$

31. $\begin{cases} 5\sqrt{x} + 2\sqrt{y} + \sqrt{z} = 22 \\ \sqrt{x} + \sqrt{y} - \sqrt{z} = 5 \\ 3\sqrt{x} - 2\sqrt{y} - 3\sqrt{z} = 10 \end{cases}$

32. $\begin{cases} \sqrt{x} + \dfrac{2}{y} - z^2 = 0 \\[2mm] 2\sqrt{x} - \dfrac{5}{y} + z^2 = 3 \\[2mm] -\sqrt{x} + \dfrac{1}{y} + 2z^2 = 7 \end{cases}$

Computer Exercises. *If the numbers above each lead entry in a matrix written in echelon form are 0, the matrix is said to be written in reduced echelon form. From such a matrix, the solution of a system of equations can often be read with no back substitution. In Exercises 33–36, use a computer and* **Algebra Pak** *to solve each system. If you choose automatic mode,* **Algebra Pak** *will put the matrix in reduced echelon form.*

33. $\begin{cases} \dfrac{1}{3}x + \dfrac{3}{4}y - \dfrac{2}{3}z = -2 \\[2mm] x + \dfrac{1}{2}y + \dfrac{1}{3}z = 1 \\[2mm] \dfrac{1}{6}x - \dfrac{1}{8}y - z = 0 \end{cases}$

34. $\begin{cases} \dfrac{1}{4}x + y + 3z = 1 \\[2mm] \dfrac{1}{2}x - 4y + 6z = -1 \\[2mm] \dfrac{1}{3}x - 2y - 2z = -1 \end{cases}$

35.
$$\begin{cases} \dfrac{1}{2}x + \dfrac{1}{4}y - z = 2 \\[2mm] \dfrac{3}{2}x + \dfrac{1}{4}y + \dfrac{1}{2}z = \dfrac{3}{2} \\[2mm] \dfrac{2}{3}x \qquad + z = -\dfrac{1}{3} \end{cases}$$

36.
$$\begin{cases} \dfrac{5}{7}x - \dfrac{1}{3}y + z = 0 \\[2mm] \dfrac{2}{7}x + y + \dfrac{1}{8}z = 9 \\[2mm] 6x + 4y - \dfrac{27}{4}z = 20 \end{cases}$$

5.3   MATRIX ALGEBRA

In this section we discuss how to add, subtract, and multiply matrices.

> **Definition.**   An **_m × n matrix_** is a rectangular array of *mn* numbers arranged in *m* rows and *n* columns.

We will use any of the following notations to denote the matrix A:

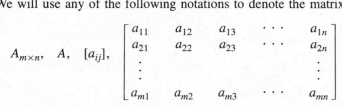

$$A_{m \times n}, \quad A, \quad [a_{ij}], \quad \begin{bmatrix} a_{11} & a_{12} & a_{13} & \cdots & a_{1n} \\ a_{21} & a_{22} & a_{23} & \cdots & a_{2n} \\ \vdots & & & & \vdots \\ a_{m1} & a_{m2} & a_{m3} & \cdots & a_{mn} \end{bmatrix}$$

The symbol a_{23} represents the entry in row two, column three, of the matrix A. Similarly, the symbol a_{ij} represents the entry in the ith row and the jth column.

Arthur Cayley (1821–1895)
Cayley was a major founder
of the theory of matrices.

> **Definition of Equal Matrices.**   If $A = [a_{ij}]$ and $B = [b_{ij}]$ are both $m \times n$ matrices, then
>
> $\qquad A = B$ if and only if $a_{ij} = b_{ij}$
>
> for all i and j, where $i = 1, 2, 3, \ldots, m$ and $j = 1, 2, 3, \ldots, n$.

The previous definition points out that two matrices must be identical to be equal. They must be the same size and have the same corresponding entries.

> **The Sum of Two Matrices.**   Let $A = [a_{ij}]$ and $B = [b_{ij}]$ be two $m \times n$ matrices. The sum, $A + B$, is the $m \times n$ matrix C found by adding the corresponding entries of matrices A and B:
>
> $\qquad A + B = C = [c_{ij}]$
>
> where $c_{ij} = a_{ij} + b_{ij}$, for $i = 1, 2, 3, \ldots, m$ and $j = 1, 2, 3, \ldots, n$.

Example 1   Add the matrices $\begin{bmatrix} 2 & 1 & 3 \\ 1 & -1 & 0 \end{bmatrix}$ and $\begin{bmatrix} 1 & -1 & 2 \\ -1 & 1 & 5 \end{bmatrix}$.

Solution   Because each matrix is 2×3, their sum is defined and can be calculated by adding their corresponding elements:

$$\begin{bmatrix} 2 & 1 & 3 \\ 1 & -1 & 0 \end{bmatrix} + \begin{bmatrix} 1 & -1 & 2 \\ -1 & 1 & 5 \end{bmatrix} = \begin{bmatrix} 2+1 & 1-1 & 3+2 \\ 1-1 & -1+1 & 0+5 \end{bmatrix}$$

$$= \begin{bmatrix} 3 & 0 & 5 \\ 0 & 0 & 5 \end{bmatrix}$$ ∎

Example 2   If possible, add the matrices $\begin{bmatrix} 2 & 4 & 3 \\ 1 & 1 & 1 \end{bmatrix}$ and $\begin{bmatrix} 1 & 2 \\ 2 & 3 \end{bmatrix}$.

Solution   The first matrix is 2×3 and the second is 2×2. Because these matrices are of different sizes, they cannot be added. ∎

Several of the field properties discussed in Chapter 1 apply to matrices also.

Theorem.   The addition of two $m \times n$ matrices is commutative.

Proof   Let $A = [a_{ij}]$ and $B = [b_{ij}]$ be $m \times n$ matrices. Then $A + B = C = [c_{ij}]$, where $c_{ij} = a_{ij} + b_{ij}$ for each $i = 1, 2, 3, \ldots, m$ and $j = 1, 2, 3, \ldots, n$. On the other hand, $B + A = D = [d_{ij}]$, where $d_{ij} = b_{ij} + a_{ij}$ for each i and j.

Because each entry in each matrix is a real number and the addition of real numbers is commutative, it follows that

$$c_{ij} = a_{ij} + b_{ij} = b_{ij} + a_{ij} = d_{ij}$$

for all i and j, $i = 1, 2, 3, \ldots, m$ and $j = 1, 2, 3, \ldots, n$.

By the definition of equality of matrices, $C = D$, and therefore $A + B = B + A$. □

Theorem.   The addition of three $m \times n$ matrices is associative.

The proof of the previous theorem is left as an exercise.

In the collection of all $m \times n$ matrices, there is a matrix called the **zero matrix**.

Definition.   Let A be any $m \times n$ matrix. There is an $m \times n$ matrix **0**, called the **zero matrix**, or the **additive identity matrix**, for which

$$A + 0 = 0 + A = A$$

The matrix **0** consists of m rows and n columns of 0's.

To illustrate the above definition, we note that the matrix

$$\begin{bmatrix} 0 & 0 & 0 \\ 0 & 0 & 0 \\ 0 & 0 & 0 \end{bmatrix}$$

is the 3×3 zero matrix, and that

$$\begin{bmatrix} 0 & 0 & 0 \\ 0 & 0 & 0 \\ 0 & 0 & 0 \end{bmatrix} + \begin{bmatrix} 1 & 2 & 3 \\ 4 & 5 & 6 \\ 7 & 8 & 9 \end{bmatrix} = \begin{bmatrix} 1 & 2 & 3 \\ 4 & 5 & 6 \\ 7 & 8 & 9 \end{bmatrix}$$

Matrices are similar to real numbers in another respect: Every matrix has an additive inverse.

> **Definition.** Any $m \times n$ matrix A has an **additive inverse**, an $m \times n$ matrix $-A$ with the property that the sum of A and $-A$ is the zero matrix:
>
> $$A + (-A) = (-A) + A = \mathbf{0}$$
>
> The entries of $-A$ are the negatives of the corresponding entries of A.

The additive inverse of the 2×3 matrix $A = \begin{bmatrix} 1 & -3 & 2 \\ 0 & 1 & -5 \end{bmatrix}$ is the matrix

$$-A = \begin{bmatrix} -1 & 3 & -2 \\ 0 & -1 & 5 \end{bmatrix}$$

because their sum is the zero matrix:

$$A + (-A) = \begin{bmatrix} 1 & -3 & 2 \\ 0 & 1 & -5 \end{bmatrix} + \begin{bmatrix} -1 & 3 & -2 \\ 0 & -1 & 5 \end{bmatrix}$$

$$= \begin{bmatrix} 1-1 & -3+3 & 2-2 \\ 0+0 & 1-1 & -5+5 \end{bmatrix}$$

$$= \begin{bmatrix} 0 & 0 & 0 \\ 0 & 0 & 0 \end{bmatrix}$$

> **The Difference of Two Matrices.** If A and B are $m \times n$ matrices, their difference, $A - B$, is the sum of A and the additive inverse of B:
>
> $$A - B = A + (-B)$$

Example 3

$$\begin{bmatrix} 2 & -5 \\ 3 & 1 \end{bmatrix} - \begin{bmatrix} 4 & -5 \\ -3 & 9 \end{bmatrix} = \begin{bmatrix} 2 & -5 \\ 3 & 1 \end{bmatrix} + \begin{bmatrix} -4 & 5 \\ 3 & -9 \end{bmatrix}$$

$$= \begin{bmatrix} -2 & 0 \\ 6 & -8 \end{bmatrix}$$ ∎

We illustrate how to find the product of two matrices by computing the product of a 2 × 3 matrix A and a 3 × 3 matrix B. The result is the 2 × 3 matrix C.

$$A \cdot B = \begin{bmatrix} 1 & 2 & 3 \\ 4 & 5 & 6 \end{bmatrix} \cdot \begin{bmatrix} a & b & c \\ d & e & f \\ g & h & i \end{bmatrix} = C$$

Each entry of matrix C is the result of a calculation that involves a row of A and a column of B. For example, the first-row, third-column entry of matrix C is found by keeping a running total of the products of corresponding entries of the first row of A and the third column of B:

$$\begin{bmatrix} 1 & 2 & 3 \\ 4 & 5 & 6 \end{bmatrix} \begin{bmatrix} a & b & c \\ d & e & f \\ g & h & i \end{bmatrix} = \begin{bmatrix} ? & ? & 1c+2f+3i \\ ? & ? & ? \end{bmatrix}$$

Similarly, the second-row, second-column entry of matrix C is formed by a calculation involving the second row of A and the second column of B.

$$\begin{bmatrix} 1 & 2 & 3 \\ 4 & 5 & 6 \end{bmatrix} \begin{bmatrix} a & b & c \\ d & e & f \\ g & h & i \end{bmatrix} = \begin{bmatrix} ? & ? & 1c+2f+3i \\ ? & 4b+5e+6h & ? \end{bmatrix}$$

To calculate the first-row, first-column entry of matrix C, we use the first row of A and the first column of B.

$$\begin{bmatrix} 1 & 2 & 3 \\ 4 & 5 & 6 \end{bmatrix} \begin{bmatrix} a & b & c \\ d & e & f \\ g & h & i \end{bmatrix} = \begin{bmatrix} 1a+2d+3g & ? & 1c+2f+3i \\ ? & 4b+5e+6h & ? \end{bmatrix}$$

The complete product C is

$$\begin{bmatrix} 1 & 2 & 3 \\ 4 & 5 & 6 \end{bmatrix} \begin{bmatrix} a & b & c \\ d & e & f \\ g & h & i \end{bmatrix} = \begin{bmatrix} 1a+2d+3g & 1b+2e+3h & 1c+2f+3i \\ 4a+5d+6g & 4b+5e+6h & 4c+5f+6i \end{bmatrix}$$

For the product $A \cdot B$ to exist, the number of columns of A must equal the number of rows of B. If the product exists, it will have as many rows as A and as many columns as B:

$$\begin{array}{ccccc} A & \cdot & B & = & C \\ m \times n & & n \times p & & m \times p \end{array}$$

These must agree.

The product is $m \times p$.

More formally, we have the following definition.

The Product of Two Matrices.   Let $A = [a_{ij}]$ be an $m \times n$ matrix and $B = [b_{ij}]$ be an $n \times p$ matrix. The product, AB, is the $m \times p$ matrix C found as follows:

$AB = C = [c_{ij}]$, where c_{ij} is the sum of the products of the corresponding entries in the ith row of A and the jth column of B, where $i = 1, 2, 3, \ldots, m$ and $j = 1, 2, 3, \ldots, p$.

Example 4   Find C if $AB = \begin{bmatrix} 1 & 2 \\ 3 & 4 \\ 5 & 6 \end{bmatrix} \begin{bmatrix} a & b \\ c & d \end{bmatrix} = C.$

Solution   Because the first matrix is 3×2 and the second matrix is 2×2, the product C exists, and it is a 3×2 matrix. The first-row, first-column entry of C is the total of the products of corresponding entries in the first row of A and the first column of B: $c_{11} = 1a + 2c$. Similarly, c_{12} is computed by using the first row of A and the second column of B: $c_{12} = 1b + 2d$. The entire product is

$$\begin{bmatrix} 1 & 2 \\ 3 & 4 \\ 5 & 6 \end{bmatrix} \begin{bmatrix} a & b \\ c & d \end{bmatrix} = \begin{bmatrix} 1a+2c & 1b+2d \\ 3a+4c & 3b+4d \\ 5a+6c & 5b+6d \end{bmatrix} \quad \blacksquare$$

Example 5   Find the product $\begin{bmatrix} 1 & -1 & 2 \\ 1 & 3 & 0 \\ 0 & 1 & 1 \end{bmatrix} \begin{bmatrix} 2 & 1 \\ 1 & 3 \\ 0 & 1 \end{bmatrix}.$

Solution   Because the matrices are 3×3 and 3×2, the product is a 3×2 matrix.

$$\begin{bmatrix} 1 & -1 & 2 \\ 1 & 3 & 0 \\ 0 & 1 & 1 \end{bmatrix} \begin{bmatrix} 2 & 1 \\ 1 & 3 \\ 0 & 1 \end{bmatrix}$$

$$= \begin{bmatrix} 1 \cdot 2 + (-1) \cdot 1 + 2 \cdot 0 & 1 \cdot 1 + (-1) \cdot 3 + 2 \cdot 1 \\ 1 \cdot 2 + 3 \cdot 1 + 0 \cdot 0 & 1 \cdot 1 + 3 \cdot 3 + 0 \cdot 1 \\ 0 \cdot 2 + 1 \cdot 1 + 1 \cdot 0 & 0 \cdot 1 + 1 \cdot 3 + 1 \cdot 1 \end{bmatrix}$$

$$= \begin{bmatrix} 1 & 0 \\ 5 & 10 \\ 1 & 4 \end{bmatrix} \quad \blacksquare$$

Example 6 Find the product $[1 \quad 2 \quad 3] \begin{bmatrix} 4 \\ 5 \\ 6 \end{bmatrix}$.

Solution Because the first matrix is 1×3 and the second matrix is 3×1, the product is a 1×1 matrix:

$$[1 \quad 2 \quad 3] \begin{bmatrix} 4 \\ 5 \\ 6 \end{bmatrix} = [1 \cdot 4 + 2 \cdot 5 + 3 \cdot 6] = [32]$$

■

Example 7 If $A = \begin{bmatrix} 1 & 1 \\ 0 & 0 \end{bmatrix}$ and $B = \begin{bmatrix} 0 & 1 \\ 0 & 1 \end{bmatrix}$, calculate AB and BA and thereby show that multiplication of matrices is not commutative.

Solution
$$AB = \begin{bmatrix} 1 & 1 \\ 0 & 0 \end{bmatrix} \begin{bmatrix} 0 & 1 \\ 0 & 1 \end{bmatrix} = \begin{bmatrix} 0 & 2 \\ 0 & 0 \end{bmatrix}$$

$$BA = \begin{bmatrix} 0 & 1 \\ 0 & 1 \end{bmatrix} \begin{bmatrix} 1 & 1 \\ 0 & 0 \end{bmatrix} = \begin{bmatrix} 0 & 0 \\ 0 & 0 \end{bmatrix}$$

Because the products are not equal, matrix multiplication is not commutative. ■

Example 8 Find values for x, y, and z such that

$$\begin{bmatrix} 1 & 2 & 3 \\ 2 & -1 & -2 \\ 1 & -3 & -3 \end{bmatrix} \begin{bmatrix} x \\ y \\ z \end{bmatrix} = \begin{bmatrix} 4 \\ 0 \\ -2 \end{bmatrix}$$

Solution Find the product of the first two matrices and set it equal to the third matrix.

$$\begin{bmatrix} 1 & 2 & 3 \\ 2 & -1 & -2 \\ 1 & -3 & -3 \end{bmatrix} \begin{bmatrix} x \\ y \\ z \end{bmatrix} = \begin{bmatrix} 1x + 2y + 3z \\ 2x - 1y - 2z \\ 1x - 3y - 3z \end{bmatrix} = \begin{bmatrix} 4 \\ 0 \\ -2 \end{bmatrix}$$

The product will be equal to the third matrix if and only if their corresponding components are equal. Set the corresponding components equal to get

$$\begin{cases} x + 2y + 3z = 4 \\ 2x - y - 2z = 0 \\ x - 3y - 3z = -2 \end{cases}$$

This system was solved in Example 2 in Section 5.2. Its solution is $x = 1$, $y = 0$, and $z = 1$. ■

The number 1 is called the *identity for multiplication* because multiplying a number by 1 does not change that number. There is a **multiplicative identity matrix** with a similar property.

Definition.   Let A be an $n \times n$ matrix. There is an $n \times n$ **identity matrix** I for which

$$AI = IA = A$$

The matrix I consists of 1s on its diagonal and 0's elsewhere.

$$I = \begin{bmatrix} 1 & 0 & 0 & \cdots & 0 \\ 0 & 1 & 0 & \cdots & 0 \\ 0 & 0 & 1 & \cdots & 0 \\ \vdots & \vdots & \vdots & & \vdots \\ 0 & 0 & 0 & \cdots & 1 \end{bmatrix}$$

Note that an identity matrix is a square matrix—it has the same number of rows and columns.

Example 9 illustrates the previous definition for the 3×3 identity matrix.

Example 9   Find   **a.** $\begin{bmatrix} 1 & 0 & 0 \\ 0 & 1 & 0 \\ 0 & 0 & 1 \end{bmatrix} \begin{bmatrix} 1 & 2 & 3 \\ 4 & 5 & 6 \\ 7 & 8 & 9 \end{bmatrix}$ and **b.** $\begin{bmatrix} 1 & 2 & 3 \\ 4 & 5 & 6 \\ 7 & 8 & 9 \end{bmatrix} \begin{bmatrix} 1 & 0 & 0 \\ 0 & 1 & 0 \\ 0 & 0 & 1 \end{bmatrix}.$

Solution   **a.** $\begin{bmatrix} 1 & 0 & 0 \\ 0 & 1 & 0 \\ 0 & 0 & 1 \end{bmatrix} \begin{bmatrix} 1 & 2 & 3 \\ 4 & 5 & 6 \\ 7 & 8 & 9 \end{bmatrix}$

$$= \begin{bmatrix} 1 \cdot 1 + 0 \cdot 4 + 0 \cdot 7 & 1 \cdot 2 + 0 \cdot 5 + 0 \cdot 8 & 1 \cdot 3 + 0 \cdot 6 + 0 \cdot 9 \\ 0 \cdot 1 + 1 \cdot 4 + 0 \cdot 7 & 0 \cdot 2 + 1 \cdot 5 + 0 \cdot 8 & 0 \cdot 3 + 1 \cdot 6 + 0 \cdot 9 \\ 0 \cdot 1 + 0 \cdot 4 + 1 \cdot 7 & 0 \cdot 2 + 0 \cdot 5 + 1 \cdot 8 & 0 \cdot 3 + 0 \cdot 6 + 1 \cdot 9 \end{bmatrix}$$

$$= \begin{bmatrix} 1 & 2 & 3 \\ 4 & 5 & 6 \\ 7 & 8 & 9 \end{bmatrix}$$

b. $\begin{bmatrix} 1 & 2 & 3 \\ 4 & 5 & 6 \\ 7 & 8 & 9 \end{bmatrix} \begin{bmatrix} 1 & 0 & 0 \\ 0 & 1 & 0 \\ 0 & 0 & 1 \end{bmatrix} = \begin{bmatrix} 1 & 2 & 3 \\ 4 & 5 & 6 \\ 7 & 8 & 9 \end{bmatrix}$ ∎

EXERCISE 5.3

In Exercises 1–8, find values of x and y, if any, that will make the two matrices equal.

1. $\begin{bmatrix} x & y \\ 1 & 3 \end{bmatrix} = \begin{bmatrix} 2 & 5 \\ 1 & 3 \end{bmatrix}$

2. $\begin{bmatrix} x & 5 \\ 3 & y \end{bmatrix} = \begin{bmatrix} 0 & 5 \\ 3 & 2 \end{bmatrix}$

3. $\begin{bmatrix} x & y \\ 1 & 3 \end{bmatrix} = \begin{bmatrix} 2 & 5 \\ 1 & 4 \end{bmatrix}$

4. $\begin{bmatrix} x & y \\ 1 & x+y \end{bmatrix} = \begin{bmatrix} 2 & 1 \\ 1 & 2 \end{bmatrix}$

5. $\begin{bmatrix} x+y & 3+x \\ -2 & 5y \end{bmatrix} = \begin{bmatrix} 3 & 4 \\ -2 & 10 \end{bmatrix}$

6. $\begin{bmatrix} x+y & x-y \\ 2x & 3y \end{bmatrix} = \begin{bmatrix} -x & x-2 \\ -y & 8-y \end{bmatrix}$

7. $\begin{bmatrix} x & 3x \\ y & x+1 \end{bmatrix} = \begin{bmatrix} y & 6 \\ 2 & 3 \end{bmatrix}$

8. $\begin{bmatrix} x \\ y \end{bmatrix} = \begin{bmatrix} 1 & 2 \\ 3 & 4 \end{bmatrix}$

In Exercises 9–18, perform the indicated operation, if possible.

9. $\begin{bmatrix} 2 & 1 & -1 \\ -3 & 2 & 5 \end{bmatrix} + \begin{bmatrix} -3 & 1 & 2 \\ -3 & -2 & -5 \end{bmatrix}$

10. $\begin{bmatrix} 3 & 1 \\ 2 & 2 \end{bmatrix} + \begin{bmatrix} 2 & 1 \\ -1 & 0 \end{bmatrix} + \begin{bmatrix} -5 & -2 \\ -1 & -2 \end{bmatrix}$

11. $\begin{bmatrix} 3 & 2 & 1 \\ -2 & 3 & -3 \\ -4 & -2 & -1 \end{bmatrix} - \begin{bmatrix} -2 & 6 & -2 \\ 5 & 7 & -1 \\ -4 & -6 & 7 \end{bmatrix}$

12. $\begin{bmatrix} -2 & 7 & -3 \\ 3 & 6 & -7 \\ -9 & -2 & -5 \end{bmatrix} + \begin{bmatrix} -5 & -4 & -3 \\ -1 & 2 & 10 \\ -1 & -3 & -4 \end{bmatrix}$

13. $\begin{bmatrix} 1 & 3 & -1 \\ 2 & 1 & 5 \\ 1 & 3 & 0 \end{bmatrix} + \begin{bmatrix} 2 \\ 0 \\ -3 \end{bmatrix}$

14. $\begin{bmatrix} 3 \\ 2 \\ 3 \end{bmatrix} - \begin{bmatrix} -3 & 5 & -6 \\ -3 & -5 & -6 \\ 4 & 6 & -6 \end{bmatrix}$

15. $[1 \quad 2 \quad 3] - [4 \quad 5 \quad 6]$

16. $\begin{bmatrix} 1 \\ 2 \\ 3 \end{bmatrix} + [4 \quad -5 \quad -6]$

17. $\begin{bmatrix} 1 & 3 & -4 \\ 2 & -1 & 3 \\ 1 & 5 & 7 \end{bmatrix} + \begin{bmatrix} 3 & 2 & -8 \\ 9 & 11 & 17 \\ 2 & 1 & 3 \end{bmatrix} - \begin{bmatrix} 1 & 3 & -5 \\ 2 & -9 & 5 \\ 3 & 10 & 11 \end{bmatrix}$

18. $\begin{bmatrix} -3 & -2 & 15 \\ 2 & -5 & 9 \end{bmatrix} - \begin{bmatrix} 3 & 2 & -15 \\ -2 & 5 & -9 \end{bmatrix} + \begin{bmatrix} 6 & 4 & -30 \\ -3 & 12 & -15 \end{bmatrix}$

In Exercises 19–30, find each product, if possible.

19. $\begin{bmatrix} 2 & 3 \\ 3 & -2 \end{bmatrix} \begin{bmatrix} 1 & 2 \\ 0 & -2 \end{bmatrix}$

20. $\begin{bmatrix} -2 & 3 \\ 3 & -2 \end{bmatrix} \begin{bmatrix} 2 & 4 \\ -5 & 7 \end{bmatrix}$

21. $\begin{bmatrix} -4 & -2 \\ 21 & 0 \end{bmatrix} \begin{bmatrix} -5 & 6 \\ 21 & -1 \end{bmatrix}$

22. $\begin{bmatrix} -5 & 4 \\ 4 & -5 \end{bmatrix} \begin{bmatrix} 6 & -2 \\ 1 & 3 \end{bmatrix}$

23. $\begin{bmatrix} 2 & 1 & 3 \\ 1 & 2 & -1 \\ 0 & 1 & 0 \end{bmatrix} \begin{bmatrix} 1 & 2 & 3 \\ 2 & -2 & 1 \\ 0 & 0 & 1 \end{bmatrix}$

24. $\begin{bmatrix} 2 & 1 & 1 \\ 1 & 1 & 2 \\ 1 & -2 & -1 \end{bmatrix} \begin{bmatrix} 1 & 2 & 3 \\ 1 & 2 & -3 \\ -1 & -1 & 3 \end{bmatrix}$

25.
$$\begin{bmatrix} 1 & -2 & -3 \\ 2 & 0 & 1 \end{bmatrix} \begin{bmatrix} 4 \\ -5 \\ -6 \end{bmatrix}$$

26.
$$\begin{bmatrix} 1 \\ -2 \\ -3 \end{bmatrix} \begin{bmatrix} 4 & -5 & -6 \end{bmatrix}$$

27.
$$\begin{bmatrix} 1 & 2 & 3 \end{bmatrix} \begin{bmatrix} 4 & 5 & 6 \\ 7 & 8 & 9 \end{bmatrix}$$

28.
$$\begin{bmatrix} 2 & 3 & 4 \\ 1 & 2 & 3 \\ -2 & 2 & 2 \end{bmatrix} \begin{bmatrix} -1 \\ 2 \\ 3 \end{bmatrix}$$

29.
$$\begin{bmatrix} 1 & 1 & 2 \\ 2 & 2 & 1 \end{bmatrix} \begin{bmatrix} 1 & 2 & 3 \\ 4 & 5 & 6 \\ 7 & 8 & 9 \end{bmatrix}$$

30.
$$\begin{bmatrix} 1 & 2 & 3 \\ 1 & 2 & 1 \\ 1 & -1 & -1 \end{bmatrix} \begin{bmatrix} 1 & 2 \\ 2 & 1 \\ 1 & 1 \end{bmatrix}$$

In Exercises 31–36, perform the indicated operations.

31.
$$\begin{bmatrix} 1 & 2 \\ 2 & 3 \end{bmatrix} \left(\begin{bmatrix} 2 & 1 & -5 \\ 1 & 1 & 2 \end{bmatrix} + \begin{bmatrix} -2 & -1 & 6 \\ 0 & -1 & -1 \end{bmatrix} \right)$$

32.
$$\begin{bmatrix} 1 & 2 \\ 2 & 3 \end{bmatrix} \begin{bmatrix} 2 & 1 & -5 \\ 1 & 1 & 2 \end{bmatrix} + \begin{bmatrix} 1 & 2 \\ 2 & 3 \end{bmatrix} \begin{bmatrix} -2 & -1 & 6 \\ 0 & -1 & -1 \end{bmatrix}$$

33.
$$\begin{bmatrix} 1 & 2 & 3 \\ 2 & 3 & 1 \\ 1 & 2 & 1 \end{bmatrix} \begin{bmatrix} 2 & 1 & 1 \\ 3 & -1 & -1 \\ 2 & -2 & 2 \end{bmatrix} + \begin{bmatrix} -2 & 3 & 4 \\ 1 & 1 & 1 \\ 0 & 1 & 0 \end{bmatrix}$$

34.
$$\begin{bmatrix} 2 & 1 & 0 \\ 1 & -2 & -1 \\ 1 & 1 & -1 \end{bmatrix} \left(\begin{bmatrix} 1 & 0 & 1 \\ 1 & 1 & 2 \\ 1 & 2 & -1 \end{bmatrix} + \begin{bmatrix} -1 & -1 & 2 \\ 0 & 0 & 1 \\ 1 & 0 & -1 \end{bmatrix} \right)$$

35.
$$\left(\begin{bmatrix} 1 & 2 \\ 2 & 3 \end{bmatrix} \begin{bmatrix} 1 \\ -3 \end{bmatrix} + \begin{bmatrix} -2 \\ 1 \end{bmatrix} \right) \left(\begin{bmatrix} 1 & 2 \end{bmatrix} \begin{bmatrix} 1 \\ -3 \end{bmatrix} + \begin{bmatrix} 4 \end{bmatrix} \right)$$

36.
$$\begin{bmatrix} 1 \\ 2 \end{bmatrix} \begin{bmatrix} -3 & -4 \end{bmatrix} - \begin{bmatrix} 0 & 3 \\ 2 & 1 \end{bmatrix} \begin{bmatrix} 2 & 0 \\ 1 & -1 \end{bmatrix}$$

In Exercises 37–44, let $A = \begin{bmatrix} 1 & 3 \\ 2 & 5 \end{bmatrix}$, $B = \begin{bmatrix} -1 \\ 3 \end{bmatrix}$, *and* $C = \begin{bmatrix} 3 & 2 \end{bmatrix}$. *Perform, if possible, the indicated operations.*

37. $A - BC$
38. $AB + B$
39. $CB - AB$
40. CAB
41. ABC
42. $CA + C$
43. $A^2 B$
44. $(BC)^2$

45. Let $A = \begin{bmatrix} 1 & 1 \\ 1 & 1 \end{bmatrix}$. Find A^7. (*Hint:* Calculate A^2 and A^3. Do you see a pattern?)

46. If $A = \begin{bmatrix} 1 & 1 \\ 0 & 1 \end{bmatrix}$, compute A^n for various values of n ($n = 2, 3, 4, \ldots$). What do you notice?

47. If $A = \begin{bmatrix} 1 & 0 \\ 1 & 1 \end{bmatrix}$, compute A^n for various values of n ($n = 2, 3, 4, \ldots$). What do you notice?

48. If $A = \begin{bmatrix} 0 & 1 \\ 1 & 0 \end{bmatrix}$, compute $A^2, A^3, A^4, \ldots$. Give a general rule for A^n, where n is a natural number.

49. If A and B are 2×2 matrices, is $(AB)^2$ equal to A^2B^2? Support your answer.

50. Let a, b, and c be real numbers. If $ab = ac$ and $a \neq 0$, then $b = c$. Find 2×2 matrices A, B, and C, where $A \neq 0$, to show that such a law does not hold for all matrices.

51. In the real number system the numbers 0 and 1 are the only numbers that equal their own squares: if $a^2 = a$, then $a = 0$ or $a = 1$. Find a 2×2 matrix A that is neither the zero matrix nor the identity matrix, such that $A^2 = A \cdot A = A$.

52. Another property of the real numbers is that, if $ab = 0$, then either $a = 0$ or $b = 0$. To show that this property is not truc for matriccs, find two nonzero 2×2 matrices, A and B, such that $AB = 0$.

53. Multiplication of three $n \times n$ matrices is associative: $(AB)C = A(BC)$. Verify this by an example chosen from the set of 2×2 matrices.

54. If A is an $m \times n$ matrix and B and C are each $n \times p$ matrices, then $A(B + C) = AB + AC$. Illustrate this with an example showing that matrix multiplication distributes over matrix addition.

55. Prove that the addition of matrices is associative.

56. If $AB = -BA$, matrices A and B are said to **anticommute**. Show that any two of the following matrices anticommute.

$$\begin{bmatrix} 0 & -i \\ i & 0 \end{bmatrix} \quad \begin{bmatrix} 1 & 0 \\ 0 & -1 \end{bmatrix} \quad \begin{bmatrix} 0 & 1 \\ 1 & 0 \end{bmatrix} \quad (i = \sqrt{-1})$$

In quantum mechanics, these matrices are important for describing electron spin.

5.4 MATRIX INVERSION

Two real numbers are called **multiplicative inverses** if their product is the multiplicative identity 1. Some matrices have multiplicative inverses also.

Definition. If A and B are $n \times n$ matrices, I is the $n \times n$ identity matrix, and

$$AB = BA = I$$

then A and B are called **multiplicative inverses**. Matrix A is the **inverse** of B, and B is the **inverse** of A.

It can be shown that the inverse of a matrix A, if it exists, is unique. The inverse of A is denoted by A^{-1}.

Example 1 If $A = \begin{bmatrix} 1 & 1 & 0 \\ 4 & 3 & 0 \\ 2 & 1 & -1 \end{bmatrix}$ and $B = \begin{bmatrix} -3 & 1 & 0 \\ 4 & -1 & 0 \\ -2 & 1 & -1 \end{bmatrix}$, show that A and B are inverses.

Solution   Multiply the matrices in each order to show that the product is the identity matrix.

$$AB = \begin{bmatrix} 1 & 1 & 0 \\ 4 & 3 & 0 \\ 2 & 1 & -1 \end{bmatrix} \begin{bmatrix} -3 & 1 & 0 \\ 4 & -1 & 0 \\ -2 & 1 & -1 \end{bmatrix}$$

$$= \begin{bmatrix} -3 + 4 & 1 - 1 & 0 \\ -12 + 12 & 4 - 3 & 0 \\ -6 + 4 + 2 & 2 - 1 - 1 & 1 \end{bmatrix} = \begin{bmatrix} 1 & 0 & 0 \\ 0 & 1 & 0 \\ 0 & 0 & 1 \end{bmatrix}$$

$$BA = \begin{bmatrix} -3 & 1 & 0 \\ 4 & -1 & 0 \\ -2 & 1 & -1 \end{bmatrix} \begin{bmatrix} 1 & 1 & 0 \\ 4 & 3 & 0 \\ 2 & 1 & -1 \end{bmatrix} = \begin{bmatrix} 1 & 0 & 0 \\ 0 & 1 & 0 \\ 0 & 0 & 1 \end{bmatrix} \blacksquare$$

If a matrix has an inverse, it is called a **nonsingular matrix**. Otherwise, it is called a **singular matrix**. The following theorem, stated without proof, provides a way of calculating the inverse of a nonsingular matrix.

Theorem.   If a sequence of elementary row operations performed on the $n \times n$ matrix A reduces A to the $n \times n$ identity matrix I, then those same row operations, performed in the same order on the identity matrix I, will transform I into A^{-1}. Furthermore, if *no* sequence of row operations will reduce A to I, then A is singular.

To use the previous theorem, we perform elementary row operations on matrix A to change it to the identity matrix I. At the same time, we perform these elementary row operations on the identity matrix I. This changes I into A^{-1}.

A notation for this process uses an n-row by $2n$-column matrix, with matrix A as the left half and matrix I as the right half. If A is nonsingular, the proper row operations performed on $[A \mid I]$ will transform it into $[I \mid A^{-1}]$.

Example 2   Find the inverse of matrix A if $A = \begin{bmatrix} 2 & -4 \\ 4 & -7 \end{bmatrix}$.

Solution   Set up a 2×4 matrix with A on the left and I on the right of the broken line:

$$[A \mid I] = \begin{bmatrix} 2 & -4 & \vdots & 1 & 0 \\ 4 & -7 & \vdots & 0 & 1 \end{bmatrix}$$

Perform row operations on the entire matrix to transform the left half into I:

$$\left(\tfrac{1}{2}\right)R1 \rightarrow R1$$
$$(-2)R1 + R2 \rightarrow R2 \qquad\qquad (2)R2 + R1 \rightarrow R1$$

$$\begin{bmatrix} 2 & -4 & \vdots & 1 & 0 \\ 4 & -7 & \vdots & 0 & 1 \end{bmatrix} \Leftrightarrow \begin{bmatrix} 1 & -2 & \vdots & \tfrac{1}{2} & 0 \\ 0 & 1 & \vdots & -2 & 1 \end{bmatrix} \Leftrightarrow \begin{bmatrix} 1 & 0 & \vdots & -\tfrac{7}{2} & 2 \\ 0 & 1 & \vdots & -2 & 1 \end{bmatrix}$$

Matrix A has been transformed into I. Thus, the right side of the previous matrix is A^{-1}. Verify this by finding AA^{-1} and $A^{-1}A$ and showing that each product is I:

$$AA^{-1} = \begin{bmatrix} 2 & -4 \\ 4 & -7 \end{bmatrix} \begin{bmatrix} -\frac{7}{2} & 2 \\ -2 & 1 \end{bmatrix} = \begin{bmatrix} 1 & 0 \\ 0 & 1 \end{bmatrix}$$

$$A^{-1}A = \begin{bmatrix} -\frac{7}{2} & 2 \\ -2 & 1 \end{bmatrix} \begin{bmatrix} 2 & -4 \\ 4 & -7 \end{bmatrix} = \begin{bmatrix} 1 & 0 \\ 0 & 1 \end{bmatrix}$$

∎

Example 3   Find the inverse of matrix A if $A = \begin{bmatrix} 1 & 1 & 0 \\ 1 & 2 & 1 \\ 2 & 3 & 2 \end{bmatrix}$.

Solution   Set up a 3×6 matrix with A on the left and I on the right of the broken line:

$$[A \mid I] = \begin{bmatrix} 1 & 1 & 0 & 1 & 0 & 0 \\ 1 & 2 & 1 & 0 & 1 & 0 \\ 2 & 3 & 2 & 0 & 0 & 1 \end{bmatrix}$$

Perform row operations on the entire matrix to transform the left half into I.

$$(-1)R1 + R2 \rightarrow R2$$
$$(-2)R1 + R3 \rightarrow R3$$

$$\begin{bmatrix} 1 & 1 & 0 & 1 & 0 & 0 \\ 1 & 2 & 1 & 0 & 1 & 0 \\ 2 & 3 & 2 & 0 & 0 & 1 \end{bmatrix} \Leftrightarrow \begin{bmatrix} 1 & 1 & 0 & 1 & 0 & 0 \\ 0 & 1 & 1 & -1 & 1 & 0 \\ 0 & 1 & 2 & -2 & 0 & 1 \end{bmatrix}$$

$$(-1)R2 + R1 \rightarrow R1$$
$$(-1)R2 + R3 \rightarrow R3$$

$$\Leftrightarrow \begin{bmatrix} 1 & 0 & -1 & 2 & -1 & 0 \\ 0 & 1 & 1 & -1 & 1 & 0 \\ 0 & 0 & 1 & -1 & -1 & 1 \end{bmatrix}$$

$$R3 + R1 \rightarrow R1$$
$$(-1)R3 + R2 \rightarrow R2$$

$$\Leftrightarrow \begin{bmatrix} 1 & 0 & 0 & 1 & -2 & 1 \\ 0 & 1 & 0 & 0 & 2 & -1 \\ 0 & 0 & 1 & -1 & -1 & 1 \end{bmatrix}$$

The left half has been transformed into the identity matrix, and the right half has become A^{-1}. Thus,

$$A^{-1} = \begin{bmatrix} 1 & -2 & 1 \\ 0 & 2 & -1 \\ -1 & -1 & 1 \end{bmatrix}$$

∎

Example 4   Find the inverse of $A = \begin{bmatrix} 1 & 2 \\ 2 & 4 \end{bmatrix}$, if possible.

Solution   Form the 2×4 matrix

$$[A \mid I] = \begin{bmatrix} 1 & 2 & \mid & 1 & 0 \\ 2 & 4 & \mid & 0 & 1 \end{bmatrix}$$

and begin to transform the left side of the matrix into the identity matrix I:

$$(-2)R1 + R2 \rightarrow R2$$

$$\begin{bmatrix} 1 & 2 & \mid & 1 & 0 \\ 2 & 4 & \mid & 0 & 1 \end{bmatrix} \Leftrightarrow \begin{bmatrix} 1 & 2 & \mid & 1 & 0 \\ 0 & 0 & \mid & -2 & 1 \end{bmatrix}$$

In obtaining the second-row, first-column position of A, the entire second row of A is "zeroed out." Because it is impossible to transform A to the identity, matrix A is singular and has no inverse.   ■

The next example shows how the inverse of a nonsingular matrix can be used to solve a system of equations.

Example 5   Solve the system $\begin{cases} x + y & = 3 \\ x + 2y + z = -2. \\ 2x + 3y + 2z = 1 \end{cases}$

Solution   This system can be written as a single equation involving three matrices.

1. $\begin{bmatrix} 1 & 1 & 0 \\ 1 & 2 & 1 \\ 2 & 3 & 2 \end{bmatrix} \begin{bmatrix} x \\ y \\ z \end{bmatrix} = \begin{bmatrix} 3 \\ -2 \\ 1 \end{bmatrix}$

The 3×3 matrix on the left is the matrix whose inverse was found in Example 3. Multiply each side of Equation 1 on the left by this inverse to obtain an equivalent system of equations. The solution of this system can be read directly from the matrix to the right of the equals sign:

$$\begin{bmatrix} 1 & -2 & 1 \\ 0 & 2 & -1 \\ -1 & -1 & 1 \end{bmatrix} \begin{bmatrix} 1 & 1 & 0 \\ 1 & 2 & 1 \\ 2 & 3 & 2 \end{bmatrix} \begin{bmatrix} x \\ y \\ z \end{bmatrix} = \begin{bmatrix} 1 & -2 & 1 \\ 0 & 2 & -1 \\ -1 & -1 & 1 \end{bmatrix} \begin{bmatrix} 3 \\ -2 \\ 1 \end{bmatrix}$$

$$\begin{bmatrix} 1 & 0 & 0 \\ 0 & 1 & 0 \\ 0 & 0 & 1 \end{bmatrix} \begin{bmatrix} x \\ y \\ z \end{bmatrix} = \begin{bmatrix} 8 \\ -5 \\ 0 \end{bmatrix}$$

$$\begin{bmatrix} x \\ y \\ z \end{bmatrix} = \begin{bmatrix} 8 \\ -5 \\ 0 \end{bmatrix}$$

The solution of this system of equations is $x = 8$, $y = -5$, $z = 0$. Verify that these results satisfy all three of the original equations.   ■

The equations of Example 5 can be thought of as the matrix equation $AX = B$, where A is the coefficient matrix,

$$A = \begin{bmatrix} 1 & 1 & 0 \\ 1 & 2 & 1 \\ 2 & 3 & 2 \end{bmatrix}$$

X is a column matrix of the variables,

$$X = \begin{bmatrix} x \\ y \\ z \end{bmatrix}$$

and B is a column matrix of the constants from the right sides of the equations,

$$B = \begin{bmatrix} 3 \\ -2 \\ 1 \end{bmatrix}$$

When each side of $AX = B$ is multiplied on the left by A^{-1}, the solution of the system appears as the column of numbers in the matrix that is the product of A^{-1} and B:

$$A^{-1}AX = A^{-1}B$$
$$IX = A^{-1}B$$
$$X = A^{-1}B$$

This method is especially useful for finding solutions of several systems of equations that differ from each other *only* in the column matrix B. If the coefficient matrix A remains unchanged from one system of equations to the next, then A^{-1} needs to be found only once. The solution of each system is found by a single matrix multiplication, $A^{-1}B$.

EXERCISE 5.4

In Exercises 1–16, find the inverse of each matrix, if possible.

1. $\begin{bmatrix} 3 & -4 \\ -2 & 3 \end{bmatrix}$ **2.** $\begin{bmatrix} 2 & 3 \\ 3 & 5 \end{bmatrix}$ **3.** $\begin{bmatrix} 3 & 7 \\ 2 & 5 \end{bmatrix}$ **4.** $\begin{bmatrix} 1 & -2 \\ 2 & -5 \end{bmatrix}$

5. $\begin{bmatrix} 1 & 2 & 3 \\ 2 & 5 & 3 \\ 1 & 0 & 8 \end{bmatrix}$ **6.** $\begin{bmatrix} 2 & 1 & -1 \\ 2 & 2 & -1 \\ -1 & -1 & 1 \end{bmatrix}$

7. $\begin{bmatrix} 3 & 2 & 1 \\ 1 & 1 & -1 \\ 4 & 3 & 1 \end{bmatrix}$ **8.** $\begin{bmatrix} -2 & 1 & -3 \\ 2 & 3 & 0 \\ 1 & 0 & 1 \end{bmatrix}$

9. $\begin{bmatrix} 1 & 3 & 5 \\ 0 & 1 & 6 \\ 1 & 4 & 11 \end{bmatrix}$ **10.** $\begin{bmatrix} 1 & 1 & 1 \\ 2 & 2 & 2 \\ 3 & 3 & 3 \end{bmatrix}$ **11.** $\begin{bmatrix} 1 & 2 & 3 \\ 0 & 1 & 2 \\ 0 & 0 & 1 \end{bmatrix}$ **12.** $\begin{bmatrix} 1 & 2 & 3 \\ 0 & 1 & 1 \\ 0 & -1 & 0 \end{bmatrix}$

13.
$$\begin{bmatrix} 1 & 6 & 4 \\ 1 & -2 & -5 \\ 2 & 4 & -1 \end{bmatrix}$$
14.
$$\begin{bmatrix} 1 & 1 & 1 \\ 1 & 0 & -1 \\ 1 & 2 & 3 \end{bmatrix}$$
15.
$$\begin{bmatrix} 1 & 2 & 3 & 4 \\ 0 & 1 & 2 & 3 \\ 0 & 0 & 1 & 2 \\ 0 & 0 & 0 & 1 \end{bmatrix}$$
16.
$$\begin{bmatrix} 1 & 0 & 0 & 0 \\ 1 & 1 & 0 & 0 \\ 1 & 1 & 1 & 0 \\ 1 & 2 & 2 & 1 \end{bmatrix}$$

In Exercises 17–26, use the method of Example 5 to solve each system of equations.

17.
$$\begin{cases} 3x - 4y = 1 \\ -2x + 3y = 5 \end{cases}$$
18.
$$\begin{cases} 2x + 3y = 7 \\ 3x + 5y = -5 \end{cases}$$
19.
$$\begin{cases} 3x + 7y = 0 \\ 2x + 5y = -10 \end{cases}$$
20.
$$\begin{cases} x - 2y = 12 \\ 2x - 5y = 13 \end{cases}$$

21.
$$\begin{cases} x + 2y + 3z = 1 \\ 2x + 5y + 3z = 3 \\ x + 8z = -2 \end{cases}$$
22.
$$\begin{cases} 2x + y - z = 3 \\ 2x + 2y - z = -1 \\ -x - y + z = 4 \end{cases}$$

23.
$$\begin{cases} 3x + 2y + z = 2 \\ x + y - z = -1 \\ 4x + 3y + z = 0 \end{cases}$$
24.
$$\begin{cases} -2x + y - 3z = 5 \\ 2x + 3y = 1 \\ x + z = -2 \end{cases}$$

25.
$$\begin{cases} x + 2y + 3z = 1 \\ 2x + 2y + 2z = 2x + y \\ z = 3 \end{cases}$$
26.
$$\begin{cases} x + 2y + 3z = 2 \\ x + y + z = x \\ x - y = x \end{cases}$$

27. If the $n \times n$ matrix A is nonsingular, and if B and C are $n \times n$ matrices such that $AB = AC$, prove that $B = C$.

28. If B is an $n \times n$ matrix that behaves as an identity ($AB = BA = A$, for any $n \times n$ matrix A), prove that $B = I$.

29. Prove that $\begin{bmatrix} a & b \\ c & d \end{bmatrix}$ has an inverse if and only if $ad - bc \neq 0$. (*Hint*: Try to find the inverse and see what happens.)

30. For what value of x will $\begin{bmatrix} 3 & 8 \\ 6 & x \end{bmatrix}$ not have a multiplicative inverse? (*Hint*: See Exercise 29.)

31. For what values of x will $\begin{bmatrix} x & 8 \\ 2 & x \end{bmatrix}$ not have a multiplicative inverse?

32. Does $(AB)^{-1} = A^{-1}B^{-1}$? Support your answer with an example chosen from 2×2 matrices.

33. Use an example chosen from 2×2 matrices to illustrate that $(AB)^{-1} = B^{-1}A^{-1}$.

34. Let A be any 3×3 matrix. Find a 3×3 matrix E such that the product EA is the result of performing the row operation $R1 \leftrightarrow R2$ on matrix A (the row operation that exchanges rows one and two of matrix A). What is E^{-1}?

35. Let A be any 3×3 matrix. Find another 3×3 matrix E such that the product EA is the result of performing the row operation $(3)R1 + R3 \rightarrow R3$ on matrix A. What is E^{-1}?

36. Let A be any 3×3 matrix. Find another 3×3 matrix E such that the product EA is the result of performing the row operation $(5)R2 \rightarrow R2$ on matrix A. What is E^{-1}?

Computer Exercises. *In Exercises 37–38, use a computer and **Algebra Pak** to find the inverse of each matrix, if it exists.*

37.
$$\begin{bmatrix} 1 & 0 & 1 & 1 \\ 0 & \frac{1}{5} & 0 & -1 \\ 0 & \frac{1}{3} & \frac{1}{3} & 1 \\ 1 & 1 & 1 & -3 \end{bmatrix}$$
38.
$$\begin{bmatrix} 1 & 1 & \frac{1}{2} & 1 \\ 0 & 1 & 1 & \frac{1}{3} \\ -2 & \frac{1}{2} & 0 & \frac{1}{3} \\ 1 & 0 & -\frac{1}{2} & \frac{2}{3} \end{bmatrix}$$

5.5   SOLUTION OF SYSTEMS OF EQUATIONS BY DETERMINANTS

There is a function, called the **determinant function**, that associates a numerical value with every square matrix. For any square matrix A, the symbol $\det(A)$ or the symbol $|A|$ represents the determinant of matrix A. We begin by defining the determinant of a 2×2 matrix.

Definition. If a, b, c, and d are numbers, then the determinant of

$$A = \begin{bmatrix} a & b \\ c & d \end{bmatrix} \text{ is}$$

$$\det(A) = \begin{vmatrix} a & b \\ c & d \end{vmatrix} = ad - bc$$

Gabriel Cramer (1704–1752) Although other mathematicians had worked with determinants, it was the work of Cramer that popularized them.

The determinant of a 2×2 matrix A is the number that is equal to the product of the entries on the major diagonal

$$\begin{vmatrix} a & b \\ c & d \end{vmatrix}$$

minus the product of the entries on the other diagonal

$$\begin{vmatrix} a & b \\ c & d \end{vmatrix}$$

Example 1

a. $\begin{vmatrix} 1 & 2 \\ 3 & 4 \end{vmatrix} = 1 \cdot 4 - 2 \cdot 3 = 4 - 6 = -2$

b. $\begin{vmatrix} -1 & -3 \\ 2 & 0 \end{vmatrix} = (-1) \cdot 0 - (-3) \cdot (2) = 0 - (-6) = 6$

c. $\begin{vmatrix} -2 & 3 \\ -\pi & \frac{1}{2} \end{vmatrix} = (-2) \cdot \left(\frac{1}{2}\right) - (3) \cdot (-\pi) = -1 + 3\pi$ ∎

To see how determinants can be used to solve a system of equations, we consider the system

$$\begin{cases} ax + by = e \\ cx + dy = f \end{cases}$$

By multiplying the first equation by d, the second equation by $-b$, and adding, the terms that involve y drop out:

$$\begin{array}{r} adx + bdy = ed \\ -bcx - bdy = -bf \\ \hline adx - bcx = ed - bf \end{array}$$

We can solve the resulting equation for x:

$$adx - bcx = ed - bf$$
$$(ad - bc)x = ed - bf$$

1. $$x = \frac{ed - bf}{ad - bc} \quad \text{provided } ad - bc \neq 0$$

Solving this system for y gives

2. $$y = \frac{af - ec}{ad - bc} \quad \text{provided } ad - bc \neq 0$$

The numerators and denominators of Equations 1 and 2 can be expressed as determinants:

$$x = \frac{\begin{vmatrix} e & b \\ f & d \end{vmatrix}}{\begin{vmatrix} a & b \\ c & d \end{vmatrix}} = \frac{ed - bf}{ad - bc} \qquad y = \frac{\begin{vmatrix} a & e \\ c & f \end{vmatrix}}{\begin{vmatrix} a & b \\ c & d \end{vmatrix}} = \frac{af - ec}{ad - bc}$$

If these formulas are compared with the original system of equations,

$$\begin{cases} ax + by = e \\ cx + dy = f \end{cases}$$

it is apparent that the denominator determinant consists of the coefficients of the variables of each equation:

$$\text{denominator determinant} = \begin{vmatrix} a & b \\ c & d \end{vmatrix}$$

Each numerator determinant is a modified copy of the denominator determinant. The column of coefficients of the variable for which we are solving is replaced with the column of constants that appears to the right of the equals signs. Thus, when solving for x, the coefficients of x (a and c) are replaced in the numerator determinant by the constants e and f:

$$x = \frac{\begin{vmatrix} e & b \\ f & d \end{vmatrix}}{\begin{vmatrix} a & b \\ c & d \end{vmatrix}} \qquad \begin{cases} ax + by = e \\ cx + dy = f \end{cases}$$

Similarly, when solving for y, the coefficients of y (b and d) are replaced in the numerator determinant by the constants e and f.

$$y = \frac{\begin{vmatrix} a & e \\ c & f \end{vmatrix}}{\begin{vmatrix} a & b \\ c & d \end{vmatrix}} \qquad \begin{cases} ax + by = e \\ cx + dy = f \end{cases}$$

The method of using determinants to solve systems of equations is called **Cramer's rule**.

Cramer's Rule for Two Equations in Two Variables.   If the system

$$\begin{cases} ax + by = e \\ cx + dy = f \end{cases}$$

has a single solution, it is given by

$$x = \frac{D_x}{D} \quad \text{and} \quad y = \frac{D_y}{D}$$

where

$$D = \begin{vmatrix} a & b \\ c & d \end{vmatrix}, \quad D_x = \begin{vmatrix} e & b \\ f & d \end{vmatrix}, \quad \text{and} \quad D_y = \begin{vmatrix} a & e \\ c & f \end{vmatrix}$$

If the denominators and the numerators of these fractions are all 0, the system is consistent but the equations are dependent.

If the denominators are 0 and any numerator is not 0, the system is inconsistent.

Example 2   Use determinants to solve the system $\begin{cases} 3x + 2y = 7 \\ -x + 5y = 9 \end{cases}$.

Solution   Use Cramer's rule:

$$x = \frac{\begin{vmatrix} 7 & 2 \\ 9 & 5 \end{vmatrix}}{\begin{vmatrix} 3 & 2 \\ -1 & 5 \end{vmatrix}} = \frac{7 \cdot 5 - 2 \cdot 9}{3 \cdot 5 - 2(-1)} = \frac{35 - 18}{15 + 2} = \frac{17}{17} = 1$$

$$y = \frac{\begin{vmatrix} 3 & 7 \\ -1 & 9 \end{vmatrix}}{\begin{vmatrix} 3 & 2 \\ -1 & 5 \end{vmatrix}} = \frac{3 \cdot 9 - 7(-1)}{3 \cdot 5 - 2(-1)} = \frac{27 + 7}{15 + 2} = \frac{34}{17} = 2$$

The solution of the given system is $x = 1$ and $y = 2$. The pair $(1, 2)$ satisfies both of the equations in the given system.   ∎

Cramer's rule can be used to solve many systems of n linear equations in n variables where $n > 2$. But to do so, we must develop a method of evaluating determinants larger than 2×2 that will give the correct solutions. Such a method is called **expansion by minors**.

> **Definition.**   Let $|A|$ be a determinant of an $n \times n$ matrix A, and let a_{ij} be the element in the ith row and jth column of A. The **minor** of a_{ij} is the determinant of the $n - 1 \times n - 1$ matrix formed by those elements of A that do not lie in row i or in column j.

To find the minor of any chosen element of a determinant, we cross out the row and the column of that element. The minor is the determinant of the square array of numbers that remains.

> **Definition.**   The **cofactor** of a_{ij} is the minor of a_{ij} if $i + j$ is even, or the negative of the minor of a_{ij} if $i + j$ is odd.

Example 3   Find the cofactor of **a.** 7   and   **b.** 2 in the matrix $\begin{bmatrix} 1 & 2 & 3 \\ 4 & 5 & 6 \\ 7 & 8 & 9 \end{bmatrix}$.

Solution   **a.** In the determinant of the 3×3 matrix A

$$\begin{vmatrix} 1 & 2 & 3 \\ 4 & 5 & 6 \\ 7 & 8 & 9 \end{vmatrix}$$

the minor of the 7 that appears in the third row, first column, is the determinant of the 2×2 matrix

$$\begin{vmatrix} 2 & 3 \\ 5 & 6 \end{vmatrix}$$

Because 7 is in the third row, first column, its row number plus its column number is even $(3 + 1 = 4)$. Thus, the cofactor of 7 is its minor:

$$\begin{vmatrix} 2 & 3 \\ 5 & 6 \end{vmatrix} = 2 \cdot 6 - 3 \cdot 5 = 12 - 15 = -3$$

b. The minor of 2, which appears in the first row, second column of A is

$$\begin{vmatrix} 4 & 6 \\ 7 & 9 \end{vmatrix} = 4 \cdot 9 - 6 \cdot 7 = 36 - 42 = -6$$

This is found by ignoring the row and column in which the 2 is located. Because the sum of the row number and column number is odd $(1 + 2 = 3)$, the co-factor of the element 2 is the *negative* of its minor. Thus, the cofactor of 2 is $-(-6) = +6$.   ■

Definition.   The value of the determinant of any square matrix is the sum of the products of the elements in the first row of the matrix and the cofactors of those elements.

It can be proved that the value of the determinant of any square matrix is the sum of the products of the elements of *any* row (or column) of the matrix and the cofactors of those elements.

Example 4   Evaluate the determinant $\begin{vmatrix} 1 & 2 & -3 \\ -1 & 0 & 1 \\ -2 & 2 & 1 \end{vmatrix}$ along   **a.** the first row,   **b.** the third row, and   **c.** the second column.

Solution   **a.** Multiply each element in the first row by its cofactor and form the sum of these three products:

$$\begin{vmatrix} 1 & 2 & -3 \\ -1 & 0 & 1 \\ -2 & 2 & 1 \end{vmatrix} = 1\begin{vmatrix} 0 & 1 \\ 2 & 1 \end{vmatrix} - 2\begin{vmatrix} -1 & 1 \\ -2 & 1 \end{vmatrix} + (-3)\begin{vmatrix} -1 & 0 \\ -2 & 2 \end{vmatrix}$$

Evaluate the three 2×2 determinants and simplify:

$$\begin{vmatrix} 1 & 2 & -3 \\ -1 & 0 & 1 \\ -2 & 2 & 1 \end{vmatrix} = 1(-2) - 2(1) - 3(-2)$$

$$= -2 - 2 + 6$$
$$= 2$$

b. Multiply each element in the third row by its cofactor and form the sum of these three products:

$$\begin{vmatrix} 1 & 2 & -3 \\ -1 & 0 & 1 \\ -2 & 2 & 1 \end{vmatrix} = -2\begin{vmatrix} 2 & -3 \\ 0 & 1 \end{vmatrix} - 2\begin{vmatrix} 1 & -3 \\ -1 & 1 \end{vmatrix} + 1\begin{vmatrix} 1 & 2 \\ -1 & 0 \end{vmatrix}$$

$$= -2(2) - 2(-2) + 1(2)$$
$$= -4 + 4 + 2$$
$$= 2$$

c. Multiply each element of the second column by its cofactor and form the sum of these three products:

$$\begin{vmatrix} 1 & 2 & -3 \\ -1 & 0 & 1 \\ -2 & 2 & 1 \end{vmatrix} = -2\begin{vmatrix} -1 & 1 \\ -2 & 1 \end{vmatrix} + 0\begin{vmatrix} 1 & -3 \\ -2 & 1 \end{vmatrix} - 2\begin{vmatrix} 1 & -3 \\ -1 & 1 \end{vmatrix}$$

$$= -2(1) + 0(-5) - 2(-2)$$
$$= -2 + 4$$
$$= 2$$

Note that the same result is obtained in each part of this example. In fact, the same result will be obtained if you expand the determinant along *any* row or column. ∎

Example 5   Evaluate the determinant $\begin{vmatrix} 0 & 0 & 2 & 0 \\ 1 & 2 & 17 & -3 \\ -1 & 0 & 28 & 1 \\ -2 & 2 & -37 & 1 \end{vmatrix}$.

Solution   Because the determinant can be evaluated along *any* row or column, choose the row or column with the most 0's. So expand this determinant along its first row. Note that three of the four minors need not be evaluated, because each will be multiplied by 0.

$$\begin{vmatrix} 0 & 0 & 2 & 0 \\ 1 & 2 & 17 & -3 \\ -1 & 0 & 28 & 1 \\ -2 & 2 & -37 & 1 \end{vmatrix} = 0\begin{vmatrix} \text{Who} \\ \text{cares?} \end{vmatrix} - 0\begin{vmatrix} \text{Who} \\ \text{cares?} \end{vmatrix} + 2\begin{vmatrix} 1 & 2 & -3 \\ -1 & 0 & 1 \\ -2 & 2 & 1 \end{vmatrix} - 0\begin{vmatrix} \text{Who} \\ \text{cares?} \end{vmatrix}$$

$$= 2(2) \qquad \text{See Example 4.}$$
$$= 4$$ ∎

Example 5 suggests the following theorem.

Theorem.   If any row or column of a square matrix consists entirely of 0's, the value of the determinant of the matrix is 0.

Proof   A determinant can be evaluated along any of its rows or columns. Evaluate the determinant along its row or column of 0's. Each entry in that row or column is 0, and the product of each entry with its cofactor is also 0. The value of the determinant is the sum of these products. Thus, the value of the determinant is 0. □

The next example illustrates how Cramer's rule can be used to solve a system of three equations in three variables.

Example 6   Use Cramer's rule to solve the following system of equations:

$$\begin{cases} 2x - y + 2z = 3 \\ x - y + z = 2 \\ x + y + 2z = 3 \end{cases}$$

Solution   Each of the values x, y, and z is the ratio of two 3×3 determinants. The denominator of each quotient is the determinant consisting of the nine coefficients of the variables. The numerators for x, y, and z are modified copies of this denominator determinant. The column of constants is substituted for the coefficients of the variable for which you are solving.

$$\begin{cases} 2x - y + 2z = 3 \\ x - y + z = 2 \\ x + y + 2z = 3 \end{cases}$$

$$x = \frac{\begin{vmatrix} 3 & -1 & 2 \\ 2 & -1 & 1 \\ 3 & 1 & 2 \end{vmatrix}}{\begin{vmatrix} 2 & -1 & 2 \\ 1 & -1 & 1 \\ 1 & 1 & 2 \end{vmatrix}} = \frac{3\begin{vmatrix} -1 & 1 \\ 1 & 2 \end{vmatrix} - (-1)\begin{vmatrix} 2 & 1 \\ 3 & 2 \end{vmatrix} + 2\begin{vmatrix} 2 & -1 \\ 3 & 1 \end{vmatrix}}{2\begin{vmatrix} -1 & 1 \\ 1 & 2 \end{vmatrix} - (-1)\begin{vmatrix} 1 & 1 \\ 1 & 2 \end{vmatrix} + 2\begin{vmatrix} 1 & -1 \\ 1 & 1 \end{vmatrix}} = \frac{2}{-1} = -2$$

$$y = \frac{\begin{vmatrix} 2 & 3 & 2 \\ 1 & 2 & 1 \\ 1 & 3 & 2 \end{vmatrix}}{\begin{vmatrix} 2 & -1 & 2 \\ 1 & -1 & 1 \\ 1 & 1 & 2 \end{vmatrix}} = \frac{2\begin{vmatrix} 2 & 1 \\ 3 & 2 \end{vmatrix} - 3\begin{vmatrix} 1 & 1 \\ 1 & 2 \end{vmatrix} + 2\begin{vmatrix} 1 & 2 \\ 1 & 3 \end{vmatrix}}{-1} = \frac{1}{-1} = -1$$

$$z = \frac{\begin{vmatrix} 2 & -1 & 3 \\ 1 & -1 & 2 \\ 1 & 1 & 3 \end{vmatrix}}{\begin{vmatrix} 2 & -1 & 2 \\ 1 & -1 & 1 \\ 1 & 1 & 2 \end{vmatrix}} = \frac{2\begin{vmatrix} -1 & 2 \\ 1 & 3 \end{vmatrix} - (-1)\begin{vmatrix} 1 & 2 \\ 1 & 3 \end{vmatrix} + 3\begin{vmatrix} 1 & -1 \\ 1 & 1 \end{vmatrix}}{-1} = \frac{-3}{-1} = 3$$

The triple $(-2, -1, 3)$ satisfies each of the equations in the given system.   ■

Cramer's rule can be used to solve larger systems of equations. A typical equation in a system of n equations in the n variables $x_1, x_2, x_3, \ldots, x_n$ has the form

$$a_1 x_1 + a_2 x_2 + \cdots + a_n x_n = c$$

where the a_i are coefficients of the variables of the system, and c represents one number in the column of constants that appears to the right of the equals signs in the system. Let D be the coefficient matrix of the system, and let D_{x_i} be the matrix formed from D by replacing the ith column of D by the column of constants from the right of the equals signs. Then, if $|D| \neq 0$, Cramer's rule provides the following unique solution:

$$
x_1 = \frac{|D_{x_1}|}{|D|}, \qquad x_2 = \frac{|D_{x_2}|}{|D|}, \qquad \ldots, \qquad x_n = \frac{|D_{x_n}|}{|D|}
$$

Type 3 row operations can be used to simplify the calculations involved in determinant expansion.

> **Theorem.** A type 3 elementary row operation performed on a square matrix does not alter the value of its determinant.

Justification. The following discussion will illustrate the previous theorem for one particular row operation on a 3×3 matrix.

Evaluating $|D|$ by expanding along the first row, we get

$$
|D| = \begin{vmatrix} a & b & c \\ d & e & f \\ g & h & i \end{vmatrix} = a \begin{vmatrix} e & f \\ h & i \end{vmatrix} - b \begin{vmatrix} d & f \\ g & i \end{vmatrix} + c \begin{vmatrix} d & e \\ g & h \end{vmatrix}
$$

If we perform the type 3 row operation $kR3 + R1 \rightarrow R1$ on matrix D, the resulting modified determinant, $|D'|$, has a new first row. We expand that determinant on its first row:

$$
|D'| = \begin{vmatrix} a+kg & b+kh & c+ki \\ d & e & f \\ g & h & i \end{vmatrix} = (a + kg) \begin{vmatrix} e & f \\ h & i \end{vmatrix} - (b + kh) \begin{vmatrix} d & f \\ g & i \end{vmatrix} + (c + ki) \begin{vmatrix} d & e \\ g & h \end{vmatrix}
$$

Using the distributive law to remove the parentheses and rearranging the terms gives

$$
|D'| = a \begin{vmatrix} e & f \\ h & i \end{vmatrix} - b \begin{vmatrix} d & f \\ g & i \end{vmatrix} + c \begin{vmatrix} d & e \\ g & h \end{vmatrix} + kg \begin{vmatrix} e & f \\ h & i \end{vmatrix} - kh \begin{vmatrix} d & f \\ g & i \end{vmatrix} + ki \begin{vmatrix} d & e \\ g & h \end{vmatrix}
$$

The first three terms of the above expansion are identical to the original determinant $|D|$. The original determinant and the modified determinant differ by the amount that is represented by the last three terms of the expansion of $|D'|$. Hence,

$$
|D'| = |D| + k \left(g \begin{vmatrix} e & f \\ h & i \end{vmatrix} - h \begin{vmatrix} d & f \\ g & i \end{vmatrix} + i \begin{vmatrix} d & e \\ g & h \end{vmatrix} \right)
$$

$$
|D'| = |D| + k(gei - ghf - hdi + hgf + idh - ige)
$$

All terms within the parentheses subtract out, and we have

$$|D'| = |D| + k \cdot 0$$
$$|D'| = |D|$$

$\square$

A similar result holds for type 3 column operations.

Theorem.   If any column of a square matrix is altered by adding to it any multiple of another column, the value of the determinant of the matrix is unchanged.

These two theorems provide a way to reduce the work involved in evaluating a large determinant.

Example 7   Evaluate $|A|$ if $|A| = \begin{vmatrix} 1 & 2 & -1 & 2 \\ 2 & 1 & 1 & 1 \\ 1 & 2 & -3 & 2 \\ 2 & -1 & -1 & 1 \end{vmatrix}$.

Solution   Expanding the given determinant along any row or column leads to four determinants involving 3×3 matrices. The row operation $(-1)R3 + R1 \rightarrow R1$ gives three 0's in the first row. The introduction of these 0's simplifies the work, because only one 3×3 determinant needs to be evaluated. Expand the new determinant along the first row.

$$|A| = \begin{vmatrix} 0 & 0 & 2 & 0 \\ 2 & 1 & 1 & 1 \\ 1 & 2 & -3 & 2 \\ 2 & -1 & -1 & 1 \end{vmatrix} = 2 \begin{vmatrix} 2 & 1 & 1 \\ 1 & 2 & 2 \\ 2 & -1 & 1 \end{vmatrix}$$

To introduce more 0's, perform the column operation $(-2)C3 + C1 \rightarrow C1$ on the 3×3 determinant $\begin{vmatrix} 2 & 1 & 1 \\ 1 & 2 & 2 \\ 2 & -1 & 1 \end{vmatrix}$ and expand that result on its first column:

$$|A| = 2 \begin{vmatrix} 0 & 1 & 1 \\ -3 & 2 & 2 \\ 0 & -1 & 1 \end{vmatrix} = 2 \left[-(-3) \begin{vmatrix} 1 & 1 \\ -1 & 1 \end{vmatrix} \right]$$

$$= 2 \cdot 3[1 - (-1)]$$
$$= 2 \cdot 3 \cdot 2$$
$$= 12$$

■

Two more theorems describe the effect of type 2 and type 1 row operations on a determinant.

Theorem.   If any row or column of a square matrix is multiplied by a constant k, the value of the determinant of the matrix is multiplied by k.

Theorem.   If two rows or columns of a square matrix are interchanged, the value of the determinant of the matrix is multiplied by -1.

EXERCISE 5.5

In Exercises 1–16, evaluate each determinant.

1. $\begin{vmatrix} 2 & 1 \\ -2 & 3 \end{vmatrix}$
   2. $\begin{vmatrix} -3 & -6 \\ 2 & -5 \end{vmatrix}$
   3. $\begin{vmatrix} 2 & -3 \\ -3 & 5 \end{vmatrix}$
   4. $\begin{vmatrix} 5 & 8 \\ -6 & -2 \end{vmatrix}$

5. $\begin{vmatrix} 2 & -3 & 5 \\ -2 & 1 & 3 \\ 1 & 3 & -2 \end{vmatrix}$
   6. $\begin{vmatrix} 1 & 3 & 1 \\ -2 & 5 & 3 \\ 3 & -2 & -2 \end{vmatrix}$

7. $\begin{vmatrix} 1 & -1 & 2 \\ 2 & 1 & 3 \\ 1 & 1 & -1 \end{vmatrix}$
   8. $\begin{vmatrix} 1 & 3 & 1 \\ 2 & 1 & -1 \\ 2 & -1 & 1 \end{vmatrix}$

9. $\begin{vmatrix} 2 & 1 & -1 \\ 1 & 3 & 5 \\ 2 & -5 & 3 \end{vmatrix}$
   10. $\begin{vmatrix} 3 & 1 & -2 \\ -3 & 2 & 1 \\ 1 & 3 & 0 \end{vmatrix}$

11. $\begin{vmatrix} 0 & 1 & -3 \\ -3 & 5 & 2 \\ 2 & -5 & 3 \end{vmatrix}$
   12. $\begin{vmatrix} 1 & -7 & -2 \\ -2 & 0 & 3 \\ -1 & 7 & 1 \end{vmatrix}$

13. $\begin{vmatrix} 1 & 2 & 1 & 3 \\ -2 & 1 & -3 & 1 \\ -1 & 0 & 1 & -2 \\ 2 & -1 & -1 & 3 \end{vmatrix}$
   14. $\begin{vmatrix} -1 & 3 & -2 & 5 \\ 2 & 1 & 0 & 1 \\ 1 & 3 & -2 & 5 \\ 2 & -1 & 0 & -1 \end{vmatrix}$

15. $\begin{vmatrix} 1 & 2 & 3 & 4 & 5 \\ 0 & 1 & 2 & 3 & 4 \\ 0 & 0 & 1 & 2 & 3 \\ 0 & 0 & 0 & 1 & 2 \\ 0 & 0 & 0 & 0 & 1 \end{vmatrix}$
   16. $\begin{vmatrix} 1 & 1 & 1 & 1 & 1 \\ 1 & 1 & 1 & 1 & 2 \\ 1 & 1 & 1 & 2 & 2 \\ 1 & 1 & 2 & 2 & 2 \\ 1 & 2 & 2 & 2 & 2 \end{vmatrix}$

In Exercises 17–28, use Cramer's rule to find the solution to each system of equations, if possible.

17. $\begin{cases} 3x + 2y = 7 \\ 2x - 3y = -4 \end{cases}$

18. $\begin{cases} x - 5y = -6 \\ 3x + 2y = -1 \end{cases}$

19. $\begin{cases} x - y = 3 \\ 3x - 7y = 9 \end{cases}$

20. $\begin{cases} 2x - y = -6 \\ x + y = 0 \end{cases}$

21. $\begin{cases} x + 2y + z = 2 \\ x - y + z = 2 \\ x + y + 3z = 4 \end{cases}$

22. $\begin{cases} x + 2y - z = -1 \\ 2x + y - z = 1 \\ x - 3y - 5z = 17 \end{cases}$

23. $\begin{cases} 2x - y + z = 5 \\ 3x - 3y + 2z = 10 \\ x + 3y + z = 0 \end{cases}$

24. $\begin{cases} x - y - z = 2 \\ x + y + z = 2 \\ -x - y + z = -4 \end{cases}$

25. $\begin{cases} \dfrac{x}{2} + \dfrac{y}{3} + \dfrac{z}{2} = 11 \\[2mm] \dfrac{x}{3} + y - \dfrac{z}{6} = 6 \\[2mm] \dfrac{x}{2} + \dfrac{y}{6} + z = 16 \end{cases}$

26. $\begin{cases} \dfrac{x}{2} + \dfrac{y}{5} + \dfrac{z}{3} = 17 \\[2mm] \dfrac{x}{5} + \dfrac{y}{2} + \dfrac{z}{5} = 32 \\[2mm] x + \dfrac{y}{3} + \dfrac{z}{2} = 30 \end{cases}$

27. $\begin{cases} 2p - q + 3r - s = 3 \\ p + q - 2s = 0 \\ 3p - r = 2 \\ p - q + 3s = 3 \end{cases}$

28. $\begin{cases} a + b + c + d = -1 \\ a + b + c + 2d = 0 \\ a + b + 2c + 3d = 1 \\ a + 2b + 3c + 4d = 0 \end{cases}$

29. Use the method of addition to solve the system $\begin{cases} ax + by = e \\ cx + dy = f \end{cases}$ for y, and thereby show that $y = \dfrac{af - ec}{ad - bc}$

30. Use an example chosen from 2×2 matrices to show that the determinant of the product of two matrices is the product of the determinants of those two matrices.

31. Find an example among 2×2 determinants to show that the determinant of a sum of two matrices is not equal to the sum of the determinants of those matrices.

32. Find a 2×2 matrix A for which $|A| \neq 0$. Then find A^{-1}, and verify that $|A^{-1}| = \dfrac{1}{|A|}$.

33. Show that $\begin{vmatrix} a & b & c \\ 0 & d & e \\ 0 & 0 & f \end{vmatrix} = adf$.

34. Show that $\begin{vmatrix} a & b & c & d \\ 0 & e & f & g \\ 0 & 0 & h & i \\ 0 & 0 & 0 & j \end{vmatrix} = aehj$.

35. Show that multiplying the first row of $\begin{vmatrix} 2 & -1 & 3 \\ 1 & 2 & -1 \\ 3 & 2 & -1 \end{vmatrix}$ by 3 multiplies the value of the determinant by 3.

36. Interchange two rows of the determinant in Exercise 35 and show that this has the effect of multiplying the value of the determinant by -1.

37. Interchange two columns of the determinant in Exercise 35 and show that this has the effect of multiplying the value of the determinant by -1.

38. Find the value of k if $\begin{vmatrix} 2a & 2b & 2c \\ 3d & 3e & 3f \\ 5g & 5h & 5i \end{vmatrix} = k \begin{vmatrix} a & b & c \\ d & e & f \\ g & h & i \end{vmatrix}$.

In Exercises 39–42, expand the determinants and solve for x.

39. $\begin{vmatrix} 3 & x \\ 1 & 2 \end{vmatrix} = \begin{vmatrix} 2 & -1 \\ x & -5 \end{vmatrix}$

40. $\begin{vmatrix} 4 & x^2 \\ 1 & -1 \end{vmatrix} = \begin{vmatrix} x & 4 \\ 2 & 3 \end{vmatrix}$

41. $\begin{vmatrix} 3 & x & 1 \\ x & 0 & -2 \\ 4 & 0 & 1 \end{vmatrix} = \begin{vmatrix} 2 & x \\ x & 4 \end{vmatrix}$

42. $\begin{vmatrix} x & -1 & 2 \\ -2 & x & 3 \\ 4 & -3 & -1 \end{vmatrix} = \begin{vmatrix} 2 & 2 \\ 5 & x \end{vmatrix}$

43. A determinant is a function that associates a number with every square matrix. Give the domain and the range of that function.

44. Use an example chosen from 2×2 matrices to show that for $n \times n$ matrices A and B, $AB \neq BA$ but $|AB| = |BA|$.

45. If A and B are matrices and if $|AB| = 0$, must $|A| = 0$ or $|B| = 0$? Support your answer.

46. If A and B are matrices and if $|AB| = 0$, must $A = \mathbf{0}$ or $B = \mathbf{0}$? Support your answer.

5.6 PARTIAL FRACTIONS

In this section, we discuss how to express a complicated fraction as the sum of several simpler fractions. This process of decomposing a fraction into **partial fractions** is used in calculus. We begin by reviewing the process of adding fractions.

Example 1   Find the sum: $\dfrac{2}{x} + \dfrac{6}{x + 1} + \dfrac{-1}{(x + 1)^2}$.

Solution   Write each fraction in a form with the least common denominator, $x(x + 1)^2$, and add:

$$\frac{2}{x} + \frac{6}{x + 1} + \frac{-1}{(x + 1)^2} = \frac{2(x + 1)^2}{x(x + 1)^2} + \frac{6x(x + 1)}{(x + 1)x(x + 1)} + \frac{-1x}{(x + 1)^2 x}$$

$$= \frac{2x^2 + 4x + 2 + 6x^2 + 6x - x}{x(x + 1)^2}$$

$$= \frac{8x^2 + 9x + 2}{x(x + 1)^2} \qquad \blacksquare$$

Example 2   Express the fraction $\dfrac{3x^2 - x + 1}{x(x - 1)^2}$ as the sum of several fractions with denominators of the smallest degree possible.

Solution   Example 1 suggests that constants A, B, and C can be found such that

$$\frac{3x^2 - x + 1}{x(x - 1)^2} = \frac{A}{x} + \frac{B}{x - 1} + \frac{C}{(x - 1)^2}$$

After you write the terms on the right side as fractions with the common denominator $x(x - 1)^2$, combine them:

$$\frac{3x^2 - x + 1}{x(x-1)^2} = \frac{A(x-1)^2}{x(x-1)^2} + \frac{Bx(x-1)}{x(x-1)(x-1)} + \frac{Cx}{(x-1)^2 x}$$

$$= \frac{Ax^2 - 2Ax + A + Bx^2 - Bx + Cx}{x(x-1)^2}$$

$$= \frac{(A+B)x^2 + (-2A - B + C)x + A}{x(x-1)^2}$$

Because the fractions are equal, the numerator $3x^2 - x + 1$ must equal the numerator $(A + B)x^2 + (-2A - B + C)x + A$. These quantities are equal provided their coefficients are equal. Thus,

$$\begin{cases} A + B & = & 3 \\ -2A - B + C & = & -1 \\ A & = & 1 \end{cases}$$

The coefficients of x^2.

The coefficients of x.

The constants.

This system of three equations in three variables can be solved by substitution. The solutions are $A = 1$, $B = 2$, and $C = 3$. Hence,

$$\frac{3x^2 - x + 1}{x(x-1)^2} = \frac{A}{x} + \frac{B}{x-1} + \frac{C}{(x-1)^2} = \frac{1}{x} + \frac{2}{x-1} + \frac{3}{(x-1)^2} \qquad \blacksquare$$

Example 3 Express the fraction $\dfrac{2x^2 + x + 1}{x^3 + x}$ as the sum of fractions with denominators of the smallest possible degree.

Solution Factoring the denominator suggests that this fraction can be expressed as the sum of two fractions, one with a denominator of x and the other with a denominator of $x^2 + 1$.

$$\frac{2x^2 + x + 1}{x(x^2 + 1)} = \frac{}{x} + \frac{}{x^2 + 1}$$

Because the denominator x of the first fraction is of first degree, its numerator must be of degree 0—that is, a constant. Because the denominator $x^2 + 1$ of the second fraction is of second degree, the numerator might be a first-degree polynomial or a constant. You can allow for both possibilities by using a numerator of $Bx + C$. If $B = 0$, then $Bx + C$ is a constant. If $B \neq 0$, then $Bx + C$ is a first-degree polynomial. Thus,

$$\frac{2x^2 + x + 1}{x(x^2 + 1)} = \frac{A}{x} + \frac{Bx + C}{x^2 + 1} = \frac{A(x^2 + 1) + (Bx + C)x}{x(x^2 + 1)}$$

$$= \frac{Ax^2 + A + Bx^2 + Cx}{x(x^2 + 1)} = \frac{(A + B)x^2 + Cx + A}{x(x^2 + 1)}$$

Equate the corresponding coefficients of the polynomials $2x^2 + x + 1$ and $(A + B)x^2 + Cx + A$ to produce the following system of equations:

$$\begin{cases} A + B = 2 \\ \quad\ C = 1 \\ \quad\ A = 1 \end{cases}$$

The solutions are $A = 1$, $B = 1$, $C = 1$. Therefore, the given fraction can be written as a sum:

$$\frac{2x^2 + x + 1}{x^3 + x} = \frac{1}{x} + \frac{x + 1}{x^2 + 1}$$

∎

The process illustrated in these examples can be summarized: Let $\frac{P(x)}{Q(x)}$ be the quotient of two polynomials with real coefficients, with the degree of $P(x)$ less than the degree of $Q(x)$. Suppose also that the fraction $\frac{P(x)}{Q(x)}$ has been simplified. The polynomial $Q(x)$ can always be factored as a product of first-degree and irreducible second-degree expressions.

If all the identical factors of $Q(x)$ are collected into single factors of the form $(ax + b)^n$ and the form $(ax^2 + bx + c)^n$, then the partial fractions required for the decomposition of $\frac{P(x)}{Q(x)}$ can be found. Each factor of $Q(x)$ of the form $(ax + b)^n$ generates a sum of n partial fractions of the form

$$\frac{A_1}{ax + b} + \frac{A_2}{(ax + b)^2} + \cdots + \frac{A_n}{(ax + b)^n}$$

where each A_i represents a constant. Each factor of the form $(ax^2 + bx + c)^n$ generates the sum of n fractions of the form

$$\frac{B_1 x + C_1}{ax^2 + bx + c} + \frac{B_2 x + C_2}{(ax^2 + bx + c)^2} + \cdots + \frac{B_n x + C_n}{(ax^2 + bx + c)^n}$$

where each B_i and C_i is a constant. After finding a least common denominator and adding the fractions, we obtain a fractional expression that must be equivalent to $\frac{P(x)}{Q(x)}$. Equating the corresponding coefficients of the numerators gives a system of linear equations that can be solved for the constants A_i, B_i, and C_i.

Example 4   What fractions should be used in the decomposition of the rational expression

$$\frac{3x^7 - 5x^5 + 3x + 2}{x^3(x - 3)(x + 2)^2(2x^2 + x + 3)^2(x^2 + 1)^3}$$

Solution   The factor x^3 in the denominator requires three possible fractions in the decomposition:

$$\frac{A}{x} + \frac{B}{x^2} + \frac{C}{x^3}$$

The factor $x - 3$ adds one more to the list:

$$\frac{A}{x} + \frac{B}{x^2} + \frac{C}{x^3} + \frac{D}{x - 3}$$

The factor $(x + 2)^2$ generates two more fractions, each with a constant as numerator:

$$\frac{A}{x} + \frac{B}{x^2} + \frac{C}{x^3} + \frac{D}{x - 3} + \frac{E}{x + 2} + \frac{F}{(x + 2)^2}$$

The factor $(2x^2 + x + 3)^2$ produces two more fractions, each requiring first-degree numerators:

$$\frac{A}{x} + \frac{B}{x^2} + \frac{C}{x^3} + \frac{D}{x - 3} + \frac{E}{x + 2} + \frac{F}{(x + 2)^2} + \frac{Gx + H}{2x^2 + x + 3}$$

$$+ \frac{Jx + K}{(2x^2 + x + 3)^2}$$

Finally, the factor $(x^2 + 1)^3$ requires three more fractions, also with first-degree numerators:

$$\frac{A}{x} + \frac{B}{x^2} + \frac{C}{x^3} + \frac{D}{x - 3} + \frac{E}{x + 2} + \frac{F}{(x + 2)^2} + \frac{Gx + H}{2x^2 + x + 3}$$

$$+ \frac{Jx + K}{(2x^2 + x + 3)^2} + \frac{Lx + M}{x^2 + 1} + \frac{Nx + P}{(x^2 + 1)^2} + \frac{Rx + S}{(x^2 + 1)^3}$$

If you find a common denominator and combine the fractions, equating the corresponding coefficients of the numerators will give 16 equations in 16 variables. These can be solved for the variables $A, B, C, \ldots, S$. ∎

Example 5 Express the fraction $\dfrac{x^2 + 4x + 2}{x^2 + x}$ as the sum of several fractions with denominators of the smallest degree possible.

Solution The method of partial fractions requires that the degree of the numerator be less than the degree of the denominator. Because the degree of both the numerator and denominator are the same in this example, you must perform a long division and express the given fraction in $quotient + \dfrac{remainder}{divisor}$ form:

$$
\begin{array}{r}
1 \\
x^2 + x \overline{\smash{)}\, x^2 + 4x + 2} \\
\underline{x^2 + x} \\
3x + 2
\end{array}
$$

Hence,

$$\frac{x^2 + 4x + 2}{x^2 + x} = 1 + \frac{3x + 2}{x^2 + x}$$

Because the degree of the numerator of the fraction $\dfrac{3x + 2}{x^2 + x}$ is less than the degree of the denominator, you can find the partial fraction decomposition of this fraction:

$$\frac{3x + 2}{x^2 + x} = \frac{3x + 2}{x(x + 1)}$$

$$= \frac{A}{x} + \frac{B}{x + 1}$$

$$= \frac{A(x + 1) + Bx}{x(x + 1)}$$

$$= \frac{(A + B)x + A}{x(x + 1)}$$

Equate the corresponding coefficients in the numerator, and solve the resulting system of equations:

$$\begin{cases} A + B = 3 \\ \quad A = 2 \end{cases}$$

The solution is $A = 2$, $B = 1$, and the decomposition of the given fraction is

$$\frac{x^2 + 4x + 2}{x^2 + x} = 1 + \frac{2}{x} + \frac{1}{x + 1} \qquad \blacksquare$$

EXERCISE 5.6

In Exercises 1–24, decompose each expression into partial fractions.

1. $\dfrac{3x + 1}{(x + 1)(x - 1)}$

2. $\dfrac{-2x + 11}{x^2 - x - 6}$

3. $\dfrac{-3x^2 + x - 5}{(x + 1)(x^2 + 2)}$

4. $\dfrac{-x^2 - 3x - 5}{x^3 + x^2 + 2x + 2}$

5. $\dfrac{-2x^2 + x - 2}{x^3 - x^2}$

6. $\dfrac{2x^2 - 7x + 2}{x(x - 1)^2}$

7. $\dfrac{2x^2 + 1}{x^4 + x^2}$

8. $\dfrac{x^2 + x + 1}{x^3}$

9. $\dfrac{5x^2 + 2x + 2}{x^3 + x}$

10. $\dfrac{-2x^3 + 7x^2 + 6}{x^2(x^2 + 2)}$

11. $\dfrac{x^3 + 4x^2 + 2x + 1}{x^4 + x^3 + x^2}$

12. $\dfrac{x^3 + 4x^2 + 3x + 6}{(x^2 + 2)(x^2 + x + 2)}$

13. $\dfrac{x^3 + 3x^2 + 6x + 6}{(x^2 + x + 5)(x^2 + 1)}$

14. $\dfrac{x^2 - 2x - 3}{(x - 1)^3}$

15. $\dfrac{x^4 - x^3 + x^2 - x + 1}{x(x^2 + 1)^2}$

16. $\dfrac{x^2 + 2}{x^3 + 3x^2 + 3x + 1}$

17. $\dfrac{x^3 + 3x^2 + 2x + 4}{(x^2 + 1)(x^2 + x + 2)}$

18. $\dfrac{4x^3 + 5x^2 + 3x + 4}{x^2(x^2 + 1)}$

19. $\dfrac{2x^4 + 6x^3 + 20x^2 + 22x + 25}{x(x^2 + 2x + 5)^2}$

20. $\dfrac{3x^3 + 5x^2 + 3x + 1}{x^2(x^2 + x + 1)}$

21. $\dfrac{2x^3 + 6x^2 + 3x + 2}{x^3 + x^2}$

22. $\dfrac{x^3}{x^2 + 3x + 2}$

23. $\dfrac{x^4 + x^3 + 3x^2 + x + 4}{(x^2 + 1)^2}$

24. $\dfrac{x^3 + 3x^2 + 2x + 1}{x^3 + x^2 + x}$

5.7   SYSTEMS OF INEQUALITIES AND LINEAR PROGRAMMING

We now consider the graphs of systems of inequalities.

Example 1   Graph the solution set of the system

$$\begin{cases} x + y \le 1 \\ 2x - y > 2 \end{cases}$$

Solution   On the same set of coordinate axes, graph each inequality. See Figure 5-3. The graph of the inequality $x + y \le 1$ includes the line graph of the equation $x + y = 1$ and all points below it. Because the boundary line is included, it is drawn as a solid line. The graph of the inequality $2x - y > 2$ contains only those points below the line graph of the equation $2x - y = 2$. Because the boundary line is not included, it is drawn as a broken line. The area that is shaded twice represents the simultaneous solutions of the given system of inequalities. Any point in the doubly shaded region has coordinates that satisfy both inequalities in the system.

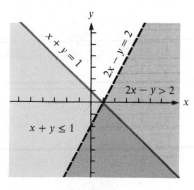

$x + y = 1 \quad 2x - y = 2$

x	y	x	y
0	1	0	−2
1	0	1	0

Figure 5-3

Example 2   Graph the solution set of the system

$$\begin{cases} y < x^2 \\ y > \dfrac{x^2}{4} - 2 \end{cases}$$

Solution   The graph of the equation $y = x^2$ is a parabola opening upward with vertex at the origin. See Figure 5-4. The points with coordinates that satisfy the inequality $y < x^2$ are those points below the parabola.

The graph of $y = \dfrac{x^2}{4} - 2$ is also a parabola opening upward. However, this time the points with coordinates that satisfy the inequality $y > \dfrac{x^2}{4} - 2$ are those points above the parabola. Thus, the graph of the solution set of this system is the shaded area between the two parabolas.

$$y = x^2 \qquad y = \frac{x^2}{4} - 2$$

x	y
0	0
1	1
−1	1
2	4
−2	4

x	y
0	−2
2	−1
−2	−1
4	2
−4	2

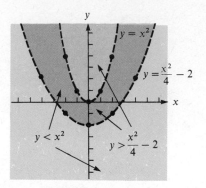

Figure 5-4

Example 3   Graph the solution set of the system

$$\begin{cases} x \geq 1 \\ y \geq x \\ 4x + 5y < 20 \end{cases}$$

Solution   The graph of the solution set of the inequality $x \geq 1$ includes those points on the graph of the equation $x = 1$ and to the right. See Figure 5-5**a**. The graph of the solution set of the inequality $y \geq x$ includes those points on the graph of the equation $y = x$ and above it. See Figure 5-5**b**. The graph of the solution set of the inequality $4x + 5y < 20$ includes those points below the line graph of the equation $4x + 5y = 20$. See Figure 5-5**c**.

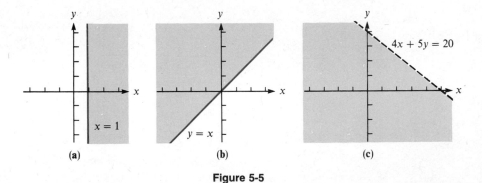

Figure 5-5

If these three graphs are merged onto a single coordinate system, the graph of the original system of inequalities includes those points within the shaded triangle together with the points on the sides of the triangle drawn as solid lines. See Figure 5-6.

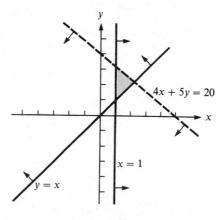

Figure 5-6   ■

Linear Programming

Systems of inequalities provide the basis for an area of applied mathematics known as **linear programming**. Linear programming is used to help answer such questions as "How can a business make as much money as possible?" or "How can I plan a nutritious menu at the least cost?" In such problems, the solution depends on certain **constraints:** The business has limited resources, and the nutritious meal must contain sufficient vitamins, minerals, and so on. Any solution that satisfies the constraints is called a **feasible solution**. In linear programming, the constraints are expressed as a system of linear inequalities, and the quantity that is to be maximized (or minimized) is expressed as a linear function of several variables.

Example 4   Many ordered pairs (x, y) satisfy each inequality in the system

$$\begin{cases} x + y \geq 1 \\ x - y \leq 1 \\ x - y \geq 0 \\ x \leq 2 \end{cases}$$

If $Z = y - 2x$, which of these pairs will produce the greatest value of Z?

Solution   Find the solution of the given system of inequalities, and find the coordinates of each corner of region R, as shown in Figure 5-7. Then rewrite the equation

$$Z = y - 2x$$

in the equivalent form

$$y = 2x + Z$$

This is the equation of a straight line with slope of 2 and y-intercept of Z. Many such lines pass through the region R. To decide which of these provides the greatest value of Z, refer to Figure 5-8 and find the line with the greatest y-intercept. It is line l passing through point P, the left-most corner of R. The coordinates of P are $(\frac{1}{2}, \frac{1}{2})$. Thus, the greatest value of Z possible (subject to the given constraints) is

$$Z = \frac{1}{2} - 2\left(\frac{1}{2}\right) = -\frac{1}{2}$$

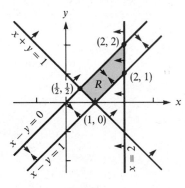

Figure 5-7

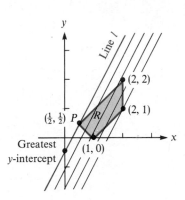

Figure 5-8   ∎

Example 4 illustrates this fact:

The maximum (or minimum) value of a linear function that is subject to the constraints of a system of linear inequalities in two variables is always attained at a corner or along an entire edge of the region R that represents the solution of the system.

Example 5   Fred and Donna are in a part-time business manufacturing clock cases. Fred must work 4 hours and Donna 2 hours to complete one case for a grandfather clock. To build one case for a wall clock, Fred must work 3 hours and Donna 4 hours. Neither partner wishes to work more than 20 hours per week. If they receive $80 for each grandfather clock and $64 for each wall clock, how many of each should they build each week to maximize their profit?

Solution   If the partners manufacture cases for x grandfather clocks and y wall clocks each week, their profit P (in dollars) is

The profit on one grandfather clock	·	the number of grandfather clocks	+	the profit on one wall clock	·	the number of wall clocks.

$$P = \boxed{}$$

$$P = 80x + 64y$$

The time requirements are summarized in the following chart:

Partner	Time for one grandfather clock	Time for one wall clock
Fred	4 hours	3 hours
Donna	2 hours	4 hours

The profit function is subject to the following constraints:

$$\begin{cases} x \geq 0 \\ y \geq 0 \\ 4x + 3y \leq 20 \\ 2x + 4y \leq 20 \end{cases}$$

The inequalities $x \geq 0$ and $y \geq 0$ state that the number of clock cases to be built cannot be negative. The inequality $4x + 3y \leq 20$ is a constraint on Fred's time because he spends 4 hours on each of the x grandfather clocks and 3 hours on each of the y wall clocks, and his total time cannot exceed 20 hours. Similarly, the inequality $2x + 4y \leq 20$ is a constraint on Donna's time.

Graph each of the constraints to find the **feasibility region R**, as in Figure 5-9. The four corners of region R have coordinates of $(0, 0)$, $(0, 5)$, $(2, 4)$, and $(5, 0)$. Substitute each of these number pairs into the equation $P = 80x + 64y$ to find the maximum profit, P.

Corner	Profit
$(0, 0)$	$P = \$80(0) + \$64(0) = \$\ \ 0$
$(0, 5)$	$P = \$80(0) + \$64(5) = \$320$
$(2, 4)$	$P = \$80(2) + \$64(4) = \$416$
$(5, 0)$	$P = \$80(5) + \$64(0) = \$400$

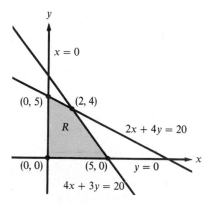

Figure 5-9

Donna and Fred will maximize their profits if they build cases for 2 grandfather clocks and 4 wall clocks each week. If they do so, they will earn $416. ∎

EXERCISE 5.7

In Exercises 1–20, graph the solution set of each system of inequalities.

1. $\begin{cases} y < 3x + 2 \\ y < -2x + 3 \end{cases}$

2. $\begin{cases} y \le x - 2 \\ y \ge 2x + 1 \end{cases}$

3. $\begin{cases} 3x + 2y > 6 \\ x + 3y \le 2 \end{cases}$

4. $\begin{cases} x + y < 2 \\ x + y \le 1 \end{cases}$

5. $\begin{cases} 3x + y \le 1 \\ -x + 2y \ge 9 \end{cases}$

6. $\begin{cases} x + 2y < 3 \\ 2x - 4y < 8 \end{cases}$

7. $\begin{cases} 2x - y > 4 \\ y < -x^2 + 2 \end{cases}$

8. $\begin{cases} x \le y^2 \\ y \ge x \end{cases}$

9. $\begin{cases} y > x^2 - 4 \\ y < -x^2 + 4 \end{cases}$

10. $\begin{cases} x \ge y^2 \\ y \ge x^2 \end{cases}$

11. $\begin{cases} 2x + 3y \le 5 \\ 3x + y \le 1 \\ x \le 0 \end{cases}$

12. $\begin{cases} 2x + y \le 2 \\ y \ge x \\ x \ge 0 \end{cases}$

13. $\begin{cases} x - y < 4 \\ y \ge 0 \\ xy = 12 \end{cases}$

14. $\begin{cases} xy \le 1 \\ x \ge 0 \\ y \ge 0 \end{cases}$

15. $\begin{cases} x \ge 0 \\ y \ge 0 \\ 9x + 3y \le 18 \\ 3x + 6y \le 18 \end{cases}$

16. $\begin{cases} x + y \ge 1 \\ x - y \le 1 \\ x - y \ge 0 \\ x \le 2 \end{cases}$

17. $\begin{cases} y < \sqrt{x} \\ x \ge 0 \end{cases}$

18. $\begin{cases} y < -\sqrt{x} \\ x \ge 0 \end{cases}$

19. $\begin{cases} |x - 2| > 3 \\ |y| > 1 \end{cases}$

20. $\begin{cases} |x + 2| < 3 \\ |y| > 2 \end{cases}$

In Exercises 21–26, maximize P subject to the given constraints.

21. $P = 2x + 3y$
$$\begin{cases} x \ge 0 \\ y \ge 0 \\ x + y \le 4 \end{cases}$$

22. $P = 3x + 2y$
$$\begin{cases} x \ge 0 \\ y \ge 0 \\ x + y \le 4 \end{cases}$$

23. $P = y + \dfrac{1}{2}x$
$$\begin{cases} 2y - x \le 1 \\ y - 2x \ge -2 \\ x \ge 0 \\ y \ge 0 \end{cases}$$

24. $P = 4y - x$
$$\begin{cases} 2y - x \le 1 \\ x \le 2 \\ x + y \ge 1 \\ y \ge 0 \end{cases}$$

25. $P = 2x + y$
$$\begin{cases} y - x \le 2 \\ 2x + 3y \le 6 \\ 3x + y \le 3 \\ y \ge 0 \end{cases}$$

26. $P = 3x - 2y$
$$\begin{cases} y - x \le 1 \\ x - y \le 1 \\ x \ge -1 \\ x \le 1 \end{cases}$$

27. Sally and Sandra each have 12 hours a week to make furniture. They obtain a $50 profit from each table and a $10 profit from each chair. Sally must work 3 hours and Sandra 2 to make a chair. Sally must work 2 hours and Sandra 6 to make a table. How many tables and how many chairs should they make each week to maximize profits?

28. Two machines, A and B, work 24 hours a day, 7 days a week, to manufacture chewing gum and bubble gum. There is a profit of $150 per case of chewing gum and $100 per case of bubble gum. To make a case of chewing gum, machine A must run 2 hours and machine B must run 8 hours. To make a case of bubble gum, machine A must run 4 hours and machine B must run 2 hours. How many cases of each should be produced each week to maximize profits?

29. A clothier stocks from 20 to 30 leather jackets and from 30 to 50 cloth jackets. In the store, there is room to stock up to 60 jackets. The profit on a leather jacket is $50, and $40 on a cloth jacket. If the clothier can sell all the jackets, how many of each should be stocked to maximize profits?

30. A health diet requires at least 16 units of vitamin C and at least 34 units of vitamin B complex. Two food supplements are available that provide these nutrients in the amounts and costs indicated in the table. How much of each should be used to minimize the cost?

Supplement	Vitamin C	Vitamin B	Cost
A	3 units per gram	2 units per gram	3¢ per gram
B	2 units per gram	6 units per gram	4¢ per gram

31. Sarah and her sister Heidi have decided to plant a garden. Sarah wants to plant strawberries; Heidi wants to plant pumpkins. Neither crop may use more than $\frac{3}{4}$ of the 40 square meters of available space. Each square meter of strawberries will earn the children $4, and each square meter of pumpkins will earn $3. The children plan to eat half of the strawberries themselves. How should the planting be divided to maximize the total income?

32. A distributor of hybrid corn has two storehouses, A and B. At A is stored 110 tons of corn; 190 tons are stored at B. Farmer X ordered 60 tons to be delivered to his farm, and farmer Y ordered 80 tons. The shipping costs appear in the following table.

Storehouse	Farmer	Costs per ton
A	X	$5
A	Y	$7
B	X	$8
B	Y	$14

How should the orders be filled to minimize the shipping costs? (*Hint*: Let x represent the number of tons shipped from A to X, and let y represent the number of tons shipped from A to Y.)

33. Bill packs two foods for his camping trip. One ounce of food X costs 35¢ and provides 150 calories and 21 units of vitamins. One ounce of food Y costs 27¢ and provides 60 calories and 42 units of vitamins. Every day, Bill needs at least 3000 calories and at least 1260 units of vitamins, but he does not want to carry more than 60 ounces of food for each day of his trip. How much of each food should Bill pack to minimize the cost?

34. To manufacture two products X and Y, three workers are scheduled as follows:

Worker	Hours required for product X	Hours required for product Y	Time available per week
A	1	4	18
B	2	3	12
C	2	1	6

The profit on product X is $40, and on product Y, $60. How many of each product should be produced to maximize the profit?

CHAPTER SUMMARY

Key Words

additive inverse of a matrix (5.3)
augmented matrix (5.2)
back substitution (5.2)
coefficient matrix (5.2)
cofactor of an element of a determinant (5.5)
consistent systems of equations (5.1)
Cramer's rule (5.5)
dependent equations (5.1)
determinant (5.5)
echelon form of a matrix (5.2)
elementary row operations (5.2)
expansion by minors (5.5)
identity matrix (5.3)

inconsistent system of equations (5.1)
inverse of a matrix (5.4)
independent equations (5.1)
linear programming (5.7)
m × n matrix (5.3)
minor of an element of a determinant (5.5)
nonsingular matrix (5.4)
partial fractions (5.6)
row equivalent matrices (5.2)
singular matrix (5.4)
triangular form of a matrix (5.2)
zero matrix (5.3)

Key Ideas

(5.1)   Systems of equations can be solved by graphing, substitution, and addition.

(5.2)   The three elementary row operations for matrices:

  1. In a type 1 row operation, any two rows of a matrix are interchanged.
  2. In a type 2 row operation, the elements of any row of a matrix are multiplied by any nonzero constant.
  3. In a type 3 row operation, any row of a matrix is altered by adding to it any multiple of another row.

To solve a system of equations by matrix methods, set up the augmented matrix of the system, and use elementary row operations to transform it into another matrix that represents a system that can be solved by back substitution.

(5:3)   Matrices are equal if and only if they are the same size and have the same corresponding entries.

Two $m \times n$ matrices are added by adding the corresponding elements of those matrices.

The product $A \cdot B$ of the $m \times n$ matrix A and the $n \times p$ matrix B is the $m \times p$ matrix C. The ith-row, jth-column entry of C is found by keeping a running total of the products of the elements in the ith row of A with the corresponding elements in the jth column of B.

Matrix addition is associative and commutative. Matrix multiplication is associative but *not* commutative.

(5.4)   The inverse (if it exists) of an $n \times n$ matrix A can be found by using elementary row operations to transform the $n \times 2n$ matrix $[A \mid I]$ into $[I \mid A^{-1}]$, where I is an identity matrix. If A cannot be transformed into I, then A is singular.

(5.5)   The determinant of matrix $A = \begin{bmatrix} a & b \\ c & d \end{bmatrix}$ is $\det(A) = |A| = ad - bc$.

The determinant of an $n \times n$ matrix A is the sum of the products of the elements of any row (or column) and the cofactors of those elements.

To solve a system $AX = B$ of n independent equations in n unknowns by Cramer's rule, form the quotients of two determinants. The denominator is the determinant of the coefficient matrix, A. The numerator is the determinant of a modification of the coefficient matrix: when solving for the ith variable, replace the ith column of A with the column of constants, B.

If two rows (or columns) of a matrix are interchanged, the value of its determinant is multiplied by -1.

If a row (or column) of a matrix is multiplied by a constant k, the value of its determinant is multiplied by k.

If a row (or column) of a matrix is altered by adding to it a multiple of another row (or column), the value of its determinant is unchanged.

(5.6) The fraction $\frac{P(x)}{Q(x)}$ can be written as the sum of several simpler fractions with denominators determined by the irreducible factors of $Q(x)$.

(5.7) Systems of inequalities in two variables can be solved by graphing. The solution is represented by a plane region with boundaries determined by graphing the inequalities as if they were equations.

The maximum and the minimum values of a linear function in two variables, subject to the constraints of a system of linear inequalities, are attained at a vertex or along an entire edge of the region determined by the system of inequalities.

REVIEW EXERCISES

In Review Exercises 1–4, solve each system of equations by the method of graphing.

1. $\begin{cases} 2x - y = -1 \\ x + y = 7 \end{cases}$

2. $\begin{cases} 5x + 2y = 1 \\ 2x - y = -5 \end{cases}$

3. $\begin{cases} y = 5x + 7 \\ x = y - 7 \end{cases}$

4. $\begin{cases} x = y + 5 \\ y = -4 + \dfrac{x}{2} \end{cases}$

In Review Exercises 5–8, solve each system of equations by substitution.

5. $\begin{cases} y = 3x + 2 \\ y = 5x \end{cases}$

6. $\begin{cases} 2y + x = 0 \\ x = y + 3 \end{cases}$

7. $\begin{cases} 2x + y = -3 \\ x - y = 3 \end{cases}$

8. $\begin{cases} \dfrac{x + y}{2} + \dfrac{x - y}{3} = 1 \\ y = 3x - 2 \end{cases}$

In Review Exercises 9–12, solve each system of equations by addition.

9. $\begin{cases} x + 5y = 7 \\ 3x + y = -7 \end{cases}$

10. $\begin{cases} 2x + 3y = 11 \\ 3x - 7y = -41 \end{cases}$

11. $\begin{cases} 2(x + y) - x = 0 \\ 3(x + y) + 2y = 1 \end{cases}$

12. $\begin{cases} \dfrac{x + y}{2} + \dfrac{x - y}{3} = \dfrac{7}{2} \\ \dfrac{x + y}{5} + \dfrac{x - y}{2} = \dfrac{5}{2} \end{cases}$

In Review Exercises 13–16, solve each system of equations by any method.

13. $\begin{cases} 3x + 2y - z = 2 \\ x + y - z = 0 \\ 2x + 3y - z = 1 \end{cases}$

14. $\begin{cases} 5x - y + z = 3 \\ 3x + y + 2z = 2 \\ x + y = 2 \end{cases}$

15. $\begin{cases} 2x - y + z = 1 \\ x - y + 2z = 3 \\ x - y + z = 1 \end{cases}$

16. $\begin{cases} x + 2y - z = -6 \\ x + y - z = -4 \\ 3y - 2z = -12 \end{cases}$

In Review Exercises 17–22, solve each system of equations by matrix methods, if possible.

17. $\begin{cases} 2x + 5y = 7 \\ 3x - y = 2 \end{cases}$

18. $\begin{cases} x + 3y - z = 8 \\ 2x + y - 2z = 11 \\ x - y + 5z = -8 \end{cases}$

19. $\begin{cases} x + 3y + z = 7 \\ 2x - y + z = 0 \\ 3x + 2y + 2z = 7 \end{cases}$

20. $\begin{cases} x + y + z = 4 \\ 3x - 2y - 2z = -3 \\ 4x - y - z = 0 \end{cases}$

21. $\begin{cases} w + x - 3y + z = 2 \\ x - y + 2z = 0 \\ w + y - z = 2 \\ w - x + 2z = -3 \end{cases}$

22. $\begin{cases} w + x + z = 3 \\ x - y + z = 3 \\ w + 2x - y + 2z = 6 \\ y - z = -2 \end{cases}$

In Review Exercises 23–32, perform the indicated matrix arithmetic.

23. $\begin{bmatrix} 3 & 2 & 1 \\ 3 & 2 & 1 \end{bmatrix} + \begin{bmatrix} -2 & 1 & 3 \\ 1 & -2 & 1 \end{bmatrix}$

24. $\begin{bmatrix} 2 & 3 & 5 \\ 1 & -2 & 4 \\ 2 & 1 & -2 \end{bmatrix} - \begin{bmatrix} 0 & -2 & 1 \\ 3 & 4 & -2 \\ 6 & -4 & 1 \end{bmatrix}$

25. $\begin{bmatrix} 2 & 3 \\ -1 & 2 \end{bmatrix} \begin{bmatrix} 1 & -2 \\ -3 & 1 \end{bmatrix}$

26. $\begin{bmatrix} -2 & 3 & 5 \\ 1 & -2 & -3 \end{bmatrix} \begin{bmatrix} 2 & 1 \\ -1 & 2 \\ -2 & 3 \end{bmatrix}$

27. $\begin{bmatrix} 1 & -3 & 2 \end{bmatrix} \begin{bmatrix} 2 \\ 1 \\ 3 \end{bmatrix}$

28. $\begin{bmatrix} 1 & -1 & -2 \\ 2 & -1 & 1 \end{bmatrix} \begin{bmatrix} 1 & 3 & -1 \\ 2 & -1 & 5 \\ 1 & -5 & 3 \end{bmatrix}$

29. $\begin{bmatrix} 1 \\ 2 \\ 1 \\ 5 \end{bmatrix} \begin{bmatrix} 2 & -1 & 1 & 3 \end{bmatrix}$

30. $\begin{bmatrix} 1 & -5 & 3 \\ 2 & 1 & -1 \end{bmatrix} \begin{bmatrix} 2 \\ -2 \\ 3 \end{bmatrix} + \begin{bmatrix} 1 & -1 \\ -1 & 3 \end{bmatrix} \begin{bmatrix} 1 \\ -2 \end{bmatrix}$

31. $\begin{bmatrix} 1 & -3 & 2 \end{bmatrix} \begin{bmatrix} 2 \\ 1 \\ -5 \end{bmatrix} + \begin{bmatrix} 1 & -3 \end{bmatrix} \begin{bmatrix} 2 \\ 5 \end{bmatrix} + \begin{bmatrix} 6 \end{bmatrix}$

32. $\left(\begin{bmatrix} 1 & -3 \\ 3 & 1 \end{bmatrix} + \begin{bmatrix} -1 & 3 \\ 1 & 1 \end{bmatrix} \right) \begin{bmatrix} 1 \\ -5 \end{bmatrix}$

In Review Exercises 33–38, find the inverse of each matrix, if possible.

33. $\begin{bmatrix} 1 & 3 \\ -3 & 5 \end{bmatrix}$
34. $\begin{bmatrix} 4 & 7 \\ 5 & 8 \end{bmatrix}$
35. $\begin{bmatrix} 1 & 3 & -5 \\ 0 & 1 & 9 \\ 0 & 0 & 1 \end{bmatrix}$
36. $\begin{bmatrix} 1 & 0 & 0 \\ 2 & 0 & -2 \\ 1 & 2 & 2 \end{bmatrix}$

37. $\begin{bmatrix} 1 & 0 & 8 \\ 3 & 7 & 6 \\ 1 & 2 & 3 \end{bmatrix}$
38. $\begin{bmatrix} -1 & 1 & 0 \\ -2 & 1 & 0 \\ 3 & -1 & -1 \end{bmatrix}$

In Review Exercises 39–40, use the inverse of the coefficient matrix to solve each system of equations.

39. $\begin{cases} 4x - y + 2z = 0 \\ x + y + 2z = 1 \\ x \quad\; + z = 0 \end{cases}$

40. $\begin{cases} w + 3x + y + 3z = 1 \\ w + 4x + y + 3z = 2 \\ x + y \quad\;\; = 1 \\ w + 2x - y + 2z = 1 \end{cases}$

In Review Exercises 41–44, evaluate each determinant.

41. $\begin{vmatrix} 3 & -2 \\ 1 & -3 \end{vmatrix}$

42. $\begin{vmatrix} 1 & 3 & -1 \\ 1 & 2 & 1 \\ 1 & 0 & 2 \end{vmatrix}$

43. $\begin{vmatrix} 1 & -2 & 3 \\ 2 & -1 & 3 \\ 1 & -1 & 0 \end{vmatrix}$

44. $\begin{vmatrix} 1 & 2 & 3 & 4 \\ -1 & 3 & -3 & 2 \\ 0 & 0 & 0 & -1 \\ 3 & 3 & 4 & 3 \end{vmatrix}$

In Review Exercises 45–48, use Cramer's rule to solve each system of equations.

45. $\begin{cases} x + 3y = -5 \\ -2x + y = -4 \end{cases}$

46. $\begin{cases} x - y + z = -1 \\ 2x - y + 3z = -4 \\ x - 3y + z = -1 \end{cases}$

47. $\begin{cases} x - 3y + z = 7 \\ x + y - 3z = -9 \\ x + y + z = 3 \end{cases}$

48. $\begin{cases} w + x - y + z = 4 \\ 2w + x \quad\;\; + z = 4 \\ x + 2y + z = 0 \\ w \quad\;\; + y + z = 2 \end{cases}$

In Review Exercises 49–52, decompose each fraction into partial fractions.

49. $\dfrac{4x^2 + 4x + 1}{x^3 + x}$
50. $\dfrac{4x^3 + 3x + x^2 + 2}{x^4 + x^2}$
51. $\dfrac{x^2 + 5}{x^3 + x^2 + 5x}$
52. $\dfrac{x^2 + 1}{(x + 1)^3}$

In Review Exercises 53–56, maximize P subject to the given conditions.

53. $P = 2x + y$
$\begin{cases} x \geq 0 \\ y \geq 0 \\ x + y \leq 3 \end{cases}$

54. $P = 2x - 3y$
$\begin{cases} x \geq 0 \\ y \leq 3 \\ x - y \leq 4 \end{cases}$

55. $P = 3x - y$
$\begin{cases} y \geq 1 \\ y \leq 2 \\ y \leq 3x + 1 \\ x \leq 1 \end{cases}$

56. $P = y - 2x$
$\begin{cases} x + y \geq 1 \\ x \leq 1 \\ y \leq \dfrac{x}{2} + 2 \\ x + y \leq 2 \end{cases}$

57. A company manufactures two fertilizers, X and Y. Each 50-pound bag of fertilizer requires three ingredients, which are available in the limited quantities shown below:

Ingredient	Number of pounds in fertilizer X	Number of pounds in fertilizer Y	Total number of pounds available
Nitrogen	6	10	20000
Phosphorus	8	6	16400
Potash	6	4	12000

The profit on each bag of fertilizer X is $6, and on each bag of Y, $5. How many bags of each product should be produced to maximize the profit?

6 Conic Sections and Quadratic Systems

The graphs of second-degree equations in x and y represent figures that have interested mathematicians since the time of the ancient Greeks. However, the equations of those graphs were not carefully studied until the seventeenth century, when René Descartes (1596–1650) and Blaise Pascal (1623–1662) began investigating them.

Definition. If A, B, C, D, E, and F are real numbers, and if at least one of A, B, and C is not 0, then

$$Ax^2 + Bxy + Cy^2 + Dx + Ey + F = 0$$

is called the **general form of a second-degree equation in x and y.**

Descartes discovered that the graphs of second-degree equations always fall into one of seven categories: a single point, a pair of lines, a circle, a parabola, an ellipse, a hyperbola, or no graph at all. These graphs are called **conic sections** because each is the intersection of a plane and a right-circular cone. See Figure 6-1.

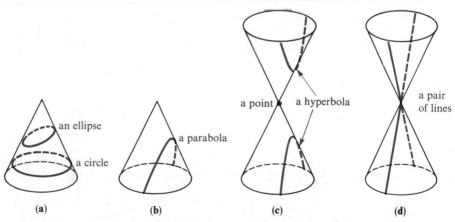

Figure 6-1

The conic sections have many practical applications. For example, the properties of parabolas are used in building flashlights, satellite antennas, and solar furnaces.

279

The orbits of the planets around the sun are ellipses. Hyperbolas are used in navigation and the design of gears.

6.1 THE CIRCLE

The most familiar of the conic sections is the circle.

> **Definition.**   A **circle** is the set of all points in a plane that are a fixed distance from a point called its **center**. The fixed distance is called the **radius of the circle**.

To find the general equation of a circle with radius r and center at the point $C(h, k)$, we must find all points $P(x, y)$ such that the length of the line segment PC is r. See Figure 6-2. We can use the distance formula to find the length of CP, which is r:

$$r = \sqrt{(x - h)^2 + (y - k)^2}$$

We square both sides to get

$$r^2 = (x - h)^2 + (y - k)^2$$

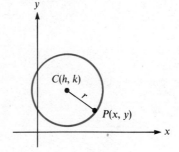

Figure 6-2

This equation is called the **standard form of the equation of a circle**.

> **Theorem.**   Any equation that can be written in the form
> $$(x - h)^2 + (y - k)^2 = r^2$$
> has a graph that is a circle with radius r and center at the point (h, k).

If $r = 0$, the circle reduces to a single point called a **point circle**. If the center of the circle is the origin, then $(h, k) = (0, 0)$, and we have the following result.

> **Theorem.**   Any equation that can be written in the form
> $$x^2 + y^2 = r^2$$
> has a graph that is a circle with radius r and with center at the origin.

Example 1   Find the equation of the circle with radius 5 and center $(3, 2)$. Express the equation in general form.

Solution   Substitute 5 for r, 3 for h, and 2 for k in the standard form of the equation of the circle and simplify:

$$(x - h)^2 + (y - k)^2 = r^2$$
$$(x - 3)^2 + (y - 2)^2 = 5^2$$
$$x^2 - 6x + 9 + y^2 - 4y + 4 = 25$$
$$x^2 + y^2 - 6x - 4y - 12 = 0$$

This final equation is a special case of the general form of a second-degree equation. The coefficient of the xy-term equals 0, and the coefficients of x^2 and y^2 are both 1.   ∎

Example 2   Find the equation of the circle with endpoints of its diameter at $(8, -3)$ and $(-4, 13)$.

Solution   First find the center (h, k) of the circle by finding the midpoint of its diameter. Use the midpoint formulas with $(x_1, y_1) = (8, -3)$ and $(x_2, y_2) = (-4, 13)$:

$$h = \frac{x_1 + x_2}{2} \qquad k = \frac{y_1 + y_2}{2}$$

$$h = \frac{8 + (-4)}{2} \qquad k = \frac{-3 + 13}{2}$$

$$= \frac{4}{2} \qquad\qquad = \frac{10}{2}$$

$$= 2 \qquad\qquad = 5$$

Thus, the center of the circle is the point $(h, k) = (2, 5)$.

To find the radius of the circle, use the distance formula to find the distance between the center and one endpoint of the diameter. Because one endpoint is $(8, -3)$, substitute 8 for x_1, -3 for y_1, 2 for x_2, and 5 for y_2 in the distance formula and simplify:

$$r = \sqrt{(x_2 - x_1)^2 + (y_2 - y_1)^2}$$
$$r = \sqrt{(2 - 8)^2 + [5 - (-3)]^2}$$
$$= \sqrt{(-6)^2 + (8)^2}$$
$$= \sqrt{36 + 64}$$
$$= \sqrt{100}$$
$$= 10$$

Thus, the radius of the circle is 10.

To find the equation of the circle with radius 10 and center at the point $(2, 5)$, substitute 2 for h, 5 for k, and 10 for r in the standard form of the equation of the circle and simplify:

$$(x - h)^2 + (y - k)^2 = r^2$$
$$(x - 2)^2 + (y - 5)^2 = 10^2$$
$$x^2 - 4x + 4 + y^2 - 10y + 25 = 100 \qquad \text{Remove parentheses.}$$
$$x^2 + y^2 - 4x - 10y - 71 = 0 \qquad \text{Simplify.} \qquad \blacksquare$$

Example 3   Graph the circle $x^2 + y^2 - 4x + 2y = 20$.

Solution   To find the coordinates of the center and the radius, write the equation in standard form by completing the square on both x and y and then simplifying:

$$x^2 + y^2 - 4x + 2y = 20$$
$$x^2 - 4x + y^2 + 2y = 20$$
$$x^2 - 4x + 4 + y^2 + 2y + 1 = 20 + 4 + 1 \qquad \begin{array}{l}\text{Add 4 and 1 to both sides} \\ \text{to complete the square.}\end{array}$$

$$(x - 2)^2 + (y + 1)^2 = 25 \qquad \begin{array}{l}\text{Factor } x^2 - 4x + 4 \text{ and} \\ y^2 + 2y + 1.\end{array}$$

$$(x - 2)^2 + [y - (-1)]^2 = 5^2$$

Note that the radius of the circle is 5 and the coordinates of its center are $h = 2$ and $k = -1$. Plot the center of the circle and construct the circle with a radius of 5 units, as shown in Figure 6-3.

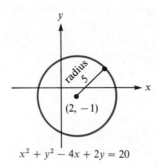

$$x^2 + y^2 - 4x + 2y = 20$$

Figure 6-3 $\blacksquare$

=========== **EXERCISE 6.1** ===========

In Exercises 1–26, write an equation for the circle with the given properties.

1. Center at the origin; $r = 1$
2. Center at the origin; $r = 4$
3. Center at $(6, 8)$; $r = 4$
4. Center at $(5, 3)$; $r = 2$
5. Center at $(-5, 3)$ and tangent to the y-axis
6. Center at $(-7, -2)$ and tangent to the x-axis
7. Center at $(3, -4)$; $r = \sqrt{2}$
8. Center at $(-9, 8)$; $r = 2\sqrt{3}$
9. Ends of diameter at $(3, -2)$ and $(3, 8)$
10. Ends of diameter at $(5, 9)$ and $(-5, -9)$
11. Ends of diameter at $(-6, 9)$ and $(-4, -7)$
12. Ends of diameter at $(17, 0)$ and $(-3, -3)$
13. Center at $(-3, 4)$ and circle passing through the origin

14. Center at (4, 0) and circle passing through the origin

15. Center at (−2, −6) and circle passing through the origin

16. Center at (−19, −13) and circle passing through the origin

17. Center at (0, −3) and circle passing through (6, 8)

18. Center at (2, 4) and circle passing through (1, 1)

19. Center at (5, 8) and circle passing through (−2, −9)

20. Center at (7, −5) and circle passing through (−3, −7)

21. Center at (−4, −2) and circle passing through (3, 5)

22. Center at (0, −7) and circle passing through (0, 7)

23. Radius of 6 and center at the intersection of $3x + y = 1$ and $-2x - 3y = 4$

24. Radius of 8 and center at the intersection of $x + 2y = 8$ and $2x - 3y = -5$

25. Radius of $\sqrt{10}$ and center at the intersection of $x - y = 12$ and $3x - y = 12$

26. Radius of $2\sqrt{2}$ and center at the intersection of $6x - 4y = 8$ and $2x + 3y = 7$

27. Can a circle with a radius of 10 have endpoints of its diameter at (6, 8) and (−2, −2)?

28. Can a circle with radius 25 have endpoints of its diameter at (0, 0) and (6, 24)?

In Exercises 29–38, graph the circle.

29. $x^2 + y^2 - 25 = 0$

30. $x^2 + y^2 - 8 = 0$

31. $(x - 1)^2 + (y + 2)^2 = 4$

32. $(x + 1)^2 + (y - 2)^2 = 9$

33. $x^2 + y^2 + 2x - 26 = 0$

34. $x^2 + y^2 - 4y = 12$

35. $9x^2 + 9y^2 - 12y = 5$

36. $4x^2 + 4y^2 + 4y = 15$

37. $4x^2 + 4y^2 - 4x + 8y + 1 = 0$

38. $9x^2 + 9y^2 - 6x + 18y + 1 = 0$

39. Write the equation of the circle passing through (0, 8), (5, 3), and (4, 6).

40. Write the equation of the circle passing through (−2, 0), (2, 8), and (5, −1).

41. Find the area of the circle $3x^2 + 3y^2 + 6x + 12y = 0$. (*Hint:* $A = \pi r^2$.)

42. Find the circumference of the circle $x^2 + y^2 + 4x - 10y - 20 = 0$. (*Hint:* $C = 2\pi r$.)

6.2 THE PARABOLA

We have encountered parabolas that open upward and downward in the discussion of quadratic functions. We now examine their equations in greater detail.

> **Definition.** A **parabola** is the set of all points in a plane such that each point in the set is equidistant from a line l, called the **directrix**, and a fixed point F, called the **focus**. The point on the parabola that is closest to the directrix is called the **vertex**. The line passing through the vertex and the focus is called the **axis**.

We will consider parabolas that open to the left, to the right, upward, and downward and that have a vertex at point (h, k). There is a standard form for the equation of each of these parabolas.

Consider the parabola in Figure 6-4, which opens to the right and has its vertex at the point $V(h, k)$. Let $P(x, y)$ be any point on the parabola. Because each point on the parabola is the same distance from the focus (point F) and from the directrix, we can let $d(DV) = d(VF) = p$, where p is some positive constant.

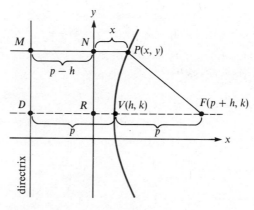

Figure 6-4

Because of the geometry of the figure,

$$d(MP) = p - h + x$$

Because of the distance formula,

$$d(PF) = \sqrt{[x - (p + h)]^2 + (y - k)^2}$$

By the definition of the parabola, $d(MP) = d(PF)$. Thus,

$$p - h + x = \sqrt{[x - (p + h)]^2 + (y - k)^2}$$
$$(p - h + x)^2 = [x - (p + h)]^2 + (y - k)^2 \qquad \text{Square both sides.}$$

Finally, we expand the expression on each side of the equation and simplify:

$$p^2 - ph + px - ph + h^2 - hx + px - hx + x^2$$
$$= x^2 - 2px - 2hx + p^2 + 2ph + h^2 + (y - k)^2$$
$$-2ph + 2px = -2px + 2ph + (y - k)^2$$
$$4px - 4ph = (y - k)^2$$
$$4p(x - h) = (y - k)^2$$

The above argument proves the following theorem.

Theorem. The standard form of the equation of a parabola with vertex at point (h, k) and opening to the right is

$$(y - k)^2 = 4p(x - h)$$

where p is the distance from the vertex to the focus.

If the parabola has its vertex at the origin, both h and k are equal to zero, and we have the following theorem.

Theorem. The standard form of the equation of a parabola with vertex at the origin and opening to the right is

$$y^2 = 4px$$

where p is the distance from the vertex to the focus.

Equations of parabolas that open to the right, left, upward, and downward are summarized in Table 6-1. If $p > 0$, then

Table 6-1

Parabola opening	Vertex at origin	Vertex at $V(h, k)$
Right	$y^2 = 4px$	$(y - k)^2 = 4p(x - h)$
Left	$y^2 = -4px$	$(y - k)^2 = -4p(x - h)$
Upward	$x^2 = 4py$	$(x - h)^2 = 4p(y - k)$
Downward	$x^2 = -4py$	$(x - h)^2 = -4p(y - k)$

Example 1 Find the equation of the parabola with vertex at the origin and focus at $(3, 0)$.

Solution Sketch the parabola as in Figure 6-5. Because the focus is to the right of the vertex, the parabola opens to the right. Because the vertex is the origin, the standard form of the equation is $y^2 = 4px$. The distance between the focus and the vertex is 3, which is p. Therefore, the equation of the parabola is $y^2 = 4(3)x$, or

$$y^2 = 12x$$

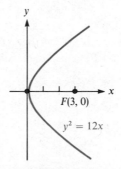

$F(3, 0)$

$y^2 = 12x$

Figure 6-5

Example 2   Find the equation of the parabola that opens upward, has vertex at the point (4, 5), and passes through the point (0, 7).

Solution   Because the parabola opens upward, use the standard form $(x - h)^2 = 4p(y - k)$. Because the point (0, 7) is on the curve, substitute 0 for x and 7 for y in the equation. Because the vertex (h, k) is (4, 5), also substitute 4 for h and 5 for k. Then solve the equation to determine p:

$$(x - h)^2 = 4p(y - k)$$
$$(0 - 4)^2 = 4p(7 - 5)$$
$$16 = 8p$$
$$2 = p$$

To find the equation of the parabola, substitute 4 for h, 5 for k, and 2 for p in the standard form of the equation and simplify:

$$(x - h)^2 = 4p(y - k)$$
$$(x - 4)^2 = 4 \cdot 2(y - 5)$$
$$(x - 4)^2 = 8(y - 5)$$

The graph of this equation appears in Figure 6-6.

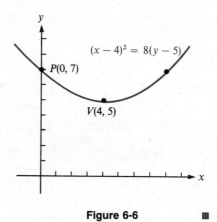

Figure 6-6   ■

Example 3   Find the equations of the two parabolas each having its vertex at (2, 4) and passing through the point (0, 0).

Solution   Sketch the two possible parabolas as shown in Figure 6-7.

Part 1.   To find the parabola that opens to the left, use the standard form of the equation $(y - k)^2 = -4p(x - h)$. Because the curve passes through the point $(x, y) = (0, 0)$ and the vertex is $(h, k) = (2, 4)$, substitute 0 for x, 0 for y, 2 for h, and 4 for k in the equation $(y - k)^2 = -4p(x - h)$ and solve for p:

$$(y - k)^2 = -4p(x - h)$$
$$(0 - 4)^2 = -4p(0 - 2)$$
$$16 = 8p$$
$$2 = p$$

Because $h = 2$, $k = 4$, and $p = 2$ and because the parabola opens to the left, its equation is

$$(y - k)^2 = -4p(x - h)$$
$$(y - 4)^2 = -4(2)(x - 2)$$
$$(y - 4)^2 = -8(x - 2)$$

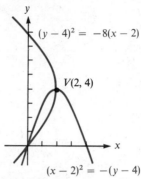

$(y - 4)^2 = -8(x - 2)$

$V(2, 4)$

$(x - 2)^2 = -(y - 4)$

Figure 6-7

Part 2. To find the equation of the parabola that opens downward, use the standard form $(x - h)^2 = -4p(y - k)$. Substitute 2 for h, 4 for k, 0 for x, and 0 for y in the equation and solve for p:

$$(x - h)^2 = -4p(y - k)$$
$$(0 - 2)^2 = -4p(0 - 4)$$
$$4 = 16p$$
$$\frac{1}{4} = p$$

Because $h = 2$, $k = 4$, and $p = \dfrac{1}{4}$, and the parabola opens downward, its equation is

$$(x - h)^2 = -4p(y - k)$$
$$(x - 2)^2 = -4\left(\frac{1}{4}\right)(y - 4)$$
$$(x - 2)^2 = -(y - 4)$$

∎

Example 4   Find the vertex and y-intercepts of the parabola $y^2 + 8x - 4y = 28$. Then graph the parabola.

Solution   Complete the square on y to write the equation in standard form:

$$y^2 + 8x - 4y = 28$$

$y^2 - 4y = -8x + 28$	Add $-8x$ to both sides.
$y^2 - 4y + 4 = -8x + 28 + 4$	Add 4 to both sides.
$(y - 2)^2 = -8(x - 4)$	Factor both sides.

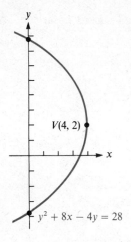

Figure 6-8

Observe that this equation represents a parabola opening to the left with vertex at (4, 2). To find the points where the graph intersects the y-axis, substitute 0 for x in the equation of the parabola, and solve for y.

$$(y - 2)^2 = -8(x - 4)$$
$$(y - 2)^2 = -8(0 - 4) \qquad \text{Substitute 0 for } x.$$
$$y^2 - 4y + 4 = 32 \qquad \text{Remove parentheses.}$$
$$y^2 - 4y - 28 = 0$$

Use the quadratic formula to determine that the roots of this quadratic equation are $y \approx 7.7$ and $y \approx -3.7$.

The points with coordinates of approximately (0, 7.7) and (0, -3.7) are on the graph of the parabola. Using this information and the knowledge that the graph opens to the left and has a vertex at (4, 2), draw the curve as shown in Figure 6-8. ∎

Example 5   A stone is thrown straight up. The equation $s = 128t - 16t^2$ expresses the height of the stone in feet t seconds after it was thrown. Find the maximum height reached by the stone.

Solution   The stone goes straight up and then straight down. The graph of $s = 128t - 16t^2$, expressing the height of the stone t seconds after it was thrown, is a parabola (see Figure 6-9). To find the maximum height reached by the stone, calculate the y-coordinate, k, of the vertex of the parabola. To find k, write the equation of the parabola, $s = 128t - 16t^2$, in standard form. To change this equation into standard form, complete the square on t:

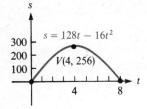

Figure 6-9

$$s = 128t - 16t^2$$
$$16t^2 - 128t = -s \qquad \text{Multiply both sides by } -1.$$
$$t^2 - 8t = \frac{-s}{16} \qquad \text{Divide both sides by 16.}$$
$$t^2 - 8t + 16 = \frac{-s}{16} + 16 \qquad \text{Add 16 to both sides.}$$
$$(t - 4)^2 = \frac{-s + 256}{16} \qquad \text{Factor } t^2 - 8t + 16 \text{ and combine terms.}$$
$$(t - 4)^2 = -\frac{1}{16}(s - 256) \qquad \text{Factor out } -\frac{1}{16}.$$

This equation indicates that the maximum height of 256 feet was reached in 4 seconds. ∎

EXERCISE 6.2

In Exercises 1–16, find the equation of each parabola.

1. Vertex at (0, 0) and focus at (0, 3)

2. Vertex at (0, 0) and focus at (0, −3)

3. Vertex at (0, 0) and focus at (3, 0)

4. Vertex at (0, 0) and focus at (−3, 0)

5. Vertex at (3, 5) and focus at (3, 2)

6. Vertex at (3, 5) and focus at (−3, 5)

7. Vertex at (3, 5) and focus at (3, −2)

8. Vertex at (3, 5) and focus at (6, 5)

9. Vertex at (2, 2) and the parabola passing through (0, 0)

10. Vertex at (−2, −2) and the parabola passing through (0, 0)

11. Vertex at (−4, 6) and the parabola passing through (0, 3)

12. Vertex at (−2, 3) and the parabola passing through (0, −3)

13. Vertex at (6, 8) and the parabola passing through (5, 10) and (5, 6)

14. Vertex at (2, 3) and the parabola passing through $(1, \frac{13}{4})$ and $(-1, \frac{21}{4})$

15. Vertex at (3, 1) and the parabola passing through (4, 3) and (2, 3)

16. Vertex at (−4, −2) and the parabola passing through (−3, 0) and $(\frac{9}{4}, 3)$

In Exercises 17–26, change each equation to standard form and graph each parabola.

17. $y = x^2 + 4x + 5$

18. $2x^2 - 12x - 7y = 10$

19. $y^2 + 4x - 6y = -1$

20. $x^2 - 2y - 2x = -7$

21. $y^2 + 2x - 2y = 5$

22. $y^2 - 4y = -8x + 20$

23. $x^2 - 6y + 22 = -4x$

24. $4y^2 - 4y + 16x = 7$

25. $4x^2 - 4x + 32y = 47$

26. $4y^2 - 16x + 17 = 20y$

27. A parabolic arch spans 30 meters and has a maximum height of 10 meters. Derive the equation of the arch using the vertex of the arch as the origin.

28. Find the maximum value of y in the parabola $x^2 + 8y - 8x = 8$.

29. A resort owner plans to build and rent n cabins for d dollars per week. The price, d, that she can charge for each cabin depends on the number of cabins she builds, where $d = -45(\frac{n}{32} - \frac{1}{2})$. Find the number of cabins that she should build to maximize her weekly income.

30. A toy rocket is s feet above the earth at the end of t seconds, where $s = -16t^2 + 80\sqrt{3}t$. Find the maximum height of the rocket.

31. An engineer plans to build a tunnel whose arch is in the shape of a parabola. The tunnel will span a two-lane highway that is 8 meters wide. To allow safe passage for most vehicles, the tunnel must be 5 meters high at a distance of 1 meter from the tunnel's edge. What will be the maximum height of the tunnel?

32. The towers of a suspension bridge are 900 feet apart and rise 120 feet above the roadway. The cable between the towers has the shape of a parabola with a vertex 15 feet above the roadway. Find the equation of the parabola with respect to the indicated coordinate system. See Illustration 1.

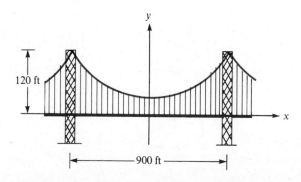

120 ft

900 ft

Illustration 1

33. A satellite antenna with a parabolic cross section is a dish 6 feet in diameter and 1 foot deep at its center. How far is the focus from the center of the dish?

34. A stone tossed upward is s feet above the earth after t seconds, where $s = -16t^2 + 128t$. Show that the stone's height x seconds *after* it is thrown is equal to its height x seconds *before* it returns to the ground.

35. Derive the standard form of the equation of a parabola that opens downward and has vertex at the origin.

36. Show that the result in Example 1 is a special case of the general form of the equation of second degree.

In Exercises 37–38, find the equation of the form $y = ax^2 + bx + c$ that determines a parabola passing through the three given points.

37. $(1, 8)$, $(-2, -1)$, and $(2, 15)$

38. $(1, -3)$, $(-2, 12)$, and $(-1, 3)$

6.3 THE ELLIPSE

A third important conic is the ellipse.

> **Definition.**   An **ellipse** is the set of all points P in a plane such that the sum of the distances from P to two other fixed points F and F' is a positive constant.

In the ellipse shown in Figure 6-10, the two fixed points F and F' are called **foci** of the ellipse, the midpoint of the chord FF' is called the **center**, the chord VV' is called the **major axis**, and each endpoint of the major axis is called a **vertex**. The chord BB', perpendicular to the major axis and passing through the center C, is called the **minor axis**.

To keep the algebra manageable, we will derive the equation of the ellipse shown in Figure 6-11, which has its center at $(0, 0)$. Because the origin is the midpoint of the chord FF', we can let $d(OF) = d(OF') = c$, where $c > 0$. Then the coordinates of point F are $(c, 0)$, and the coordinates of F' are $(-c, 0)$. We also let $P(x, y)$ be any point on the ellipse.

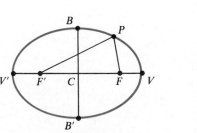

Figure 6-10

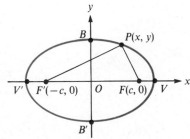

Figure 6-11

The definition of an ellipse requires that the sum of $d(F'P)$ and $d(PF)$ be a positive constant, which we will call $2a$. Thus,

1. $d(F'P) + d(PF) = 2a$

We use the distance formula to compute the lengths of $F'P$ and PF:

$$d(F'P) = \sqrt{[x - (-c)]^2 + y^2}$$
$$d(PF) = \sqrt{(x - c)^2 + y^2}$$

and substitute these values into Equation 1 to obtain

$$\sqrt{[x - (-c)]^2 + y^2} + \sqrt{(x - c)^2 + y^2} = 2a$$

or

$$\sqrt{[x + c]^2 + y^2} = 2a - \sqrt{(x - c)^2 + y^2}$$

We square both sides of this equation and simplify to get

$$(x + c)^2 + y^2 = 4a^2 - 4a\sqrt{(x - c)^2 + y^2} + [(x - c)^2 + y^2]$$
$$x^2 + 2cx + c^2 + y^2 = 4a^2 - 4a\sqrt{(x - c)^2 + y^2} + x^2 - 2cx + c^2 + y^2$$
$$4cx = 4a^2 - 4a\sqrt{(x - c)^2 + y^2}$$
$$cx = a^2 - a\sqrt{(x - c)^2 + y^2}$$
$$cx - a^2 = -a\sqrt{(x - c)^2 + y^2}$$

We square both sides again and simplify to obtain

$$c^2x^2 - 2a^2cx + a^4 = a^2[(x - c)^2 + y^2]$$
$$c^2x^2 - 2a^2cx + a^4 = a^2(x^2 - 2cx + c^2 + y^2)$$
$$c^2x^2 - 2a^2cx + a^4 = a^2x^2 - 2a^2cx + a^2c^2 + a^2y^2$$
$$c^2x^2 + a^4 = a^2x^2 + a^2c^2 + a^2y^2$$
$$a^4 - a^2c^2 = a^2x^2 - c^2x^2 + a^2y^2$$

2. $a^2(a^2 - c^2) = (a^2 - c^2)x^2 + a^2y^2$

Because the shortest path between two points is a line segment, $d(F'P) + d(PF) > d(F'F)$. Therefore, $2a > 2c$. This implies that $a > c$ and that $a^2 - c^2$ is a positive number, which we will call b^2. Letting $b^2 = a^2 - c^2$ and substituting into Equation 2, we have

$$a^2b^2 = b^2x^2 + a^2y^2$$

Dividing both sides of this equation by a^2b^2 gives the standard form of the equation for an ellipse with center at the origin and major axis on the x-axis:

$$\frac{x^2}{a^2} + \frac{y^2}{b^2} = 1 \qquad \text{where } a > b > 0$$

To find the coordinates of the vertices V and V', we substitute 0 for y and solve for x:

$$\frac{x^2}{a^2} + \frac{y^2}{b^2} = 1$$

$$\frac{x^2}{a^2} + \frac{0^2}{b^2} = 1$$

$$\frac{x^2}{a^2} = 1$$

$$x^2 = a^2$$

$$x = a \qquad \text{or} \qquad x = -a$$

Thus, the coordinates of V are $(a, 0)$, and the coordinates of V' are $(-a, 0)$. In other words, a is the distance between the center of the ellipse, $(0, 0)$, and either of its vertices, and the center of the ellipse is the midpoint of the major axis.

To find the coordinates of B and B', we substitute 0 for x and solve for y:

$$\frac{x^2}{a^2} + \frac{y^2}{b^2} = 1$$

$$\frac{0^2}{a^2} + \frac{y^2}{b^2} = 1$$

$$y^2 = b^2$$

$$y = b \qquad \text{or} \qquad y = -b$$

Thus, the coordinates of B are $(0, b)$, and the coordinates of B' are $(0, -b)$. The distance between the center of the ellipse and either endpoint of the minor axis is b.

Theorem.   The standard form of the equation of an ellipse with center at the origin and major axis on the x-axis is

$$\frac{x^2}{a^2} + \frac{y^2}{b^2} = 1 \qquad \text{where } a > b > 0$$

If the major axis of an ellipse with center at $(0, 0)$ lies on the y-axis, the standard form of the equation of the ellipse is

$$\frac{y^2}{a^2} + \frac{x^2}{b^2} = 1 \qquad \text{where } a > b > 0$$

In either case, the length of the major axis is $2a$, and the length of the minor axis is $2b$.

If we develop the equation of the ellipse with center at (h, k), we obtain the following results.

Theorem.   The standard form of the equation of an ellipse with center at (h, k) and major axis parallel to the x-axis is

$$\frac{(x - h)^2}{a^2} + \frac{(y - k)^2}{b^2} = 1 \qquad \text{where } a > b > 0$$

If the major axis of an ellipse with center at (h, k) is parallel to the y-axis, the standard form of the equation of the ellipse is

$$\frac{(y - k)^2}{a^2} + \frac{(x - h)^2}{b^2} = 1 \qquad \text{where } a > b > 0$$

In either case, the length of the major axis is $2a$, and the length of the minor axis is $2b$.

Example 1   Find the equation of the ellipse with center at the origin, major axis of length 6 units located on the x-axis, and minor axis of length 4 units.

Solution   Because the center of the ellipse is the origin and the length of the major axis is 6, $a = 3$ and the coordinates of the vertices of the ellipse are $(3, 0)$ and $(-3, 0)$, as shown in Figure 6-12.

Because the length of the minor axis is 4, the value of b is 2 and the coordinates of B and B' are $(0, 2)$ and $(0, -2)$. To find the desired equation, substitute 3 for a and 2 for b in the standard form of the equation of an ellipse with center at the origin and major axis on the x-axis. Then simplify the equation:

$$\frac{x^2}{a^2} + \frac{y^2}{b^2} = 1$$

$$\frac{x^2}{3^2} + \frac{y^2}{2^2} = 1$$

$$\frac{x^2}{9} + \frac{y^2}{4} = 1$$

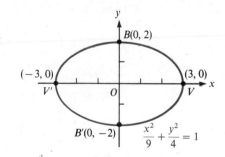

Figure 6-12   ■

Example 2   Find the equation of the ellipse with focus $(0, 3)$ and vertices V and V' at $(3, 3)$ and $(-5, 3)$.

Solution   Because the midpoint of the major axis is the center of the ellipse, the coordinates of the center are $(-1, 3)$. Look at Figure 6-13 and note that the major axis is parallel to the x-axis. The standard form of the equation to use is

$$\frac{(x - h)^2}{a^2} + \frac{(y - k)^2}{b^2} = 1 \qquad \text{where } a > b > 0$$

The distance between the center of the ellipse and a vertex is $a = 4$; the distance between the focus and the center is $c = 1$. In the ellipse, $b^2 = a^2 - c^2$. From this equation, compute b^2:

$$b^2 = a^2 - c^2$$
$$= 4^2 - 1^2$$
$$= 15$$

To find the equation of the ellipse, substitute -1 for h, 3 for k, 16 for a^2, and 15 for b^2 in the standard form of the equation for an ellipse and simplify:

$$\frac{(x - h)^2}{a^2} + \frac{(y - k)^2}{b^2} = 1$$

$$\frac{[x - (-1)]^2}{16} + \frac{(y - 3)^2}{15} = 1$$

$$\frac{(x + 1)^2}{16} + \frac{(y - 3)^2}{15} = 1$$

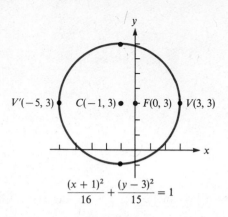

$$\frac{(x + 1)^2}{16} + \frac{(y - 3)^2}{15} = 1$$

Figure 6-13

Example 3   The orbit of the earth is approximately an ellipse, with the sun at one focus. The ratio of c to a (called the *eccentricity* of the ellipse) is about $\frac{1}{62}$, and the length of the major axis is approximately 186,000,000 miles. How does the earth get close to the sun?

Solution   Assume that this ellipse has its center at the origin and vertices V' and V at $(-93,000,000, 0)$ and $(93,000,000, 0)$, as shown in Figure 6-14. This implies that $a = 93,000,000$.

$$\frac{c}{a} = \frac{1}{62} \quad \text{or} \quad c = \frac{1}{62}a$$

Because $a = 93,000,000$,

$$c = \frac{1}{62}(93,000,000)$$
$$= 1,500,000$$

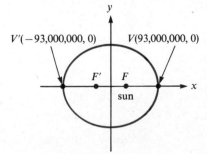

Figure 6-14

$d(FV)$ represents the shortest possible distance between the earth and the sun. (You'll be asked to prove this in the exercises.) Thus,

$$d(FV) = a - c = 93{,}000{,}000 - 1{,}500{,}000 = 91{,}500{,}000 \text{ miles}$$

The earth's point of closest approach to the sun (called *perihelion*) is approximately 91.5 million miles.   ■

Example 4   Graph the ellipse $\dfrac{(x + 2)^2}{4} + \dfrac{(y - 2)^2}{9} = 1$.

Solution   The center of the ellipse is at $(-2, 2)$, and the major axis is parallel to the y-axis. Because $a = 3$, the vertices are 3 units above and below the center at points $(-2, 5)$ and $(-2, -1)$. Because $b = 2$, the endpoints of the minor axis are 2 units to the right and left of the center at points $(0, 2)$ and $(-4, 2)$. Using these four points as guides, sketch the ellipse, as shown in Figure 6-15.

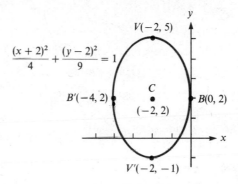

Figure 6-15   ■

Example 5   Graph the equation $4x^2 + 9y^2 - 16x - 18y = 11$.

Solution   Write the equation in standard form by completing the square on x and y as follows:

$$4x^2 + 9y^2 - 16x - 18y = 11$$
$$4x^2 - 16x + 9y^2 - 18y = 11$$
$$4(x^2 - 4x) + 9(y^2 - 2y) = 11$$
$$\mathbf{4}(x^2 - 4x + \mathbf{4}) + 9(y^2 - 2y + \mathbf{1}) = 11 + \mathbf{16} + \mathbf{9}$$
$$4(x - 2)^2 + 9(y - 1)^2 = 36$$
$$\frac{(x - 2)^2}{9} + \frac{(y - 1)^2}{4} = 1$$

You can now see that the graph of the given equation is an ellipse with center at $(2, 1)$ and major axis parallel to the x-axis. Because $a = 3$, the vertices are at $(-1, 1)$ and $(5, 1)$. Because $b = 2$, the endpoints of the minor axis are at $(2, -1)$ and $(2, 3)$. Using these four points as guides, sketch the ellipse, as shown in Figure 6-16.

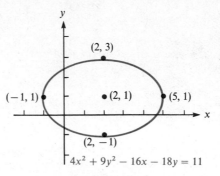

$$4x^2 + 9y^2 - 16x - 18y = 11$$

Figure 6-16

EXERCISE 6.3

In Exercises 1–6, write the equation of the ellipse that has its center at the origin.

1. Focus at $(3, 0)$ and a vertex at $(5, 0)$

2. Focus at $(0, 4)$ and a vertex at $(0, 7)$

3. Focus at $(0, 1)$; $\frac{4}{3}$ is one-half the length of the minor axis.

4. Focus at $(1, 0)$; $\frac{4}{3}$ is one-half the length of the minor axis.

5. Focus at $(0, 3)$ and major axis equal to 8

6. Focus at $(5, 0)$ and major axis equal to 12

In Exercises 7–16, write the equation of each ellipse.

7. Center at $(3, 4)$; $a = 3$, $b = 2$; the major axis is parallel to the y-axis.

8. Center at $(3, 4)$; the curve passes through $(3, 10)$ and $(3, -2)$; $b = 2$.

9. Center at $(3, 4)$; $a = 3$, $b = 2$; the major axis is parallel to the x-axis.

10. Center at $(3, 4)$; the curve passes through $(8, 4)$ and $(-2, 4)$; $b = 2$.

11. Foci at $(-2, 4)$ and $(8, 4)$; $b = 4$

12. Foci at $(-8, 5)$ and $(4, 5)$; $b = 3$

13. Vertex at $(6, 4)$ and foci at $(-4, 4)$ and $(4, 4)$

14. Center at $(-4, 5)$; $\frac{c}{a} = \frac{1}{3}$; vertex at $(-4, -1)$

15. Foci at $(6, 0)$ and $(-6, 0)$; $\frac{c}{a} = \frac{3}{5}$

16. Vertices at $(2, 0)$ and $(-2, 0)$; $\frac{2b^2}{a} = 2$

In Exercises 17–24, graph each ellipse.

17. $\dfrac{x^2}{25} + \dfrac{y^2}{49} = 1$

18. $4x^2 + y^2 = 4$

19. $\dfrac{x^2}{16} + \dfrac{(y + 2)^2}{36} = 1$

20. $(x - 1)^2 + \dfrac{4y^2}{25} = 4$

21. $x^2 + 4y^2 - 4x + 8y + 4 = 0$

22. $x^2 + 4y^2 - 2x - 16y = -13$

23. $16x^2 + 25y^2 - 160x - 200y + 400 = 0$

24. $3x^2 + 2y^2 + 7x - 6y = -1$

25. The moon has an orbit that is an ellipse with the earth at one focus. If the major axis of the orbit is 378,000 miles and the ratio of c to a is approximately $\frac{11}{200}$, how far does the moon get from the earth? (This farthest point in an orbit is called the *apogee*.)

26. An arch is a semiellipse 10 meters wide and 5 meters high. Write the equation of the ellipse if the ellipse is centered at the origin.

27. A track is built in the shape of an ellipse and has a maximum length of 100 meters and a maximum width of 60 meters. Write the equation of the ellipse and find its focal width. That is, find the length of a chord that is perpendicular to the major axis and that passes through either focus of the ellipse.

28. An arch has the shape of a semiellipse and has a maximum height of 5 meters. The foci are on the ground with a distance between them of 24 meters. Find the total distance from one focus to any point on the arch and back to the other focus.

29. Consider the ellipse in Illustration 1. If F is a focus of the ellipse and B is an endpoint of the minor axis, use the distance formula to prove that the length of segment FB is a. (*Hint:* Remember that in an ellipse $a^2 - c^2 = b^2$.)

30. Consider the ellipse in Illustration 1. If F is a focus of the ellipse and P is any point on the ellipse, use the distance formula to show that the length of segment FP is $a - \frac{c}{a}x$. (*Hint:* Remember that in an ellipse $a^2 - c^2 = b^2$.)

31. Consider the ellipse in Illustration 2. Chord AA' passes through the focus F and is perpendicular to the major axis. Show that the length of AA' (called the **focal width** of the ellipse) is $\frac{2b^2}{a}$.

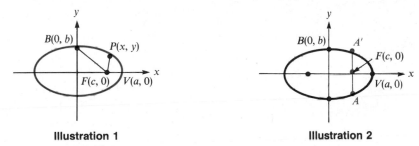

Illustration 1   Illustration 2

32. Prove that segment FV in Example 3 does represent the shortest distance between the earth and the sun. (*Hint:* You might find the result of Exercise 30 helpful.)

33. The ends of a piece of string 6 meters long are attached to two thumbtacks that are 2 meters apart. A pencil catches the loop and draws it tight. As the pencil is moved about the thumbtacks (always keeping the tension), an ellipse is produced with the thumbtacks as foci. Write the equation of the ellipse. (*Hint:* You'll have to establish a coordinate system.)

34. Prove that $a > b$ in the development of the standard form of the equation of an ellipse.

35. Show that the expansion of the standard equation of an ellipse is a special case of the general second-degree equation.

36. The distance between point $P(x, y)$ and the point $(0, 2)$ is $\frac{1}{3}$ of the distance of point P from the line $y = 18$. Find the equation of the curve on which point P lies.

6.4 THE HYPERBOLA

The definition of the hyperbola is similar to the definition of the ellipse, except that we demand a constant *difference* of $2a$ instead of a constant sum.

> **Definition.**   A **hyperbola** is the set of all points P in a plane such that the difference of the distances from point P to two other points in the plane, F and F', is a positive constant.

Points F and F' (see Figure 6-17) are called the **foci** of the hyperbola, and the midpoint of chord FF' is called the **center** of the hyperbola. The points V and V', where the hyperbola intersects the line segment FF', are called the **vertices** of the hyperbola, and the line segment VV' is called the **transverse axis**.

As with the ellipse, we will develop the equation of the hyperbola centered at the origin. Because the origin is the midpoint of chord FF', we can let $d(F'O) = d(OF) = c > 0$. Therefore, F is at $(c, 0)$ and F' is at $(-c, 0)$. The definition requires that $|d(F'P) - d(PF)| = 2a$, where $2a$ is a positive constant. Using the distance formula to compute the lengths of $F'P$ and PF gives

$$d(F'P) = \sqrt{[x - (-c)]^2 + y^2}$$

$$d(PF) = \sqrt{(x - c)^2 + y^2}$$

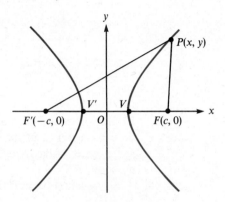

Figure 6-17

Substituting these values into the equation $d(F'P) - d(PF) = 2a$ gives

$$\sqrt{(x + c)^2 + y^2} - \sqrt{(x - c)^2 + y^2} = 2a$$

or

$$\sqrt{(x + c)^2 + y^2} = 2a + \sqrt{(x - c)^2 + y^2}$$

Squaring both sides of this equation and simplifying gives

$$(x + c)^2 + y^2 = 4a^2 + 4a\sqrt{(x - c)^2 + y^2} + (x - c)^2 + y^2$$

$$x^2 + 2cx + c^2 + y^2 = 4a^2 + 4a\sqrt{(x - c)^2 + y^2} + x^2 - 2cx + c^2 + y^2$$

$$4cx = 4a^2 + 4a\sqrt{(x - c)^2 + y^2}$$

$$cx - a^2 = a\sqrt{(x - c)^2 + y^2}$$

Squaring both sides again and simplifying gives

$$c^2x^2 - 2a^2cx + a^4 = a^2(x^2 - 2cx + c^2 + y^2)$$

$$c^2x^2 - 2a^2cx + a^4 = a^2x^2 - 2a^2cx + a^2c^2 + a^2y^2$$

$$c^2x^2 + a^4 = a^2x^2 + a^2c^2 + a^2y^2$$

1. $(c^2 - a^2)x^2 - a^2y^2 = a^2(c^2 - a^2)$

Because $c > a$ (you will be asked to prove this in the exercises), $c^2 - a^2$ is a positive number. Thus, we can let $b^2 = c^2 - a^2$ and substitute b^2 for $c^2 - a^2$ in Equation 1 to get

$$b^2x^2 - a^2y^2 = a^2b^2$$

Dividing both sides of the previous equation by a^2b^2 gives the standard form of the equation for a hyperbola with center at the origin and foci on the x-axis:

$$\frac{x^2}{a^2} - \frac{y^2}{b^2} = 1$$

If $y = 0$, the previous equation becomes

$$\frac{x^2}{a^2} = 1 \qquad \text{or} \qquad x^2 = a^2$$

Solving this equation for x gives

$$x = a \qquad \text{or} \qquad x = -a$$

This implies that the coordinates of V and V' are $(a, 0)$ and $(-a, 0)$ and that the distance between the center of the hyperbola and either vertex is a. This, in turn, implies that the center of the hyperbola is the midpoint of the segment $V'V$ as well as of the segment FF'.

If $x = 0$, the equation becomes

$$\frac{-y^2}{b^2} = 1 \qquad \text{or} \qquad y^2 = -b^2$$

Because this equation has no real solutions, the hyperbola cannot intersect the y-axis. These results suggest the following theorem.

Theorem. The standard form of the equation of a hyperbola with center at the origin and foci on the x-axis is

$$\frac{x^2}{a^2} - \frac{y^2}{b^2} = 1$$

The standard form of the equation of a hyperbola with center at the origin and foci on the y-axis is

$$\frac{y^2}{a^2} - \frac{x^2}{b^2} = 1$$

As with the ellipse, the standard equation of the hyperbola can be developed with center at (h, k). We state the results without proof.

Theorem.   The standard form of the equation of a hyperbola with center at (h, k) and foci on a line parallel to the x-axis is

$$\frac{(x - h)^2}{a^2} - \frac{(y - k)^2}{b^2} = 1$$

The standard form of the equation of a hyperbola with center at (h, k) and foci on a line parallel to the y-axis is

$$\frac{(y - k)^2}{a^2} - \frac{(x - h)^2}{b^2} = 1$$

Example 1   Write the equation of the hyperbola with vertices $(3, -3)$ and $(3, 3)$ and with a focus at $(3, 5)$.

Solution   First, plot the vertices and focus, as shown in Figure 6-18. Note that the foci lie on a vertical line. Therefore, the standard form to use is

$$\frac{(y - k)^2}{a^2} - \frac{(x - h)^2}{b^2} = 1$$

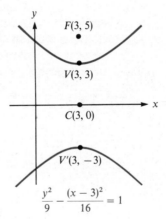

Figure 6-18

Because the center of the hyperbola is midway between the vertices V and V', the center is point $(3, 0)$, so $h = 3$, and $k = 0$. The distance between the vertex and the center of the hyperbola is $a = 3$, and the distance between the focus and the center is $c = 5$. Also, in a hyperbola, $b^2 = c^2 - a^2$. Therefore, $b^2 = 5^2 - 3^2 = 16$. Substituting the values for h, k, a^2, and b^2 into the standard form of the equation gives the desired result:

$$\frac{(y - 0)^2}{9} - \frac{(x - 3)^2}{16} = 1$$

$$\frac{y^2}{9} - \frac{(x - 3)^2}{16} = 1$$

■

Asymptotes of a Hyperbola

The values of a and b play an important role in graphing hyperbolas. To see their significance, we consider the hyperbola

$$\frac{x^2}{a^2} - \frac{y^2}{b^2} = 1$$

with the center at the origin and vertices at $V(a, 0)$ and $V'(-a, 0)$. We plot points V, V', $B(0, b)$, and $B'(0, -b)$ and form rectangle $RSQP$, called the **fundamental rectangle**, as shown in Figure 6-19. The extended diagonals of this rectangle are asymptotes of the hyperbola. In the exercises, you will be asked to show that the equations of these two lines are

$$y = \frac{b}{a}x \qquad \text{and} \qquad y = -\frac{b}{a}x$$

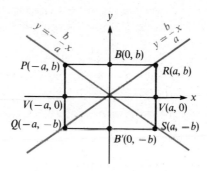

Figure 6-19

To show that the extended diagonals are asymptotes of the hyperbola, we solve the equation

$$\frac{x^2}{a^2} - \frac{y^2}{b^2} = 1$$

for y and modify its form:

$$\frac{x^2}{a^2} - \frac{y^2}{b^2} = 1$$

$$b^2x^2 - a^2y^2 = a^2b^2 \qquad \qquad \text{Multiply both sides by } a^2b^2.$$

$$y^2 = \frac{b^2x^2 - a^2b^2}{a^2} \qquad \qquad \text{Add } -b^2x^2 \text{ to both sides and divide both sides by } -a^2.$$

$$y^2 = \frac{b^2x^2}{a^2}\left(1 - \frac{a^2}{x^2}\right) \qquad \qquad \text{Factor out } b^2x^2 \text{ in the numerator.}$$

$$y = \pm\frac{bx}{a}\sqrt{1 - \frac{a^2}{x^2}} \qquad \qquad \text{Take the square root of both sides.}$$

If $|x|$ grows large without bound, the fraction $\dfrac{a^2}{x^2}$ in the previous equation approaches 0 and $\sqrt{1 - \dfrac{a^2}{x^2}}$ approaches 1. Hence, the hyperbola approaches the lines

$$y = \frac{b}{a}x \qquad \text{and} \qquad y = -\frac{b}{a}x$$

This fact makes it easy to sketch a hyperbola. We convert the equation into standard form, find the coordinates of its vertices, and plot them. Then we construct the fundamental rectangle and its extended diagonals. Using the vertices and the asymptotes as guides, we make a quick and relatively accurate sketch, as in Figure 6-20. The segment BB' is called the **conjugate axis** of the hyperbola.

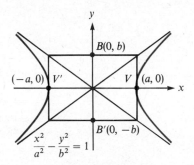

Figure 6-20

Example 2   Graph the hyperbola $x^2 - y^2 - 2x + 4y = 12$.

Solution   First complete the square on x and y to convert the equation into standard form:

$$x^2 - 2x - y^2 + 4y = 12$$
$$x^2 - 2x - (y^2 - 4y) = 12$$
$$x^2 - 2x + 1 - (y^2 - 4y + 4) = 12 + 1 - 4$$
$$(x - 1)^2 - (y - 2)^2 = 9$$
$$\frac{(x - 1)^2}{9} - \frac{(y - 2)^2}{9} = 1$$

From the standard form of the equation of a hyperbola, observe that the center is $(1, 2)$, that $a = 3$ and $b = 3$, and that the vertices are on a line segment parallel to the x-axis, as shown in Figure 6-21. Therefore, the vertices V and V' are 3 units to the right and left of the center and have coordinates of $(4, 2)$ and $(-2, 2)$. Points B and B', 3 units above and below the center, have coordinates $(1, 5)$ and $(1, -1)$. After plotting points V, V', B, and B', construct the fundamental rectangle and its extended diagonals. Using the vertices as points on the hyperbola and the extended diagonals as asymptotes, sketch the graph.

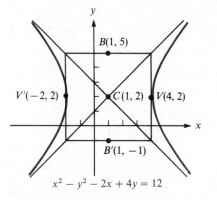

Figure 6-21

This discussion of the hyperbola has considered only cases where the segment that joins the foci is horizontal or vertical. However, there are hyperbolas where this is not true. For example, the graph of the equation $xy = 4$ is a hyperbola with vertices at $(2, 2)$ and $(-2, -2)$, as shown in Figure 6-22.

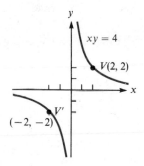

Figure 6-22

EXERCISE 6.4

In Exercises 1–12, write the equation of each hyperbola.

1. Vertices at $(5, 0)$ and $(-5, 0)$ and focus at $(7, 0)$

2. Focus at $(3, 0)$, vertex at $(2, 0)$, and center at $(0, 0)$

3. Center at $(2, 4)$; $a = 2$, $b = 3$; transverse axis is horizontal

4. Center at $(-1, 3)$, vertex at $(1, 3)$, and focus at $(2, 3)$

5. Center at $(5, 3)$, vertex at $(5, 6)$, hyperbola passes through $(1, 8)$

6. Foci at $(0, 10)$ and $(0, -10)$; $\dfrac{c}{a} = \dfrac{5}{4}$

7. Vertices at $(0, 3)$ and $(0, -3)$; $\dfrac{c}{a} = \dfrac{5}{3}$

8. Focus at $(4, 0)$, vertex at $(2, 0)$, and center at the origin

9. Center at $(1, -3)$; $a^2 = 4$, $b^2 = 16$

10. Center at $(1, 4)$, focus at $(7, 4)$, and vertex at $(3, 4)$

11. Center at the origin; hyperbola passes through points $(4, 2)$ and $(8, -6)$

12. Center at $(3, -1)$, y-intercept of -1, x-intercept of $3 + \dfrac{3\sqrt{5}}{2}$

In Exercises 13–16, find the area of the fundamental rectangle of each hyperbola.

13. $4(x - 1)^2 - 9(y + 2)^2 = 36$ **14.** $x^2 - y^2 - 4x - 6y = 6$

15. $x^2 + 6x - y^2 + 2y = -11$ **16.** $9x^2 - 4y^2 = 18x + 24y + 63$

In Exercises 17–20, write the equation of each hyperbola.

17. Center at $(-2, -4)$; $a = 2$; area of fundamental rectangle is 36 square units

18. Center at $(3, -5)$; $b = 6$; area of fundamental rectangle is 24 square units

19. One vertex at $(6, 0)$, one end of conjugate axis at $(0, \frac{5}{4})$

20. One vertex at $(3, 0)$, one focus at $(-5, 0)$, center at $(0, 0)$

In Exercises 21–30, graph the hyperbola.

21. $\dfrac{x^2}{9} - \dfrac{y^2}{4} = 1$ **22.** $\dfrac{y^2}{4} - \dfrac{x^2}{9} = 1$

23. $4x^2 - 3y^2 = 36$ **24.** $3x^2 - 4y^2 = 36$

25. $y^2 - x^2 = 1$ **26.** $9(y + 2)^2 - 4(x - 1)^2 = 36$

27. $4x^2 - 2y^2 + 8x - 8y = 8$ **28.** $x^2 - y^2 - 4x - 6y = 6$

29. $y^2 - 4x^2 + 6y + 32x = 59$ **30.** $x^2 + 6x - y^2 + 2y = -11$

In Exercises 31–32, graph each hyperbola by plotting points.

31. $-xy = 6$ **32.** $xy = 20$

In Exercises 33–36, find the equation of each curve on which point P lies.

33. The difference of the distances between $P(x, y)$ and the points $(-2, 1)$ and $(8, 1)$ is 6.

34. The difference of the distances between $P(x, y)$ and the points $(3, -1)$ and $(3, 5)$ is 5.

35. The distance between point $P(x, y)$ and the point $(0, 3)$ is $\frac{3}{2}$ of the distance between P and the line $y = -2$.

36. The distance between point $P(x, y)$ and the point $(5, 4)$ is $\frac{5}{3}$ of the distance between P and the line $x = -3$.

37. Prove that $c > a$ for a hyperbola with center at the origin and line segment FF' on the x-axis.

38. Show that the equations of the extended diagonals of the fundamental rectangle of the hyperbola with equation $\dfrac{x^2}{a^2} - \dfrac{y^2}{b^2} = 1$ are $y = \frac{b}{a}x$ and $y = -\frac{b}{a}x$.

39. Show that the expansion of the standard form of the equation of a hyperbola is a special case of the general equation of second degree with $B = 0$.

6.5   SOLVING SIMULTANEOUS SECOND-DEGREE EQUATIONS

We now discuss techniques for solving systems of two equations in two variables where at least one of the equations is of second degree.

Example 1   Solve the following system of equations by graphing:

$$\begin{cases} x^2 + y^2 = 25 \\ 2x + y = 10 \end{cases}$$

Solution   The graph of the equation $x^2 + y^2 = 25$ is a circle with center at the origin and radius of 5. The graph of the equation $2x + y = 10$ is a straight line. Depending on whether the line is a secant (intersecting the circle at two points) or a tangent (intersecting the circle at one point) or does not intersect the circle at all, there are two, one, or no solutions to the system, respectively. After graphing the circle and the line, as shown in Figure 6-23, note that there are two intersection points, P and P', with the coordinates of (3, 4) and (5, 0). Thus, the solutions to the given system of equations are

$$\begin{cases} x = 3 \\ y = 4 \end{cases} \quad \text{and} \quad \begin{cases} x = 5 \\ y = 0 \end{cases}$$

Verify that these are *exact* solutions.

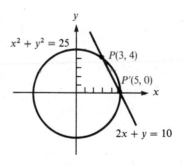

Figure 6-23

Graphical solutions of systems of equations usually give only approximate solutions. Algebraic methods can be used to find exact solutions.

Example 2   Solve the following system of equations algebraically.

$$\begin{cases} x^2 + y^2 = 25 \\ 2x + y = 10 \end{cases}$$

Solution This system contains one equation of second degree and another of first degree. Solve systems of this kind by the substitution method. Solving the linear equation for y gives

$$2x + y = 10$$
$$y = -2x + 10$$

Substitute the expression $-2x + 10$ for y in the second-degree equation, and solve the resulting quadratic equation for x:

$x^2 + y^2 = 25$	
$x^2 + (-2x + 10)^2 = 25$	
$x^2 + 4x^2 - 40x + 100 = 25$	Remove parentheses.
$5x^2 - 40x + 75 = 0$	Combine terms.
$x^2 - 8x + 15 = 0$	Divide both sides by 5.
$(x - 5)(x - 3) = 0$	Factor $x^2 - 8x + 15 = 0$.
$x - 5 = 0 \quad$ or $\quad x - 3 = 0$	
$x = 5 \quad | \quad x = 3$	

Because $y = -2x - 10$, if $x = 5$ then $y = 0$, and if $x = 3$ then $y = 4$. The two solutions are

$$\begin{cases} x = 5 \\ y = 0 \end{cases} \quad \text{or} \quad \begin{cases} x = 3 \\ y = 4 \end{cases}$$

■

Example 3 Solve the following system of equations algebraically.

$$\begin{cases} 4x^2 + 9y^2 = 5 \\ \qquad y = x^2 \end{cases}$$

Solution Solve this system by substitution.

$4x^2 + 9y^2 = 5$	
$4y + 9y^2 = 5$	Substitute y for x^2.
$9y^2 + 4y - 5 = 0$	Add -5 to both sides.
$(9y - 5)(y + 1) = 0$	Factor $9y^2 + 4y - 5$.
$9y - 5 = 0 \quad$ or $\quad y + 1 = 0$	
$y = \dfrac{5}{9} \quad \Big| \quad y = -1$	

Because $y = x^2$, the values of x are found by solving the equations

$$x^2 = \frac{5}{9} \quad \text{and} \quad x^2 = -1$$

Because the equation $x^2 = -1$ has no real solutions, this possibility is discarded. The solutions of the equation $x^2 = \frac{5}{9}$ are

$$x = \frac{\sqrt{5}}{3} \quad \text{or} \quad x = \frac{-\sqrt{5}}{3}$$

Thus, the solutions of the system are

$$\left(\frac{\sqrt{5}}{3}, \frac{5}{9}\right) \quad \text{and} \quad \left(\frac{-\sqrt{5}}{3}, \frac{5}{9}\right)$$ ∎

Example 4   Solve the following system of equations algebraically.

$$\begin{cases} 3x^2 + 2y^2 = 36 \\ 4x^2 - y^2 = 4 \end{cases}$$

Solution   In this system, both equations are of second degree and in the form $ax^2 + by^2 = c$. Solve systems like this by eliminating one of the variables by addition. Copy the first equation and multiply the second equation by 2 to obtain the equivalent system of equations

$$\begin{cases} 3x^2 + 2y^2 = 36 \\ 8x^2 - 2y^2 = 8 \end{cases}$$

Add the equations to eliminate the variable y, and solve the resulting equation for x:

$$11x^2 = 44$$
$$x^2 = 4$$
$$x = 2 \quad \text{or} \quad x = -2$$

To find y, substitute 2 for x and then -2 for x in the first equation and proceed as follows:

For $x = 2$	For $x = -2$
$3x^2 + 2y^2 = 36$	$3x^2 + 2y^2 = 36$
$3(2)^2 + 2y^2 = 36$	$3(-2)^2 + 2y^2 = 36$
$12 + 2y^2 = 36$	$12 + 2y^2 = 36$
$2y^2 = 24$	$2y^2 = 24$
$y^2 = 12$	$y^2 = 12$
$y = +\sqrt{12} \quad \text{or} \quad y = -\sqrt{12}$	$y = +\sqrt{12} \quad \text{or} \quad y = -\sqrt{12}$
$y = 2\sqrt{3} \quad \mid \quad y = -2\sqrt{3}$	$y = 2\sqrt{3} \quad \mid \quad y = -2\sqrt{3}$

The four solutions of this system are

$$(2, 2\sqrt{3}), \quad (2, -2\sqrt{3}), \quad (-2, 2\sqrt{3}), \quad \text{and} \quad (-2, -2\sqrt{3})$$ ∎

EXERCISE 6.5

In Exercises 1–10, solve each system of equations by graphing.

1. $\begin{cases} 8x^2 + 32y^2 = 256 \\ x = 2y \end{cases}$

2. $\begin{cases} x^2 + y^2 = 2 \\ x + y = 2 \end{cases}$

3. $\begin{cases} x^2 + y^2 = 90 \\ y = x^2 \end{cases}$

4. $\begin{cases} x^2 + y^2 = 5 \\ x + y = 3 \end{cases}$

5. $\begin{cases} x^2 + y^2 = 25 \\ 12x^2 + 64y^2 = 768 \end{cases}$

6. $\begin{cases} x^2 + y^2 = 13 \\ y = x^2 - 1 \end{cases}$

7. $\begin{cases} x^2 - 13 = -y^2 \\ y = 2x - 4 \end{cases}$

8. $\begin{cases} x^2 + y^2 = 20 \\ y = x^2 \end{cases}$

9. $\begin{cases} x^2 - 6x - y = -5 \\ x^2 - 6x + y = -5 \end{cases}$

10. $\begin{cases} x^2 - y^2 = -5 \\ 3x^2 + 2y^2 = 30 \end{cases}$

In Exercises 11–36, solve each system of equations algebraically for real values of x and y.

11. $\begin{cases} 25x^2 + 9y^2 = 225 \\ 5x + 3y = 15 \end{cases}$

12. $\begin{cases} x^2 + y^2 = 20 \\ y = x^2 \end{cases}$

13. $\begin{cases} x^2 + y^2 = 2 \\ x + y = 2 \end{cases}$

14. $\begin{cases} x^2 + y^2 = 36 \\ 49x^2 + 36y^2 = 1764 \end{cases}$

15. $\begin{cases} x^2 + y^2 = 5 \\ x + y = 3 \end{cases}$

16. $\begin{cases} x^2 - x - y = 2 \\ 4x - 3y = 0 \end{cases}$

17. $\begin{cases} x^2 + y^2 = 13 \\ y = x^2 - 1 \end{cases}$

18. $\begin{cases} x^2 + y^2 = 25 \\ 2x^2 - 3y^2 = 5 \end{cases}$

19. $\begin{cases} x^2 + y^2 = 30 \\ y = x^2 \end{cases}$

20. $\begin{cases} 9x^2 - 7y^2 = 81 \\ x^2 + y^2 = 9 \end{cases}$

21. $\begin{cases} x^2 + y^2 = 13 \\ x^2 - y^2 = 5 \end{cases}$

22. $\begin{cases} 2x^2 + y^2 = 6 \\ x^2 - y^2 = 3 \end{cases}$

23. $\begin{cases} x^2 + y^2 = 20 \\ x^2 - y^2 = -12 \end{cases}$

24. $\begin{cases} xy = -\dfrac{9}{2} \\ 3x + 2y = 6 \end{cases}$

25. $\begin{cases} y^2 = 40 - x^2 \\ y = x^2 - 10 \end{cases}$

26. $\begin{cases} x^2 - 6x - y = -5 \\ x^2 - 6x + y = -5 \end{cases}$

27. $\begin{cases} y = x^2 - 4 \\ x^2 - y^2 = -16 \end{cases}$

28. $\begin{cases} 6x^2 + 8y^2 = 182 \\ 8x^2 - 3y^2 = 24 \end{cases}$

29. $\begin{cases} x^2 - y^2 = -5 \\ 3x^2 + 2y^2 = 30 \end{cases}$

30. $\begin{cases} \dfrac{1}{x} + \dfrac{1}{y} = 5 \\ \dfrac{1}{x} - \dfrac{1}{y} = -3 \end{cases}$

31. $\begin{cases} \dfrac{1}{x} + \dfrac{2}{y} = 1 \\ \dfrac{2}{x} - \dfrac{1}{y} = \dfrac{1}{3} \end{cases}$

32. $\begin{cases} \dfrac{1}{x} + \dfrac{3}{y} = 4 \\ \dfrac{2}{x} - \dfrac{1}{y} = 7 \end{cases}$

33. $\begin{cases} 3y^2 = xy \\ 2x^2 + xy - 84 = 0 \end{cases}$

34. $\begin{cases} x^2 + y^2 = 10 \\ 2x^2 - 3y^2 = 5 \end{cases}$

35. $\begin{cases} xy = \dfrac{1}{6} \\ y + x = 5xy \end{cases}$

36. $\begin{cases} xy = \dfrac{1}{12} \\ y + x = 7xy \end{cases}$

37. The area of a rectangle is 63 square centimeters, and its perimeter is 32 centimeters. Find the dimensions of the rectangle.

38. The product of two integers is 32 and their sum is 12. Find the integers.

39. The sum of the squares of two numbers is 221, and the sum of the numbers is 212 less. Find the numbers.

40. Grant receives $225 annual income from one investment. Jeff invested $500 more than Grant, but at an annual rate of 1% less. Jeff's annual income is $240. What is the amount and rate of Grant's investment?

41. Carol receives $67.50 annual income from one investment. John invested $150 more than Carol at an annual rate of $1\frac{1}{2}\%$ more. John's annual income is $94.50. What is the amount and rate of Carol's investment? (*Hint:* There are two answers.)

42. Jim drove 306 miles. Jim's brother made the same trip at a speed 17 miles per hour slower than Jim did and required an extra $1\frac{1}{2}$ hours. What was Jim's rate and time?

6.6 TRANSLATION OF COORDINATE AXES

The graph of the equation

$$(x - 3)^2 + (y - 1)^2 = 4$$

is a circle with radius 2 and with center at the point (3, 1). See Figure 6-24. If we were to shift the black xy-coordinate system 3 units to the right and 1 unit up, we would establish the colored $x'y'$-coordinate system. With respect to this new $x'y'$-system, the center of the circle is the origin and its equation is

$$x'^2 + y'^2 = 4$$

In this section, we will discuss how to change the equation of a graph by shifting the position of the x- and y-axes. A shift to the left, right, up, or down is called a **translation of the coordinate axes**.

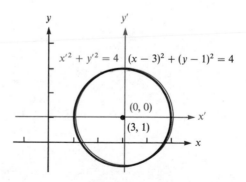

Figure 6-24

Figure 6-25 shows both an xy- and an $x'y'$-coordinate system. The colored x'- and y'-axes are parallel to the black x- and y-axes, respectively, and the unit distance on each is the same. The origin of the $x'y'$-system is the point O' with $x'y'$-coordinates of (0, 0) and with xy-coordinates of (h, k). The $x'y'$-system is called a **translated coordinate system**.

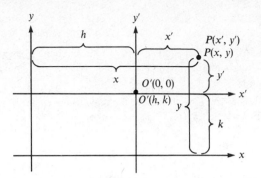

Figure 6-25

Relative to the xy-system in Figure 6-25, the coordinates of point P are (x, y). Relative to the $x'y'$-system, the coordinates of point P are (x', y'). By the geometry of the figure,

1. $\begin{cases} x = x' + h \\ y = y' + k \end{cases}$ or   **2.** $\begin{cases} x' = x - h \\ y' = y - k \end{cases}$

Equations 1 and Equations 2 are called the **translation-of-axes formulas**. They enable us to determine the coordinates of any point with respect to any translated coordinate system.

Example 1   The $x'y'$-coordinates of point P in Figure 6-26 are $(-3, -2)$, and the xy-coordinates of point O' are $(2, 1)$. Find the xy-coordinates of point P.

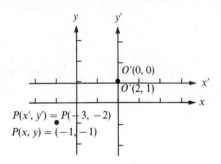

Figure 6-26

Solution   Because you want to find xy-coordinates, use Equations 1 with $(h, k) = (2, 1)$ and $(x', y') = (-3, -2)$:

$$x = x' + h \qquad y = y' + k$$
$$x = -3 + 2 \qquad y = -2 + 1$$
$$= -1 \qquad\qquad = -1$$

The xy-coordinates of point P are $(-1, -1)$.   ∎

Example 2   The xy-coordinates of point O' in Figure 6-27 are $(-3, -5)$, and the xy-coordinates of point Q are $(1, -2)$. Find the $x'y'$-coordinates of point Q.

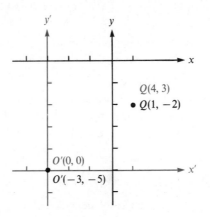

Figure 6-27

Solution   Because you want to find $x'y'$-coordinates, use Equations 2 with $(h, k) = (-3, -5)$ and $(x, y) = (1, -2)$:

$$x' = x - h \qquad y' = y - k$$
$$x' = 1 - (-3) \qquad y' = -2 - (-5)$$
$$= 4 \qquad\qquad = 3$$

The $x'y'$-coordinates of point Q are $(4, 3)$.   ■

Example 3   The xy-coordinates of point O' in Figure 6-28 are $(1, 5)$. Find the equation of the line determined by $y = 2x + 3$ in the variables x' and y'.

Solution   Because the xy-coordinates of point O' are $(h, k) = (1, 5)$, substitute 1 for h and 5 for k in Equations 1 to obtain

$$x = x' + h \qquad y = y' + k$$
$$x = x' + 1 \qquad y = y' + 5$$

Then substitute $x' + 1$ for x and $y' + 5$ for y in the equation $y = 2x + 3$ and simplify:

$$y = 2x + 3$$
$$y' + 5 = 2(x' + 1) + 3$$
$$y' + 5 = 2x' + 2 + 3$$
$$y' + 5 = 2x' + 5$$
$$y' = 2x'$$

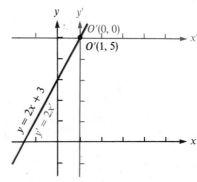

Figure 6-28

In the variables x' and y', the equation of the line is $y' = 2x'$. Note that graphing $y = 2x + 3$ with respect to the black xy-coordinate system gives the same line as graphing $y' = 2x'$ with respect to the colored $x'y'$-coordinate system. ■

Example 4   Find the equation of the parabola $y + 3 = (x - 2)^2$ with respect to a translated coordinate system with origin at $O'(2, -3)$. Graph the resulting equation with respect to the translated coordinate system.

Solution   Because the origin is translated to the point $(h, k) = (2, -3)$, substitute 2 for h and -3 for k in the translation-of-axes formulas:

$$x = x' + h \qquad y = y' + k$$
$$x = x' + 2 \qquad y = y' + (-3)$$
$$y = y' - 3$$

To obtain the equation of the same parabola with respect to the translated axes, substitute $x' + 2$ for x and $y' - 3$ for y in the given equation:

$$y + 3 = (x - 2)^2$$
$$(y' - 3) + 3 = [(x' + 2) - 2]^2$$
$$y' = x'^2 \qquad\qquad \text{Simplify.}$$

The graph of the equation $y' = x'^2$ is a parabola with vertex at the origin of the $x'y'$-system. See Figure 6-29.

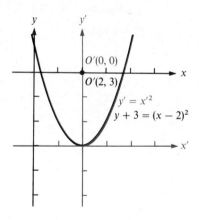

Figure 6-29   ■

Example 5   To what point should the origin of the xy-coordinate system be translated to remove the first-degree terms in the equation below?

$$4x^2 + y^2 + 8x - 6y + 9 = 0$$

Graph the resulting equation with respect to the translated coordinate system.

Solution   Begin by completing the square in both x and y and simplifying:

$$4x^2 + y^2 + 8x - 6y + 9 = 0$$

$$4x^2 + 8x + y^2 - 6y = -9 \qquad \text{Rearrange terms and add } -9 \text{ to both sides.}$$

$$4(x^2 + 2x) + y^2 - 6y = -9 \qquad \text{Factor 4 from } 4x^2 + 8x.$$

$$4(x^2 + 2x + 1) + y^2 - 6y + 9 = -9 + 4 + 9 \qquad \text{Add 4 and 9 to both sides.}$$

$$4(x + 1)^2 + (y - 3)^2 = 4 \qquad \text{Factor both trinomials.}$$

$$\frac{(x + 1)^2}{1} + \frac{(y - 3)^2}{4} = 1 \qquad \text{Divide both sides by 4.}$$

Let the origin of the $x'y'$-coordinate system be the point $O'(h, k) = O'(-1, 3)$. Then, by the translation-of-axes formulas, you have $x' = x + 1$ and $y' = y - 3$. Substitute x' for $x + 1$ and y' for $y - 3$ to obtain the equation

$$\frac{(x + 1)^2}{1} + \frac{(y - 3)^2}{4} = 1$$

$$\frac{x'^2}{1} + \frac{y'^2}{4} = 1$$

Note that this final equation contains no first-degree terms. Its graph is an ellipse centered at the origin of an $x'y'$-coordinate system whose origin has been translated to the point $(-1, 3)$ of the xy-system. See Figure 6-30.

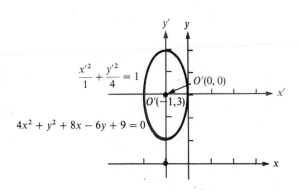

Figure 6-30

Example 6   Show that the slope of a nonvertical line is not affected by a translation of axes.

Solution   In the xy-system, the slope-intercept form of the equation of a nonvertical line with slope m is

$$y = mx + b$$

Use the translation-of-axes formulas to find the equation of the *same* line with respect to an $x'y'$-system whose origin has been translated to the point (h, k) of the xy-system. Substitute $x' + h$ for x and $y' + k$ for y and simplify. Proceed as follows:

$$y = mx + b$$
$$y' + k = m(x' + h) + b$$
$$y' + k = mx' + mh + b \qquad \text{Remove parentheses.}$$
$$y' = mx' + mh + b - k \qquad \text{Add } -k \text{ to both sides.}$$

This final equation is in slope-intercept form and represents a line with y'-intercept of $mh + b - k$ and a slope of m. Thus, the translation does not change the slope of the line. ∎

EXERCISE 6.6

In Exercises 1–4, the origin of the $x'y'$-system is at the point $(1, -3)$ of the xy-system. Find the xy-coordinates of the point whose $x'y'$-coordinates are given.

1. $P(2, 4)$ **2.** $Q(-1, 3)$ **3.** $R(0, 0)$ **4.** $S(1, -3)$

In Exercises 5–8, the origin of the $x'y'$-system is at the point $(-2, 4)$ of the xy-system. Find the $x'y'$-coordinates of the point whose xy-coordinates are given.

5. $P(2, -4)$ **6.** $Q(-2, 4)$ **7.** $R(0, 0)$ **8.** $S(4, 2)$

In Exercises 9–12, the origin of the $x'y'$-system is at the point $(0, -5)$ of the xy-system. Express each equation in terms of the variables x' and y'. Draw a sketch that shows the relation of the graphs of the equations to both coordinate systems.

9. $y = 3x - 2$ **10.** $y = -3x - 4$ **11.** $y = x^2 - 5$ **12.** $x = y^2 + 10y + 25$

In Exercises 13–16, the origin of the $x'y'$-system is at the point $(3, -2)$ of the xy-system. Express each equation in terms of the variables x' and y'. Draw a sketch that shows the relation of the graphs of the equations to both coordinate systems.

13. $x^2 + y^2 = 9$ **14.** $2x^2 + 3y^2 = 12$ **15.** $x^2 - 6x + 7 = y$ **16.** $y^2 + 4y + 7 = x$

In Exercises 17–22, determine the point to which the origin of the $x'y'$-system should be translated to eliminate the first-degree terms of the given equation. Draw a sketch that shows the relation of the graphs of the equations to both coordinate systems.

17. $x^2 + y^2 + 4x - 10y - 6 = 0$ **18.** $x^2 + y^2 - 8x - 2y + 1 = 0$

19. $2x^2 + y^2 + 4x + 2y - 1 = 0$ **20.** $4x^2 + 9y^2 + 24x - 18y + 9 = 0$

21. $x^2 - y^2 - 6x - 4y + 4 = 0$ **22.** $4x^2 - 9y^2 + 8x - 36y = 68$

23. Show that the radius of a circle is not affected by a translation of axes.

24. Show that no translation of axes will remove the xy-term of the equation $xy = 1$.

In Exercises 25–26, suppose that the equation $Ax^2 + Bxy + Cy^2 + Dx + Ey + F = 0$ is changed into $A'x'^2 + B'x'y' + C'y'^2 + D'x' + E'y' + F' = 0$ by substituting $x' + h$ for x and $y' + k$ for y and then simplifying.

25. Show that $A + C = A' + C'$.   **26.** Show that $B^2 - 4AC = B'^2 - 4A'C'$.

CHAPTER SUMMARY

Key Words

center of a circle (6.1)
center of an ellipse (6.3)
center of a hyperbola (6.4)
circle (6.1)
conjugate axis of a hyperbola (6.4)
directrix of a parabola (6.2)
ellipse (6.3)
foci of an ellipse (6.3)
foci of a hyperbola (6.4)
focus of a parabola (6.2)

fundamental rectangle (6.4)
hyperbola (6.4)
major axis of an ellipse (6.3)
minor axis of an ellipse (6.3)
parabola (6.2)
translation of axes (6.6)
transverse axis of a hyperbola (6.4)
vertices of an ellipse (6.3)
vertices of a hyperbola (6.4)
vertex of a parabola (6.2)

Key Ideas

(6.1) Any equation that can be written in the form

$$(x - h)^2 + (y - k)^2 = r^2$$

has a graph that is a circle with center at point $C(h, k)$ and with a radius of r.

Any equation that can be written in the form

$$x^2 + y^2 = r^2$$

has a graph that is a circle with center at the origin and with radius r.

(6.2) The standard forms of the equations of parabolas that open to the right, left, upward, and downward are as follows (consider $p > 0$):

Parabola opening	Vertex at origin	Vertex at $V(h, k)$
Right	$y^2 = 4px$	$(y - k)^2 = 4p(x - h)$
Left	$y^2 = -4px$	$(y - k)^2 = -4p(x - h)$
Upward	$x^2 = 4py$	$(x - h)^2 = 4p(y - k)$
Downward	$x^2 = -4py$	$(x - h)^2 = -4p(y - k)$

(6.3) The standard form of the equation of an ellipse with center at the origin and major axis on the x-axis is

$$\frac{x^2}{a^2} + \frac{y^2}{b^2} = 1 \quad \text{where } a > b > 0$$

If the major axis of the ellipse with center at the origin lies on the y-axis, the standard form of the equation of an ellipse is

$$\frac{y^2}{a^2} + \frac{x^2}{b^2} = 1 \quad \text{where } a > b > 0$$

In either case, the length of the major axis is $2a$, and the length of the minor axis is $2b$.

The standard form of the equation of an ellipse with center at (h, k) and major axis parallel to the x-axis is

$$\frac{(x - h)^2}{a^2} + \frac{(y - k)^2}{b^2} = 1 \qquad \text{where } a > b > 0$$

If the major axis of an ellipse with center (h, k) is parallel to the y-axis, the standard form of the equation of the ellipse is

$$\frac{(y - k)^2}{a^2} + \frac{(x - h)^2}{b^2} = 1 \qquad \text{where } a > b > 0$$

In either case, the length of the major axis is $2a$, and the length of the minor axis is $2b$.

(6.4)   The standard form of the equation of a hyperbola with center at the origin and foci on the x-axis is

$$\frac{x^2}{a^2} - \frac{y^2}{b^2} = 1$$

The standard form of the equation of a hyperbola with center at the origin and foci on the y-axis is

$$\frac{y^2}{a^2} - \frac{x^2}{b^2} = 1$$

The standard form of the equation of a hyperbola with center at (h, k) and foci on a line parallel to the x-axis is

$$\frac{(x - h)^2}{a^2} - \frac{(y - k)^2}{b^2} = 1$$

The standard form of the equation of a hyperbola with center at (h, k) and foci on a line parallel to the y-axis is

$$\frac{(y - k)^2}{a^2} - \frac{(x - h)^2}{b^2} = 1$$

The extended diagonals of the fundamental rectangle are asymptotes of the graph of a hyperbola.

(6.5)   Good estimates for solutions to systems of simultaneous second-degree equations can be found by graphing.

Exact solutions to systems of simultaneous second-degree equations can often be found by algebraic techniques.

(6.6)   Equations in two variables x and y can often be simplified by using the translation-of-axes formulas

$$\begin{cases} x = x' + h \\ y = y' + k \end{cases} \qquad \text{or} \qquad \begin{cases} x' = x - h \\ y' = y - k \end{cases}$$

REVIEW EXERCISES

1. Write the equation of the circle with center at the origin and passing through point $(5, 5)$.

2. Write the equation of the circle with center at the origin and passing through point $(6, 8)$.

$x_1 \; y_1 \qquad x_2 \; y_2$

3. Write the equation of a circle with endpoints of its diameter at $(-2, 4)$ and $(12, 16)$.

4. Write the equation of a circle with endpoints of its diameter at $(-3, -6)$ and $(7, 10)$.

5. Write in standard form the equation of the circle $x^2 + y^2 - 6x + 4y = 3$ and graph the circle.

6. Write in standard form the equation of the circle $x^2 + 4x + y^2 - 10y = -13$ and graph the circle.

7. Write the equation of the parabola with vertex at the origin and curve passing through $(-8, 4)$ and $(-8, -4)$.

8. Write the equation of the parabola with vertex at the origin and curve passing through $(-8, 4)$ and $(8, 4)$.

9. Write the equation of the parabola $y = ax^2 + bx + c$ in standard form to show that the x-coordinate of the vertex of the parabola is $-\frac{b}{2a}$.

10. Find the equation of the parabola with vertex at $(-2, 3)$, curve passing through point $(-4, -8)$, and opening downward.

11. Graph $x^2 - 4y - 2x + 9 = 0$.   **12.** Graph $y^2 - 6y = 4x - 13$.

13. Graph $y^2 - 4x - 2y + 13 = 0$.

14. Write the equation of the ellipse with center at the origin, major axis that is horizontal and 12 units long, and minor axis 8 units long.

15. Write the equation of the ellipse with center at point $(-2, 3)$ and curve passing through points $(-2, 0)$ and $(2, 3)$.

16. Graph $4x^2 + y^2 - 16x + 2y = -13$.

17. Graph $x^2 + 9y^2 - 6x - 18y + 9 = 0$.

18. Write the equation of the hyperbola with vertices at points $(-3, 3)$ and $(3, 3)$ and a focus at point $(5, 3)$.

19. Graph $9x^2 - 4y^2 - 16y - 18x = 43$.   **20.** Graph $-2xy = 9$.

21. Solve the following system of equations by graphing:
$$\begin{cases} 3x^2 + y^2 = 52 \\ x^2 - y^2 = 12 \end{cases}$$

22. Solve the system in Review Exercise 21 algebraically.

23. Solve the following system of equations by graphing:
$$\begin{cases} x^2 + y^2 = 16 \\ -\sqrt{3}y + 4\sqrt{3} = 3x \end{cases}$$

24. Solve the system in Review Exercise 23 algebraically.

25. Solve the following system of equations by graphing:
$$\begin{cases} \dfrac{x^2}{16} + \dfrac{y^2}{12} = 1 \\ x^2 - \dfrac{y^2}{3} = 1 \end{cases}$$

26. Solve the system in Review Exercise 25 algebraically.

In Review Exercises 27–30, the origin of the $x'y'$-system is at the point $(2, -3)$ of the xy-system. Express each equation in terms of the variables x' and y'. Draw a sketch that shows the relation of the graphs to both coordinate systems.

27. $x^2 + y^2 = 25$ **28.** $x^2 - y^2 = 4$

29. $4x^2 + 9y^2 = 36$ **30.** $x^2 - 4x - 3y = 5$

In Review Exercises 31–32, determine the point at which the origin of the $x'y'$-system should be located to eliminate the first-degree terms of the given equation. Draw a sketch that shows the relation of the graphs of the equations to both coordinate systems.

31. $x^2 + y^2 - 6x - 4y + 12 = 0$ **32.** $4x^2 - y^2 - 8x + 4y = 4$

7 Theory of Equations

Some equations of degree greater than 2 can be solved by factoring. For example, the equation

$$x^3 - 3x^2 + 2x = 0$$

can be solved as follows:

$x^3 - 3x^2 + 2x = 0$	
$x(x^2 - 3x + 2) = 0$	Factor out x.
$x(x - 1)(x - 2) = 0$	Factor $x^2 - 3x + 2$.
$x = 0$ or $x - 1 = 0$ or $x - 2 = 0$	Set each factor equal to 0.
$x = 1$ $x = 2$	

The solution set of the given equation is $\{0, 1, 2\}$.

In this chapter we will discuss how to solve more complicated polynomial equations.

7.1 THE FACTOR AND REMAINDER THEOREMS

A **polynomial equation** is an equation that can be written in the form $P(x) = 0$, where

$$P(x) = a_n x^n + a_{n-1} x^{n-1} + \cdots + a_1 x + a_0$$

is a polynomial of degree n (n is a natural number). A **zero of the polynomial** $P(x)$ is any number r for which $P(r) = 0$. It follows that a zero of the polynomial $P(x)$ is a solution or root of the equation $P(x) = 0$. For example, 2 is a zero of the polynomial

$$P(x) = x^2 - 3x + 2$$

because

$$P(2) = 2^2 - 3(2) + 2$$
$$= 4 - 6 + 2$$
$$= 0$$

Because 2 is a zero of $x^2 - 3x + 2$, it is also a root of the polynomial equation

$$x^2 - 3x + 2 = 0$$

Before attempting to find zeros of more complicated polynomials, we need to know whether a given polynomial even has a zero. This question was answered by Karl Friedrich Gauss (1777–1855) when he proved the **fundamental theorem of algebra**.

> **The Fundamental Theorem of Algebra.**   If $P(x)$ is a polynomial with positive degree, then $P(x)$ has at least one zero.

The fundamental theorem points out that polynomials such as

$$2x + 3 \qquad \text{and} \qquad 32.75x^{1984} + ix^3 - (2 + i)x - 5$$

all have zeros. It may be difficult to find the zeros, and we may have to settle for approximations of the zeros, but the zeros do exist.

There is a relationship between a zero r of a polynomial $P(x)$ and the results of a long division of $P(x)$ by the binomial $x - r$. This relationship can be illustrated by an example.

Example 1   Let $P(x) = 3x^3 - 5x^2 + 3x - 10$.

a. Find $P(1)$ and $P(-2)$.

b. Divide $P(x)$ by $x - 1$ and by $x + 2$.

Solution   **a.** $P(1) = 3(1)^3 - 5(1)^2 + 3(1) - 10$ $P(-2) = 3(-2)^3 - 5(-2)^2 + 3(-2) - 10$
$\qquad\qquad = 3 - 5 + 3 - 10$ $\qquad\qquad\qquad\qquad = 3(-8) - 5(4) + 3(-2) - 10$
$\qquad\qquad = -9$ $\qquad\qquad\qquad\qquad\qquad = -60$

b.

$$
\begin{array}{r}
3x^2 - 2x\ + 1 \\
x - 1{\overline{\smash{\big)}\,3x^3 - 5x^2 + 3x - 10}} \\
\underline{3x^3 - 3x^2} \\
-2x^2 + 3x \\
\underline{-2x^2 + 2x} \\
+\ x - 10 \\
\underline{x - 1} \\
-\ 9
\end{array}
$$

$$
\begin{array}{r}
3x^2 - 11x\ +\ 25 \\
x + 2{\overline{\smash{\big)}\,3x^3 -\ 5x^2 +\ 3x - 10}} \\
\underline{3x^3 +\ 6x^2} \\
-11x^2 +\ 3x \\
\underline{-11x^2 - 22x} \\
25x - 10 \\
\underline{25x + 50} \\
-\ 60
\end{array}
$$

Note the results of Example 1. When $P(x)$ was divided by $x - 1$, the remainder was $P(1)$, or -9. When $P(x)$ was divided by $x - (-2)$, or $x + 2$, the remainder was $P(-2)$, or -60. These results are not coincidental. The following theorem, called the **remainder theorem**, asserts that the division of any polynomial $P(x)$ by the binomial $x - r$ yields $P(r)$ as the remainder.

> **The Remainder Theorem.**   If $P(x)$ is a polynomial, r is a real or complex number, and $P(x)$ is divided by $x - r$, then the remainder is $P(r)$.

Proof   To divide $P(x)$ by $x - r$, we must find a quotient $Q(x)$ and a remainder $R(x)$ such that

$$\text{dividend} = \text{divisor} \cdot \text{quotient} + \text{remainder}$$
$$\downarrow \qquad\qquad \downarrow \qquad\qquad \downarrow \qquad\qquad\qquad \downarrow$$
$$P(x) \quad = (x - r) \cdot \quad Q(x) \quad + \quad R(x)$$

Furthermore, the degree of the remainder $R(x)$ must be less than the degree of the divisor $x - r$. Because the divisor is of degree 1, the remainder must be a constant R. The expression $P(x) = (x - r)Q(x) + R$ indicates that the polynomial on the left side of the equation is the same as the polynomial on the right. In particular, the values that these polynomials assume for any replacement of the variable x must be equal. Replacing x with the number r, we have $P(r) = (r - r)Q(r) + R$. Because $(r - r) = 0$, it follows that $P(r) = R$; that is, the value of the polynomial $P(x)$ attained at $x = r$ is the remainder produced by dividing $P(x)$ by $x - r$. The proof is complete.   □

The **factor theorem**, a corollary to the remainder theorem, applies when the remainder R is 0.

> **The Factor Theorem.**   Let $P(x)$ be any polynomial, and let r be a real or complex number. Then $P(r) = 0$ if and only if $x - r$ is a factor of $P(x)$.

Proof   First, assume that $P(r) = 0$ and prove that $x - r$ is a factor of $P(x)$. Divide $P(x)$ by $x - r$. The remainder theorem asserts that the remainder must be $P(r)$. But $P(r)$, by assumption, is 0. Hence, $x - r$ is a factor of $P(x)$.

Conversely, assume that $x - r$ is a factor of $P(x)$ and prove that $P(r) = 0$. Because $x - r$ is a factor of $P(x)$, dividing $P(x)$ by $x - r$ gives a remainder of 0. The remainder theorem asserts that this remainder is $P(r)$. Hence, $P(r) = 0$.   □

Example 2   Let $P(x) = 3x^3 - 5x^2 + 3x - 10$. Show that $P(2) = 0$, and use the factor theorem to factor $P(x)$.

Solution   Calculate $P(2)$:

$$P(x) = 3x^3 - 5x^2 + 3x - 10$$
$$P(2) = 3(2)^3 - 5(2)^2 + 3(2) - 10 \qquad \text{Substitute 2 for } x.$$
$$= 3 \cdot 8 - 5 \cdot 4 + 6 - 10$$
$$= 0$$

Because $P(2) = 0$, it follows (by the factor theorem) that $x - 2$ is a factor of $P(x)$. To determine the other factor of $P(x)$, divide $P(x)$ by $x - 2$:

$$
\begin{array}{r}
3x^2 + x + 5 \\
x - 2\overline{)3x^3 - 5x^2 + 3x - 10} \\
\underline{3x^3 - 6x^2} \\
x^2 + 3x \\
\underline{x^2 - 2x} \\
5x - 10 \\
\underline{5x - 10} \\
0
\end{array}
$$

Thus, $P(x)$ factors as

$$P(x) = (x - 2)(3x^2 + x + 5) \qquad \blacksquare$$

Example 3   Solve the equation $3x^3 - 5x^2 + 3x - 10 = 0$.

Solution   This equation is related to the polynomial of Example 2, so you can use the work already done there:

$$3x^3 - 5x^2 + 3x - 10 = 0$$
$$(x - 2)(3x^2 + x + 5) = 0$$

To solve for x, set each factor equal to 0 and apply the quadratic formula to the equation $3x^2 + x + 5 = 0$. The complete solution set is

$$\left\{ 2, \quad -\frac{1}{6} + \frac{\sqrt{59}}{6}i, \quad -\frac{1}{6} - \frac{\sqrt{59}}{6}i \right\} \qquad \blacksquare$$

Example 4   Find a polynomial $P(x)$ that has zeros of 2, 3, and -5.

Solution   By the factor theorem, if 2, 3, and -5 are zeros of $P(x)$, then $x - 2$, $x - 3$, and $x - (-5)$ are all factors of $P(x)$. Hence,

$$
\begin{aligned}
P(x) &= (x - 2)(x - 3)(x + 5) \\
&= (x^2 - 5x + 6)(x + 5) \\
&= x^3 - 19x + 30
\end{aligned}
$$

The polynomial $P(x) = x^3 - 19x + 30$ has zeros of 2, 3, and -5. $\qquad \blacksquare$

Example 5   Is $x + 2$ a factor of the polynomial $P(x) = x^4 - 7x^2 - 6x$?

Solution   By the factor theorem, $x + 2$ will be a factor of $P(x)$ if -2 is a zero of $P(x)$. So evaluate $P(-2)$ and see if it is a zero of $x^4 - 7x^2 - 6x$:

$$
\begin{aligned}
P(x) &= x^4 - 7x^2 - 6x \\
P(-2) &= (-2)^4 - 7(-2)^2 - 6(-2) \\
&= 16 - 28 + 12 \\
&= 0
\end{aligned}
$$

Because -2 is a zero of the polynomial $P(x)$, then $x - (-2)$, or $x + 2$, is a factor of $P(x) = x^4 - 7x^2 - 6x$. ∎

Example 6   Find the three cube roots of -1.

Solution   The three cube roots of -1 are solutions of the equation $x^3 = -1$, or $x^3 + 1 = 0$. One of the cube roots of -1 is the value -1. Because of the factor theorem, $x - (-1)$ must be a factor of $x^3 + 1$. Hence, $x + 1$ divides $x^3 + 1$. Obtain the other factor of $x^3 + 1$ by long division.

$$
\begin{array}{r}
x^2 - x + 1 \\
x + 1{\overline{\smash{\big)}\,x^3 \qquad\quad + 1}} \\
\underline{x^3 + x^2} \\
-x^2 \\
\underline{-x^2 - x} \\
x + 1 \\
\underline{x + 1} \\
0
\end{array}
$$

Thus, the equation $x^3 + 1 = 0$ factors as

$$(x + 1)(x^2 - x + 1) = 0$$

Setting each factor equal to 0 and solving for x gives the three cube roots of -1:

$$x = -1, \qquad x = \frac{1}{2} + \frac{\sqrt{3}}{2}i, \qquad x = \frac{1}{2} - \frac{\sqrt{3}}{2}i$$

∎

EXERCISE 7.1

In Exercises 1–6, let $P(x) = 2x^4 - 2x^3 + 5x^2 - 1$. Evaluate the polynomial by substituting the given value of x into the polynomial and simplifying. Then use the remainder theorem to evaluate the polynomial.

1. $P(2)$ **2.** $P(-1)$ **3.** $P(0)$ **4.** $P(1)$ **5.** $P(-4)$ **6.** $P(4)$

7. Let $P(x) = x^5 - 1$. Use the remainder theorem to find $P(2)$.

8. Let $P(x) = x^5 + 4x^2 - 1$. Use the remainder theorem to find $P(-3)$.

9. Let $P(x) = -x^5 + 6x^3 - 20x + 1$. Use the remainder theorem to find $P(-3)$.

10. Let $P(x) = 3x^4 - 2x^3 + 7x^2 - 5x + 3$. Use the remainder theorem to find $P(2)$.

In Exercises 11–18, use the factor theorem to decide whether each statement is true. If not, so indicate.

11. $x - 1$ is a factor of $x^7 - 1$.

12. $x - 2$ is a factor of $x^3 - x^2 + 2x - 8$.

13. $x - 1$ is a factor of $3x^5 + 4x^2 - 7$.

14. $x + 1$ is a factor of $3x^5 + 4x^2 - 7$.

15. $x + 3$ is a factor of $2x^3 - 2x^2 + 1$.

16. $x - 3$ is a factor of $3x^5 - 3x^4 + 5x^2 - 13x - 6$.

17. $x - 1$ is a factor of $x^{1984} - x^{1776} + x^{1492} - x^{1066}$.

18. $x + 1$ is a factor of $x^{1984} + x^{1776} - x^{1492} - x^{1066}$.

19. Completely solve $x^3 + 3x^2 - 13x - 15 = 0$, given that $x = -1$ is a root.

20. Completely solve $x^4 + 4x^3 - 10x^2 - 28x - 15 = 0$, given that $x = -1$ is a double root.

21. Completely solve $x^4 - 2x^3 - 2x^2 + 6x - 3 = 0$, given that $x = 1$ is a double root.

22. Completely solve $x^5 + 4x^4 + 4x^3 - x^2 - 4x - 4 = 0$, given that $x = -2$ is a double root.

In Exercises 23–32, find a polynomial of lowest degree that has the indicated zeros.

23. $1, 1, 1$ 24. $1, 0, -1$ 25. $2, 4, 5$ 26. $7, 6, 3$

27. $-1, 1, -\sqrt{2}, \sqrt{2}$ 28. $0, 0, 0, \sqrt{3}, -\sqrt{3}$ 29. $\sqrt{2}, i, -i$ 30. $i, i, 2$

31. $1 + i, 1 - i, 0$ 32. $2 + i, 2 - i, i$

In Exercises 33–36, find the three cube roots of each number.

33. 1 34. 64 35. -125 36. -216

37. Completely solve $x^4 - 5x^3 + 7x^2 - 5x + 6 = 0$, given that $x = 3$ and $x = 2$ are roots.

38. Completely solve $x^4 + 2x^3 - 3x^2 - 4x + 4 = 0$, given that $x = 1$ and $x = -2$ are roots.

39. Completely solve $x^4 - 2x^3 - 9x^2 + 2x + 8 = 0$, given that $x = 4$ and $x = -1$ are roots.

40. If 0 is a zero of $P(x) = a_n x^n + a_{n-1} x^{n-1} + \cdots + a_1 x + a_0$, what is a_0?

41. If 0 occurs as a zero twice in the polynomial $P(x) = a_n x^n + a_{n-1} x^{n-1} + \cdots + a_1 x + a_0$, what is a_1?

42. Explain why the fundamental theorem of algebra guarantees that every polynomial equation of positive degree has at least one root.

43. Explain why the fundamental theorem of algebra and the factor theorem guarantee that an nth-degree polynomial equation has n roots.

44. The fundamental theorem of algebra demands that a polynomial be of positive degree. Would the theorem still be true if the polynomial were of degree 0? Explain.

7.2 SYNTHETIC DIVISION

A shortcut, called **synthetic division**, can be used to divide a higher-degree polynomial written in descending powers of x by a binomial of the form $x - r$. To see how this method works, we consider the long division of $2x^3 + 4x^2 - 3x + 10$ by $x - 3$:

$$
\begin{array}{r}
2x^2 + 10x + 27 \\
x - 3 \overline{)\, 2x^3 + 4x^2 - 3x + 10} \\
\underline{2x^3 - 6x^2} \\
10x^2 - 3x \\
\underline{10x^2 - 30x} \\
27x + 10 \\
\underline{27x - 81} \\
\text{(remainder)} \ 91
\end{array}
$$

$$
\begin{array}{r}
2 + 10 + 27 \\
1 - 3 \overline{)\, 2 + 4 - 3 + 10} \\
\underline{2 - 6} \\
10 - 3 \\
\underline{10 - 30} \\
27 + 10 \\
\underline{27 - 81} \\
\text{(remainder)} \ 91
\end{array}
$$

On the left is the complete long division. On the right is a modified version of the long division in which the variables have been removed. We can shorten the work on the right even further by omitting the numbers printed in color:

$$
\begin{array}{r}
2 + 10 + 27 \\
\hline
-3\overline{)2 + \ \ 4 - \ \ 3 + 10} \\
- \ \ 6 \\
\hline
10 \\
- \ 30 \\
\hline
27 \\
- \ 81 \\
\hline
\end{array}
$$

(remainder) 91

We can then compress the work vertically to obtain

$$
\begin{array}{r}
2 + 10 + 27 \\
\hline
-3\overline{)2 + \ \ 4 - \ \ 3 + 10} \\
- \ 6 - 30 - 81 \\
\hline
10 \quad 27 \quad 91
\end{array}
$$

There is no reason why the quotient, represented by the numbers 2, 10, and 27, must appear above the long division symbol. If we write the 2 on the bottom line, the bottom line gives both the coefficients of the quotient and the remainder. The top line can be eliminated, and the division now appears as

$$
\begin{array}{r|rrrr}
-3 & 2 & +4 & -3 & +10 \\
   &   & -6 & -30 & -81 \\
\hline
   & 2 & 10 & 27 & 91
\end{array}
$$

The bottom line was obtained by subtracting the middle line from the top line. If we replace the -3 in the divisor with a 3, the signs of every entry in the middle line are reversed in the division process. Then, the bottom line can be obtained by addition. Thus, we have this final form of the synthetic division:

$$
\begin{array}{r|rrrr}
3 & 2 & 4 & -3 & 10 \\
  &   & 6 & 30 & 81 \\
\hline
  & 2 & 10 & 27 & 91
\end{array}
$$ The coefficients of the dividend.

The coefficients of the quotient and the remainder.

Thus,

$$
\frac{2x^3 + 4x^2 - 3x + 10}{x - 3} = 2x^2 + 10x + 27 + \frac{91}{x - 3}
$$

Example 1 Use synthetic division to divide $3x^4 - 8x^3 + 10x + 3$ by $x - 2$.

Solution Begin by writing the coefficients of the dividend and the 2 from the divisor in the following form:

$$\underline{2\rvert\ \ 3 \ \ -8 \ \ \mathbf{0} \ \ 10 \ \ 3}$$

Write 0 for the coefficient of the missing term of x^2.

Then follow these steps:

$$
\begin{array}{r|rrrrr}
2 & 3 & -8 & 0 & 10 & 3 \\
& \downarrow \\
\hline
& 3
\end{array}
$$

Begin by bringing down the 3.

$$
\begin{array}{r|rrrrr}
2 & 3 & -8 & 0 & 10 & 3 \\
& & 6 \\
\hline
& 3 & -2
\end{array}
$$

Multiply 2 and 3 and add the product to -8 to get -2.

$$
\begin{array}{r|rrrrr}
2 & 3 & -8 & 0 & 10 & 3 \\
& & 6 & -4 \\
\hline
& 3 & -2 & -4
\end{array}
$$

Multiply 2 and -2 and add the product to 0 to get -4.

$$
\begin{array}{r|rrrrr}
2 & 3 & -8 & 0 & 10 & 3 \\
& & 6 & -4 & -8 \\
\hline
& 3 & -2 & -4 & 2
\end{array}
$$

Multiply 2 and -4 and add the product to 10 to get 2.

$$
\begin{array}{r|rrrrr}
2 & 3 & -8 & 0 & 10 & 3 \\
& & 6 & -4 & -8 & 4 \\
\hline
& 3 & -2 & -4 & 2 & 7
\end{array}
$$

Multiply 2 and 2 and add the product to 3 to get 7.

Thus,

$$\frac{3x^4 - 8x^3 + 10x + 3}{x - 2} = 3x^3 - 2x^2 - 4x + 2 + \frac{7}{x - 2}$$ ∎

Example 2 Use synthetic division to find $P(-2)$ if $P(x) = 5x^3 + 3x^2 - 21x - 1$.

Solution Because of the remainder theorem, $P(-2)$ will be the remainder when $P(x)$ is divided by $x - (-2)$. Perform the division as follows:

$$
\begin{array}{r|rrrr}
-2 & 5 & 3 & -21 & -1 \\
& & -10 \\
\hline
& 5 & -7
\end{array}
\qquad
\begin{array}{r|rrrr}
-2 & 5 & 3 & -21 & -1 \\
& & -10 & 14 \\
\hline
& 5 & -7 & -7
\end{array}
\qquad
\begin{array}{r|rrrr}
-2 & 5 & 3 & -21 & -1 \\
& & -10 & 14 & 14 \\
\hline
& 5 & -7 & -7 & 13
\end{array}
$$

Because the remainder is 13, $P(-2) = 13$. ∎

Example 3   Find $P(i)$, where $i = \sqrt{-1}$ and $P(x) = x^3 - x^2 + x - 1$.

Solution   Use synthetic division.

$$
\begin{array}{r|rrrr}
i & 1 & -1 & +1 & -1 \\
  &   & i & -1-i & +1 \\
\hline
  & 1 & i-1 & -i & 0
\end{array}
$$

Because the remainder is 0, $P(i) = 0$, and i is a zero of $P(x)$.   ■

Example 4   Graph $y = x^3 + x^2 - 2x$.

Solution   Let $x = 0$ and find the y-intercept; if $x = 0$, then $y = 0$. Then use synthetic division to help find coordinates of other points on the graph. For example, if $x = 1$, then

$$
\begin{array}{r|rrrr}
1 & 1 & 1 & -2 & 0 \\
  &   & 1 & 2 & 0 \\
\hline
  & 1 & 2 & 0 & 0
\end{array}
$$

The point with coordinates $(1, 0)$ lies on the graph. As another example, if $x = -1$, then

$$
\begin{array}{r|rrrr}
-1 & 1 & 1 & -2 & 0 \\
   &   & -1 & 0 & 2 \\
\hline
   & 1 & 0 & -2 & 2
\end{array}
$$

The point with coordinates $(-1, 2)$ lies on the graph. These coordinates and others that satisfy the equation appear in the graph in Figure 7-1.

$y = x^3 + x^2 - 2x$

x	y
0	0
1	0
−1	2
2	8
−2	0
3	30
−3	−12

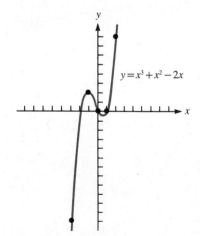

Figure 7-1   ■

Example 5   Graph $y = x^4 - 5x^2 + 4$.

Solution   Because x appears only with even powers, the graph is symmetric about the y-axis. Because $y = 4$ when $x = 0$, the graph passes through the point $(0, 4)$. Use synthetic division to help find coordinates of other points on the graph. For example, if $x = -3$, then

$$
\begin{array}{r|rrrrr}
-3 & 1 & 0 & -5 & 0 & 4 \\
   &   & -3 & 9 & -12 & 36 \\
\hline
   & 1 & -3 & 4 & -12 & 40
\end{array}
$$

The graph is shown in Figure 7-2.

$$y = x^4 - 5x^2 + 4$$

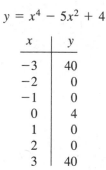

x	y
-3	40
-2	0
-1	0
0	4
1	0
2	0
3	40

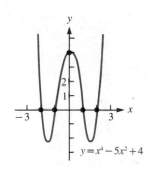

Figure 7-2

EXERCISE 7.2

In Exercises 1–8, assume that $P(x) = 5x^3 + 2x^2 - x + 1$. Use synthetic division to find each value of $P(x)$.

1. $P(2)$ **2.** $P(-2)$ **3.** $P(-5)$ **4.** $P(3)$

5. $P(0)$ **6.** $P(5)$ **7.** $P(-i)$ **8.** $P(i)$

In Exercises 9–16, assume that $P(x) = 2x^4 - x^2 + 2$. Use synthetic division to find each value of $P(x)$.

9. $P(1)$ **10.** $P(-1)$ **11.** $P(-2)$ **12.** $P(3)$

13. $P(\frac{1}{2})$ **14.** $P(\frac{1}{3})$ **15.** $P(i)$ **16.** $P(-i)$

In Exercises 17–24, assume that $P(x) = x^4 - 8x^3 + 8x + 14x^2 - 15$. Write the terms of $P(x)$ in descending powers of x and use synthetic division to find each value of $P(x)$.

17. $P(1)$ **18.** $P(0)$ **19.** $P(-3)$ **20.** $P(-1)$

21. $P(3)$ **22.** $P(5)$ **23.** $P(2)$ **24.** $P(-5)$

In Exercises 25–32, assume that $P(x) = 8 - 8x^2 + x^5 - x^3$. Write the terms of $P(x)$ in descending powers of x and use synthetic division to find each value of $P(x)$.

25. $P(0)$ **26.** $P(1)$ **27.** $P(2)$ **28.** $P(-2)$

29. $P(i)$ **30.** $P(-i)$ **31.** $P(-2i)$ **32.** $P(2i)$

In Exercises 33–38, use synthetic division to express $P(x) = 3x^3 - 2x^2 - 6x - 4$ in the form (divisor)(quotient) + remainder for each of the given divisors.

33. $x + 1$ **34.** $x - 1$ **35.** $x - 2$ **36.** $x + 2$

37. x **38.** $x - 7$

In Exercises 39–44, use synthetic division to perform each division.

39. $(7x^3 - 3x^2 - 5x + 1) \div (x + 1)$ **40.** $(2x^3 + 4x^2 - 3x + 8) \div (x - 3)$

41. $\dfrac{4x^4 - 3x^3 - x + 5}{x - 3}$ **42.** $\dfrac{x^4 + 5x^3 - 2x^2 + x - 1}{x + 1}$

43. $\dfrac{3x^5 - 768x}{x - 4}$ **44.** $\dfrac{x^5 - 4x^2 + 4x + 4}{x + 3}$

In Exercises 45–48, use synthetic division to find each power. (Hint: $y = 3^5$ is equivalent to $y = x^5$ when $x = 3$.)

45. 3^5 **46.** 4^5 **47.** 2^7 **48.** 3^6

In Exercises 49–56, use synthetic division to help graph each equation.

49. $y = x^3 - 4x$ **50.** $y = -x^3 + 4x$ **51.** $y = x^3 - x^2 - 2x$ **52.** $y = x^3 - x^2 - 6x$

53. $y = -x^4 + x^2$ **54.** $y = x^5 - 2x^3$

55. $y = x^5 - 3x^4 - 5x^3 + 15x^2 + 4x - 12$ **56.** $y = x^4 + 2x^3 - 5x^2 - 6x$

57. Let $P(x) = 5x^3 - 10x^2 + kx - 9$. For what value of k will the number 3 be a zero of $P(x)$?

58. Let $P(x) = 3x^3 + 8x^2 + kx - 6$. For what values of k will the number -2 be a zero of $P(x)$?

59. Use a calculator and synthetic division to find $P(1.3)$ where $P(x) = 2.5x^3 - 0.78x^2 - 2.7x + 4.3$.

60. Use a calculator and synthetic division to find $P(0.13)$ where $P(x) = 2.1x^3 - 1.2x^2 + 3.5x - 1.8$.

7.3 DESCARTES' RULE OF SIGNS AND BOUNDS ON ROOTS

The remainder theorem and synthetic division provide a way of verifying that a particular number is a root of a polynomial equation, but they do not provide the solutions. Selecting numbers at random, checking to see if they work, and hoping for the best is not an efficient technique! Some guidelines are needed to indicate how many solutions to expect, the kind of solutions to expect, and where they can be found. This section presents several theorems that provide such guidelines. The first theorem tells how many solutions to expect when solving a polynomial equation.

> **Theorem.**   If multiple roots are counted individually, the polynomial equation $P(x) = 0$ with complex coefficients (which include real coefficients) and degree $n > 0$ has exactly n roots among the complex numbers.

Proof   Let $P(x)$ be a polynomial of degree n. The fundamental theorem of algebra asserts that $P(x)$ has a zero, r_1. Therefore, the equation $P(x) = 0$ has r_1 as a root. The factor theorem guarantees that $x - r_1$ is a factor of $P(x)$. Thus,

$$P(x) = (x - r_1)Q_1(x)$$

If the lead coefficient of the nth-degree polynomial $P(x)$ is a_n, then $Q_1(x)$ is a polynomial of degree $n - 1$ whose lead coefficient is also a_n. The fundamental theorem of algebra asserts that $Q_1(x)$ also has a zero, which we call r_2. According to the factor theorem, $x - r_2$ is a factor of $Q_1(x)$, and

$$P(x) = (x - r_1)(x - r_2)Q_2(x)$$

where $Q_2(x)$ is a polynomial of degree $n - 2$ with lead coefficient a_n. This process can continue only to n factors of the form $x - r_i$; the final quotient $Q_n(x)$ is a polynomial of degree $n - n$, or degree 0. A polynomial of degree 0 with lead coefficient a_n is simply the constant a_n. The original polynomial $P(x)$ factors completely as

$$P(x) = a_n(x - r_1)(x - r_2)(x - r_3) \cdots (x - r_n)$$

Each of the n values r_i is a zero of $P(x)$ and a root of the equation $P(x) = 0$. There are no other roots, because no single factor in this product becomes 0 for any value of x not included in the list $r_1, r_2, r_3, \ldots, r_n$.

The theorem is proved. $\square$

The r_i in the above proof need not be different. Any root that occurs k times is called a **root of multiplicity k**.

The next theorem points out a pattern in the complex roots of real polynomial equations.

Theorem. If a polynomial equation $P(x) = 0$ with real coefficients has a complex root $a + bi$, with $b \neq 0$, then the conjugate $a - bi$ is a root also. (This theorem is often stated as "complex roots of real polynomial equations occur in conjugate pairs.")

The proof of this theorem is omitted.

Example 1   Form a fourth-degree equation with real coefficients and a double root of i.

Solution   Because i is a root of multiplicity two and a fourth-degree equation must have four roots, two more roots are needed. According to the previous theorem, the missing roots must be the conjugates of the given roots. Thus, the complete solution set is

$$\{i, i, -i, -i\}$$

The equation is

$$(x - i)(x + i)(x - i)(x + i) = 0$$
$$(x^2 + 1)(x^2 + 1) = 0$$
$$x^4 + 2x^2 + 1 = 0$$

∎

Example 2   Can a quadratic equation have a double root of i? If so, find such an equation.

Solution   Complex roots of a quadratic equation with *real* coefficients will be conjugates. A quadratic equation can have two nonreal, nonconjugate roots if the equation has coefficients that are not real. If i is a solution of multiplicity two, the equation is

$$(x - i)(x - i) = 0$$
$$x^2 - 2ix - 1 = 0$$ ∎

Descartes' Rule of Signs

René Descartes is credited with a theorem known as **Descartes' rule of signs**, which enables us to estimate the number of positive, negative, and nonreal roots of a polynomial equation.

If a polynomial is written in descending powers of x and we scan it from left to right, a "variation in sign" occurs whenever successive terms have opposite signs. For example,

$$P(x) = 3x^5 - 2x^4 - 5x^3 + x^2 - x - 9$$

has three variations in sign, and

$$P(-x) = 3(-x)^5 - 2(-x)^4 - 5(-x)^3 + (-x)^2 - (-x) - 9$$
$$= -3x^5 - 2x^4 + 5x^3 + x^2 + x - 9$$

has two variations in sign.

Descartes' Rule of Signs.   If $P(x)$ is a polynomial with real coefficients, the number of positive roots of the polynomial equation $P(x) = 0$ is either equal to the number of variations in sign of $P(x)$ or less than that by an even number.

The number of negative roots of $P(x) = 0$ is either equal to the number of variations in sign of $P(-x)$ or less than that by an even number.

The proof of this theorem is omitted.

As an example of Descartes' rule of signs, we consider the polynomial equation

$$P(x) = x^8 + x^6 + x^4 + x^2 + 1 = 0$$

Because $P(x)$ is of eighth degree, the equation has eight roots. Descartes' rule of signs can be used to show that none of these roots are real numbers. Because there are no variations in sign for $P(x)$, none of the roots can be positive. Because there are no variations in sign for $P(-x)$, none of the roots can be negative. Because $P(0) = 1$, zero is not a root. Thus, all eight roots are complex numbers with nonzero imaginary parts, and they occur in four conjugate pairs.

Example 3   Discuss the possibilities for the roots of $3x^3 - 2x^2 + x - 5 = 0$.

Solution   Let $P(x) = 3x^3 - 2x^2 + x - 5$. There are three variations in sign of $P(x)$, so there could be three positive solutions or only one (because 1 is less than 3 by an even number). Because $P(-x) = -3x^3 - 2x^2 - x - 5$ has no variations in sign, there are no negative roots. Furthermore, 0 is not a root.

If there are three positive roots, all the roots are accounted for; there would be no other possibilities. If there is only one positive root, the remaining two roots must be nonreal. The following list indicates these two possibilities.

Number of positive roots	Number of negative roots	Number of nonreal roots
3	0	0
1	0	2

The number of nonreal roots is the number needed to bring the total number of roots up to three.   ■

Example 4   Discuss the possibilities for the roots of $5x^5 - 3x^3 - 2x^2 + x - 1 = 0$.

Solution $P(x)$ has three variations in sign; there are either three positive solutions or only one. Because $P(-x) = -5x^5 + 3x^3 - 2x^2 - x - 1$ has two variations in sign, there are either two negative roots or none. Each line of the following list indicates a possible combination of positive, negative, and nonreal roots.

Number of positive roots	Number of negative roots	Number of nonreal roots
1	0	4
3	0	2
1	2	2
3	2	0

Note that in each case the number of nonreal roots is even. This is expected, because this polynomial has real coefficients and its complex roots must occur in conjugate pairs.   ■

Bounds for Roots

A final theorem provides a way of finding **bounds** for the roots, enabling us to concentrate our efforts on those regions where roots can be found. This theorem is also presented without proof.

Theorem.   Let the lead coefficient of the polynomial $P(x)$ with real coefficients be positive, and do a synthetic division of the coefficients of $P(x)$ by the positive number c. If each term in the last row is nonnegative, then no number greater than c can be a root of $P(x) = 0$. (c is an *upper bound* of the real roots.)

If $P(x)$ is divided synthetically by a negative number d, and the signs in the last row alternate,* then no value less than d can be a root of $P(x) = 0$. (d is a *lower bound* of the real roots.)

Example 5   Establish best integer bounds for the roots of $18x^3 - 3x^2 - 37x + 12 = 0$.

Solution   Try several synthetic divisions, looking for a nonnegative last row (if you synthetically divide by a positive number) or the alternating-sign last row (if you synthetically divide by a negative number). Trying 1 first gives

$$
\begin{array}{r|rrrr}
1 & 18 & -3 & -37 & 12 \\
  &   & 18 & 15 & -22 \\
\hline
  & +18 & +15 & -22 & -10
\end{array}
$$

Because some of the signs in the last row are negative, 1 is not an upper bound of the roots of the equation. Now try 2.

$$
\begin{array}{r|rrrr}
2 & 18 & -3 & -37 & 12 \\
  &   & 36 & 66 & 58 \\
\hline
  & +18 & +33 & +29 & +70
\end{array}
$$

Because the last row is entirely positive, no number greater than 2 can be a root. Thus, the smallest or best integer upper bound of the roots is 2.

Now try a negative divisor such as -2.

$$
\begin{array}{r|rrrr}
-2 & 18 & -3 & -37 & 12 \\
   &   & -36 & 78 & -82 \\
\hline
   & +18 & -39 & +41 & -70
\end{array}
$$

The alternating signs in the last row indicate that no number less than -2 can be a root. Try -1 next.

$$
\begin{array}{r|rrrr}
-1 & 18 & -3 & -37 & 12 \\
   &   & -18 & 21 & 16 \\
\hline
   & +18 & -21 & -16 & +28
\end{array}
$$

Because the signs in the last row do not alternate, -1 is not a lower bound. Thus, the largest or best integer lower bound is -2.

All of the real roots of this equation lie between -2 and 2.   ■

---

*If a 0 appears in the third row, that 0 can be assigned either a plus or a minus sign to help the signs alternate.

Example 6   Find bounds for the roots of $x^5 - x^4 - 3x^3 + 3x^2 - 10x + 10 = 0$.

Solution   Perform several divisions and watch for the desired nonnegative last row (if synthetically dividing by a positive number) or the alternating-sign last row (if synthetically dividing by a negative number):

$$
\begin{array}{r|rrrrrr}
-3 & 1 & -1 & -3 & 3 & -10 & 10 \\
   &   & -3 & 12 & -27 & 72 & -186 \\
\hline
   & +1 & -4 & +9 & -24 & +62 & -176 \\
\end{array}
$$

$$
\begin{array}{r|rrrrrr}
-2 & 1 & -1 & -3 & 3 & -10 & 10 \\
   &   & -2 & 6 & -6 & 6 & 8 \\
\hline
   & +1 & -3 & +3 & -3 & -4 & +18 \\
\end{array}
$$

Note the alternation of signs in the synthetic division by -3. You can conclude that -3 is a lower bound; that is, the equation has no roots less than -3. That claim cannot be made for -2, however, because the signs do not alternate.

Now look for an upper bound by dividing synthetically by various positive values:

$$
\begin{array}{r|rrrrrr}
3 & 1 & -1 & -3 & 3 & -10 & 10 \\
  &   & 3 & 6 & 9 & 36 & 78 \\
\hline
  & +1 & +2 & +3 & +12 & +26 & +88 \\
\end{array}
$$

$$
\begin{array}{r|rrrrrr}
2 & 1 & -1 & -3 & 3 & -10 & 10 \\
  &   & 2 & 2 & -2 & 2 & -16 \\
\hline
  & +1 & +1 & -1 & +1 & -8 & -6 \\
\end{array}
$$

When you divide synthetically by 3, you obtain a completely positive last row, which indicates that 3 is an upper bound. You cannot say the same thing for 2, because its last row contains negative numbers.

EXERCISE 7.3

1. How many roots does $x^{10} = 1$ have?
2. How many roots does $x^{40} = 1$ have?
3. One root of $x(3x^4 - 2) = 12x$ is 0. How many other roots are there?
4. One root of $3x^2(x^7 - 14x + 3) = 0$ is 0. How many other roots are there?

In Exercises 5–18, use Descartes' rule of signs to find the number of possible positive, negative, and nonreal roots of each equation. **Do not attempt to find the roots.**

5. $3x^3 + 5x^2 - 4x + 3 = 0$
6. $3x^3 - 5x^2 - 4x - 3 = 0$
7. $2x^3 + 7x^2 + 5x + 4 = 0$
8. $-2x^3 - 7x^2 - 5x - 4 = 0$
9. $8x^4 = -5$
10. $-3x^3 = -5$
11. $x^4 + 8x^2 - 5x = 10$
12. $5x^7 + 3x^6 - 2x^5 + 3x^4 + 9x^3 + x^2 + x + 1 = 0$
13. $-x^{10} - x^8 - x^6 - x^4 - x^2 - 1 = 0$
14. $x^{10} + x^8 + x^6 + x^4 + x^2 + 1 = 0$
15. $x^9 + x^7 + x^5 + x^3 + x = 0$ (Is 0 a root?)
16. $-x^9 - x^7 - x^5 - x^3 = 0$ (Is 0 a root?)
17. $-2x^4 - 3x^2 + 2x + 3 = 0$
18. $-7x^5 - 6x^4 + 3x^3 - 2x^2 + 7x - 4 = 0$

In Exercises 19–26, find the best integer bounds for the roots of each equation.

19. $x^2 - 5x - 6 = 0$ **20.** $6x^2 + x - 1 = 0$

21. $6x^2 - 13x - 110 = 0$ **22.** $3x^2 + 12x + 24 = 0$

23. $x^5 + x^4 - 8x^3 - 8x^2 + 15x + 15 = 0$ **24.** $12x^3 + 20x^2 - x - 6 = 0$

25. $2x^3 + 9x^2 - 5x = 41$ **26.** $x^4 - 34x^2 = -225$

27. Prove that any odd-degree polynomial equation with real coefficients must have at least one real root.

28. If a, b, c, and d are positive numbers, prove that $ax^4 + bx^2 + cx - d = 0$ has exactly two nonreal roots.

7.4 RATIONAL ROOTS OF POLYNOMIAL EQUATIONS

This section considers a method for actually finding the rational roots of polynomial equations with integral coefficients.

Theorem.   If the polynomial equation

$$P(x) = a_n x^n + a_{n-1} x^{n-1} + a_{n-2} x^{n-2} + \cdots + a_1 x + a_0 = 0$$

has integral coefficients and the rational number p/q (in lowest terms) is a root of the equation, then p is a factor of the constant term a_0, and q is a factor of the lead coefficient a_n.

Proof   Let p/q be a rational root in lowest terms of the equation $P(x) = 0$. The equation is satisfied by p/q:

$$a_n\left(\frac{p}{q}\right)^n + a_{n-1}\left(\frac{p}{q}\right)^{n-1} + a_{n-2}\left(\frac{p}{q}\right)^{n-2} + \cdots + a_1\left(\frac{p}{q}\right) + a_0 = 0$$

By multiplying both sides of the equation by the lowest common denominator q^n, we clear the equation of fractions. (Remember that p, q, and each of the a_i are integers.)

1. $a_n p^n + a_{n-1} p^{n-1} q + a_{n-2} p^{n-2} q^2 + \cdots + a_1 p q^{n-1} + a_0 q^n = 0$

Note that all the terms but the last share a common factor of p. We rewrite the equation in the form

$$p(a_n p^{n-1} + a_{n-1} p^{n-2} q + a_{n-2} p^{n-3} q^2 + \cdots + a_1 q^{n-1}) = -a_0 q^n$$

Because p is a factor of the left side, it must also be a factor of the right side. It cannot be a factor of q^n, because the fraction p/q is in lowest terms, and p and q share no common factor. Therefore, p and q^n share no common factors either. It follows that p must be a factor of a_0.

We return to Equation 1, note that all terms but the first share a common factor of q, and rewrite Equation 1 as

$$q(a_{n-1} p^{n-1} + a_{n-2} p^{n-2} q + a_{n-3} p^{n-3} q^2 + \cdots + a_0 q^{n-1}) = -a_n p^n$$

Now q is a factor of the left side and must therefore be a factor of the right side as well. Because q cannot be a factor of p^n, it must be a factor of a_n.

The theorem is proved. $\square$

Example 1 Find the only possible rational roots of the equation

$$\frac{1}{2}x^4 + \frac{2}{3}x^3 + 3x^2 - \frac{3}{2}x + 3 = 0$$

Solution The previous theorem applies to polynomial equations with *integral* coefficients. To clear this equation of its fractional coefficients, multiply both sides by 6 to get

$$3x^4 + 4x^3 + 18x^2 - 9x + 18 = 0$$

The only possible numerators available for a rational root are the factors of the constant term 18: ± 1, ± 2, ± 3, ± 6, ± 9, and ± 18. The only possible denominators are the factors of the lead coefficient 3: ± 1 and ± 3. You can form the list of possible rational solutions by listing all the combinations of values from these two sets.

$$\pm\frac{1}{1}, \ \pm\frac{2}{1}, \ \pm\frac{3}{1}, \ \pm\frac{6}{1}, \ \pm\frac{9}{1}, \ \pm\frac{18}{1}, \ \pm\frac{1}{3}, \ \pm\frac{2}{3}, \ \pm\frac{3}{3}, \ \pm\frac{6}{3}, \ \pm\frac{9}{3}, \ \pm\frac{18}{3}$$

Several of these are duplicates, so you can condense the list to obtain

Possible Rational Roots

$$\pm 1, \ \pm 2, \ \pm 3, \ \pm 6, \ \pm 9, \ \pm 18, \ \pm\frac{1}{3}, \ \pm\frac{2}{3}$$

$\blacksquare$

Example 2 Prove that $\sqrt{2}$ is irrational.

Solution $\sqrt{2}$ is a real root of the polynomial equation $x^2 - 2 = 0$. Any rational solution of the equation must have a numerator of either ± 1 or ± 2 and a denominator of ± 1. The only possible rational solutions, therefore, are ± 1 and ± 2, but none of these satisfies the equation. Therefore, the solution $\sqrt{2}$ must be irrational. $\blacksquare$

Example 3 Solve the polynomial equation $P(x) = 2x^3 + 3x^2 - 8x + 3 = 0$.

Solution Because the equation is of third degree, it must have three roots. According to Descartes' rule of signs, there are two possible combinations of positive, negative, and nonreal roots. They are summarized as follows:

Number of positive roots	Number of negative roots	Number of nonreal roots
2	1	0
0	1	2

The only possible rational roots are

$$\pm\frac{3}{1},\ \pm\frac{1}{1},\ \pm\frac{3}{2},\ \pm\frac{1}{2}$$

or

$$-3,\ -\frac{3}{2},\ -1,\ -\frac{1}{2},\ \frac{1}{2},\ 1,\ \frac{3}{2},\ 3$$

Check each one, crossing out those that do not satisfy the equation. Start, for example, with $x = \frac{3}{2}$.

$$\frac{3}{2} \begin{array}{|rrrr} 2 & 3 & -8 & 3 \\ & 3 & 9 & \frac{3}{2} \\ \hline 2 & 6 & 1 & \frac{9}{2} \end{array}$$

Because the remainder is not 0, the number $\frac{3}{2}$ is not a root and can be crossed off the list. Because the last row in the synthetic division is entirely positive, $\frac{3}{2}$ is an upper bound. Thus, 3 cannot be a root either, and you can cross it off the list as well:

$$-3,\ -\frac{3}{2},\ -1,\ -\frac{1}{2},\ \frac{1}{2},\ 1,\ \not{\frac{3}{2}},\ \not{3}$$

Now try $x = \frac{1}{2}$:

$$\frac{1}{2} \begin{array}{|rrrr} 2 & 3 & -8 & 3 \\ & 1 & 2 & -3 \\ \hline 2 & 4 & -6 & 0 \end{array}$$

Because the remainder is 0, the number $\frac{1}{2}$ is a root.

Thus, the binomial $x - \frac{1}{2}$ is a factor of $P(x)$, and any remaining roots must be supplied by the remaining factor, which is the quotient $2x^2 + 4x - 6$. The other roots can be found by solving the equation $2x^2 + 4x - 6 = 0$ or the equation

$$x^2 + 2x - 3 = 0 \qquad \text{Divide both sides by 2.}$$

This equation, called the **depressed equation**, is a quadratic equation that can be solved by factoring:

$$x^2 + 2x - 3 = 0$$
$$(x - 1)(x + 3) = 0$$
$$x - 1 = 0 \qquad \text{or} \qquad x + 3 = 0$$
$$x = 1 \qquad \qquad \qquad x = -3$$

The solution set is $\{\frac{1}{2},\ 1,\ -3\}$.   ■

Example 4   Solve the equation $x^7 - 2x^6 - 5x^5 + 6x^4 - x^3 + 2x^2 + 5x - 6 = 0$.

Solution Being of seventh degree, the equation must have seven roots. According to Descartes' rule of signs, there are six possible combinations of positive, negative, and nonreal roots for this equation.

Number of positive roots	Number of negative roots	Number of nonreal roots
5	2	0
3	2	2
1	2	4
5	0	2
3	0	4
1	0	6

The only possible rational roots are

$$-6, -3, -2, -1, 1, 2, 3, 6$$

Check each one, crossing off those that do not satisfy the equation. Begin with -3:

$$
\begin{array}{r|rrrrrrrr}
-3 & 1 & -2 & -5 & 6 & -1 & 2 & 5 & -6 \\
   &   & -3 & 15 & -30 & 72 & -213 & 633 & -1914 \\
\hline
   & 1 & -5 & 10 & -24 & 71 & -211 & 638 & -1920
\end{array}
$$

Because the last number in the synthetic division is not 0, -3 is not a root and can be crossed off the list. Because the last row is alternately positive and negative, -3 is a lower bound. Thus, you can cross off -6 as well:

$$-\cancel{6}, -\cancel{3}, -2, -1, 1, 2, 3, 6$$

Now try -2:

$$
\begin{array}{r|rrrrrrrr}
-2 & 1 & -2 & -5 & 6 & -1 & 2 & 5 & -6 \\
   &   & -2 & 8 & -6 & 0 & 2 & -8 & 6 \\
\hline
   & 1 & -4 & 3 & 0 & -1 & 4 & -3 & 0
\end{array}
$$

Because the remainder is 0, -2 is a root.

This root is negative, so you can revise the chart of positive/negative/nonreal possibilities. (Until now, 0 negative roots was a possibility.)

Number of positive roots	Number of negative roots	Number of nonreal roots
5	2	0
3	2	2
1	2	4

Because -2 is a root, the factor theorem asserts that $x - (-2)$, or $x + 2$, is a factor of $P(x)$. Any remaining roots can be found by solving the depressed equation

$$x^6 - 4x^5 + 3x^4 - x^2 + 4x - 3 = 0$$

Because the constant term of this equation is different from the constant term of the original equation, you can cross off some other possible rational roots. The numbers 2 and 6 must go because neither is a factor of the constant, 3. The number -2 cannot be a root a second time because -2 is not a factor of 3. The list of candidates is now

$$-6, \; -3, \; -2, \; -1, \; 1, \; 2, \; 3, \; 6$$

The only solution found thus far is -2. Try -1 next, since you know there must be one more negative root. The coefficients are those of the depressed equation. (Don't forget the missing x^3.)

$$
\begin{array}{r|rrrrrr}
-1 & 1 & -4 & 3 & 0 & -1 & 4 & -3 \\
   &   & -1 & 5 & -8 & 8 & -7 & 3 \\
\hline
   & 1 & -5 & 8 & -8 & 7 & -3 & \;0
\end{array}
$$

Because the remainder is 0, the number -1 is another root. The roots found so far are -1 and -2. The root -1 cannot appear again because there can be only two negative roots, and you have found them both. The current list of candidates is now

$$-6, \; -3, \; -2, \; -1, \; 1, \; 2, \; 3, \; 6$$

Other roots can be found by solving the depressed equation

$$x^5 - 5x^4 + 8x^3 - 8x^2 + 7x - 3 = 0$$

Try 1 next:

$$
\begin{array}{r|rrrrrr}
1 & 1 & -5 & 8 & -8 & 7 & -3 \\
  &   & 1 & -4 & 4 & -4 & 3 \\
\hline
  & 1 & -4 & 4 & -4 & 3 & \;0
\end{array}
$$

The solution 1 joins the growing list of solutions. To see if 1 is a multiple solution, try it again in the depressed equation.

$$
\begin{array}{r|rrrrr}
1 & 1 & -4 & 4 & -4 & 3 \\
  &   & 1 & -3 & 1 & -3 \\
\hline
  & 1 & -3 & 1 & -3 & \;0
\end{array}
$$

Again, 1 is a root. Will it work a third time?

$$
\begin{array}{r|rrrr}
1 & 1 & -3 & 1 & -3 \\
  &   & 1 & -2 & -1 \\
\hline
  & 1 & -2 & -1 & \;-4
\end{array}
$$

No, 1 is only a double root.

Solutions found thus far are -2, -1, 1, and 1. Now try 3:

$$
\begin{array}{r|rrrr}
3 & 1 & -3 & 1 & -3 \\
  &   & 3 & 0 & 3 \\
\hline
  & 1 & 0 & 1 & 0
\end{array}
$$

The solution 3 is added to the list of roots. So far the roots are -2, -1, 1, 1, and 3.

The depressed equation is now the quadratic equation $x^2 + 1 = 0$, which can be solved as follows:

$$x^2 + 1 = 0$$
$$x^2 = -1$$
$$x = i \quad \text{or} \quad x = -i$$

Because i and $-i$ are solutions, the complete solution set of the original seventh-degree equation is

$$\{-2, -1, 1, 1, 3, i, -i\}$$

Note that this list contains three positives, two negatives, and two conjugate complex numbers. This combination was one of the predicted possibilities. ■

EXERCISE 7.4

In Exercises 1–30, find all roots for each equation.

1. $x^3 - 5x^2 - x + 5 = 0$

2. $x^3 + 7x^2 - x - 7 = 0$

3. $x^3 - 2x^2 - 9x + 18 = 0$

4. $x^3 + 3x^2 - 4x - 12 = 0$

5. $x^3 - 2x^2 - x + 2 = 0$

6. $x^3 + 2x^2 - x - 2 = 0$

7. $x^4 - 10x^3 + 35x^2 - 50x + 24 = 0$

8. $x^4 + 4x^3 + 6x^2 + 4x + 1 = 0$

9. $x^4 + 3x^3 - 13x^2 - 9x + 30 = 0$

10. $x^4 - 8x^3 + 14x^2 + 8x - 15 = 0$

11. $x^5 + 3x^4 - 5x^3 - 15x^2 + 4x + 12 = 0$

12. $x^5 - 3x^4 - 16x + 48 = 0$

13. $x^7 - 12x^5 + 48x^3 - 64x = 0$

14. $x^7 + 7x^6 + 21x^5 + 35x^4 + 35x^3 + 21x^2 + 7x + 1 = 0$

15. $3x^3 - 2x^2 + 12x - 8 = 0$

16. $4x^4 - 8x^3 - x^2 + 8x - 3 = 0$

17. $3x^4 - 14x^3 + 11x^2 + 16x - 12 = 0$

18. $2x^4 - x^3 - 2x^2 - 4x - 40 = 0$

19. $12x^4 + 20x^3 - 41x^2 + 20x - 3 = 0$

20. $4x^5 - 12x^4 + 15x^3 - 45x^2 - 4x + 12 = 0$

21. $6x^5 - 7x^4 - 48x^3 + 81x^2 - 4x - 12 = 0$

22. $36x^4 - x^2 + 2x - 1 = 0$

23. $30x^3 - 47x^2 - 9x + 18 = 0$

24. $20x^3 - 53x^2 - 27x + 18 = 0$

25. $15x^3 - 61x^2 - 2x + 24 = 0$

26. $12x^4 + x^3 + 42x^2 + 4x - 24 = 0$

27. $x^3 - \dfrac{4}{3}x^2 - \dfrac{13}{3}x - 2 = 0$

28. $x^3 - \dfrac{19}{6}x^2 + \dfrac{1}{6}x + 1 = 0$

29. $x^{-5} - 8x^{-4} + 25x^{-3} - 38x^{-2} + 28x^{-1} - 8 = 0$

30. $1 - x^{-1} - x^{-2} - 2x^{-3} = 0$

31. If n is an even positive integer and c is a positive constant, prove that the equation $x^n + c = 0$ has no real roots.

32. If n is an even positive integer and c is a positive constant, prove that the equation $x^n - c = 0$ has exactly two real roots.

7.5   IRRATIONAL ROOTS OF POLYNOMIAL EQUATIONS

First-degree equations are easy to solve, and quadratic equations can be solved by the quadratic formula. There are also formulas for solving general third- and fourth-degree polynomial equations, although these formulas are complicated.

However, there are no explicit algebraic formulas for solving polynomial equations of degree 5 or greater. This fact was proven for fifth-degree equations by the Norwegian mathematician Niels Henrik Abel (1802–1829) and for equations of degree greater than 5 by the French mathematician Evariste Galois (1811–1832). To solve a polynomial equation of high degree with integer coefficients, we could use the methods of Section 7.4 to find the rational roots. Once they were found, however, the remaining depressed equation would have to be a first- or second-degree equation, or else we could not finish. The purpose of this section is to provide techniques for approximating the real roots of equations of high degree, even though the exact roots cannot be found. The following theorem provides one method for locating a root.

Theorem.   Let $P(x)$ be a polynomial with real coefficients. If $P(a)$ and $P(b)$ have opposite signs, there is at least one number r between a and b for which $P(r) = 0$.

Justification.   A proof of this theorem requires the use of calculus. The theorem becomes plausible if we consider the graph of the polynomial $y = P(x)$. Graphs of polynomials are *continuous* curves, a technical term that means, roughly, that they can be drawn without lifting the pencil from the paper. If $P(a)$ and $P(b)$ have opposite signs, the points $A(a, P(a))$ and $B(b, P(b))$ on the graph of $y = P(x)$ lie on opposite sides of the x-axis. Because the continuous curve joining A and B has no gaps, it must cross the x-axis at least once. The point of crossing, $x = r$, is a zero of $P(x)$, and a solution of the equation $P(x) = 0$.   □

The previous theorem provides a method for finding roots of $P(x) = 0$ to any degree of accuracy desired. Suppose we find, by trial and error, the numbers x_L and x_R (for left and right) that straddle a root; that is, $x_L < x_R$ and $P(x_L)$ and $P(x_R)$ have opposite signs. For purposes of discussion, $P(x_L)$ will be negative and $P(x_R)$ will be positive. We compute a number c that is halfway between x_L and x_R (c is the average of x_L and x_R), and then evaluate $P(c)$. If $P(c)$ is 0, we've found a root. More likely, however, $P(c)$ will not be 0.

If $P(c)$ is negative, the root, r, lies between c and x_R, as shown in Figure 7-3. In such a case, let c become the *new* x_L, and repeat the procedure.

If $P(c)$ is positive, however, the root lies between x_L and c, as shown in Figure 7-4. In this case, let c become the new x_R, and repeat the process.

At any stage in this procedure, the root is contained between the current numbers x_L and x_R. If the original bounds were, say, 1 unit apart, after 10 repetitions of this procedure, the root would be contained between bounds that were 2^{-10}, or about

0.001, units apart. After 20 repetitions, the bounds would be only 0.000001 units apart. The actual zero of $P(x)$ would be within 0.000001 of either x_L or x_R. This procedure is called **binary chopping**.

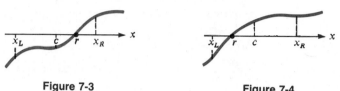

Figure 7-3   Figure 7-4

Example 1   Find $\sqrt{2}$ to two decimal places.

Solution $\sqrt{2}$ is a root of the polynomial equation $P(x) = x^2 - 2 = 0$. Note that the values of $P(1) = 1^2 - 2 = -1$ and $P(2) = 2^2 - 2 = 2$ have opposite signs. Set x_L equal to 1 and x_R equal to 2 and compute the midpoint, $c = \dfrac{1 + 2}{2} = 1.5$.

Because $P(c) = P(1.5) = 0.25$ is a positive number, c becomes the new x_R and we calculate a new c. Tabulating the information in a chart helps keep things straight.

Step	x_L	c	x_R	$P(x_L)$	$P(c)$	$P(x_R)$
0	1	1.5	2	negative	positive	positive
1	1	1.25	1.5	negative	negative	positive

At this point, $P(c)$ and $P(x_R)$ are of opposite signs, so c becomes the new x_L. The process continues, but without a hand calculator it can be difficult.

Step	x_L	c	x_R	$P(x_L)$	$P(c)$	$P(x_n)$
0	1	1.5	2	negative	positive	positive
1	1	1.25	1.5	negative	negative	positive
2	1.25	1.375	1.5	negative	negative	positive
3	1.375	1.4375	1.5	negative	positive	positive
4	1.375	1.40625	1.4375	negative	negative	positive
5	1.40625	1.421875	1.4375	negative	positive	positive
6	1.40625	1.4140625	1.421875	negative	negative	positive
7	1.4140625	1.41796875	1.421875	negative	positive	positive
8	1.4140625	1.416015625	1.41796875	negative	positive	positive

In Step 7, the bounds x_R and x_L agree to only one decimal place; our approximation of $\sqrt{2}$ is 1.4. In Step 8, x_R and x_L agree to two decimal places. Thus, $\sqrt{2} \approx 1.41$.   ∎

Gottfried Wilhelm von Leibnitz (1646–1716) Although principally known as one of the inventors of the calculus, Leibnitz also developed the binary numeration system, which is basic to modern computers.

Procedures that produce estimates that converge on a problem's solution are known as **iterative procedures**; they are ideal for solving equations by computer. Binary chopping is only one of many such iterative processes. Several others are based on **Newton's method**, a process developed by one of the inventors of the calculus, Sir Isaac Newton (1642–1727). They converge more quickly to the correct answer than does binary chopping.

EXERCISE 7.5

In Exercises 1–10, show that each equation has at least one real root between the numbers specified.

1. $2x^2 + x - 3 = 0$; $-2, -1$

2. $2x^3 + 17x^2 + 31x - 20 = 0$; $-1, 2$

3. $3x^3 - 11x^2 - 14x = 0$; $4, 5$

4. $2x^3 - 3x^2 + 2x - 3 = 0$; $1, 2$

5. $x^4 - 8x^2 + 15 = 0$; $1, 2$

6. $x^4 - 8x^2 + 15 = 0$; $2, 3$

7. $30x^3 + 10 = 61x^2 + 39x$; $2, 3$

8. $30x^3 + 10 = 61x^2 + 39x$; $-1, 0$

9. $30x^3 + 10 = 61x^2 + 39x$; $0, 1$

10. $5x^3 - 9x^2 - 4x + 9 = 0$; $-1, 2$

11. Use binary chopping to evaluate $\sqrt{3}$ to two decimal places.

12. Use binary chopping to evaluate $\sqrt[3]{53}$ to two decimal places.

13. Find a positive root of $x^4 - x^2 - 6 = 0$ to one decimal place.

14. Find a negative root of $2x^4 + x^2 - 6 = 0$ to one decimal place.

15. Find a negative root of $3x^4 + 3x^3 - x^2 - 4x - 4 = 0$ to one decimal place.

16. Find a positive root of $x^5 + x^4 - 4x^3 - 4x^2 - 5x - 5 = 0$ to one decimal place.

*In Exercises 17–22, use a computer and the binary chopping option of **Algebra Pak** to find all real solutions of each equation. Select nonautomatic scaling and begin with intervals that contain a zero of the polynomial.*

17. $x^2 - 3 = 0$ (Find $\sqrt{3}$)

18. $x^3 - 3 = 0$ (Find $\sqrt[3]{3}$)

19. $2x^3 - x^2 - 10x + 5 = 0$

20. $x^3 - 5x^2 + 6 = 0$

21. $x^6 - 6x^4 + 11x^2 - 6 = 0$

22. $4x^3 - 7x + 3 = 0$

CHAPTER SUMMARY

Key Words

binary chopping (7.5)

bounds on roots (7.3)

depressed equation (7.4)

Descartes' rule of signs (7.3)

factor theorem (7.1)

fundamental theorem of algebra (7.1)

remainder theorem (7.1)

roots of multiplicity k (7.3)

synthetic division (7.2)

zeros of a polynomial (7.1)

Key Ideas

(7.1)   Every polynomial with complex coefficients (which include real coefficients) and positive degree has a complex zero.

The Remainder Theorem. If $P(x)$ is any polynomial and r is any number, and $P(x)$ is divided by $x - r$, the remainder is $P(r)$.

The Factor Theorem. Let $P(x)$ be any polynomial and let r be any number. Then, $P(r) = 0$ if and only if $x - r$ is a factor of $P(x)$.

(7.2) Synthetic division can be used to find the quotient and remainder when a polynomial is to be divided by a binomial of the form $x - r$.

(7.3) If multiple roots are counted individually, a polynomial equation $P(x) = 0$ with complex coefficients (which includes real coefficients) and degree $n > 0$ has exactly n roots among the complex numbers.

Complex roots of polynomial equations with real coefficients occur in conjugate pairs.

Descartes' Rule of Signs. If $P(x)$ is a polynomial with real coefficients, the number of positive roots of the polynomial equation $P(x) = 0$ is either equal to the number of variations in sign of $P(x)$ or less than that by an even number.

The number of negative roots of $P(x) = 0$ is either equal to the number of variations in sign of $P(-x)$ or less than that by an even number.

Let the lead coefficient of the polynomial $P(x)$ with real coefficients be positive, and do a synthetic division of the coefficients of $P(x)$ by the positive number c. If none of the terms in the last row is negative, then c is an upper bound for the real roots of $P(x) = 0$.

If $P(x)$ is divided synthetically by a negative number d, and the signs in the last row alternate, then d is a lower bound for the real roots of $P(x) = 0$.

(7.4) If the polynomial equation $P(x) = a_n x^n + a_{n-1} x^{n-1} + a_{n-2} x^{n-2} + \cdots + a_1 x + a_0 = 0$ has integral coefficients, and the rational number p/q (reduced to lowest terms) is a root of the equation, then p is a factor of the constant term a_0, and q is a factor of the lead coefficient a_n.

(7.5) Let $P(x)$ be a polynomial with real coefficients. If $P(a)$ and $P(b)$ have opposite signs, then there is at least one number r between a and b for which $P(r) = 0$.

REVIEW EXERCISES

In Review Exercises 1–4, let $P(x) = 4x^4 + 2x^3 - 3x - 2$. Use synthetic division to evaluate the polynomial for the given value.

1. $P(0)$ **2.** $P(2)$ **3.** $P(-3)$ **4.** $P(\frac{1}{2})$

In Review Exercises 5–8, use the factor theorem to decide whether each statement is true. If not, so indicate.

5. $x - 2$ is a factor of $x^3 + 4x^2 - 2x + 4$.

6. $x + 3$ is a factor of $2x^4 + 10x^3 + 4x^2 + 7x + 21$.

7. $x - 5$ is a factor of $x^5 - 3125$.

8. $x - 6$ is a factor of $x^5 - 6x^4 - 4x + 24$.

9. Find the three cube roots of -64.

10. Find the three cube roots of 343.

11. Find the polynomial of lowest degree with zeros of -1, 2, and $\frac{3}{2}$.

12. Find the polynomial equation of lowest degree with roots of 1, -3, and $\frac{1}{2}$.

13. Use synthetic division to find the quotient when the polynomial $3x^4 + 2x^2 + 3x + 7$ is divided by $x - 3$.

14. Use synthetic division to find the quotient when the polynomial $5x^5 - 4x^4 + 3x^3 - 2x^2 + x - 1$ is divided by $x + 2$.

15. How many roots does the polynomial equation $3x^6 - 4x^5 + 3x + 2 = 0$ have?

16. How many roots does the equation $x^{1984} - 1 = 0$ have?

In Review Exercises 17–20, use Descartes' rule of signs to find the number of possible positive, negative, and nonreal roots.

17. $3x^4 + 2x^3 - 4x + 2 = 0$

18. $4x^5 + 3x^4 + 2x^3 + x^2 + x = 7$

19. $x^4 + x^2 + 24{,}567 = 0$

20. $-x^7 - 5 = 0$

21. Find all roots of the equation $2x^3 + 17x^2 + 41x + 30 = 0$.

22. Find all roots of the equation $3x^3 + 2x^2 + 2x = 1$.

23. Show that $5x^3 + 37x^2 + 59x + 18 = 0$ has a root between $x = 0$ and $x = -1$.

24. Show that $6x^3 - x^2 - 10x - 3 = 0$ has a root between $x = 1$ and $x = 2$.

25. Use binary chopping to find $\sqrt{7}$ to the nearest hundredth.

26. Use binary chopping to find an approximation of the root of the equation $0 = 3x - 1$. What is the exact root?

8 Exponential and Logarithmic Functions

In this chapter, we discuss two functions that are important in certain applications of mathematics. The *exponential function* can be used, for example, to compute compound interest or to provide a model for population growth and radioactivity. The *logarithmic function* can be used to simplify calculations, measure the acidity of a solution or the intensity of an earthquake, or determine safe noise levels for factory workers.

8.1 EXPONENTIAL FUNCTIONS

In the discussion of exponential functions, we will consider expressions such as 3^x where x is a *real* number. Because we have only defined 3^x where x is a *rational* number, we must now give meaning to 3^x where x is an *irrational* number.

To this end, we consider the expression $3^{\sqrt{2}}$, where $\sqrt{2}$ is the irrational number $1.414213562 \ldots$. Because $1 < \sqrt{2} < 2$, it can be shown that $3^1 < 3^{\sqrt{2}} < 3^2$. Similarly, $1.4 < \sqrt{2} < 1.5$, so $3^{1.4} < 3^{\sqrt{2}} < 3^{1.5}$.

The value of $3^{\sqrt{2}}$ is bounded by two numbers involving only rational powers of 3, as shown in the following list. As the list continues, $3^{\sqrt{2}}$ gets squeezed into a smaller and smaller interval:

$$3^1 = 3 \qquad < 3^{\sqrt{2}} < 9 \qquad = 3^2$$
$$3^{1.4} \approx 4.656 \qquad < 3^{\sqrt{2}} < 5.196 \qquad \approx 3^{1.5}$$
$$3^{1.41} \approx 4.7070 \qquad < 3^{\sqrt{2}} < 4.7590 \qquad \approx 3^{1.42}$$
$$3^{1.414} \approx 4.727695 < 3^{\sqrt{2}} < 4.732892 \approx 3^{1.415}$$

There is exactly one real number that is larger than any of the increasing numbers on the left of the previous list and less than all the decreasing numbers on the right. By definition, that number is $3^{\sqrt{2}}$.

To find an approximation for $3^{\sqrt{2}}$, we can press the following keys on a calculator:

$$3 \;\boxed{y^x}\; 2 \;\boxed{\sqrt{}}\; \boxed{=}$$

The display will show 4.7288044. Thus,

$$3^{\sqrt{2}} \approx 4.7288044$$

It can be shown that all of the familiar properties of exponents hold for irrational exponents.

In general, if b is a *positive* real number and x is *any* real number, then the exponential expression b^x represents a unique positive real number. If $b > 0$ and $b \neq 1$, the function defined by the equation $y = f(x) = b^x$ is called an **exponential function**.

Definition.   The **exponential function with base b** is defined by the equation

$$y = f(x) = b^x$$

where $b > 0$ and $b \neq 1$.

The domain of the exponential function with base b is the set of real numbers. Its range is the interval $(0, \infty)$ of positive real numbers.

Example 1   Graph the exponential functions $y = 2^x$ and $y = 7^x$.

Solution   Calculate several pairs (x, y) that satisfy each equation. Plot the points and join them with a smooth curve. The graph of $y = 2^x$ appears in Figure 8-1**a**, and the graph of $y = 7^x$ appears in Figure 8-1**b**.

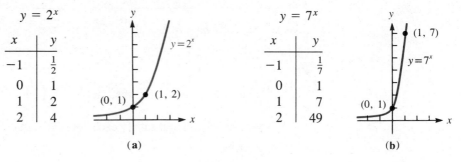

$y = 2^x$	
x	y
-1	$\frac{1}{2}$
0	1
1	2
2	4

$y = 7^x$	
x	y
-1	$\frac{1}{7}$
0	1
1	7
2	49

(a)   (b)

Figure 8-1

Note that $y = 2^x$ and $y = 7^x$ are *increasing functions*, that each graph passes through the point $(0, 1)$, and that the x-axis is an asymptote of each graph. Note also that the graph of $y = 2^x$ passes through the point $(1, 2)$, and that of $y = 7^x$ passes through the point $(1, 7)$.   ∎

Example 2   Graph the exponential functions $y = \left(\frac{1}{2}\right)^x$ and $y = \left(\frac{1}{7}\right)^x$.

Solution   Calculate and plot several pairs (x, y) that satisfy each equation. The graph of $y = \left(\frac{1}{2}\right)^x$ appears in Figure 8-2**a** and the graph of $y = \left(\frac{1}{7}\right)^x$ in Figure 8-2**b**.

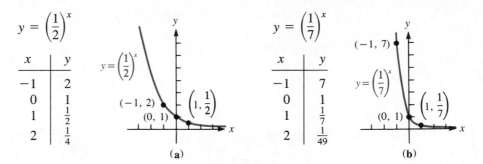

Figure 8-2

Note that $y = (\frac{1}{2})^x$ and $y = (\frac{1}{7})^x$ are *decreasing* functions, that each graph passes through the point $(0, 1)$, and that the x-axis is an asymptote of each graph. Note also that the graph of $y = (\frac{1}{2})^x$ passes through the point $(1, \frac{1}{2})$, and that of $y = (\frac{1}{7})^x$ passes through $(1, \frac{1}{7})$. ∎

Examples 1 and 2 suggest that an exponential function with base b is either increasing (for $b > 1$) or decreasing (for $0 < b < 1$). Thus, distinct real numbers x will determine distinct values b^x. The exponential function, therefore, is one-to-one. This fact is the basis of an important property of exponential expressions.

> **Theorem.** If $b > 0$, $b \neq 1$ and $b^r = b^s$, then $r = s$.

Example 3 On the same set of coordinate axes, graph $y = \left(\frac{3}{2}\right)^x$ and $y = \left(\frac{2}{3}\right)^x$.

Solution Plot several pairs (x, y) that satisfy each equation and draw each graph as in Figure 8-3.

$y = \left(\frac{3}{2}\right)^x$

x	y
-2	$\frac{4}{9}$
-1	$\frac{2}{3}$
0	1
1	$\frac{3}{2}$
2	$\frac{9}{4}$

$y = \left(\frac{2}{3}\right)^x$

x	y
-2	$\frac{9}{4}$
-1	$\frac{3}{2}$
0	1
1	$\frac{2}{3}$
2	$\frac{4}{9}$

Figure 8-3

Note that $\frac{3}{2}$ and $\frac{2}{3}$ are reciprocals of each other and that the graphs are reflections of each other in the y-axis. This follows from the properties of exponents. If the number x in the first equation $y = (\frac{3}{2})^x$ is replaced with $-x$, the result is the second equation, $y = (\frac{2}{3})^x$.

$$y = \left(\frac{3}{2}\right)^x$$

$$y = \left(\frac{3}{2}\right)^{-x} \qquad \text{Replace } x \text{ with } -x.$$

$$= \left[\left(\frac{3}{2}\right)^{-1}\right]^x$$

$$= \left(\frac{2}{3}\right)^x$$

The symmetry discussed in this example is not the same as the y-axis symmetry discussed in Chapter 3. There, you considered curves that are reflections of themselves in the y-axis. Here, *two* curves are reflections of each other. ∎

We summarize the properties of the exponential function with base b as follows:

The domain of the exponential function is the set of real numbers, and its range is the set of positive real numbers.

If $b > 1$, then $y = b^x$ defines an *increasing* function.

If $0 < b < 1$, then $y = b^x$ defines a *decreasing* function.

The graph of $y = b^x$ passes through the points $(0, 1)$ and $(1, b)$.

The x-axis is an asymptote of the graph of $y = b^x$.

The graphs of $y = b^x$ and $y = b^{-x}$ are reflections of each other in the y-axis.

The exponential function defined by $y = b^x$ is one-to-one; that is,

if $b^r = b^s$, then $r = s$

In Section 3.5, we saw that, for $k > 0$, the graph of

1. $y = f(x) + k$ **3.** $y = f(x - k)$
2. $y = f(x) - k$ **4.** $y = f(x + k)$

is identical to the graph of $y = f(x)$ except that it is translated k units

1. upward   **3.** to the right
2. downward   **4.** to the left

The same principles apply to the graphs of exponential functions.

Example 4   Graph the functions defined by   **a.** $y = 2^x - 3$,   **b.** $y = 2^{x+2}$, and
c. $y = 2(3^{x/2})$.

Solution   **a.** The graph of $y = 2^x - 3$ is identical to the graph of $y = 2^x$, except that it is
shifted 3 units downward. The graph is shown in Figure 8-4**a**. The graph of
$y = 2^x$ is included for reference.

b. The graph of $y = 2^{x+2}$ is identical to the graph of $y = 2^x$, except that it is
shifted 2 units to the left. The graph is shown in Figure 8-4**b**. The graph of
$y = 2^x$ is included for reference.

c. The graph of $y = 2(3^{x/2})$ has the general shape of an exponential function. To
determine its exact shape, plot several pairs (x, y) that satisfy the equation and
join them with a smooth curve. The graph appears in Figure 8-4**c**.

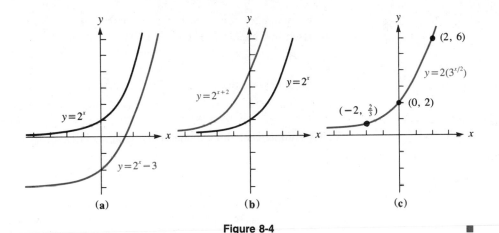

Figure 8-4

Applications of Exponential Expressions

A mathematical description of an observed event is called a **model** of that event.
Many observed events that change with time can be modeled by functions defined
by equations of the form

$$y = f(t) = ab^{kt} \qquad \text{Remember that } ab^{kt} \text{ means } a(b^{kt}).$$

where a, b, and k are constants, and t represents time. If f is an increasing function
such as the one in Example 4**c**, then y is said to **grow exponentially**. If f is a
decreasing function, then y **decays exponentially**.

A decreasing function determined by an equation of the form $y = ab^{kt}$ models
a process called **radioactive decay**. The atomic structure of radioactive material
changes as the material emits radiation. Uranium, for example, changes, or decays,
into thorium, then into radium, and eventually into lead.

Experiments have determined the time it takes for *half* of a given amount of
radioactive material to decompose. This time, called the **half-life**, is constant for

any given substance. The amount A of radioactive material present decays exponentially according to the model

Radioactive Decay Formula

$$A = A_0 2^{-t/h}$$

where t is time, A_0 is the amount present at $t = 0$, and h is the material's half-life.

Example 5   The half-life of radium is approximately 1600 years. How much of a 1-gram sample will remain after 660 years?

Solution   In this example, $A_0 = 1$, $h = 1600$, and $t = 660$. Substitute these three values into the equation $A = A_0 2^{-t/h}$ and simplify:

$$A = A_0 2^{-t/h}$$
$$A = 1 \cdot 2^{-660/1600}$$
$$= 1 \cdot 2^{-0.4125} \qquad \text{Use a calculator.}$$
$$\approx 0.75$$

After 660 years, approximately 0.75 gram of radium will remain.   ■

Another example of exponential growth is **compound interest**. If the interest earned on money in a savings account is allowed to accumulate in the account, then that interest also earns interest. The amount in the account grows exponentially according to the equation

Compound Interest Formula

$$A = A_0 \left(1 + \frac{r}{k}\right)^{kt}$$

where A represents the amount in the account after t years, with interest paid k times a year at an annual rate of r on an initial deposit A_0.

Example 6   If $1000 is deposited in an account that earns 12% interest, compounded quarterly, how much will be in the account after 20 years?

Solution   Calculate A using the formula

$$A = A_0 \left(1 + \frac{r}{k}\right)^{kt}$$

with $A_0 = 1000$, $r = 0.12$, and $t = 20$. Because quarterly interest payments occur four times a year, $k = 4$.

$$A = A_0 \left(1 + \frac{r}{k}\right)^{kt}$$
$$A = 1000 \left(1 + \frac{0.12}{4}\right)^{4 \cdot 20}$$
$$= 1000(1.03)^{80}$$
$$= 10,640.89 \qquad \text{Use a calculator.}$$

In 20 years, the account will contain $10,640.89.   ■

EXERCISE 8.1

In Exercises 1–8, graph each exponential function.

1. $y = 3^x$ **2.** $y = 5^x$ **3.** $y = \left(\dfrac{1}{5}\right)^x$ **4.** $y = \left(\dfrac{1}{3}\right)^x$

5. $y = -2^x$ **6.** $y = -3^x$ **7.** $y = \left(\dfrac{3}{4}\right)^x$ **8.** $y = \left(\dfrac{4}{3}\right)^x$

In Exercises 9–20, graph the function defined by each equation.

9. $y = 3^x - 1$ **10.** $y = 2^x + 3$ **11.** $y = 2^x + 1$ **12.** $y = 4^x - 4$

13. $y = 3^{x-1}$ **14.** $y = 2^{x+3}$ **15.** $y = 3^{x+1} - 1$ **16.** $y = 3^{x-1} + 1$

17. $y = 5(2^x)$ **18.** $y = 2(5^x)$ **19.** $y = 2^{|x|}$ **20.** $y = 2^{-x^2}$

In Exercises 21–26, find the value of b, if any, that would cause the graph of $y = b^x$ to look like the graph indicated.

21.
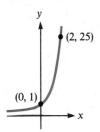

22.
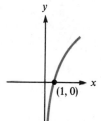

23.
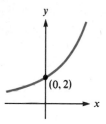

24.
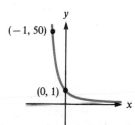

25.
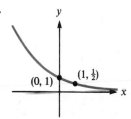

26.
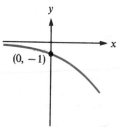

27. A radioactive material decays according to the formula $A = A_0\left(\frac{2}{3}\right)^t$, where A_0 is the amount present initially and t is measured in years. What amount will be present in 5 years?

28. Tritium, a radioactive isotope of hydrogen, has a half-life of 12.4 years. Of an initial sample of 0.05 gram, how much will remain after 100 years?

29. The half-life of radioactive carbon-14 is 5700 years. How much of an initial sample will remain after 3000 years?

30. The **biological half-life** of the asthma medication theophylline is 4.5 hours for smokers and 8 hours for nonsmokers. Twelve hours after administering equal doses, what is the ratio of drug retained in a smoker's system to that in a nonsmoker's?

In Exercises 31–34, assume that there are no deposits or withdrawals.

31. An initial deposit of $500 earns 10% interest, compounded quarterly. How much will be in the account in 10 years?

32. An initial deposit of $1000 earns 12% interest, compounded monthly. How much will be in the account in $4\frac{1}{2}$ years?

33. If $1 had been invested in 1776 at 5% interest, compounded annually, what would it be worth on January 1, 2076?

34. Some financial institutions pay daily interest, compounded by the **360/365 method**, by using the formula

$$A = A_0\left(1 + \frac{r}{360}\right)^{365t} \qquad (t \text{ is in years})$$

Using this method, what will an initial investment of $1000 be worth in 5 years, assuming a 12% annual interest rate?

35. A colony of 6 million bacteria is growing in a culture medium. The population P after t hours is modeled by the formula $P = (6 \times 10^6)(2.3)^t$. What is the population after 4 hours?

36. The population of North Rivers is growing exponentially according to the model $P = 375(1.3)^t$, where t is measured in years from the present date. What will be the population in 3 years?

37. A bacteria culture grows exponentially according to the model $P = P_0 2^{t/24}$, where P_0 is the initial population and t is measured in hours. By what factor will it have increased in 36 hours?

38. The charge remaining in a battery is decreasing exponentially according to the formula $C = C_0(0.7)^t$, where C is the charge remaining after t days, and C_0 is the initial charge. If a charge of 2.471×10^{-5} coulombs remains after 7 days, what was the battery's initial charge?

39. To have P dollars available in n years, A dollars can be invested now in an account paying interest at an annual rate i, compounded annually. Show that

$$A = P(1 + i)^{-n}$$

40. Atmospheric pressure P (in pounds per square inch) is an exponential function of the altitude a (in feet above sea level) given by

$$P = 14.7(2^{-0.000056a})$$

At what altitude will the atmospheric pressure be half that at sea level?

8.2   BASE-e EXPONENTIAL FUNCTIONS

Leonhard Euler
(1707–1783) Euler was one of the most prolific mathematicians of all time. He did much of his work after he became blind.

In mathematical models of natural events, one number appears often as the base of an exponential function. This number is denoted by the letter e, the symbol first used by Leonhard Euler (1707–1783). We introduce this important number by recalling the compound interest formula

$$A = A_0\left(1 + \frac{r}{k}\right)^{kt}$$

and allowing k, representing the number of times per year that interest is compounded, to become very large. To see what happens, we let $k = rp$, where p is a new variable, and proceed as follows:

$$A = A_0\left(1 + \frac{r}{k}\right)^{kt}$$

$$= A_0\left(1 + \frac{r}{rp}\right)^{rpt} \qquad \text{Substitute } rp \text{ for } k.$$

$$= A_0\left(1 + \frac{1}{p}\right)^{rpt} \qquad \text{Simplify } \frac{r}{rp}.$$

$$= A_0\left[\left(1 + \frac{1}{p}\right)^p\right]^{rt} \qquad \text{Remember that } (x^m)^n = x^{mn}.$$

Because r is a positive constant and $k = rp$, it follows that as k becomes very large, so does p. The question of what happens to the value of A becomes tied to the question: What happens to the value of $(1 + 1/p)^p$ as p becomes very large? Some results calculated for increasing values of p appear in Table 8-1.

Table 8.1

p	$\left(1 + \dfrac{1}{p}\right)^p$
1	2
10	2.5937425
100	2.7048138
1,000	2.7169239
1,000,000	2.7182805
1,000,000,000	2.7182818

The results in the table suggest that as p increases, the value of $(1 + 1/p)^p$ approaches a fixed number. This number is e, an irrational number with a decimal representation of 2.71828182845904. . . .

If interest on an amount A_0 is compounded more and more often, the number p grows large without bound, and the formula

$$A = A_0\left[\left(1 + \frac{1}{p}\right)^p\right]^{rt}$$

becomes

Continuous Compound Interest Formula

$$A = A_0 e^{rt}$$

When the amount invested grows exponentially according to the formula $A = A_0 e^{rt}$, interest is said to be **compounded continuously**.

Example 1   If \$1000 accumulates interest at an annual rate of 12% compounded continuously, how much money will be in the account in 20 years?

Solution

$$A = A_0 e^{rt}$$
$$A = 1000 e^{(0.12)(20)} \qquad \text{Substitute 1000 for } A_0, \, 0.12 \text{ for } r, \text{ and 20 for } t.$$
$$\approx 1000(11.02318) \qquad \text{Use a calculator.}$$
$$\approx 11{,}023.18$$

In 20 years, the account will contain \$11,023.18. ∎

The exponential function $y = e^x$ is so important that it is often called *the* exponential function. In some textbooks, on many calculators, and in several computer languages, the exponential function is denoted by the letters exp. Thus,

$$\exp(x) = e^x$$

Example 2   Graph the exponential function.

Solution   Use a calculator to find several pairs (x, y) that satisfy the equation $y = e^x$. Plot them and join them with a smooth curve. The graph appears in Figure 8-5.

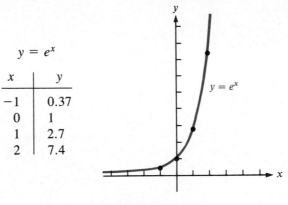

$$y = e^x$$

x	y
-1	0.37
0	1
1	2.7
2	7.4

Figure 8-5

Example 3   Graph $y = 3e^{-x/2}$.

Solution   Plot several pairs (x, y) that satisfy the equation $y = 3e^{-x/2}$ and join them with a smooth curve. The graph appears in Figure 8-6.

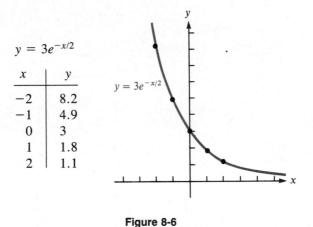

$$y = 3e^{-x/2}$$

x	y
-2	8.2
-1	4.9
0	3
1	1.8
2	1.1

Figure 8-6

An equation based on the exponential function provides a model for **population growth**. Many factors—such as birth and death rates, immigration, pollution, diet, wars, plagues, and famines—affect the growth of a population. Models of population growth that account for several factors are very complex. A simpler model, called the **Malthusian model of population growth**, assumes a constant birth rate B and constant death rate D and incorporates no other factors. In that model, the population P grows exponentially according to the formula

Population Growth Formula

$$P = P_0 e^{kt}$$

where P_0 is the population at $t = 0$, and $k = B - D$. If t is measured in years, k is called the **annual growth rate**.

Example 4   The annual birth rate in a certain country is 19 per 1000 population, and the death rate is 7 per 1000. What would the Malthusian model predict the population of the country to be in 50 years if the current population is 2.3 million?

Solution   Use the Malthusian model of population growth

$$P = P_0 e^{kt}$$

The number k is the difference between the birth and death rates. The birth rate, B, is $\frac{19}{1000}$, or 0.019. The death rate, D, is $\frac{7}{1000}$, or 0.007. Thus

$$k = B - D$$
$$k = 0.019 - 0.007$$
$$= 0.012$$

Substitute 2.3×10^6 for P_0, 50 for t, and 0.012 for k in the equation $P = P_0 e^{kt}$ and simplify:

$$P = P_0 e^{kt}$$
$$P = (2.3 \times 10^6)e^{(0.012)(50)}$$
$$= (2.3 \times 10^6)(1.82)$$
$$= 4.2 \times 10^6$$

After 50 years, the population will exceed 4 million.   ∎

Example 5   A population of 1000 bacteria doubles in 8 hours. Assuming the Malthusian model, what will be the population in 12 hours?

Solution   The population P grows according to the formula $P = P_0 e^{kt}$. Let $P_0 = 1000$, $P = 2000$, and $t = 8$. Then proceed as follows:

$$P = P_0 e^{kt}$$

$2000 = 1000e^{k8}$	Substitute 2000 for P, 1000 for P_0, and 8 for t.
$2 = e^{k \cdot 8}$	Divide both sides by 1000.
$2^{1/8} = (e^{k \cdot 8})^{1/8}$	Raise both sides to the $\frac{1}{8}$ power.
$2^{1/8} = e^k$	

We know that the population grows according to the formula

$P = 1000e^{kt}$	
$= 1000(2^{1/8})^t$	Substitute $2^{1/8}$ for e^k.
$= 1000(2^{t/8})$	

To find the population after 12 hours, substitute 12 for t and simplify:

$$P = 1000(2^{t/8})$$
$$= 1000(2^{12/8})$$
$$= 1000(2^{3/2})$$
$$\approx 1000(2.8284) \qquad \text{Use a calculator.}$$
$$\approx 2800$$

After 12 hours, there are approximately 2800 bacteria.   ■

EXERCISE 8.2

In Exercises 1–8, graph the function defined by each equation. **Use a calculator.**

1. $y = -e^x$
2. $y = e^{-x}$
3. $y = e^{-0.5x}$
4. $y = -e^{2x}$

5. $y = 2e^{-x}$
6. $y = -3e^x$
7. $y = e^x + 1$
8. $y = 2 - e^x$

In Exercises 9–16, tell whether the graph of $y = e^x$ could look like the graph indicated.

9.

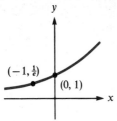

10.

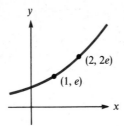

11.

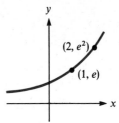

12.

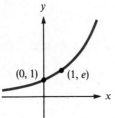

13.

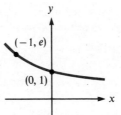

14.

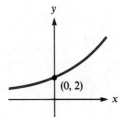

15.

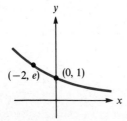

16.

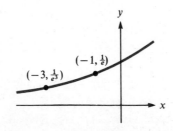

In Exercises 17–21, assume that there are no deposits or withdrawals.

17. An initial investment of $5000 earns 11.2% interest, compounded continuously. What will the investment be worth in 12 years?

18. An initial deposit of $2000 earns 8% interest, compounded continuously. How much will be in the account in 15 years?

19. An account now contains $11,180. It has been accumulating interest at 13%, compounded continuously, for 7 years. What was the initial deposit?

20. An account now contains $3610. It has been accumulating interest at $10\frac{1}{2}$%, compounded continuously. How much was in the account 1 year ago?

21. An initial deposit grows at a continuously compounded annual rate of 14%. If $5000 is in the account after 2 years, how much will be in the account after 6 years?

22. The growth of a population is modeled by

$$P = 173e^{0.03t}$$

How large will the population be when $t = 20$?

23. The decline of a population is modeled by

$$P = 1.2 \times 10^6 e^{-0.008t}$$

How large will the population be when $t = 30$?

24. The world population is approximately 5 billion and is growing at an annual rate of 1.9%. Assuming a Malthusian growth model, what will be the world's population in 30 years?

25. Assuming a Malthusian model and an annual growth rate of 1.9%, by what factor will the earth's current population increase in 50 years?

26. A country's population is now 2×10^5 people and is expected to double every 20 years. Assuming a Malthusian model, what will be the population in 35 years?

27. The population of a small town is now 140 and is expected to grow exponentially, tripling every 15 years. Assuming a Malthusian model, what can the city planners expect the population to be in 5 years?

28. The amount A of a drug remaining in a person's system after t hours is given by the formula

$$A = A_0 e^{kt}$$

where A_0 is the initial dose. After 2.3 hours, one-half of an initial dose of triazolam, a drug for treating insomnia, will remain. What percent will remain after 24 hours?

29. If $e^{t+3} = ke^t$, determine k.

30. If $e^{3t} = k^t$, determine k.

31. On a sheet of graph paper, graph the function $y = \frac{1}{2}(e^x + e^{-x})$ for values of x between -2 and 2. You'll need to calculate and plot about five or six points before joining them with a smooth curve. The graph looks like a parabola, but it is not. It is called a **catenary** and is important in the design of power distribution networks because it represents the shape of a cable drooping between its supporting poles.

32. The value of e can be calculated to any degree of accuracy by adding the first several terms of the following list.

$$1, 1, \frac{1}{2}, \frac{1}{2 \cdot 3}, \frac{1}{2 \cdot 3 \cdot 4}, \cdots, \frac{1}{2 \cdot 3 \cdots \cdots n - 1}, \cdots$$

The more terms that are added, the closer the sum is to the actual value of e. Calculate an approximation of e by adding the first eight values in the preceding list. To how many decimal places is your sum accurate?

33. The **hyperbolic cosine** function, cosh x, and the **hyperbolic sine** function, sinh x, are defined by the equations

$$\cosh x = \frac{e^x + e^{-x}}{2} \qquad \text{and} \qquad \sinh x = \frac{e^x - e^{-x}}{2}$$

Show that $(\cosh x)^2 - (\sinh x)^2 = 1$.

34. Show that $\sinh 2x = 2 \sinh x \cosh x$. See Exercise 33.

35. Show that $\cosh 2x = (\cosh x)^2 + (\sinh x)^2$. See Exercise 33.

36. The **hyperbolic tangent**, tanh x, is defined by the equation

$$\tanh x = \frac{e^x - e^{-x}}{e^x + e^{-x}}$$

Show that

$$\frac{1 + \tanh x}{2} = \frac{1}{1 + e^{-2x}}$$

8.3   LOGARITHMIC FUNCTIONS

Because an exponential function defined by $y = b^x$ is one-to-one, it has an inverse that is defined by the equation $x = b^y$. To express this inverse function in the form $y = f^{-1}(x)$, we must solve the equation $x = b^y$ for y. To do so, we need the following definition.

> **Definition.**   The **logarithmic function with base b** is defined by the equation
>
> $$y = \log_b x$$
>
> where $b > 0$ and $b \neq 1$. This equation is equivalent to the exponential equation
>
> $$x = b^y$$
>
> The domain of the logarithmic function is the interval $(0, \infty)$ of positive real numbers. Its range is the set of real numbers.

Because the function $y = \log_b x$ is the inverse of the one-to-one exponential function $y = b^x$, the function $y = \log_b x$ is one-to-one also.

The previous definition implies that any pair (x, y) that satisfies the equation $x = b^y$ also satisfies the equation $y = \log_b x$ (or $\log_b x = y$). Thus,

$$\log_5 25 = 2 \quad \text{because} \quad 25 = 5^2$$

$$\log_7 1 = 0 \quad \text{because} \quad 1 = 7^0$$

$$\log_{16} 4 = \frac{1}{2} \quad \text{because} \quad 4 = 16^{1/2}$$

and

$$\log_2 \frac{1}{8} = -3 \quad \text{because} \quad \frac{1}{8} = 2^{-3}$$

Note that in each case the logarithm of a number is an exponent: $\log_b x$ *is the power to which b is raised to get x.*

Because the domain of the logarithmic function is the set of positive real numbers, it is impossible to find the logarithm of 0 or of a negative number.

Example 1 Find the value of y in each equation: **a.** $\log_5 1 = y$, **b.** $\log_2 8 = y$, and **c.** $\log_7 \frac{1}{7} = y$.

Solution **a.** Change the equation $\log_5 1 = y$ into the equivalent exponential form $1 = 5^y$. Because $1 = 5^0$, it follows that $y = 0$. Hence, $\log_5 1 = 0$.

b. $\log_2 8 = y$ is equivalent to $8 = 2^y$. Because $8 = 2^3$, it follows that $y = 3$. Hence, $\log_2 8 = 3$.

c. $\log_7 \frac{1}{7} = y$ is equivalent to $\frac{1}{7} = 7^y$. Because $\frac{1}{7} = 7^{-1}$, it follows that $y = -1$. Hence, $\log_7 \frac{1}{7} = -1$. ∎

Example 2 Find the value of a in each equation: **a.** $\log_3 \frac{1}{9} = a$, **b.** $\log_a 32 = 5$, and **c.** $\log_9 a = -\frac{1}{2}$.

Solution **a.** $\log_3 \frac{1}{9} = a$ is equivalent to $\frac{1}{9} = 3^a$. Because $\frac{1}{9} = 3^{-2}$, it follows that $a = -2$.

b. $\log_a 32 = 5$ is equivalent to $a^5 = 32$. Because $2^5 = 32$, it follows that $a = 2$.

c. $\log_9 a = -\frac{1}{2}$ is equivalent to $9^{-1/2} = a$. Because $9^{-1/2} = \frac{1}{3}$, it follows that $a = \frac{1}{3}$. ∎

Example 3 Graph the logarithmic functions **a.** $y = \log_2 x$ and **b.** $y = \log_{1/2} x$.

Solution **a.** The equation $y = \log_2 x$ is equivalent to the equation $x = 2^y$. Calculate and plot pairs (x, y) that satisfy the equation $x = 2^y$ and connect them with a smooth curve. The graph appears in Figure 8-7**a**.

b. Rewrite $y = \log_{1/2} x$ as $x = (\frac{1}{2})^y$ and proceed as in part **a**. The graph appears in Figure 8-7**b**.

$y = \log_2 x$

x	y
$\frac{1}{4}$	-2
$\frac{1}{2}$	-1
1	0
2	1
4	2
8	3

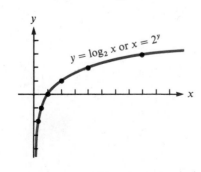

(a)

$y = \log_{1/2} x$

x	y
8	-3
4	-2
2	-1
1	0
$\frac{1}{2}$	1
$\frac{1}{4}$	2

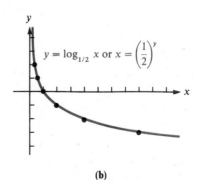

(b)

Figure 8-7 ∎

Example 4   Graph the functions defined by   **a.** $y = 3 + \log_2 x$ and   **b.** $y = \log_2 (x - 1)$.

Solution   **a.** The graph of $y = 3 + \log_2 x$ is identical to the graph of $y = \log_2 x$ (see Figure 8-7**a**), except that it is shifted 3 units upward. The graph is shown in Figure 8-8**a**.

b. The graph of $y = \log_2 (x - 1)$ is identical to the graph of $y = \log_2 x$, except that it is shifted 1 unit to the right. The graph is shown in Figure 8-8**b**.

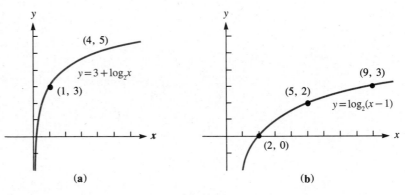

(a)   (b)

Figure 8-8   ∎

Just as $y = e^x$ represents an important exponential function, $y = \log_e x$, denoted as $y = \ln x$, represents an important logarithmic function.

Example 5   Graph the function determined by $y = \ln x$.

Solution   The equation $y = \ln x$ is equivalent to the equation $x = e^y$. Calculate and plot pairs (x, y) that satisfy the equation $x = e^y$ and connect them with a smooth curve. The graph appears in Figure 8-9.

$y = \ln x$

x	y
$\dfrac{1}{e} \approx 0.4$	-1
1	0
$e \approx 2.7$	1
$e^2 \approx 7.4$	2

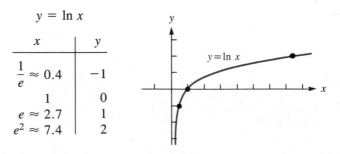

Figure 8-9   ∎

Examples 3–5 suggest that the graphs of logarithmic functions are similar to those in Figure 8-10. If $b > 1$, the logarithmic function is *increasing* as in Figure 8-10**a**, and if $0 < b < 1$, the logarithmic function is *decreasing* as in Figure 8-10**b**. Note that each graph of $y = \log_b x$ passes through the points $(1, 0)$ and $(b, 1)$ and that the y-axis is an asymptote to the curve.

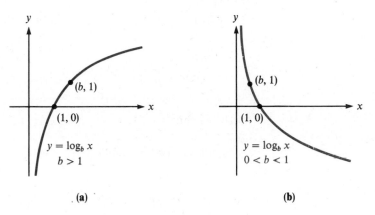

(a)   **(b)**

Figure 8-10

The exponential and logarithmic functions to base b are inverses of each other and, therefore, have symmetry about the line $y = x$. The graphs of $y = \log_b x$ and $y = b^x$ appear as in Figure 8-11**a** if $b > 1$ and as in Figure 8-11**b** if $0 < b < 1$.

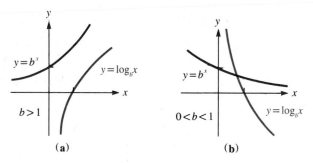

(a)   **(b)**

Figure 8-11

Because the logarithmic function is one-to-one, we have the following property of logarithms:

Theorem.   If $\log_b r = \log_b s$, then $r = s$.

Several other properties of logarithms can be found by expressing the properties of exponents in logarithmic form.

Properties of Logarithms.   If M, N, p, and b are positive numbers, and $b \neq 1$, then

1. $\log_b 1 = 0$ **2.** $\log_b b = 1$

3. $\log_b b^x = x$ **4.** $b^{\log_b x} = x$

5. $\log_b MN = \log_b M + \log_b N$ **6.** $\log_b \dfrac{M}{N} = \log_b M - \log_b N$

7. $\log_b M^p = p \log_b M$

Proof   Properties 1 through 4 follow directly from the definition of logarithms. To prove Property 5, we let $x = \log_b M$ and $y = \log_b N$. Using the definition of logarithms, we write these equations in the form

$$M = b^x \qquad \text{and} \qquad N = b^y$$

We multiply equal quantities by equal quantities to get

$$MN = b^x b^y$$

or

$$MN = b^{x+y}$$

By the definition of a logarithm, this equation is equivalent to

$$\log_b MN = x + y$$

We substitute the values of x and y to complete the proof of Property 5.

$$\log_b MN = \log_b M + \log_b N$$

To prove Property 7, we let $x = \log_b M$, write this equation in exponential form, and raise both sides of the resulting equation to the pth power.

$$M = b^x$$
$$(M)^p = (b^x)^p$$
$$M^p = b^{px}$$

Using the definition of logarithms in reverse gives

$$\log_b M^p = px$$

Substituting the value for x completes the proof.

$$\log_b M^p = p \log_b M \qquad \qquad \Box$$

The proof of Property 6 is similar to the proof of Property 5 and is left as an exercise.

Property 5 of logarithms asserts that the logarithm of the *product* of two numbers is equal to the *sum* of their logarithms. The logarithm of a *sum* or a *difference* usually does not simplify. In general,

$$\log_b(M + N) \neq \log_b M + \log_b N$$

Similarly,

$$\log_b(M - N) \neq \log_b M - \log_b N$$

Property 6 of logarithms asserts that the logarithm of the *quotient* of two numbers is equal to the *difference* of their logarithms. The logarithm of a quotient is not the quotient of the logarithms:

$$\log_b \frac{M}{N} \neq \frac{\log_b M}{\log_b N}$$

Example 6 Simplify each expression: **a.** $\log_3 1$, **b.** $\log_4 4$, **c.** $\log_7 7^3$, and **d.** $b^{\log_b 3}$.

Solution **a.** By Property 1, $\log_3 1 = 0$.
b. By Property 2, $\log_4 4 = 1$.
c. By Property 3, $\log_7 7^3 = 3$.
d. By Property 4, $b^{\log_b 3} = 3$. ■

The properties of logarithms are often used to expand or condense a logarithmic expression, as in the following two examples.

Example 7 Assume that x, y, z, and b are positive numbers. Use the properties of logarithms to write each expression in terms of the logarithms of x, y, and z:

a. $\log_b \frac{xy}{z}$, **b.** $\log_b(x^3y^2z)$, and **c.** $\log_b \frac{y^2\sqrt{z}}{x}$.

Solution **a.** $\log_b \dfrac{xy}{z} = \log_b(xy) - \log_b z$ Use Property 6.

$= \log_b x + \log_b y - \log_b z$ Use Property 5.

b. $\log_b(x^3y^2z) = \log_b x^3 + \log_b y^2 + \log_b z$ Use Property 5 twice.

$= 3 \log_b x + 2 \log_b y + \log_b z$ Use Property 7 twice.

c. $\log_b \dfrac{y^2\sqrt{z}}{x} = \log_b(y^2\sqrt{z}) - \log_b x$ Use Property 6.

$= \log_b y^2 + \log_b \sqrt{z} - \log_b x$ Use Property 5.

$= \log_b y^2 + \log_b z^{1/2} - \log_b x$ Write $\sqrt{z}$ as $z^{1/2}$.

$= 2 \log_b y + \dfrac{1}{2}\log_b z - \log_b x$ Use Property 7 twice. ■

Example 8   Assume that x, y, z, and b are positive numbers and $b \neq 1$. Use the properties of logarithms to write each expression as the logarithm of a single quantity:

a. $2 \log_b x + \dfrac{1}{3}\log_b y$ and   **b.** $\dfrac{1}{2}\log_b(x - 2) - \log_b y + 3 \log_b z$

Solution   **a.** $2 \log_b x + \dfrac{1}{3}\log_b y = \log_b x^2 + \log_b y^{1/3}$ Use Property 7 twice.

$$= \log_b(x^2 y^{1/3})$$ Use Property 5.

$$= \log_b(x^2 \sqrt[3]{y})$$ Write $y^{1/3}$ as $\sqrt[3]{y}$.

b. $\dfrac{1}{2}\log_b(x - 2) - \log_b y + 3 \log_b z$

$$= \log_b(x - 2)^{1/2} - \log_b y + \log_b z^3$$ Use Property 7 twice.

$$= \log_b \frac{(x - 2)^{1/2}}{y} + \log_b z^3$$ Use Property 6.

$$= \log_b \frac{z^3 \sqrt{x - 2}}{y}$$ Use Property 5 and write $(x - 2)^{1/2}$ as $\sqrt{x - 2}$.   ∎

Example 9   Given that $\log_{10} 2 \approx 0.3010$ and $\log_{10} 3 \approx 0.4771$, find approximate values for
a. $\log_{10} 18$ and   **b.** $\log_{10} 2.5$.

Solution   **a.** $\log_{10} 18 = \log_{10}(2 \cdot 3^2)$ Factor 18.

$$= \log_{10} 2 + \log_{10} 3^2$$ Use Property 5.

$$= \log_{10} 2 + 2 \log_{10} 3$$ Use Property 7.

$$\approx 0.3010 + 2(0.4771)$$ Substitute the value of each logarithm.

$$\approx 1.2552$$ Simplify.

b. $\log_{10} 2.5 = \log_{10}\left(\dfrac{5}{2}\right)$ Write 2.5 as $\frac{5}{2}$.

$$= \log_{10} 5 - \log_{10} 2$$ Use Property 6.

$$= \log_{10} \frac{10}{2} - \log_{10} 2$$ Write 5 as $\frac{10}{2}$.

$$= \log_{10} 10 - \log_{10} 2 - \log_{10} 2$$ Use Property 6.

$$= 1 - 2 \log_{10} 2$$ Use Property 2 and combine terms.

$$\approx 1 - 2(0.3010)$$ Substitute 0.3010 for $\log_{10} 2$.

$$\approx 0.3980$$ Simplify.   ∎

EXERCISE 8.3

In Exercises 1–8, write each equation in exponential form.

1. $\log_3 81 = 4$ **2.** $\log_7 7 = 1$ **3.** $\log_{1/2} \frac{1}{8} = 3$ **4.** $\log_{1/5} 1 = 0$

5. $\log_4 \frac{1}{64} = -3$ **6.** $\log_6 \frac{1}{36} = -2$ **7.** $\log_x y = z$ **8.** $\log_m n = \frac{1}{2}$

In Exercises 9–16, write each equation in logarithmic form.

9. $8^2 = 64$ **10.** $10^3 = 1000$ **11.** $4^{-2} = \frac{1}{16}$ **12.** $3^{-4} = \frac{1}{81}$

13. $\left(\frac{1}{2}\right)^{-5} = 32$ **14.** $\left(\frac{1}{3}\right)^{-3} = 27$ **15.** $x^y = z$ **16.** $m^n = p$

In Exercises 17–40, find the value of x. A calculator is of no value.

17. $\log_2 8 = x$ **18.** $\log_{1/2} \frac{1}{8} = x$ **19.** $\log_{1/2} 8 = x$ **20.** $\log_{25} 5 = x$

21. $\log_5 25 = x$ **22.** $\log_8 x = 2$ **23.** $\log_x 8 = 3$ **24.** $\log_7 x = 0$

25. $\log_7 x = 1$ **26.** $\log_4 x = \frac{1}{2}$ **27.** $\log_x \frac{1}{16} = -2$ **28.** $\log_{125} x = \frac{2}{3}$

29. $\log_{100} \frac{1}{1000} = x$ **30.** $\log_{5/2} \frac{4}{25} = x$ **31.** $\log_{27} 9 = x$ **32.** $\log_{12} x = 0$

33. $\log_x 5^3 = 3$ **34.** $\log_x 5 = 1$ **35.** $\log_x \frac{9}{4} = 2$ **36.** $\log_x \frac{\sqrt{3}}{3} = \frac{1}{2}$

37. $\log_{\sqrt{3}} x = -4$ **38.** $\log_\pi x = 3$ **39.** $\log_{2\sqrt{2}} x = 2$ **40.** $\log_4 8 = x$

In Exercises 41–48, graph the function determined by each equation.

41. $y = \log_5 x$ **42.** $y = \log_{1/5} x$ **43.** $y = 2 + \log_{1/3} x$ **44.** $y = \log_3(x + 2)$

45. $y = \ln x^2$ **46.** $y = \ln 2x$ **47.** $y = (\ln x) - 1$ **48.** $y = \ln(x - 1)$

In Exercises 49–54, graph each pair of equations on one set of coordinate axes.

49. $y = \log_3 x$ and $y = \log_3(3x)$ **50.** $y = \log_3 x$ and $y = \log_3\left(\frac{x}{3}\right)$

51. $y = \log_2 x$ and $y = \log_2(x + 1)$ **52.** $y = \log_2 x$ and $y = \log_2(-x)$

53. $y = \log_5 x$ and $y = 5^x$ **54.** $y = \ln x$ and $y = e^x$

In Exercises 55–60, find the value of b, if any, that would cause the graph of $y = \log_b x$ to look like the graph indicated.

55.

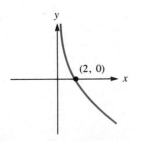

56.

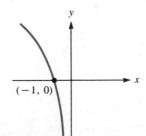

57.

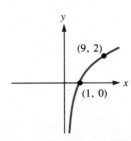

58.

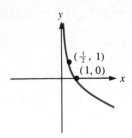

$(\frac{1}{2}, 1)$
$(1, 0)$

59.

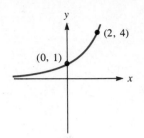

$(2, 4)$
$(0, 1)$

60.

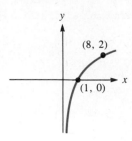

$(8, 2)$
$(1, 0)$

In Exercises 61–64, tell whether the graph of $y = \ln x$ could look like the graph indicated.

61.

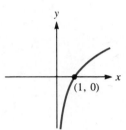

$(1, 0)$

62.

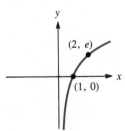

$(2, e)$
$(1, 0)$

63.

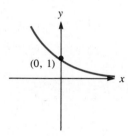

$(0, 1)$

64.

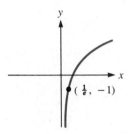

$(\frac{1}{e}, -1)$

In Exercises 65–72, assume that x, y, z, and b are positive numbers. Use the properties of logarithms to write each expression in terms of the logarithms of x, y, and z.

65. $\log_b xyz$

66. $\log_b \dfrac{x}{yz}$

67. $\log_b \left(\dfrac{x}{y}\right)^2$

68. $\log_b (xz)^{1/3}$

69. $\log_b x\sqrt{z}$

70. $\log_b \dfrac{\sqrt[3]{x}}{\sqrt[3]{yz}}$

71. $\log_b \sqrt[4]{\dfrac{x^3 y^2}{z^4}}$

72. $\log_b x \sqrt{\dfrac{\sqrt{y}}{z}}$

In Exercises 73–80, assume that x, y, and z are positive numbers. Use the properties of logarithms to write each expression as the logarithm of a single quantity.

73. $\log_b(x + 1) - \log_b x$

74. $\log_b x + \log_b(x + 2) - \log_b 8$

75. $2 \log_b x + \dfrac{1}{3} \log_b y$

76. $-2 \log_b x - 3 \log_b y + \log_b z$

77. $-3 \log_b x - 2 \log_b y + \dfrac{1}{2} \log_b z$

78. $3 \log_b(x + 1) - 2 \log_b(x + 2) + \log_b x$

79. $\log_b\left(\dfrac{x}{z} + x\right) - \log_b\left(\dfrac{y}{z} + y\right)$

80. $\log_b(xy + y^2) - \log_b(xz + yz) + \log_b x$

In Exercises 81–102, tell if the given statement is true. If it is not true, so indicate.

81. $\log_b ab = \log_b a + 1$

82. $\log_b \frac{1}{a} = -\log_b a$

83. $\log_b 0 = 1$

84. $\log_b 2 = \log_2 b$

85. $\log_b(x + y) \neq \log_b x + \log_b y$

86. $\log_b xy = (\log_b x)(\log_b y)$

87. If $\log_a b = c$, then $\log_b a = c$.

88. If $\log_a b = c$, then $\log_b a = \frac{1}{c}$.

89. $\log_7 7^7 = 7$

90. $7^{\log_7 7} = 7$

91. $\log_b(-x) = -\log_b x$

92. If $\log_b a = c$, then $\log_b a^p = pc$.

93. $\frac{\log_b A}{\log_b B} = \log_b A - \log_b B$

94. $\log_b(A - B) = \frac{\log_b A}{\log_b B}$

95. $\log_b \frac{1}{5} = -\log_b 5$

96. $3 \log_b \sqrt[3]{a} = \log_b a$

97. $\frac{1}{3} \log_b a^3 = \log_b a$

98. A logarithm cannot be negative.

99. $\log_{10} 10^3 = 3(10^{\log_{10} 3})$

100. If x lies between 0 and 1, $\log_b x$ is negative.

101. $\log_{4/3} y = -\log_{3/4} y$

102. $\log_b y + \log_{1/b} y = 0$

In Exercises 103–112, assume that $\log_{10} 4 \approx 0.6021$, $\log_{10} 7 \approx 0.8451$, and $\log_{10} 9 \approx 0.9542$. Use these values and the properties of logarithms to find the approximate value of each quantity.

103. $\log_{10} 28$

104. $\log_{10} \frac{7}{4}$

105. $\log_{10} 2.25$

106. $\log_{10} 36$

107. $\log_{10} \frac{63}{4}$

108. $\log_{10} \frac{4}{63}$

109. $\log_{10} 252$

110. $\log_{10} 49$

111. $\log_{10} 112$

112. $\log_{10} 324$

113. Prove Property 6 of logarithms:
$$\log_b \frac{M}{N} = \log_b M - \log_b N$$

114. Show that $-\log_b x = \log_{1/b} x$.

115. Show that $e^{x \ln a} = a^x$.

116. Show that $e^{\ln x} = x$.

117. Show that $\ln(e^x) = x$.

118. If $\log_b 3x = 1 + \log_b x$, what is b?

119. Explain why it is impossible to calculate $\ln(\log 0.9)$.

120. Explain why it is impossible to calculate $\log_b(\ln 1)$.

8.4 APPLICATIONS OF LOGARITHMS

Before the widespread use of calculators, logarithms provided the only practical way to simplify many difficult computations. To use logarithms, mathematicians had to rely on extensive tables. Today, however, logarithms of numbers are easy to find with a calculator.

For computational purposes, base-10 logarithms are the most convenient. For this reason, base-10 logarithms have been called **common logarithms**. In this book, if the base b is not indicated in the notation $\log x$, always assume that b is 10:

$$\log A \quad \text{means} \quad \log_{10} A$$

Because the number e appears often in mathematical models of events in nature, base-e logarithms are called **natural logarithms**. They are also called **Napierian logarithms** after John Napier (1550–1617). As we have seen, natural logarithms are usually denoted by the symbol $\ln x$ rather than $\log_e x$.

$$\ln x \quad \text{means} \quad \log_e x$$

Example 1 Use a calculator to find   **a.** $\log 2.34$ and   **b.** $\ln 2.34$

Solution **a.** To find $\log 2.34$, enter the number 2.34 and press the $\boxed{\log}$ key. (You may have to press a $\boxed{2\text{nd}}$ function key first.) The display should read .3692158574. Hence,

$$\log 2.34 \approx 0.3692$$

b. To find $\ln 2.34$, enter the number 2.34 and press the $\boxed{\ln x}$ key. The display should read .850150929. Hence,

$$\ln 2.34 \approx 0.8502$$ ∎

Example 2 Find the value of x in each equation: **a.** $\log x = 0.7482$,   **b.** $\ln x = 1.335$,   and   **c.** $\ln x = \log 5.5$.

Solution **a.** $\log x = 0.7482$ is equivalent to $10^{0.7482} = x$. To find x, enter the number 10, press the $\boxed{y^x}$ key, enter the number .7482, and press $\boxed{=}$. The display reads 5.6001544. Hence,

$$x \approx 5.6$$

If your calculator has a $\boxed{10^x}$ key, simply enter .7482 and press it to get the same result. (You might have to press a $\boxed{2\text{nd}}$ function key.)

b. $\ln x = 1.335$ is equivalent to $e^{1.335} = x$. To find x, enter the number 1.335 and press the $\boxed{e^x}$ key. The display reads 3.79999595. Hence,

$$x \approx 3.8$$

c. $\ln x = \log 5.5$ is equivalent to $e^{\log 5.5} = x$. To find x, enter 5.5 and press $\boxed{\log}$ followed by $\boxed{e^x}$. The display reads 2.0966958. Hence,

$$x \approx 2.1$$ ∎

Example 3 Use a calculator to verify Property 5 of logarithms by showing that

$$\ln[(3.7)(15.9)] = \ln 3.7 + \ln 15.9$$

Solution   Calculate the left- and the right-hand sides of the equation separately, and compare the results. To calculate $\ln[(3.7)(15.9)]$, enter the number 3.7, press $\boxed{\times}$, enter 15.9, and press $\boxed{=}$. Then press $\boxed{\ln x}$. The display should read 4.0746519.

To calculate $\ln 3.7 + \ln 15.9$, enter the number 3.7, press $\boxed{\ln x}$, press $\boxed{+}$, enter 15.9, press $\boxed{\ln x}$, and press $\boxed{=}$. The display should also read 4.0746519. Because the left- and the right-hand sides are equal, the equation is verified.   ∎

Applications of Base-10 Logarithms

Example 4   In chemistry, common logarithms are used to express the acidity of solutions. The more acidic a solution, the greater the concentration of hydrogen ions. This concentration is indicated indirectly by the **pH scale**, or **hydrogen-ion index**. The pH of a solution is defined by the equation

$$pH = -\log[H^+]$$

where $[H^+]$ is the hydrogen-ion concentration in gram-ions per liter. Pure water has a few free hydrogen ions—$[H^+]$ is approximately 10^{-7} gram-ions per liter. The pH of pure water is

$$
\begin{aligned}
pH &= -\log 10^{-7} \\
&= -(-7) \log 10 &&\text{Use Property 7.} \\
&= -(-7) &&\text{Use Property 2.} \\
&= 7
\end{aligned}
$$

Seawater has a pH of approximately 8.5, and its hydrogen-ion concentration is found by solving the equation $8.5 = -\log_{10}[H^+]$ for $[H^+]$.

$$
\begin{aligned}
8.5 &= -\log[H^+] \\
-8.5 &= \log[H^+] \\
[H^+] &= 10^{-8.5} &&\text{Change the equation from logarithmic to exponential form.}
\end{aligned}
$$

Use a calculator to find that

$$[H^+] \approx 3.2 \times 10^{-9} \text{ gram-ions per liter.}$$ ∎

John Napier (1550–1617) Napier invented a device, called **Napier's rods**, that did multiplications mechanically. His device was a forerunner of modern-day computers.

Example 5   In electrical engineering, common logarithms are used to express the voltage gain (or loss) of an electronic device such as an amplifier or a length of transmission line. The unit of gain (or loss), called the **decibel**, is defined by a logarithmic relation. If E_O is the output voltage of a device, and E_I is the input voltage, the decibel voltage gain is defined as

$$\textbf{Decibel voltage gain} = 20 \log \frac{E_O}{E_I}$$

If, for example, the input to an amplifier is 0.5 volt and the output is 40 volts, the decibel voltage gain is calculated by substituting these values into the formula:

$$\text{Decibel voltage gain} = 20 \log \frac{E_O}{E_I}$$

$$\text{Decibel voltage gain} = 20 \log \frac{40}{0.5}$$

$$= 20 \log 80$$

$$\approx 20(1.9031)$$

$$\approx 38$$

The amplifier provides a 38-decibel voltage gain.   ■

Example 6   In seismology, common logarithms are used to measure the intensity of earthquakes on the **Richter scale**. The intensity R is given by

$$R = \log \frac{A}{P}$$

where A is the amplitude of the tremor (measured in micrometers) and P is the period of the tremor (the time of one oscillation of the earth's surface, measured in seconds). To calculate the intensity of an earthquake with an amplitude of 10,000 micrometers (1 centimeter) and a period of 0.1 second, substitute 10,000 for A and 0.1 for P in the formula and simplify:

$$R = \log \frac{A}{P}$$

$$R = \log \frac{10,000}{0.1}$$

$$= \log 100,000$$

$$= \log 10^5$$

$$= 5 \log 10 \qquad \text{Use Property 7.}$$

$$= 5 \qquad \text{Use Property 2.}$$

The earthquake measures 5 on the Richter scale.   ■

Applications of Base-e Logarithms

Example 7   In electronics, a mathematical model of the time required for charging a battery uses the natural logarithm function. A battery charges at a rate that depends on how close it is to being fully charged; it charges fastest when it is most discharged. The charge C at any instant t is modeled by the formula

$$C = M(1 - e^{-kt})$$

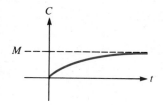

Figure 8-12

where M is the theoretical maximum charge that the battery can hold and k is a positive constant that depends on the battery and the charger. Plotting the variable C against t gives a curve like that in Figure 8-12. Notice that the full charge M is never attained; the actual charge can come very close to M, however, if the battery is charged long enough. To determine how long it will take a battery to reach a given charge C, solve the equation $C = M(1 - e^{-kt})$ for t as follows:

$$C = M(1 - e^{-kt})$$

$$\frac{C}{M} = 1 - e^{-kt} \qquad \text{Divide both sides by } M.$$

$$\frac{C}{M} - 1 = -e^{-kt} \qquad \text{Add } -1 \text{ to both sides.}$$

$$1 - \frac{C}{M} = e^{-kt} \qquad \text{Multiply both sides by } -1.$$

$$\ln\left(1 - \frac{C}{M}\right) = -kt \qquad \begin{array}{l}\text{Change the exponential equation} \\ \text{to logarithmic form.}\end{array}$$

$$-\frac{1}{k}\ln\left(1 - \frac{C}{M}\right) = t \qquad \text{Multiply both sides by } -\frac{1}{k}.$$

The formula that determines the time t required to charge a battery to a given level C is

$$t = -\frac{1}{k}\ln\left(1 - \frac{C}{M}\right)$$

■

Example 8 In physiology, experiments suggest that the relationship of loudness and intensity of sound is a logarithmic one known as the **Weber-Fechner law**: The apparent loudness L of a sound is proportional to the natural logarithm of its actual intensity I. In symbols,

$$L = k \ln I$$

For example, what actual increase in intensity will cause a doubling of the apparent loudness? If the original loudness is L_0, caused by an actual intensity I_0, then $L_0 = k \ln I_0$. To double the apparent loudness, multiply both sides of the equation by 2 and use Property 7 of logarithms:

$$2L_0 = 2k \ln I_0$$

$$= k \ln(I_0)^2$$

Thus, to double the apparent volume of a sound, the actual intensity must be squared.

■

Example 9 If a population grows exponentially at an annual rate r, the time t required for the population to double is called the **doubling time** and is given by the formula

$$t = \frac{\ln 2}{r}$$

The world's population is growing at the rate of approximately 2% per year. If this growth rate continues, when will the population be double its present size?

Solution   Because the population is growing at the rate 2%, substitute $r = 0.02$ into the formula and determine t.

$$t = \frac{\ln 2}{r}$$

$$t = \frac{\ln 2}{0.02}$$

$$\approx \frac{0.69315}{0.02}$$

$$\approx 31$$

At the current rate of growth, the world population is doubling every 31 years. ∎

EXERCISE 8.4

In Exercises 1–12, use a calculator to find the value of the variable. Express all answers to four decimal places.

1. $\log 3.25 = x$ 2. $\log 0.57 = y$ 3. $\ln 0.93 = y$ 4. $\ln 7.39 = x$

5. $\log(\ln 1.7) = x$ 6. $\ln(\log 9.8) = y$ 7. $\ln y = 4.24$ 8. $\log y = 0.926$

9. $\log x = -3.71$ 10. $\ln y = -0.28$ 11. $\log x = \ln 8$ 12. $\ln y = \log 7$

In Exercises 13–18, use a calculator to verify each equation.

13. $\log[(3.7)(2.9)] = \log 3.7 + \log 2.9$

14. $\ln \frac{9.3}{2.1} = \ln 9.3 - \ln 2.1$

15. $\ln(3.7)^3 = 3 \ln 3.7$

16. $\log 3.2 = \frac{\ln 3.2}{\ln 10}$

17. $\log \sqrt{14.1} = \frac{1}{2} \log 14.1$

18. $\ln 9.7 = \frac{\log 9.7}{\log 2.71828}$

19. Find the pH of a solution with a hydrogen-ion concentration of 1.7×10^{-5} gram-ions per liter.

20. What is the hydrogen-ion concentration of a saturated solution of calcium hydroxide whose pH is 13.2?

21. The pH of apples can range from 2.9 to 3.3. What is the range in the hydrogen-ion concentration?

22. The hydrogen-ion concentration of sour pickles is 6.31×10^{-4}. What is the pH?

23. The decibel voltage gain of an amplifier is 29. If the output is 20 volts, what is the input voltage?

24. The decibel voltage gain of an amplifier is 35. If the input signal is 0.05 volt, what is the output voltage?

25. The power output (or input) of an amplifier is directly proportional to the square of the voltage output (or input). Show that the formula for decibel voltage gain is

$$\text{Decibel voltage gain} = 10 \log \frac{P_O}{P_I}$$

where P_O is the power output and P_I is the power input.

26. An amplifier produces an output of 30 watts when driven by an input signal of 0.1 watt. What is the amplifier's voltage gain? See Exercise 25.

27. An earthquake has an amplitude of 5000 micrometers and a period of 0.2 second. What does it measure on the Richter scale?

28. An earthquake with amplitude of 8000 micrometers measures 6 on the Richter scale. What is its period?

29. An earthquake with a period of $\frac{1}{4}$ second measures 4 on the Richter scale. What is its amplitude?

30. By what factor must the period of an earthquake change to increase its severity by 1 point on the Richter scale? Assume that the amplitude remains constant.

31. If a battery can reach half of its full charge in 6 hours, how long will it take the battery to reach a 90% charge? Assume that the battery was fully discharged when it began charging.

32. A battery reaches 80% of a full charge in 8 hours. If it started charging when it was fully discharged, how long did it take to reach a 40% charge?

33. If the intensity of a sound is doubled, what is the apparent change in loudness?

34. What change in intensity of sound will cause an apparent tripling of the loudness?

35. A town's population grows at the rate of 12% per year. How long will it take the population to double?

36. A population growing at an annual rate r will triple in a time t given by the formula

$$t = \frac{\ln 3}{r}$$

How long will it take the population of the town in Exercise 35 to triple?

37. In business, equipment is often depreciated using the double-declining balance method. A piece of equipment with a life expectancy of N years, costing C dollars, will depreciate to a value V dollars in n years, where n is given by the formula

$$n = \frac{\log V - \log C}{\log\left(1 - \frac{2}{N}\right)}$$

A computer with an expected life of 5 years costs $37,000. It has depreciated to $8000. How old is it?

38. A typewriter worth $470 when new has an expected life of 12 years. It is now worth $189. How old is the typewriter? See Exercise 37.

39. If P dollars are invested at the end of each year in an annuity earning annual interest at the rate r, then the amount in the account will be A dollars after n years, where

$$n = \frac{\log\left[\frac{Ar}{P} + 1\right]}{\log(1 + r)}$$

If $1000 is invested each year in an annuity bearing annual interest of 12%, when will the account contain $20,000?

40. If $5000 is invested each year in an annuity bearing annual interest of 8%, when will the account contain $50,000? See Exercise 39.

41. Use the formula $P = P_0 e^{rt}$ to verify that P will become double the initial population P_0 when $t = \frac{\ln 2}{r}$.

42. Use the formula $P = P_0 e^{rt}$ to verify that P will become triple the initial population P_0 when $t = \frac{\ln 3}{r}$.

43. Show that the equation $t = -\dfrac{1}{k} \ln\left(1 - \dfrac{C}{M}\right)$ can be written in the form $t = \ln\left(\dfrac{M}{M - C}\right)^{1/k}$.

44. One form of the **logistic function** is given by the equation

$$y = \frac{1}{1 + e^{-2x}}$$

Determine the y-intercept of its graph.

8.5   EXPONENTIAL AND LOGARITHMIC EQUATIONS

An **exponential equation** is one that contains the variable in an exponent. A **logarithmic equation** is one that involves logarithms of expressions that contain the variable.

Example 1   Solve the exponential equation $3^x = 5$.

Solution   Because the logarithms of equal numbers are equal, you can take the common logarithm of both sides of the equation. Property 7 of logarithms then provides a means for moving the variable x from its position as an exponent to a position as a factor.

$$3^x = 5$$

$$\log 3^x = \log 5 \qquad \text{Take the common logarithm of both sides.}$$

$$x \log 3 = \log 5 \qquad \text{Use Property 7 of logarithms.}$$

1. $$x = \frac{\log 5}{\log 3} \qquad \text{Divide both sides by } \log 3.$$

$$\approx \frac{0.6990}{0.4771} \qquad \text{Substitute values for } \log 5 \text{ and } \log 3.$$

$$\approx 1.465$$

Thus, $x \approx 1.465$.

A careless reading of Equation 1 can lead to a common error. Because $\log \dfrac{A}{B} = \log A - \log B$, you may think that the expression $\dfrac{\log 5}{\log 3}$ also involves subtraction. It does not. The expression $\dfrac{\log 5}{\log 3}$ calls for division.   ∎

Example 2   Solve the exponential equation $6^{x-3} = 2^x$.

Solution

$$6^{x-3} = 2^x$$

$$\log 6^{x-3} = \log 2^x$$ Take the common logarithm of both sides.

$$(x - 3) \log 6 = x \log 2$$ Use Property 7 of logarithms.

$$x \log 6 - 3 \log 6 = x \log 2$$ Remove parentheses.

$$x \log 6 - x \log 2 = 3 \log 6$$ Add 3 log 6 and $-x$ log 2 to both sides.

$$x(\log 6 - \log 2) = 3 \log 6$$ Factor out x from the left side.

$$x = \frac{3 \log 6}{\log 6 - \log 2}$$ Divide both sides by log 6 − log 2.

$$x \approx 4.893$$ Substitute values for log 6 and log 2 and simplify.   ■

Example 3   Solve the logarithmic equation $\log x + \log(x - 3) = 1$.

Solution

$$\log x + \log(x - 3) = 1$$

$$\log x(x - 3) = 1$$ Use Property 5 of logarithms.

$$x(x - 3) = 10^1$$ Use the definition of logarithm to change the equation to exponential form.

$$x^2 - 3x - 10 = 0$$ Remove parentheses and add -10 to both sides.

$$(x + 2)(x - 5) = 0$$ Factor $x^2 - 3x - 10$.

$$x + 2 = 0 \quad \text{or} \quad x - 5 = 0$$

$$x = -2 \quad | \quad x = 5$$

Check:

The number -2 is not a solution because it does not satisfy the equation (a negative number does not have a logarithm). Check the remaining number 5:

$$\log x + \log(x - 3) = 1$$

$$\log 5 + \log(5 - 3) \overset{?}{=} 1$$ Substitute 5 for x.

$$\log 5 + \log 2 \overset{?}{=} 1$$

$$\log(5 \cdot 2) \overset{?}{=} 1$$ Use Property 5 of logarithms.

$$\log 10 \overset{?}{=} 1$$

$$1 = 1$$ Use Property 2 of logarithms.

The solution 5 does check.   ■

Example 4   Solve the logarithmic equation $\log_b(3x + 2) - \log_b(2x - 3) = 0$.

Solution

$$\log_b(3x + 2) - \log_b(2x - 3) = 0$$

$$\log_b(3x + 2) = \log_b(2x - 3) \qquad \text{Add } \log_b(2x - 3) \text{ to both sides.}$$

$$3x + 2 = 2x - 3 \qquad \text{If } \log_b r = \log_b s, \text{ then } r = s.$$

$$x = -5 \qquad \text{Add } -2x - 2 \text{ to both sides.}$$

Check:

$$\log_b(3x + 2) - \log_b(2x - 3) = 0$$

$$\log_b[3(-5) + 2] - \log_b[2(-5) - 3] \overset{?}{=} 0$$

$$\log_b(-13) - \log_b(-13) \overset{?}{=} 0$$

Because the logarithm of a negative number does not exist, the number -5 is not a solution. The given equation has no solutions.   ∎

If we know the base-a logarithm of a number, we can find the logarithm of that number to some other base b by using a formula called the **change-of-base formula**.

The Change-of-Base Formula.

$$\log_b x = \frac{\log_a x}{\log_a b}$$

Proof   We begin with the equation $\log_b x = y$ and proceed as follows:

1. $$\log_b x = y$$

$$b^y = x \qquad \text{Change the equation from logarithmic to exponential form.}$$

$$\log_a b^y = \log_a x \qquad \text{Take the base-}a \text{ logarithm of both sides.}$$

$$y \log_a b = \log_a x \qquad \text{Use Property 7 of logarithms.}$$

$$y = \frac{\log_a x}{\log_a b} \qquad \text{Divide both sides by } \log_a b.$$

$$\log_b x = \frac{\log_a x}{\log_a b} \qquad \text{Refer to Equation 1 and substitute } \log_b x \text{ for } y.$$ □

If we know logarithms to base a (for example, $a = 10$), we can find the logarithm of x to a new base b. To do so, we divide the base-a logarithm of x by the base-a logarithm of b.

Example 5   Use the change-of-base formula to find $\log_3 5$.

Solution   Use the change-of-base formula with $b = 3$, $a = 10$, and $x = 5$:

$$\log_b x = \frac{\log_a x}{\log_a b}$$

$$\log_3 5 = \frac{\log_{10} 5}{\log_{10} 3} \qquad \text{Substitute 3 for } b, \text{ 10 for } a, \text{ and 5 for } x.$$

$$\approx \frac{0.6990}{0.4771} \qquad \text{Substitute values for log 5 and log 3.}$$

$$\approx 1.465 \qquad\qquad\qquad\qquad\qquad\qquad\qquad ■$$

Applications of Exponential and Logarithmic Equations

Example 6   When a living organism dies, the oxygen/carbon dioxide cycle common to all living things ceases and carbon-14, a radioactive isotope with a half-life of 5700 years, is no longer absorbed. By measuring the amount of carbon-14 present in an ancient object, archeologists can estimate the object's age and answer questions such as the following: How old is a wooden statue that contains only $\frac{1}{3}$ of its original carbon-14 content?

Solution   The amount A of radioactive material present at time t is given by the model

$$A = A_0 2^{-t/h}$$

where A_0 is the amount present initially and h is the material's half-life.

To determine the time t when A is $\frac{1}{3}$ of A_0, substitute $A_0/3$ for A and 5700 for h and solve for t:

$$A = A_0 2^{-t/h}$$

$$\frac{A_0}{3} = A_0 2^{-t/5700} \qquad \text{Substitute } \frac{A_0}{3} \text{ for } A \text{ and 5700 for } h.$$

$$1 = 3 \cdot 2^{-t/5700} \qquad \text{Multiply both sides by } \frac{3}{A_0}.$$

$$\log 1 = \log(3 \cdot 2^{-t/5700}) \qquad \text{Take the common logarithm of both sides.}$$

$$0 = \log 3 + \log 2^{-t/5700} \qquad \text{Use Properties 1 and 5 of logarithms.}$$

$$-\log 3 = -\frac{t}{5700} \log 2 \qquad \begin{array}{l}\text{Add } -\log 3 \text{ to both sides and use} \\ \text{Property 7 of logarithms.}\end{array}$$

$$t = 5700\left(\frac{\log 3}{\log 2}\right) \qquad \text{Multiply both sides by } -\frac{5700}{\log 2}.$$

$$\approx 9034.29$$

The wooden statue is approximately 9000 years old.   ■

Example 7   When there is sufficient food supply and space, populations of living organisms tend to increase exponentially according to the Malthusian population growth model

$$P = P_0 e^{kt}$$

where P_0 is the initial population (at $t = 0$), and k depends on the rate of growth.

The bacteria population in a laboratory culture increased from an initial population of 500 to 1500 in 3 hours. Determine the time it will take the population to reach 10,000.

Solution

$$P = P_0 e^{kt}$$

$1500 = 500(e^{k \cdot 3})$ Substitute 1500 for P, 500 for P_0, and 3 for t.

$3 = e^{3k}$ Divide both sides by 500.

$3k = \ln 3$ Change the equation from exponential to logarithmic form.

$k = \dfrac{\ln 3}{3}$ Divide both sides by 3.

To find when the population will reach 10,000, substitute 10,000 for P, 500 for P_0, and $\dfrac{\ln 3}{3}$ for k in the equation $P = P_0 e^{kt}$ and solve for t:

$$P = P_0 e^{kt}$$

$10,000 = 500 e^{[(\ln 3)/3]t}$

$20 = e^{[(\ln 3)/3]t}$ Divide both sides by 500.

$\left(\dfrac{\ln 3}{3}\right)t = \ln 20$ Change the equation from exponential to logarithmic form.

$t = \dfrac{3 \ln 20}{\ln 3}$ Multiply both sides by $\dfrac{3}{\ln 3}$.

≈ 8.18

The culture will reach the 10,000 mark in approximately 8 hours.   ∎

EXERCISE 8.5

In Exercises 1–20, solve each exponential equation.

1. $4^x = 5$ **2.** $7^x = 12$ **3.** $13^{x-1} = 2$ **4.** $5^{x+1} = 3$

5. $2^{x+1} = 3^x$ **6.** $5^{x-3} = 3^{2x}$ **7.** $2^x = 3^x$ **8.** $3^{2x} = 4^x$

9. $7^{x^2} = 10$ **10.** $8^{x^2} = 11$ **11.** $8^{x^2} = 9^x$ **12.** $5^{x^2} = 2^{5x}$

13. $4^{x+2} - 4^x = 15$ (*Hint:* $4^{x+2} = 4^x 4^2$)   **14.** $3^{x+3} + 3^x = 84$

15. $2(3^x) = 6^{2x}$ **16.** $2(3^{x+1}) = 3(2^{x-1})$

17. $2^{2x} - 10(2^x) + 16 = 0$ (*Hint:* Let $y = 2^x$)   **18.** $3^{2x} - 10(3^x) + 9 = 0$

19. $2^{2x+1} - 2^x = 1$ (*Hint:* $2^{a+b} = 2^a \cdot 2^b$)   **20.** $3^{2x+1} - 10(3^x) + 3 = 0$

In Exercises 21–42, solve each logarithmic equation.

21. $\log(2x - 3) = \log(x + 4)$ **22.** $\log(3x + 5) - \log(2x + 6) = 0$

23. $\log \dfrac{4x + 1}{2x + 9} = 0$ **24.** $\log \dfrac{5x + 2}{2(x + 7)} = 0$

25. $\log x^2 = 2$ **26.** $\log x^3 = 3$

27. $\log x + \log(x - 48) = 2$

28. $\log x + \log(x + 9) = 1$

29. $\log x + \log(x - 15) = 2$

30. $\log x + \log(x + 21) = 2$

31. $\log(x + 90) = 3 - \log x$

32. $\log(x - 6) - \log(x - 2) = \log \dfrac{5}{x}$

33. $\log(x - 1) - \log 6 = \log(x - 2) - \log x$

34. $\log(2x - 3) - \log(x - 1) = 0$

35. $\log x^2 = (\log x)^2$

36. $\log(\log x) = 1$

37. $\log_3 x = \log_3 \left(\dfrac{1}{x}\right) + 4$

38. $\log_5(7 + x) + \log_5(8 - x) - \log_5 2 = 2$

39. $2 \log_2 x = 3 + \log_2(x - 2)$

40. $2 \log_3 x - \log_3(x - 4) = 2 + \log_3 2$

41. $\log(7y + 1) = 2 \log(y + 3) - \log 2$

42. $2 \log(y + 2) = \log(y + 2) - \log 12$

In Exercises 43–46, find the logarithm with the indicated base.

43. $\log_3 7$

44. $\log_7 3$

45. $\log_{\sqrt{2}} \sqrt{5}$

46. $\log_\pi e$

47. The half-life of tritium is 12.4 years. How long will it take for 25% of a sample of tritium to decompose?

48. In 2 years, 20% of a newly discovered radioactive element decays. What is its half-life?

49. An isotope of thorium, ^{227}Th, has a half-life of 18.4 days. How long will it take 80% of a sample to decompose?

50. An isotope of lead, ^{201}Pb, has a half-life of 8.4 hours. How many hours ago was there 30% more of the substance?

51. A parchment fragment is found in a newly discovered ancient tomb. It contains 60% of the carbon-14 that it is assumed to have had initially. Approximately how old is the fragment?

52. Only 10% of the carbon-14 in a small wooden bowl remains. How old is the bowl?

53. If $500 is deposited into an account paying 12% interest, compounded semiannually, how long will it take for the account to increase to $800? How long will it take if the interest is compounded continuously?

54. If $1300 is deposited into an account paying 14% interest, compounded quarterly, how long will it take to increase the amount to $2100?

55. A sum of $5000 deposited in an account grows to $7000 in 5 years. Assuming annual compounding, what interest rate is paid?

56. A quick rule of thumb for determining how long it takes an investment to double is known as the *rule of seventy:* Divide 70 by the rate (as a percent). At 5%, for example, it requires $\frac{70}{5} = 14$ years to double the capital. At 7%, it takes $\frac{70}{7} = 10$ years. Why does this formula work?

57. A bacteria culture grows according to the formula

$$P = P_0 \, a^t$$

If it takes 5 days for the culture to triple in size, how long does it take to double in size?

58. The intensity I of light a distance x meters beneath the surface of a lake decreases exponentially. If the light intensity at 6 meters is 70% of the intensity at the surface, at what depth will the intensity be 20%?

8.6 LOGARITHMIC CALCULATIONS (OPTIONAL)

Before calculators, logarithms provided the only reasonable way of performing certain calculations. In this section, we will show how to use logarithms as a computational aid. We begin by discussing the use of Table B in Appendix III.

Example 1   Use Table B to find the base-10 logarithm of 2.71.

Solution   To find log 2.71, run your finger down the left column of Table B until you reach 2.7. (A portion of Table B is reproduced below.) Then slide your finger to the right until you reach entry .4330, which is in the column headed with 1. The number 0.4330 is the logarithm of 2.71. To verify that this is true, use a calculator to show that $10^{0.4330} \approx 2.71$.

N	0	1	2	3	4	5	6	7	8	9
2.6	.4150	.4166	.4183	.4200	.4216	.4232	.4249	.4265	.4281	.4298
2.7	.4314	**.4330**	.4346	.4362	.4378	.4393	.4409	.4425	.4440	.4456
2.8	.4472	.4487	.4502	.4518	.4533	.4548	.4564	.4579	.4594	.4609

■

Example 2   Use Table B to find the base-10 logarithm of 2,710,000.

Solution   The logarithm of 2,710,000 cannot be found directly from the table. However, by using properties of logarithms, you can determine the logarithm of 2,710,000 as follows:

$$2,710,000 = 2.71 \times 10^6 \qquad \text{Write 2,710,000 in scientific notation.}$$

$$\begin{aligned}
\log 2,710,000 &= \log(2.71 \times 10^6) &&\text{Take the logarithm of both sides.} \\
&= \log 2.71 + \log 10^6 &&\text{Use Property 5 of logarithms.} \\
&= \log 2.71 + 6 \log 10 &&\text{Use Property 7 of logarithms.} \\
&= \log 2.71 + 6 &&\text{Use Property 2 of logarithms.} \\
&\approx 0.4330 + 6 &&\text{Substitute 0.4330 for log 2.71.} \\
&\approx 6.4330 &&\text{Simplify.}
\end{aligned}$$

Hence, log 2,710,000 $\approx$ 6.4330. To verify that this is true, use a calculator to show that $10^{6.4330} \approx 2,710,000$.   ■

If a common logarithm is written as the sum of an integer and a positive decimal between 0 and 1, the positive decimal is called the **mantissa**, and the integer is called the **characteristic**. The characteristic of the value 0.4330 that was obtained in Example 1 is 0, and the mantissa is 0.4330. The characteristic of the value of 6.4330 that was obtained in Example 2 is 6, and the mantissa is 0.4330.

Example 3   Find log 0.000271.

Solution   Proceed as in Example 2.

$$0.000271 = 2.71 \times 10^{-4} \qquad \text{Write 0.000271 in scientific notation.}$$

$$\log 0.000271 = \log(2.71 \times 10^{-4}) \qquad \text{Take the logarithm of both sides.}$$
$$= \log 2.71 + \log 10^{-4} \qquad \text{Use Property 5 of logarithms.}$$
$$= \log 2.71 - 4 \log 10 \qquad \text{Use Property 7 of logarithms.}$$
$$= \log 2.71 - 4 \qquad \text{Use Property 2 of logarithms.}$$
$$\approx 0.4330 - 4 \qquad \text{Substitute 0.4330 for log 2.71.}$$

Hence, $\log 0.000271 \approx 0.4330 - 4$. The mantissa of $\log 0.000271$ is the positive decimal .4330, and the characteristic is -4. If the characteristic and the mantissa are combined, we have $\log 0.000271 \approx -3.5670$. To verify that this is true, use a calculator to show that $10^{-3.5670} \approx 0.000271$. ∎

From Examples 1, 2, and 3, it is apparent that the mantissa 0.4330 is determined by the digits 271, and the characteristic is determined by the location of the decimal point. The characteristic of the logarithm is the exponent of 10 used when expressing that number in scientific notation.

Example 4   If $\log N = -2.1180$, find N.

Solution   The negative number -2.1180 cannot be found in Table B. However, if you add and subtract 3 from -2.1180, you can establish a positive mantissa:

$$\log N = -2.1180$$
$$= (-2.1180 + 3) - 3 \qquad \text{Add and subtract 3.}$$
$$= 0.8820 - 3$$

The mantissa of this logarithm is 0.8820. Determine from Table B that $\log 7.62 \approx 0.8820$ or, equivalently, that $10^{0.8820} \approx 7.62$. To place the decimal point in the answer, write the equation $\log N \approx 0.8820 - 3$ in exponential form and proceed as follows:

$$\log N \approx 0.8820 - 3$$
$$N \approx 10^{0.8820 - 3} \qquad \begin{array}{l}\text{Change the equation from logarithmic} \\ \text{to exponential form.}\end{array}$$
$$\approx 10^{0.8820} \cdot 10^{-3} \qquad \text{Use the rule } x^{m+n} = x^m x^n.$$
$$\approx 7.62 \cdot 10^{-3} \qquad \text{Substitute 7.62 for } 10^{0.8820}.$$
$$\approx 0.00762 \qquad \text{Write the number in standard form.}$$

Thus, $N \approx 0.00762$. The number 0.00762 is called the **antilogarithm** of -2.1180. ∎

Linear Interpolation

Table B in Appendix III gives values of N accurate to three significant digits and values of the mantissas accurate to four significant digits. There is a process, called **linear interpolation**, that allows us to extend these estimates by one additional digit.

The method of linear interpolation is based on the fact that a small part of the graph of the logarithmic function appears to be a straight line. For example, the logarithmic curve in Figure 8-13 appears to be straight between points A and B. If AB is assumed to be a straight line, we can set up proportions involving points on the line AB. The following examples illustrate how.

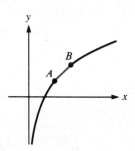

Figure 8-13

Example 5   Use Table B to find log 3.974.

Solution   Table B does not give the value of log 3.974, but it does give the values for log 3.970 and log 3.980:

$$
\begin{array}{ll}
\text{10} & \text{4} \\
\text{thousandths} & \text{thousandths}
\end{array}
\left[
\begin{array}{l}
\text{log } 3.970 = 0.5988 \\
\text{log } 3.974 = ? \\
\text{log } 3.980 = 0.5999
\end{array}
\right]
\begin{array}{ll}
x & 11 \\
\text{ten-} & \text{ten-} \\
\text{thousandths} & \text{thousandths}
\end{array}
$$

Note that the difference between 3.970 and 3.980 is 10 thousandths, and that the difference between 3.970 and 3.974 is 4 thousandths. Also note that the difference between 0.5988 and 0.5999 is 11 ten-thousandths. If you assume that the graph of $y = \log x$ is a straight line between $x = 3.97$ and $x = 3.98$, then you can set up and solve the following proportion:

$$\frac{4 \text{ thousandths}}{10 \text{ thousandths}} = \frac{x \text{ ten-thousandths}}{11 \text{ ten-thousandths}}$$

$$\frac{4}{10} = \frac{x}{11}$$

$$10x = 44$$

$$x \approx 4$$

To get a good estimate of log 3.974, you must add 4 ten-thousandths to 0.5988. Hence,

$$\log 3.974 \approx 0.5992$$ ■

Example 6   Use linear interpolation to find N, where $\log N = 0.1514$.

Solution   You cannot find 0.1514 in the body of Table B, but you can find two consecutive values, 0.1492 and 0.1523, that straddle 0.1514. Hence, you can form the following chart:

10 thousandths	*x* thousandths	log 1.410 = 0.1492	22 ten- thousandths	31 ten- thousandths
		log *N* = 0.1514		
		log 1.420 = 0.1523		

Set up the following proportion, and solve for *x*:

$$\frac{x}{10} = \frac{22}{31}$$

$$x = \frac{10 \cdot 22}{31}$$

$$x \approx 7$$

A good approximation for N is found by adding 7 thousandths to 1.410. Thus, $N \approx 1.417$.   ∎

Example 7   Use logarithms to calculate $\dfrac{\sqrt{31.7}}{3.974}$.

Solution   First determine log 31.7 and log 3.974.

$\log 31.7 = \log(3.17 \times 10^1)$	Write 31.7 in scientific notation.
$= \log 3.17 + \log 10^1$	Use Property 5 of logarithms.
$\approx 0.5011 + 1 \log 10$	Find log 3.17 and use Property 7 of logarithms.
≈ 1.5011	Use Property 2 of logarithms and simplify.

Finding log 3.974 requires interpolation. Refer to Example 5 to see that

$$\log 3.974 \approx 0.5992$$

Now form the equation

$$N = \frac{\sqrt{31.7}}{3.974}$$

and take the common logarithm of both sides:

$\log N = \log \dfrac{\sqrt{31.7}}{3.974}$	
$= \log \sqrt{31.7} - \log 3.974$	Use Property 6 of logarithms.
$= \log(31.7)^{1/2} - \log 3.974$	Write $\sqrt{31.7}$ as $(31.7)^{1/2}$.
$= \dfrac{1}{2}\log 31.7 - \log 3.974$	Use Property 7 of logarithms.
$\approx \dfrac{1}{2}(1.5011) - 0.5992$	Substitute the values of log 31.7 and log 3.974.
≈ 0.1514	

Finding the value of N requires linear interpolation. Refer to Example 6 to see that $N \approx 1.417$. Thus,

$$\frac{\sqrt{31.7}}{3.974} \approx 1.417$$

■

Before calculators, mathematicians also relied on tables to find base-e logarithms. The next example uses a table of base-e logarithmic values.

Example 8   Use Table C in Appendix III to find the value of ln 2.34.

Solution   Look up 2.3 in the left column of Table C. (A portion of Table C is reproduced below.) Follow that row to the column headed by 4. From that position in the table, you can read that ln 2.34 ≈ 0.8502.

N	0	1	2	3	**4**	5	6	7	8	9
2.2	.7885	.7930	.7975	.8020	.8065	.8109	.8154	.8198	.8242	.8286
2.3	.8329	.8372	.8416	.8459	**.8502**	.8544	.8587	.8629	.8671	.8713
2.4	.8755	.8796	.8838	.8879	.8920	.8961	.9002	.9042	.9083	.9123

■

EXERCISE 8.6

In Exercises 1–8, use Table B in Appendix III to find each logarithm.

1. log 5.97   **2.** log 3.15   **3.** log 4.23   **4.** log 9.83

5. log 432,000   **6.** log 57,900,000   **7.** log 0.00137   **8.** log 0.0823

In Exercises 9–14, use Table B in Appendix III to find the value of N.

9. log N = 0.4969   **10.** log N = 0.8785   **11.** log N = 3.9232   **12.** log N = 4.6149

13. log N = −2.5467   **14.** log N = −4.4377

In Exercises 15–18, use linear interpolation and Table B in Appendix III to find each logarithm.

15. log 6.894   **16.** log 37.43   **17.** log 0.003456   **18.** log 0.04376

In Exercises 19–24, use linear interpolation and Table B in Appendix III to find N to four digits.

19. log N = 0.6315   **20.** log N = 0.0437

21. log N = 3.2036   **22.** log N = 0.8508 − 4

23. log N = −2.1134   **24.** log N = −0.4467

In Exercises 25–30, use Table B in Appendix III to calculate the approximate value of each indicated number. Do not use linear interpolation.

25. $\sqrt[3]{0.007}$ **26.** $(0.012)^{-0.03}$ **27.** $(1.05)^{25}$ **28.** $(\log 4.1)^{2.4}$

29. $4.3^{-5.2} + 3.1^{1.3}$ **30.** $(2.3 + 1.79)^{-0.157}$

In Exercises 31–34, use Table B in Appendix III to calculate each indicated value. Use linear interpolation.

31. $(34.41)(0.4455)$ **32.** $(0.0004519)^{2.5}$ **33.** $\sqrt[15]{38,670}$ **34.** $\dfrac{(8.034)(32.6)}{\sqrt{3.869}}$

In Exercises 35–38, use Table C in Appendix III to find x.

35. $\ln 4.65 = x$ **36.** $\ln 2.93 = x$ **37.** $\ln x = 2.0451$ **38.** $\ln x = 1.1969$

In Exercises 39–42, use Table C in Appendix III and the fact that $\ln 10 \approx 2.3026$ to calculate each value.

39. $\ln 29.4$ **40.** $\ln 751$ **41.** $\ln 0.00823$ **42.** $\ln 0.436$

CHAPTER SUMMARY

Key Words

change-of-base formula   (8.5)
characteristic   (8.6)
common logarithm   (8.4)
e   (8.2)
exponential decay   (8.1)
exponential function   (8.1)
exponential growth   (8.1)

exponential equation   (8.5)
linear interpolation   (8.6)
logarithmic equation   (8.5)
logarithmic function   (8.3)
mantissa   (8.6)
Napierian logarithm   (8.4)
natural logarithm   (8.4)

Key Ideas

(8.1)   The exponential function $y = b^x$, where $b > 0$, $b \neq 1$, and x is any real number, is one-to-one. Its domain is the set of real numbers, and its range is the set of positive real numbers.

(8.2) $e = 2.7182818 \ldots$

The exponential function $y = e^x$ is one-to-one. Its domain is the set of real numbers, and its range is the set of positive real numbers.

(8.3)   The logarithmic function $y = \log_b x$, where $b > 0$, $b \neq 1$, and x is a positive real number, is one-to-one. Its domain is the set of positive real numbers, and its range is the set of all real numbers.

The equation $y = \log_b x$ is equivalent to the equation $x = b^y$.

Logarithms of negative numbers do not exist.

The functions defined by $y = \log_b x$ and $y = b^x$ are inverse functions.

If $\log_b r = \log_b s$, then $r = s$.

Properties of logarithms:

1. $\log_b 1 = 0$ **2.** $\log_b b = 1$

3. $\log_b b^x = x$ **4.** $b^{\log_b x} = x$

5. $\log_b MN = \log_b M + \log_b N$ **6.** $\log_b \dfrac{M}{N} = \log_b M - \log_b N$

7. $\log_b M^p = p \log_b M$

(8.4)   Common logarithms are base-10 logarithms.

Natural logarithms are base-e logarithms.

(8.5)   **Change-of-base formula:** $\log_b y = \dfrac{\log_a y}{\log_a b}$

(8.6)   Logarithms can be used to simplify certain arithmetic calculations.

REVIEW EXERCISES

In Review Exercises 1–4, graph the function defined by each equation.

1. $y = \left(\dfrac{6}{5}\right)^x$ **2.** $y = e^x$ **3.** $y = \log x$ **4.** $y = \ln x$

In Review Exercises 5–8, graph each pair of equations on one set of coordinate axes.

5. $y = \left(\dfrac{1}{3}\right)^x$ and $y = \log_{1/3} x$ **6.** $y = \left(\dfrac{2}{5}\right)^x$ and $y = \log_{2/5} x$

7. $y = 4^x$ and $y = \log_4 x$ **8.** $y = 3^x$ and $y = \log_3 x$

In Review Exercises 9–30, solve each equation for x.

9. $\log_2 x = 3$ **10.** $\log_3 x = -2$ **11.** $\log_x 9 = 2$ **12.** $\log_x 0.125 = -3$

13. $\log_7 7 = x$ **14.** $\log_3 \sqrt{3} = x$ **15.** $\log_8 \sqrt{2} = x$ **16.** $\log_6 36 = x$

17. $\log_{1/3} 9 = x$ **18.** $\log_{1/2} 1 = x$ **19.** $\log_x 3 = \dfrac{1}{3}$ **20.** $\log_x 25 = -2$

21. $\log_2 x = 5$ **22.** $\log_{\sqrt{3}} x = 4$ **23.** $\log_{\sqrt{3}} x = 6$ **24.** $\log_{0.1} 10 = x$

25. $\log_x 2 = -\dfrac{1}{3}$ **26.** $\log_x 32 = 5$ **27.** $\log_{0.25} x = -1$ **28.** $\log_{0.125} x = -\dfrac{1}{3}$

29. $\log_{\sqrt{2}} 32 = x$ **30.** $\log_{\sqrt{5}} x = -4$

31. Write the following expression in terms of the logarithms of x, y, and z: $\log_b \sqrt{\dfrac{x}{yz^2}}$

32. Write the following expression as the logarithm of a single quantity:

$$\dfrac{1}{2}\log_b x + 3\log_b y - 7\log_b z$$

In Review Exercises 33–34, assume that $\log a = 0.6$, $\log b = 0.36$, *and* $\log c = 2.4$. *Find the value of each expression.*

33. $\log \dfrac{ac}{b}$ **34.** $\log a^2 b^{1/2}$

In Review Exercises 35–44, solve for x, if possible.

35. $3^x = 7$ **36.** $1.2 = (3.4)^{5.6x}$

37. $2^x = 3^{x-1}$ **38.** $\log x + \log(29 - x) = 2$

39. $\log_2 x + \log_2(x - 2) = 3$ **40.** $\log_2(x + 2) + \log_2(x - 1) = 2$

41. $e^{x \ln 2} = 9$ **42.** $\ln x = \ln(x - 1)$

43. $\ln x = \ln(x - 1) + 1$

44. $\ln x = \log_{10} x$ (*Hint:* Use the change-of-base formula.)

45. A wooden statue excavated from the sands of Egypt has a carbon-14 content that is $\frac{2}{3}$ of that found in living wood. The half-life of carbon-14 is 5700 years. How old is the statue?

46. The pH of grapefruit juice is approximately 3.1. What is its hydrogen-ion concentration?

47. Some chemistry texts define the pH of a solution as the common logarithm of the reciprocal of the hydrogen-ion concentration:

$$pH = \log_{10} \frac{1}{[H^+]}$$

Show that this definition is equivalent to the one given in the text.

48. What is the half-life of a radioactive material if $\frac{1}{3}$ of it decays in 20 years?

In Review Exercises 49–52, use Table B in Appendix III and perform the indicated calculations. Use linear interpolation if necessary.

49. $(34.5)(0.236)$ **50.** $\sqrt[5]{456{,}000}$ **51.** $\dfrac{(0.00235)^3}{(0.00896)^2}$ **52.** $\dfrac{(3.476)(0.003456)}{3.45}$

9 Natural Number Functions and Probability

Karl Friedrich Gauss
(1777–1855) Many people
consider Gauss to be the
greatest mathematician of all
time. He often is called the
"prince of the
mathematicians."

The brilliant German mathematician Karl Friedrich Gauss (1777–1855) was once a student in the class of a very strict teacher. To keep the class busy one day, the teacher asked the students to add together all of the natural numbers from 1 through 100. Gauss immediately wrote the sum on his slate and put it on the teacher's desk.

Gauss's solution was relatively simple. He recognized that in the sum $1 + 2 + 3 + \cdots + 98 + 99 + 100$, the first number (1) added to the last number (100) was 101. Similarly, the second number (2) added to the second from the last number (99) was 101. The third number (3) added to the third from the last number (98) was 101 also. Gauss realized that this pattern continued, and because there were fifty pairs of numbers there were fifty sums of 101. He multiplied 101 by 50 and obtained the correct answer of 5050.

This story illustrates a group of problems involving long strings of numbers called *sequences*. We begin a discussion of sequences by considering a method of proof that is used to prove sequence formulas. This method of proof, called **mathematical induction**, was first used extensively by Giuseppe Peano (1858–1932).

9.1 MATHEMATICAL INDUCTION

Suppose we ask a theatergoer whether everyone in line for a movie gained admittance. The theatergoer answers that "the first person in line was admitted." Does this response answer our question? Certainly more than one person was admitted, but this does not mean that everyone was admitted. As a matter of fact, we know little more now than before we asked the question.

Meeting a second theatergoer, we ask the question again. This time the answer is "they promised that, if anyone was admitted, the person next in line would also be admitted." On the basis of this response we know that, if anyone was admitted, the person behind was admitted also. But a promise that begins with "if someone is admitted" does not guarantee that anyone actually was admitted.

However, when we consider both the first and second answers, we know that everyone in line gained admittance. The first theatergoer said that the first person got in; the second theatergoer said that, if anyone got in, the next person in line got in also. Because the first person was admitted, the second person was admitted also. And, if the second person was admitted, then so was the third. This pattern would have continued until everyone was admitted to the theater.

This situation is very similar to a children's game played with dominoes. The dominoes are placed on end, fairly close together, in a row. When the first domino is pushed over, it falls against the second, knocking it down. The second domino, in turn, knocks down the third, which topples the fourth, and so on until all of the dominoes fall. Two things must happen to guarantee that all of the dominoes fall: (1) The first domino must be knocked over, and (2) every domino that falls must topple the next one. When both of these conditions are met, it is certain that all of the dominoes will fall.

The preceding examples illustrate the basic idea that underlies the principle of mathematical induction.

The Axiom of Mathematical Induction. If a statement involving the natural number n has the two properties that

1. the statement is true for $n = 1$, and
2. the statement is true for $n = k + 1$ whenever it is true for $n = k$,

then the statement is true for all natural numbers.

The axiom of mathematical induction provides a method for proving many theorems. Note that any such proof by induction involves two parts: We must first show that the formula is true for the natural number 1, and then show that, *if* the formula is true for any natural number k, then it also is true for the natural number $k + 1$. A proof by induction is complete only if both of the required properties are established.

Let us return to Gauss's problem of finding the sum of the first 100 natural numbers. There is a formula for finding the sum of the first n natural numbers:

$$1 + 2 + 3 + \cdots + n = \frac{n(n + 1)}{2}$$

We will prove this formula by using the axiom of mathematical induction.

Example 1 Use mathematical induction to prove that the formula

$$1 + 2 + 3 + \cdots + n = \frac{n(n + 1)}{2}$$

is true for every natural number n.

Solution The proof has two parts.

Part 1. Verify that the formula is true for the value $n = 1$. Substituting $n = 1$ into the term n on the left side of the equation yields a single term, the number 1.

Substituting the number 1 for n on the right side, the formula becomes

$$1 = \frac{n(n + 1)}{2}$$

$$1 = \frac{(1)(1 + 1)}{2}$$

$$1 = 1$$

Thus, the formula is true for $n = 1$, and Part 1 of the proof is complete.

Part 2.   Assume that the given formula is true when n is replaced by *some* natural number k. By this assumption, called the **induction hypothesis**, you accept that

1. $1 + 2 + 3 + \cdots + k = \dfrac{k(k + 1)}{2}$

is a true statement. The plan is to show that the given formula is true for the next natural number, $k + 1$. Do this by verifying the equation

2. $1 + 2 + 3 + \cdots + k + (k + 1) = \dfrac{(k + 1)[(k + 1) + 1]}{2}$

obtained from the given formula by replacing n with $k + 1$.

Compare the left sides of Equations 1 and 2 and note that the left side of Equation 2 contains an extra term of $k + 1$. Hence, add $k + 1$ to both sides of Equation 1 (which was assumed to be true) to obtain the equation

$$1 + 2 + 3 + \cdots + k + (k + 1) = \frac{k(k + 1)}{2} + (k + 1)$$

Because both terms on the right side of this equation have a common factor of $k + 1$, the right side factors and the equation can be rewritten as follows:

$$1 + 2 + 3 + \cdots + k + (k + 1) = (k + 1)\left(\frac{k}{2} + 1\right)$$

$$= (k + 1)\left(\frac{k + 2}{2}\right)$$

$$= \frac{(k + 1)(k + 2)}{2}$$

$$= \frac{(k + 1)[(k + 1) + 1]}{2}$$

This final result is Equation 2. Because the truth of Equation 1 implies the truth of Equation 2, Part 2 of the proof is complete. Parts 1 and 2 together establish that the formula

$$1 + 2 + 3 + \cdots + n = \frac{n(n + 1)}{2}$$

is true for any natural number n.   ■

Here is a brief overview of Example 1.

1. Did the first domino fall? That is, is the formula

$$1 + 2 + 3 + \cdots + n = \frac{n(n + 1)}{2}$$

true for $n = 1$? Yes, Part 1 verified this.

2. Will toppling any domino knock over the next domino? If the given formula is true for the value $n = k$, is it also true for the value $n = k + 1$? Yes, Part 2 of the proof verified this.

Because both of the induction requirements were verified, the formula is true for all natural numbers n.

Example 2   Use mathematical induction to prove the formula

$$1 + 5 + 9 + \cdots + (4n - 3) = n(2n - 1)$$

for all natural numbers n.

Solution   The proof has two parts.

Part 1.   First verify the formula for the value $n = 1$. Substituting the value $n = 1$ into the term $4n - 3$ on the left side of the formula gives the single term 1. After substituting the same value into the right side, the equation becomes

$$1 = 1[2(1) - 1]$$
$$1 = 1$$

Thus, the formula is true for $n = 1$, and Part 1 of the proof is complete.

Part 2.   The induction hypothesis is the assumption that the formula is true for $n = k$. Hence, you assume that

$$1 + 5 + 9 + \cdots + (4k - 3) = k(2k - 1)$$

is a true statement. Because the truth of this assumption must guarantee the truth of the formula for $k + 1$ terms, add the $(k + 1)$th term to both sides of the induction hypothesis formula. In this example, the terms on the left side increase by 4, so the $(k + 1)$th term is $(4k - 3) + 4$, or $4k + 1$. Adding $4k + 1$ to both sides of the induction hypothesis formula gives

$$1 + 5 + 9 + \cdots + (4k - 3) + (4k + 1) = k(2k - 1) + (4k + 1)$$

Simplify the right side and rewrite the equation as follows:

$$1 + 5 + 9 + \cdots + (4k - 3) + [4(k + 1) - 3] = 2k^2 + 3k + 1$$
$$= (k + 1)(2k + 1)$$
$$= (k + 1)[2(k + 1) - 1]$$

Because the above equation has the same form as the given formula, except that $k + 1$ appears in place of n, the truth of the formula for $n = k$ implies the truth of the formula for $n = k + 1$. Part 2 of the proof is complete.

Because both of the induction requirements have been verified, the given formula is proved for all natural numbers.   ∎

Example 3   Prove that $\dfrac{1}{2} + \dfrac{1}{4} + \dfrac{1}{8} + \cdots + \dfrac{1}{2^n} < 1$.

Solution   The proof is by induction.

Part 1.   Verify the formula for $n = 1$. Substituting 1 for n on the left side of the inequality gives $\frac{1}{2} < 1$. Thus, the formula is true for $n = 1$, and Part 1 of the proof is complete.

Part 2.   The induction hypothesis is the assumption that the inequality is true for $n = k$. Thus, assume that

$$\frac{1}{2} + \frac{1}{4} + \frac{1}{8} + \cdots + \frac{1}{2^k} < 1$$

Multiply both sides of the above inequality by $\frac{1}{2}$ to obtain the inequality

$$\frac{1}{2}\left(\frac{1}{2} + \frac{1}{4} + \frac{1}{8} + \cdots + \frac{1}{2^k}\right) < 1\left(\frac{1}{2}\right)$$

or

$$\frac{1}{4} + \frac{1}{8} + \frac{1}{16} + \cdots + \frac{1}{2^{k+1}} < \frac{1}{2}$$

Now add $\frac{1}{2}$ to both sides of this inequality to obtain

$$\frac{1}{2} + \frac{1}{4} + \frac{1}{8} + \frac{1}{16} + \cdots + \frac{1}{2^{k+1}} < \frac{1}{2} + \frac{1}{2}$$

or

$$\frac{1}{2} + \frac{1}{4} + \frac{1}{8} + \frac{1}{16} + \cdots + \frac{1}{2^{k+1}} < 1$$

The resulting inequality is the same as the original except that $k + 1$ appears in place of n. The truth of the inequality for $n = k$ implies the truth of the inequality for $n = k + 1$. Part 2 of the proof is complete.

Because both of the induction requirements have been verified, this inequality is true for all natural numbers.   ∎

There are statements that are not true when $n = 1$, but that are true for all natural numbers equal to or greater than some given natural number, say q. In these cases, verify the given statements for $n = q$ in Part 1 of the induction proof. After establishing Part 2 of the induction proof, the given statement is proved for all natural numbers that are greater than or equal to q.

EXERCISE 9.1

In Exercises 1–4, verify each given formula for $n = 1, 2, 3$, and 4.

1. $5 + 10 + 15 + \cdots + 5n = \dfrac{5n(n + 1)}{2}$

2. $1^2 + 2^2 + 3^2 + \cdots + n^2 = \dfrac{n(n + 1)(2n + 1)}{6}$

3. $7 + 10 + 13 + \cdots + (3n + 4) = \dfrac{n(3n + 11)}{2}$

4. $1(3) + 2(4) + 3(5) + \cdots + n(n + 2) = \dfrac{n}{6}(n + 1)(2n + 7)$

In Exercises 5–20, prove each of the following formulas by mathematical induction, if possible.

5. $2 + 4 + 6 + \cdots + 2n = n(n + 1)$

6. $1 + 3 + 5 + \cdots + (2n - 1) = n^2$

7. $3 + 7 + 11 + \cdots + (4n - 1) = n(2n + 1)$

8. $4 + 8 + 12 + \cdots + 4n = 2n(n + 1)$

9. $10 + 6 + 2 + \cdots + (14 - 4n) = 12n - 2n^2$

10. $8 + 6 + 4 + \cdots + (10 - 2n) = 9n - n^2$

11. $2 + 5 + 8 + \cdots + (3n - 1) = \dfrac{n(3n + 1)}{2}$

12. $3 + 6 + 9 + \cdots + 3n = \dfrac{3n(n + 1)}{2}$

13. $1^2 + 2^2 + 3^2 + \cdots + n^2 = \dfrac{n(n + 1)(2n + 1)}{6}$

14. $1 + 2 + 3 + \cdots + (n - 1) + n + (n - 1) + \cdots + 3 + 2 + 1 = n^2$

15. $\dfrac{1}{3} + 2 + \dfrac{11}{3} + \cdots + \left(\dfrac{5}{3}n - \dfrac{4}{3}\right) = n\left(\dfrac{5}{6}n - \dfrac{1}{2}\right)$

16. $\dfrac{1}{1 \cdot 2} + \dfrac{1}{2 \cdot 3} + \dfrac{1}{3 \cdot 4} + \cdots + \dfrac{1}{n(n + 1)} = \dfrac{n}{n + 1}$

17. $\dfrac{1}{2} + \dfrac{1}{4} + \dfrac{1}{8} + \cdots + \left(\dfrac{1}{2}\right)^n = 1 - \left(\dfrac{1}{2}\right)^n$

18. $\dfrac{1}{3} + \dfrac{2}{9} + \dfrac{4}{27} + \cdots + \dfrac{1}{3}\left(\dfrac{2}{3}\right)^{n-1} = 1 - \left(\dfrac{2}{3}\right)^n$

19. $2^0 + 2^1 + 2^2 + 2^3 + \cdots + 2^{n-1} = 2^n - 1$

20. $1^3 + 2^3 + 3^3 + \cdots + n^3 = \left[\dfrac{n(n + 1)}{2}\right]^2$

21. Prove that $x - y$ is a factor of $x^n - y^n$. (*Hint:* Consider subtracting and adding xy^k to the binomial $x^{k+1} - y^{k+1}$.)

22. Prove that $n < 2^n$.

23. There are $180°$ in the sum of the angles of any triangle. Prove by induction that $(n - 2)180°$ gives the sum of the angles of any simple polygon when n is the number of sides of that polygon. (*Hint:* If a polygon has $k + 1$ sides, it has $k - 2$ sides plus three more sides.)

24. Consider the equation $1 + 3 + 5 + \cdots + 2n - 1 = 3n - 2$.
   a. Is the equation true for $n = 1$?
   b. Is the equation true for $n = 2$?
   c. Is the equation true for all natural numbers n?

25. If $1 + 2 + 3 + \cdots + n = \dfrac{n}{2}(n + 1) + 1$ were true for $n = k$, show that it would be true for $n = k + 1$. Is it true for $n = 1$?

26. Prove that $n + 1 = 1 + n$ for each natural number n.

27. If n is any natural number, prove that $7^n - 1$ is divisible by 6.

28. Prove that $1 + 2n < 3^n$ for $n > 1$.

29. Prove that, if r is a real number where $r \neq 1$, then $1 + r + r^2 + \cdots + r^n = \dfrac{1 - r^{n+1}}{1 - r}$.

30. The expression a^m where m is a natural number was defined in Section 1.3. An alternative definition of a^m, useful in proofs by induction, is (Part 1) $a^1 = a$ and (Part 2) $a^{m+1} = a^m \cdot a$. Use mathematical induction on n to prove the familiar law of exponents, $a^m a^n = a^{m+n}$.

9.2   THE BINOMIAL THEOREM

In this section, we will develop a general formula for raising binomials to positive integral powers. We begin by considering the following binomial expansions:

$$(a + b)^0 = 1$$
$$(a + b)^1 = a + b$$
$$(a + b)^2 = a^2 + 2ab + b^2$$
$$(a + b)^3 = a^3 + 3a^2b + 3ab^2 + b^3$$
$$(a + b)^4 = a^4 + 4a^3b + 6a^2b^2 + 4ab^3 + b^4$$
$$(a + b)^5 = a^5 + 5a^4b + 10a^3b^2 + 10a^2b^3 + 5ab^4 + b^5$$
$$(a + b)^6 = a^6 + 6a^5b + 15a^4b^2 + 20a^3b^3 + 15a^2b^4 + 6ab^5 + b^6$$

Four patterns are apparent in the above expansions.

1. Each expansion has one more term than the power of the binomial.
2. The degree of each term in each expansion equals the exponent of the binomial.
3. The first term in each expansion is a raised to the power of the binomial.
4. The exponents of a decrease by one in each successive term, and the exponents of b, beginning with b^0 in the first term, increase by one in each successive term.

Blaise Pascal (1623–1662)
Pascal made several contributions to the field of probability.

To make another pattern apparent, we write the coefficients of each of the binomial expansions in a triangular array:

$$
\begin{array}{ll}
(a + b)^0 & \qquad\qquad\quad 1 \\
(a + b)^1 & \qquad\qquad\; 1 \quad 1 \\
(a + b)^2 & \qquad\qquad 1 \quad 2 \quad 1 \\
(a + b)^3 & \qquad\quad\; 1 \quad 3 \quad 3 \quad 1 \\
(a + b)^4 & \qquad\; 1 \quad 4 \quad 6 \quad 4 \quad 1 \\
(a + b)^5 & \quad\; \mathbf{1} \quad \mathbf{5} \quad \mathbf{10} \quad \mathbf{10} \quad 5 \quad 1 \\
(a + b)^6 & 1 \quad 6 \quad 15 \quad \mathbf{20} \quad 15 \quad 6 \quad 1
\end{array}
$$

In this triangular array, each entry other than the 1's is the sum of the closest pair of numbers in the line immediately above it. For example, the 6 in the bottom row is the sum of the 1 and the 5 above it, and the 20 is the sum of the two 10's above it.

The triangular array, named after the French mathematician Blaise Pascal (1623–1662), continues with the same pattern forever. The next two lines are

$(a + b)^7$ 1   7   21   35   35   21   7   1

$(a + b)^8$ 1   8   28   56   70   56   28   8   1

Example 1   Expand $(x + y)^6$.

Solution   The first term in the expansion is x^6, and the exponents of x decrease by one in each successive term. The y will appear in the second term, and the exponents of y will increase in each successive term, concluding when the term y^6 is reached. The variables in the expansion are

$$x^6 \quad x^5y \quad x^4y^2 \quad x^3y^3 \quad x^2y^4 \quad xy^5 \quad y^6$$

Using Pascal's triangle, you can find the coefficients of these variables. Because the binomial is raised to the 6th power, choose the row in Pascal's triangle whose second entry is 6. The coefficients of the variables are the numbers in that row:

$$1 \quad 6 \quad 15 \quad 20 \quad 15 \quad 6 \quad 1$$

Putting these two pieces of information together, the expansion is

$$(x + y)^6 = x^6 + 6x^5y + 15x^4y^2 + 20x^3y^3 + 15x^2y^4 + 6xy^5 + y^6 \qquad ■$$

Example 2   Expand $(x - y)^6$.

Solution   To expand $(x - y)^6$, rewrite the binomial in the form

$$[x + (-y)]^6$$

and refer to Example 1. The expansion is

$$[x + (-y)]^6 = x^6 + 6x^5(-y) + 15x^4(-y)^2 + 20x^3(-y)^3 + 15x^2(-y)^4$$
$$+ 6x(-y)^5 + (-y)^6$$
$$= x^6 - 6x^5y + 15x^4y^2 - 20x^3y^3 + 15x^2y^4 - 6xy^5 + y^6$$

In general, in the binomial expansion of $(x - y)^n$, the sign of the first term, x^n, is $+$, the sign of the second term is $-$, and the signs continue to alternate.   ■

Another method for expanding a binomial, called the **binomial theorem**, uses **factorial notation**.

Definition.   The symbol $n!$ (read either as "n factorial" or as "factorial n") is defined as

$$n! = n(n - 1)(n - 2)(n - 3) \cdots (3)(2)(1)$$

where n is a natural number.

Example 3   Evaluate   **a.** 3!,   **b.** 6!,   and   **c.** 10!.

Solution   **a.** $3! = 3 \cdot 2 \cdot 1 = 6$

b. $6! = 6 \cdot 5 \cdot 4 \cdot 3 \cdot 2 \cdot 1 = 720$

c. $10! = 10 \cdot 9 \cdot 8 \cdot 7 \cdot 6 \cdot 5 \cdot 4 \cdot 3 \cdot 2 \cdot 1 = 3{,}628{,}800$ ■

There are two fundamental properties of factorials.

Property 1.   By definition, $0! = 1$.
Property 2. $n \cdot (n - 1)! = n!$

Example 4   Show that   **a.** $6 \cdot 5! = 6!$ and that   **b.** $8 \cdot 7! = 8!$.

Solution   **a.** $6 \cdot 5! = 6(5 \cdot 4 \cdot 3 \cdot 2 \cdot 1) = 6 \cdot 5 \cdot 4 \cdot 3 \cdot 2 \cdot 1 = 6!$

b. $8 \cdot 7! = 8(7 \cdot 6 \cdot 5 \cdot 4 \cdot 3 \cdot 2 \cdot 1) = 8 \cdot 7 \cdot 6 \cdot 5 \cdot 4 \cdot 3 \cdot 2 \cdot 1 = 8!$ ■

We now state the binomial theorem.

The Binomial Theorem.   If n is any positive integer, then

$$(a + b)^n = a^n + \frac{n!}{1!(n - 1)!}a^{n-1}b + \frac{n!}{2!(n - 2)!}a^{n-2}b^2$$

$$+ \frac{n!}{3!(n - 3)!}a^{n-3}b^3 + \cdots + \frac{n!}{r!(n - r)!}a^{n-r}b^r + \cdots + b^n$$

A proof of the binomial theorem appears in Appendix I.

In the binomial theorem, the exponents of the variables in each term on the right side follow familiar patterns: The sum of the exponents of a and b in each term is n, the exponents of a decrease, and the exponents of b increase. Only the method of finding the coefficients is different. Except for the first and last term, $n!$ is the numerator of each coefficient. If the exponent of b is r in a particular term, the two factors $r!$ and $(n - r)!$ form the denominator of the fractional coefficient.

Example 5   Use the binomial theorem to expand $(a + b)^5$.

Solution   Substituting directly into the binomial theorem gives

$$(a + b)^5 = a^5 + \frac{5!}{1!(5 - 1)!}a^4b + \frac{5!}{2!(5 - 2)!}a^3b^2 + \frac{5!}{3!(5 - 3)!}a^2b^3$$

$$+ \frac{5!}{4!(5 - 4)!}ab^4 + b^5$$

$$= a^5 + \frac{5 \cdot 4!}{1 \cdot 4!}a^4b + \frac{5 \cdot 4 \cdot 3!}{2 \cdot 1 \cdot 3!}a^3b^2 + \frac{5 \cdot 4 \cdot 3!}{3! \cdot 2 \cdot 1}a^2b^3$$

$$+ \frac{5 \cdot 4!}{4! \cdot 1}ab^4 + b^5$$

$$= a^5 + 5a^4b + 10a^3b^2 + 10a^2b^3 + 5ab^4 + b^5$$

Note that the coefficients in this example are the same numbers that appear in the sixth row of Pascal's triangle (the row whose second entry is 5). ■

Example 6 Find the expansion of $(2x - 3y)^4$.

Solution Note that $(2x - 3y)^4 = (2x + [-3y])^4$. To find the expansion of $(2x + [-3y])^4$, let $a = 2x$ and $b = -3y$. Then find the expansion of $(a + b)^4$. Substituting 4 for n in the binomial theorem gives

$$(a + b)^4 = a^4 + \frac{4!}{1!(4 - 1)!}a^3b + \frac{4!}{2!(4 - 2)!}a^2b^2 + \frac{4!}{3!(4 - 3)!}ab^3 + b^4$$

$$= a^4 + \frac{4 \cdot 3!}{3!}a^3b + \frac{4 \cdot 3 \cdot 2!}{2 \cdot 1 \cdot 2!}a^2b^2 + \frac{4 \cdot 3!}{3!}ab^3 + b^4$$

$$= a^4 + 4a^3b + 6a^2b^2 + 4ab^3 + b^4$$

In this expansion, substitute $2x$ for a and $-3y$ for b, and simplify to obtain

$$(2x - 3y)^4 = (2x)^4 + 4(2x)^3(-3y) + 6(2x)^2(-3y)^2 + 4(2x)(-3y)^3 + (-3y)^4$$

$$= 16x^4 - 96x^3y + 216x^2y^2 - 216xy^3 + 81y^4$$ ■

Finding a Particular Term of a Binomial Expansion

Suppose we wish to find the fifth term of the expansion of $(a + b)^{11}$. It is possible to raise the binomial $a + b$ to the 11th power and then look at the fifth term, but that would be tedious. However, this task is easy if we use the binomial theorem.

Example 7 Find the fifth term of the expansion of $(a + b)^{11}$.

Solution The exponent of b in the fifth term of this expansion is 4, because the exponent for b is always 1 less than the number of the term. Because the exponent of b added to the exponent of a must equal 11, the exponent of a must be 7. The variables of the fifth term appear as a^7b^4.

Because of the binomial theorem, the number in the numerator of the coefficient is $n!$, which in this case is 11!. The factors in the denominator are 4! and $(11 - 4)!$. Thus, the complete fifth term of the expansion of $(a + b)^{11}$ is

$$\frac{11!}{4!7!}a^7b^4 = \frac{11 \cdot 10 \cdot 9 \cdot 8 \cdot 7!}{4 \cdot 3 \cdot 2 \cdot 1 \cdot 7!}a^7b^4 = 330a^7b^4$$ ■

Example 8 Find the sixth term of the expansion of $(a + b)^9$.

Solution In the sixth term, the exponent of b is 5, and the exponent of a is $9 - 5$ or 4. The factors in the denominator of the coefficient are 5! and $(9 - 5)!$, and 9! is the numerator. Thus, the sixth term of the expansion of $(a + b)^9$ is

$$\frac{9!}{5!(9 - 5)!}a^4b^5 = \frac{9 \cdot 8 \cdot 7 \cdot 6 \cdot 5!}{5!4!}a^4b^5 = \frac{9 \cdot 8 \cdot 7 \cdot 6}{4 \cdot 3 \cdot 2 \cdot 1}a^4b^5 = 126a^4b^5 \quad \blacksquare$$

Example 9 Find the third term of the expansion of $(3x - 2y)^6$.

Solution Use the binomial theorem to find the third term in the expansion of $(a + b)^6$:

$$\frac{6!}{2!(6 - 2)!}a^4b^2 = \frac{6 \cdot 5 \cdot 4!}{2 \cdot 1 \cdot 4!}a^4b^2 = 15a^4b^2$$

Replacing a with $3x$ and b with $-2y$ in the term $15a^4b^2$ gives the third term of the expansion of $(3x - 2y)^6$.

$$15a^4b^2 = 15(3x)^4(-2y)^2$$
$$= 15(3)^4(-2)^2x^4y^2$$
$$= 4860x^4y^2 \quad \blacksquare$$

EXERCISE 9.2

In Exercises 1–10, evaluate each expression.

1. 4!

2. $-5!$

3. $3! \cdot 5!$

4. $0! \cdot 7!$

5. $6! + 6!$

6. $5! - 2!$

7. $\dfrac{9!}{12!}$

8. $\dfrac{8!}{5!}$

9. $\dfrac{18!}{6!(18 - 6)!}$

10. $\dfrac{15!}{9!(15 - 9)!}$

In Exercises 11–22, use the binomial theorem to expand each binomial.

11. $(a + b)^4$

12. $(a + b)^3$

13. $(a - b)^5$

14. $(x - y)^4$

15. $(2x - y)^3$

16. $(x + 2y)^5$

17. $(2x + y)^4$

18. $(2x - y)^4$

19. $(4x - 3y)^4$

20. $(5x + 2y)^5$

21. $(6x - 5)^3$

22. $\left(\dfrac{x}{2} + \dfrac{1}{3}\right)^4$

In Exercises 23–38, find the required term in the expansion of the given expression.

23. $(a + b)^4$; third term

24. $(a - b)^4$; second term

25. $(a + b)^7$; fifth term

26. $(a + b)^5$; fourth term

27. $(a - b)^5$; sixth term

28. $(a + b)^{12}$; twelfth term

29. $(x - y)^8$; seventh term

30. $(2x - y)^4$; second term

31. $(\sqrt{2}x + y)^5$; third term

32. $(\sqrt{2}x - 3y)^5$; third term

33. $(x + 2y)^9$; eighth term

34. $(3x - 5y)^6$; fourth term

35. $(a + b)^r$; fourth term

36. $(a - b)^r$; fifth term

37. $(a + b)^n$ rth term

38. $(a + b)^n$; $(r + 1)$th term

39. Find the sum of the numbers in each row of the first 10 rows of Pascal's triangle. What is the pattern?

40. Show that the sum of the coefficients in the binomial expansion of $(x + y)^n$ is 2^n. (*Hint:*   Let $x = y = 1$.)

41. Find the constant term in the expansion of $\left(a - \dfrac{1}{a} \right)^{10}$.

42. Find the coefficient of x^5 in the expansion of $\left(x + \dfrac{1}{x} \right)^9$.

43. Find the coefficient of x^8 in the expansion of $\left(\sqrt{x} + \dfrac{1}{2x} \right)^{25}$.

44. Find the constant term in the expansion of $\left(\dfrac{1}{a} + a \right)^8$.

9.3   SEQUENCES, SERIES, AND SUMMATION NOTATION

In the chapter introduction, we referred to a group of problems involving long strings of numbers called *sequences*. To study sequences in detail, we must formally define a **sequence**.

> **Definition.**   A **sequence** is a function whose domain is the set of natural numbers.

Since a sequence is a function whose domain is the set of natural numbers, we can write its values as a list of numbers. For example, if n is a natural number, the function defined by $f(n) = 2n - 1$ generates the list

$$1, 3, 5, \ldots , 2n - 1, \ldots$$

It is common to call such a list, as well as the function, a sequence. The number 1 is the first term of this sequence, the number 3 is the second term, and the expression $2n - 1$ represents the **general**, or **nth term** of the sequence. Likewise, if n is a natural number, then the function defined by $f(n) = 3n^2 + 1$ generates the list

$$4, 13, 28, \ldots , 3n^2 + 1, \ldots$$

The number 4 is the first term, 13 is the second term, 28 is the third term, and $3n^2 + 1$ is the general term.

Because the domain of any sequence is the infinite set of natural numbers, the sequence itself is an unending list of numbers. Note that a constant function such as $g(n) = 1$ is a sequence also; it generates the list $1, 1, 1, \ldots$.

Many times, sequences do not lend themselves to functional notation, because it is difficult or even impossible to write the general term—the expression that shows how the terms are constructed. In such cases, if there is a pattern that is assumed to be continued, it is acceptable simply to list several terms of the sequence. Some examples of sequences follow:

$$1^2, 2^2, 3^2, \ldots, n^2, \ldots$$

$$3, 9, 19, 33, \ldots, 2n^2 + 1, \ldots$$

$$1, 3, 6, 10, 15, 21, \ldots, \frac{n(n + 1)}{2}, \ldots$$

$$1, 1, 2, 3, 5, 8, 13, 21, \ldots \text{ (Fibonacci sequence)}$$

$$2, 3, 5, 7, 11, 13, 17, 19, 23, \ldots \text{ (prime numbers)}$$

Leonardo Fibonacci (late 12th and early 13th cent.) In his work *Liber abaci*, Fibonacci was among the first to advocate the adoption of Arabic numerals, the numerals that we use today.

The fourth example listed is called the **Fibonacci sequence**, after the twelfth-century mathematician Leonardo of Pisa—known to his friends as Fibonacci. After the two 1's in the Fibonacci sequence, each term is the sum of the two terms that immediately precede it. The Fibonacci sequence occurs in the study of botany, for example, in the growth patterns of certain plants.

Sometimes a sequence is defined by giving its first term, a_1, and a rule showing how to obtain the $(n + 1)$th term, a_{n+1}, from the nth term, a_n. Such a definition is called a **recursive definition**. For example, the information

$$a_1 = 5 \qquad \text{and} \qquad a_{n+1} = 3a_n - 2$$

defines a sequence recursively. To find the first five terms of this sequence, we proceed as follows:

$$a_1 = 5$$
$$a_2 = 3(a_1) - 2 = 3(5) - 2 = 13$$
$$a_3 = 3(a_2) - 2 = 3(13) - 2 = 37$$
$$a_4 = 3(a_3) - 2 = 3(37) - 2 = 109$$
$$a_5 = 3(a_4) - 2 = 3(109) - 2 = 325$$

To add the terms of a sequence, we replace each comma between the terms with a plus sign, forming what is called a **series**. Because each sequence is infinite, the number of terms in the series associated with it is infinite also. Two examples of infinite series are

$$1^2 + 2^2 + 3^2 + \cdots + n^2 + \cdots$$

and

$$1 + 2 + 3 + 5 + 8 + 13 + 21 + \cdots$$

If the signs between successive terms of an infinite series alternate, the series is often called an **alternating infinite series**. Two examples of alternating infinite series are

$$-3 + 6 - 9 + 12 - \cdots + (-1)^n 3n + \cdots$$

and

$$2 - 4 + 8 - 16 + \cdots + (-1)^{n+1} 2^n + \cdots$$

Summation Notation

There is a shorthand method of indicating the sum of the first n terms, or the **nth partial sum** of a sequence. This method, called **summation notation**, involves the symbol Σ, which is a capital sigma in the Greek alphabet. The expression

$$\sum_{n=1}^{3} (2n^2 + 1)$$

designates the sum of the three terms obtained if we successively substitute the natural numbers 1, 2, and 3 for n in the expression $2n^2 + 1$. Hence,

$$\sum_{n=1}^{3} (2n^2 + 1) = [2(1)^2 + 1] + [2(2)^2 + 1] + [2(3)^2 + 1]$$
$$= 3 + 9 + 19$$
$$= 31$$

Example 1 Evaluate $\sum_{n=1}^{4} (n^2 - 1)$.

Solution In this example, n is said to run from 1 to 4. Hence, substitute 1, 2, 3, and 4 for n in the expression $n^2 - 1$, and find the sum of the resulting values:

$$\sum_{n=1}^{4} (n^2 - 1) = (1^2 - 1) + (2^2 - 1) + (3^2 - 1) + (4^2 - 1)$$
$$= 0 + 3 + 8 + 15$$
$$= 26 \qquad \blacksquare$$

Example 2 Evaluate $\sum_{n=3}^{5} (3n + 2)$.

Solution In this example, n runs from 3 to 5. Hence, substitute 3, 4, and 5 for n in the expression $3n + 2$ and find the sum of the resulting values:

$$\sum_{n=3}^{5} (3n + 2) = [3(3) + 2] + [3(4) + 2] + [3(5) + 2]$$
$$= 11 + 14 + 17$$
$$= 42 \qquad \blacksquare$$

The following theorems give three properties of summations. The first theorem states that the summation of a constant as k runs from 1 to n is n times that constant.

Theorem. If c is a constant, then $\sum_{k=1}^{n} c = nc$.

Proof Because c is a constant, each term is c for each value for k as k runs from 1 to n.

$$\sum_{k=1}^{n} c = \overbrace{c + c + c + c + \cdots + c}^{n \text{ number of } c\text{'s}} = nc$$

$\square$

Example 3   Evaluate $\sum_{n=1}^{5} 13$.

Solution $\sum_{n=1}^{5} 13 = 13 + 13 + 13 + 13 + 13$

$$= 5(13)$$

$$= 65$$ ∎

The next theorem states that a constant factor can be brought outside a summation sign.

Theorem.   If c is a constant, then $\sum_{k=1}^{n} cf(k) = c \sum_{k=1}^{n} f(k)$.

Proof $\sum_{k=1}^{n} cf(k) = cf(1) + cf(2) + cf(3) + \cdots + cf(n)$

$$= c[f(1) + f(2) + f(3) + \cdots + f(n)] \qquad \text{Factor out } c.$$

$$= c \sum_{k=1}^{n} f(k) \qquad \qquad \qquad \qquad \qquad \square$$

Example 4   Show that $\sum_{k=1}^{3} 5k^2 = 5 \sum_{k=1}^{3} k^2$.

Solution $\sum_{k=1}^{3} 5k^2 = 5(1)^2 + 5(2)^2 + 5(3)^2$

$$= 5 + 20 + 45$$

$$= 70$$

$$5 \sum_{k=1}^{3} k^2 = 5[(1)^2 + (2)^2 + (3)^2]$$

$$= 5[1 + 4 + 9]$$

$$= 5(14)$$

$$= 70$$

Thus, the quantities are equal.   ∎

The final theorem states that the summation of a sum is equal to the sum of the summations.

Theorem. $\sum_{k=1}^{n} [f(k) + g(k)] = \sum_{k=1}^{n} f(k) + \sum_{k=1}^{n} g(k)$

Proof

$$\sum_{k=1}^{n} [f(k) + g(k)] = [f(1) + g(1)] + [f(2) + g(2)]$$

$$+ [f(3) + g(3)] + \cdots + [f(n) + g(n)]$$

$$= [f(1) + f(2) + f(3) + \cdots + f(n)]$$

$$+ [g(1) + g(2) + g(3) + \cdots + g(n)]$$

$$= \sum_{k=1}^{n} f(k) + \sum_{k=1}^{n} g(k) \qquad \Box$$

Example 5 Show that $\displaystyle\sum_{k=1}^{3} (k + k^2) = \sum_{k=1}^{3} k + \sum_{k=1}^{3} k^2$.

Solution

$$\sum_{k=1}^{3} (k + k^2) = (1 + 1^2) + (2 + 2^2) + (3 + 3^2)$$

$$= 2 + 6 + 12$$

$$= 20$$

$$\sum_{k=1}^{3} k + \sum_{k=1}^{3} k^2 = (1 + 2 + 3) + (1^2 + 2^2 + 3^2)$$

$$= 6 + 14$$

$$= 20 \qquad \blacksquare$$

Example 6 Evaluate $\displaystyle\sum_{k=1}^{5} (2k - 1)^2$ directly. Then expand the binomial, apply the previous theorems, and evaluate the expression again.

Solution *Part 1.* $\displaystyle\sum_{k=1}^{5} (2k - 1)^2 = 1 + 9 + 25 + 49 + 81 = 165$

Part 2. $\displaystyle\sum_{k=1}^{5} (2k - 1)^2 = \sum_{k=1}^{5} (4k^2 - 4k + 1)$

$$= \sum_{k=1}^{5} 4k^2 + \sum_{k=1}^{5} (-4k) + \sum_{k=1}^{5} 1 \qquad \begin{array}{l}\text{The summation of a}\\ \text{sum is the sum of}\\ \text{the summations.}\end{array}$$

$$= 4\sum_{k=1}^{5} k^2 - 4\sum_{k=1}^{5} k + \sum_{k=1}^{5} 1 \qquad \begin{array}{l}\text{Bring the constant}\\ \text{factors outside the}\\ \text{summation sign.}\end{array}$$

$$= 4\sum_{k=1}^{5} k^2 - 4\sum_{k=1}^{5} k + 5 \qquad \begin{array}{l}\text{The summation of a}\\ \text{constant as } k \text{ runs}\\ \text{from 1 to 5 is 5 times}\\ \text{that constant.}\end{array}$$

$$= 4(1 + 4 + 9 + 16 + 25) - 4(1 + 2 + 3 + 4 + 5) + 5$$

$$= 4(55) - 4(15) + 5$$

$$= 220 - 60 + 5$$

$$= 165$$

Note that the sum is 165, regardless of the method used. $\qquad \blacksquare$

EXERCISE 9.3

1. Write the first eight terms of the sequence defined by the function $f(n) = 5n(n - 1)$.

2. Write the first six terms of the sequence defined by the function $f(n) = n\left(\dfrac{n - 1}{2}\right)\left(\dfrac{n - 2}{3}\right)$.

In Exercises 3–8, write the fifth term in each sequence.

3. 1, 6, 11, 16, . . .

4. 1, 8, 27, 64, . . .

5. $a, a + d, a + 2d, a + 3d, \ldots$

6. $a, ar, ar^2, ar^3, \ldots$

7. 1, 3, 6, 10, . . .

8. 20, 17, 13, 8, . . .

In Exercises 9–16, find the sum of the first five terms of the sequence with the given general term.

9. n

10. $2k$

11. 3

12. $4k^0$

13. $2\left(\dfrac{1}{3}\right)^n$

14. $(-1)^n$

15. $3n - 2$

16. $2k + 1$

In Exercises 17–24, a sequence is defined recursively. Find the first four terms of each sequence.

17. $a_1 = 3$ and $a_{n+1} = 2a_n + 1$

18. $a_1 = -5$ and $a_{n+1} = -a_n - 3$

19. $a_1 = -4$ and $a_{n+1} = \dfrac{a_n}{2}$

20. $a_1 = 0$ and $a_{n+1} = 2a_n^2$

21. $a_1 = k$ and $a_{n+1} = a_n^2$

22. $a_1 = 3$ and $a_{n+1} = ka_n$

23. $a_1 = 8$ and $a_{n+1} = \dfrac{2a_n}{k}$

24. $a_1 = m$ and $a_{n+1} = \dfrac{(a_n)^2}{m}$

In Exercises 25–28, tell whether each series is an alternating infinite series.

25. $-1 + 2 - 3 + \cdots + (-1)^n n + \cdots$

26. $a + \dfrac{a}{b} + \dfrac{a}{b^2} + \cdots + a\left(\dfrac{1}{b}\right)^{n-1} + \cdots ; b = 4$

27. $a + a^2 + a^3 + \cdots + a^n + \cdots ; a = 3$

28. $a + a^2 + a^3 + \cdots + a^n + \cdots ; a = -2$

In Exercises 29–40, evaluate each sum.

29. $\displaystyle\sum_{k=1}^{5} 2k$

30. $\displaystyle\sum_{k=3}^{6} 3k$

31. $\displaystyle\sum_{k=3}^{4} (-2k^2)$

32. $\displaystyle\sum_{k=1}^{100} 5$

33. $\displaystyle\sum_{k=1}^{5} (3k - 1)$

34. $\displaystyle\sum_{n=2}^{5} (n^2 + 3n)$

35. $\displaystyle\sum_{k=1}^{1000} \dfrac{1}{2}$

36. $\displaystyle\sum_{x=4}^{5} \dfrac{2}{x}$

37. $\displaystyle\sum_{x=3}^{4} \dfrac{1}{x}$

38. $\displaystyle\sum_{x=2}^{6} (3x^2 + 2x) - 3\displaystyle\sum_{x=2}^{6} x^2$

39. $\displaystyle\sum_{x=1}^{4} (4x + 1)^2 - \displaystyle\sum_{x=1}^{4} (4x - 1)^2$

40. $\displaystyle\sum_{x=0}^{10} (2x - 1)^2 + 4\displaystyle\sum_{x=0}^{10} x(1 - x)$

In Exercises 41–43, use mathematical induction to prove each statement.

41. $\displaystyle\sum_{k=1}^{n} (4k - 3) = n(2n - 1)$

42. $\displaystyle\sum_{k=1}^{n} (5k - 3) = \dfrac{n(5n - 1)}{2}$

43. $\displaystyle\sum_{k=1}^{n} (6k + 4) = n(3n + 7)$

44. Construct an example to disprove the proposition that the summation of a product is the product of the summations. In other words, prove that

$$\sum_{k=1}^{n} f(k)g(k) \neq \sum_{k=1}^{n} f(k) \sum_{k=1}^{n} g(k)$$

9.4 ARITHMETIC AND GEOMETRIC PROGRESSIONS

Some important sequences are called **progressions**. One of these is the arithmetic progression.

Definition.   An **arithmetic progression** is a sequence of the form

$$a,\ a + d,\ a + 2d,\ a + 3d,\ \ldots,\ a + (n - 1)d,\ \ldots$$

where a is the first term, $a + (n - 1)d$ is the nth term, and d is the common difference.

In this definition, note that the second term of the progression has an addend of d, the third term has an addend of $2d$, the fourth term has an addend of $3d$, and so on. This is why the nth term has an addend of $(n - 1)d$.

Example 1   For an arithmetic progression that has a first term of 7 and a common difference of 5, write the first six terms and the 21st term of the progression.

Solution   Because the first term, a, is 7 and the common difference, d, is 5, the first six terms are

$$7,\quad 7 + 5,\quad 7 + 2(5),\quad 7 + 3(5),\quad 7 + 4(5),\quad 7 + 5(5)$$

or

$$7,\quad 12,\quad 17,\quad 22,\quad 27,\quad 32$$

The nth term is $a + (n - 1)d$. Because you are looking for the 21st term, $n = 21$.

$$n\text{th term} = a + (n - 1)d$$
$$21\text{st term} = 7 + (21 - 1)5$$
$$= 7 + (20)5$$
$$= 107$$

The 21st term is 107.   ■

Example 2   For an arithmetic progression with the first three terms 2, 6, and 10, find the 98th term.

Solution   In this example, $a = 2$, $n = 98$, and $d = 6 - 2 = 10 - 6 = 4$. The nth term is given by the formula $a + (n - 1)d$. Therefore, the 98th term is

$$n\text{th term} = a + (n - 1)d$$
$$98\text{th term} = 2 + (98 - 1)4$$
$$= 2 + (97)4$$
$$= 390$$

∎

Numbers inserted between a first and last term to form a segment of an arithmetic progression are called **arithmetic means**. In this type of problem, the last term, l, is considered the nth term:

$$l = a + (n - 1)d$$

Example 3   Insert three arithmetic means between the numbers -3 and 12.

Solution   Begin by finding the common difference d. In this example, the first term is -3 and the last term is 12. Because you are inserting three terms, the total number of terms is five. Thus, $a = -3$, $l = 12$, and $n = 5$. The formula for the nth (or last) term is

$$l = a + (n - 1)d$$

Substituting 12 for l, -3 for a, and 5 for n in the formula and solving for d gives

$$12 = -3 + (5 - 1)d$$
$$15 = 4d$$
$$\frac{15}{4} = d$$

Once the common difference has been found, the arithmetic means are the second, third, and fourth terms of the arithmetic progression with a first term of -3 and a fifth term of 12:

$$a + d = -3 + \frac{15}{4} = \frac{3}{4}$$
$$a + 2d = -3 + \frac{30}{4} = 4\frac{1}{2}$$
$$a + 3d = -3 + \frac{45}{4} = 8\frac{1}{4}$$

The three arithmetic means are $\frac{3}{4}$, $4\frac{1}{2}$, and $8\frac{1}{4}$.

∎

The formula stated in the following theorem gives the sum of the first n terms of an arithmetic progression.

Theorem. The formula

$$S_n = \frac{n(a + l)}{2}$$

gives the sum of the first n terms of an arithmetic progression. In this formula, a is the first term, l is the last (or nth) term, and n is the number of terms.

Proof We write the first n terms of an arithmetic progression (letting S_n represent their sum), rewrite the same sum in reverse order, and add the equations together term by term:

$$S_n = a + (a + d) + \cdots + [a + (n - 2)d] + [a + (n - 1)d]$$
$$S_n = [a + (n - 1)d] + [a + (n - 2)d] + \cdots + (a + d) + a$$

$$2S_n = [2a + (n - 1)d] + [2a + (n - 1)d] + \cdots + [2a + (n - 1)d] + [2a + (n - 1)d]$$

Because there are n equal terms on the right side of the previous equation,

$$2S_n = n[2a + (n - 1)d]$$

or

$$2S_n = n\{a + [a + (n - 1)d]\}$$

Because $a + (n - 1)d = l$, we make that substitution in the right side of the previous equation and divide both sides by 2 to obtain

$$S_n = \frac{n(a + l)}{2}$$

The theorem is proved. (Exercise 45 will ask you to prove this theorem again using mathematical induction.) □

Example 4 Find the sum of the first 30 terms of the arithmetic progression 5, 8, 11, ...

Solution In this example, $a = 5$, $n = 30$, $d = 3$, and $l = 5 + 29(3) = 92$. Substituting these values into the formula $S_n = n(a + l)/2$ and simplifying gives

$$S_{30} = \frac{30(5 + 92)}{2} = 15(97) = 1455$$

The sum of the first 30 terms is 1455. ■

Geometric Progressions

Another important sequence is the **geometric progression**.

> **Definition.** A **geometric progression** is a sequence of the form
>
> $$a, ar, ar^2, ar^3, \ldots, ar^{n-1}, \ldots$$
>
> where a is the first term, ar^{n-1} is the nth term, and r is the common ratio.

In this definition, note that the second term of the progression has a factor of r^1, the third term has a factor of r^2, the fourth term has a factor of r^3, and so on. This explains why the nth term has a factor of r^{n-1}.

Example 5 For a geometric progression with a first term of 3 and a common ratio of 2, write the first six terms and the 15th term of the progression.

Solution Write the first six terms of the geometric progression:

$$3, \quad 3(2), \quad 3(2)^2, \quad 3(2)^3, \quad 3(2)^4, \quad 3(2)^5$$

or

$$3, \quad 6, \quad 12, \quad 24, \quad 48, \quad 96$$

To obtain the 15th term, substitute 15 for n, 3 for a, and 2 for r in the formula for the nth term:

$$n\text{th term} = ar^{n-1}$$
$$15\text{th term} = 3(2)^{15-1}$$
$$= 3(2)^{14}$$
$$= 3(16{,}384)$$
$$= 49{,}152$$

■

Example 6 For a geometric progression with the first three terms 9, 3, and 1, find the eighth term.

Solution In this example, $a = 9$, $r = \frac{1}{3}$, $n = 8$, and the nth term is ar^{n-1}. To obtain the eighth term, substitute these values into the expression for the nth term:

$$n\text{th term} = ar^{n-1}$$
$$8\text{th term} = 9\left(\frac{1}{3}\right)^{8-1}$$
$$= 9\left(\frac{1}{3}\right)^{7}$$
$$= \frac{1}{243}$$

■

As with arithmetic progressions, numbers may be inserted between a first and last term to form a segment of a geometric progression. The numbers inserted are called **geometric means**. In this type of problem, the last term, l, is considered to be the nth term: $l = ar^{n-1}$.

Example 7   Insert two geometric means between the numbers 4 and 256.

Solution   Begin by finding the common ratio. The first term, a, is 4. Because 256 is to be the fourth term, $n = 4$ and $l = 256$. Substituting these values into the formula for the nth term of a geometric progression, and solving for r gives

$$a r^{n-1} = l$$
$$4 r^{4-1} = 256$$
$$r^3 = 64$$
$$r = 4$$

The common ratio is 4. The two geometric means are the second and third terms of the geometric progression:

$$ar = 4 \cdot 4 = 16$$
$$ar^2 = 4 \cdot 4^2 = 4 \cdot 16 = 64$$

The first four terms of the geometric progression are 4, 16, 64, and 256; 16 and 64 are geometric means between 4 and 256.   ■

There is a formula that gives the sum of the first n terms of a geometric progression.

Theorem.   The formula

$$S_n = \frac{a - ar^n}{1 - r} \quad (r \neq 1)$$

gives the sum of the first n terms of a geometric progression. In this formula, S_n is the sum, a is the first term, r is the common ratio, and n is the number of terms.

Proof   We write out the sum of the first n terms of the geometric progression:

1. $S_n = a + ar + ar^2 + \cdots + ar^{n-3} + ar^{n-2} + ar^{n-1}$

Multiplying both sides of this equation by r gives

2. $S_n r = \quad ar + ar^2 + \quad \cdots \quad + ar^{n-2} + ar^{n-1} + ar^n$

We now subtract Equation 2 from Equation 1 and solve for S_n:

$$S_n - S_n r = a - ar^n$$
$$S_n(1 - r) = a - ar^n$$
$$S_n = \frac{a - ar^n}{1 - r}$$

The theorem is proved. (Exercise 46 will ask you to prove this theorem using mathematical induction.)   □

Example 8   Find the sum of the first six terms of the geometric progression 8, 4, 2, ...

Solution   In this example, $a = 8$, $n = 6$, and $r = \frac{1}{2}$. Substituting these values into the formula for the sum of the first n terms of a geometric progression gives

$$S_n = \frac{a - ar^n}{1 - r}$$

$$S_6 = \frac{8 - 8\left(\frac{1}{2}\right)^6}{1 - \frac{1}{2}}$$

$$= 2\left(\frac{63}{8}\right)$$

$$= \frac{63}{4}$$

The sum of the first six terms is $\frac{63}{4}$.   ■

Infinite Geometric Progressions

Under certain conditions, it is possible to find the sum of all the terms in an infinite geometric progression. To define this sum, we consider the geometric progression

$$a, ar, ar^2, \ldots$$

The first partial sum, S_1, of the progression is a. Hence,

$$S_1 = a$$

The second partial sum, S_2, of this progression is $a + ar$. Hence,

$$S_2 = a + ar$$

In general, the nth partial sum, S_n, of this progression is

$$S_n = a + ar + ar^2 + \cdots + ar^{n-1}$$

If the nth partial sum, S_n, of an infinite geometric progression approaches some number S as n becomes large without bound, then S is called the **sum of the infinite geometric progression**. The symbol $\sum_{n=1}^{\infty} ar^{n-1}$, where ∞ is the symbol for infinity, denotes the sum, S, of the infinite geometric progression; $S = \sum_{n=1}^{\infty} ar^{n-1}$, provided that sum exists.

To develop a formula for finding the sum of all the terms in an infinite geometric progression, we consider the formula

$$S_n = \frac{a - ar^n}{1 - r} \quad (r \neq 1)$$

If $|r| < 1$ and a is a constant, then the term ar^n, or $a(r^n)$, approaches 0 as n becomes large without bound. Hence, when n is very large, the value of ar^n is extremely small, and the term ar^n in the formula can be ignored. This argument leads to the following theorem.

Theorem. If $|r| < 1$, then the sum of the terms of an infinite geometric progression is given by the formula

$$S = \frac{a}{1 - r}$$

where a is the first term and r is the common ratio.

Example 9 Change $0.\overline{4}$ to a common fraction.

Solution Write the decimal as an infinite geometric series and find its sum:

$$S = \frac{4}{10} + \frac{4}{100} + \frac{4}{1000} + \frac{4}{10,000} + \cdots$$

$$S = \frac{4}{10} + \frac{4}{10}\left(\frac{1}{10}\right) + \frac{4}{10}\left(\frac{1}{10}\right)^2 + \frac{4}{10}\left(\frac{1}{10}\right)^3 + \cdots$$

Because the common ratio is $\frac{1}{10}$ and $\left|\frac{1}{10}\right| < 1$, use the formula for the sum of an infinite geometric series:

$$S = \frac{a}{1 - r} = \frac{\dfrac{4}{10}}{1 - \dfrac{1}{10}} = \frac{\dfrac{4}{10}}{\dfrac{9}{10}} = \frac{4}{9}$$

Long division will verify that $\frac{4}{9} = 0.\overline{4}$. ∎

EXERCISE 9.4

In Exercises 1–6, write the first six terms of the arithmetic progressions with the given properties.

1. $a = 1$ and $d = 2$

2. $a = -12$ and $d = -5$

3. $a = 5$ and the third term is 2.

4. $a = 4$ and the fifth term is 12.

5. The seventh term is 24, and the common difference is $\frac{5}{2}$.

6. The 20th term is -49, and the common difference is -3.

In Exercises 7–10, find the sum of the first n terms of each arithmetic progression.

7. $5 + 7 + 9 + \cdots$ (to 15 terms)

8. $-3 + (-4) + (-5) + \cdots$ (to 10 terms)

9. $\displaystyle\sum_{n=1}^{20} \left(\frac{3}{2}n + 12\right)$

10. $\displaystyle\sum_{n=1}^{10} \left(\frac{2}{3}n + \frac{1}{3}\right)$

11. In an arithmetic progression, the 25th term is 10 and the common difference is $\frac{1}{2}$. Find the sum of the first 30 terms.

12. In an arithmetic progression, the 15th term is 86 and the first term is 2. Find the sum of the first 100 terms.

13. If the fifth term of an arithmetic progression is 14 and the second term is 5, find the 15th term.

14. Can an arithmetic progression have a first term of 4, a 25th term of 126, and a common difference of $4\frac{1}{4}$? If not, explain why.

15. Insert three arithmetic means between 10 and 20.

16. Insert five arithmetic means between 5 and 15.

17. Insert four arithmetic means between -7 and $\frac{2}{3}$.

18. Insert three arithmetic means between -11 and -2.

In Exercises 19–26, write the first four terms of each geometric progression with the given properties.

19. $a = 10$ and $r = 2$

20. $a = -3$ and $r = 2$

21. $a = -2$ and $r = 3$

22. $a = 64$ and $r = \frac{1}{2}$

23. $a = 3$ and $r = \sqrt{2}$

24. $a = 2$ and $r = \sqrt{3}$

25. $a = 2$, and the fourth term is 54.

26. The third term is 4, and $r = \frac{1}{2}$.

In Exercises 27–32, find the sum of the indicated terms of each geometric progression.

27. $4, 8, 16, \ldots$ (to 5 terms)

28. $9, 27, 81, \ldots$ (to 6 terms)

29. $2, -6, 18, \ldots$ (to 10 terms)

30. $\frac{1}{8}, \frac{1}{4}, \frac{1}{2}, \ldots$ (to 12 terms)

31. $\displaystyle\sum_{n=1}^{6} 3\left(\frac{3}{2}\right)^{n-1}$

32. $\displaystyle\sum_{n=1}^{6} 12\left(-\frac{1}{2}\right)^{n-1}$

In Exercises 33–36, find the sum of each infinite geometric progression.

33. $6 + 4 + \frac{8}{3} + \cdots$

34. $8 + 4 + 2 + 1 + \cdots$

35. $\displaystyle\sum_{n=1}^{\infty} 12\left(-\frac{1}{2}\right)^{n-1}$

36. $\displaystyle\sum_{n=1}^{\infty} \left(\frac{1}{3}\right)^{n-1}$

37. Insert three positive geometric means between 10 and 20.

38. Insert five geometric means between -5 and 5, if possible.

39. Insert four geometric means between 2 and 2048.

40. Insert three geometric means between 162 and 2. (There are two possibilities.)

In Exercises 41–44, change each decimal to a common fraction.

41. $0.\overline{5}$

42. $0.\overline{6}$

43. $0.\overline{25}$

44. $0.\overline{37}$

45. Use mathematical induction to prove the formula for finding the sum of the first n terms of an arithmetic progression.

46. Use mathematical induction to prove the formula for finding the sum of the first n terms of a geometric progression.

47. If Justin earns 1¢ on the first day of May, 2¢ on the second day, 4¢ on the third day, and the pay continues to double each day throughout the month, what will his total earnings be for the month?

48. A single arithmetic mean between two numbers is called *the* arithmetic mean of the two numbers. Similarly, a single geometric mean between two numbers is called *the* geometric mean of the two numbers. Find the arithmetic mean and the geometric mean between the numbers 4 and 64. Which is larger, the arithmetic or the geometric mean?

49. Use the definitions in Exercise 48 to compute the arithmetic mean and the geometric mean between $\frac{1}{2}$ and $\frac{7}{8}$. Which is larger, the arithmetic or geometric mean?

50. If a and b are positive numbers and $a \neq b$, prove that their arithmetic mean is greater than their geometric mean. (*Hint*:   See Exercises 48 and 49.)

51. Find the indicated sum: $\displaystyle\sum_{n=1}^{100} \frac{1}{n(n+1)}$. (*Hint*:   Use partial fractions first.)

52. Find the indicated sum: $\displaystyle\sum_{k=1}^{100} \ln\left(\frac{k}{k+1}\right)$.

9.5   APPLICATIONS OF PROGRESSIONS

The following examples illustrate some applications of arithmetic and geometric progressions.

Example 1   A town with a population of 3500 people has a predicted growth rate of 6% over the preceding year for the next 20 years. How many people are expected to live in the town 20 years from now?

Solution   Let p_0 be the initial population of the town. After 1 year, there will be a different population, p_1. The initial population (p_0) plus the growth (the product of p_0 and the rate of growth, r) will equal the new population after 1 year (p_1):

$$p_1 = p_0 + p_0 r = p_0(1 + r)$$

The population of the town at the end of 2 years will be p_2, and

$$p_2 = p_1 + p_1 r$$
$$p_2 = p_1(1 + r)$$
$$p_2 = p_0(1 + r)(1 + r)$$
$$p_2 = p_0(1 + r)^2$$

The population at the end of the third year will be $p_3 = p_0(1 + r)^3$. Writing the terms in a sequence yields

$$p_0, \quad p_0(1 + r), \quad p_0(1 + r)^2, \quad p_0(1 + r)^3, \quad p_0(1 + r)^4, \ldots$$

This is a geometric progression with p_0 as the first term and $1 + r$ as the common ratio. Recall that the nth term is given by the formula $l = ar^{n-1}$. In this example, $p_0 = 3500$, $1 + r = 1.06$, and (because the population after 20 years will be the value of the 21st term of the geometric progression) $n = 21$. The population after 20 years is $p = 3500(1.06)^{20}$. Use a calculator to find that $p \approx 11{,}225$. ∎

Example 2   A woman deposits $2500 in a bank at 9% annual interest, compounded daily. If the investment is left untouched for 6 years, how much money will be in the account?

Solution   This problem is similar to that of Example 1. Let the initial amount in the account be A_0. At the end of the first day, the amount in the account is

$$A_1 = A_0 + A_0\left(\frac{r}{365}\right) = A_0\left(1 + \frac{r}{365}\right)$$

The amount in the bank after the second day is

$$A_2 = A_1 + A_1\left(\frac{r}{365}\right) = A_1\left(1 + \frac{r}{365}\right) = A_0\left(1 + \frac{r}{365}\right)^2$$

Just as in Example 1, the amounts in the account each day form a geometric progression.

$$A_0, \quad A_0\left(1 + \frac{r}{365}\right), \quad A_0\left(1 + \frac{r}{365}\right)^2, \quad A_0\left(1 + \frac{r}{365}\right)^3, \ldots$$

where A_0 is the initial deposit and r is the annual rate of interest.

Because the interest is compounded daily for 6 years, the amount in the bank at the end of 6 years will be the 2191th term $(6 \cdot 365 + 1)$ of the progression. The amount in the account at the end of 6 years is

$$A_{2191} = 2500\left(1 + \frac{0.09}{365}\right)^{2190}$$

Use a calculator to find that $A_{2191} \approx \$4289.73$.   ■

Example 3   The equation $S = 16t^2$ represents the distance in feet, S, that an object will fall in t seconds. After 1 second, the object has fallen 16 feet. After 2 seconds, the object has fallen 64 feet. After 3 seconds, the object has fallen 144 feet. In other words, the object fell 16 feet during the first second, 48 feet during the next second, and 80 feet during the third second. Thus, the sequence 16, 48, 80, . . . represents the distance an object will fall during the first second, second second, third second, and so forth. Find the distance the object falls during the 12th second.

Solution   The sequence 16, 48, 80, . . . is an arithmetic progression with $a = 16$ and $d = 32$. To find the 12th term, substitute these values into the formula $l = a + (n - 1)d$ and simplify:

$$l = a + (n - 1)d$$
$$l = 16 + 11(32)$$
$$l = 16 + 352$$
$$l = 368$$

During the 12th second, the object falls 368 feet.   ■

Example 4   A pump can remove 20% of the gas in a container with each stroke. What percentage of the gas will remain in the container after six strokes?

Solution   Let V represent the volume of the container. Because each stroke of the pump removes 20% of the gas, 80% of the gas remains after each stroke, and we have the sequence

$$V, \quad 0.80V, \quad 0.80(0.80V), \quad 0.80[0.80(0.80V)], \; \ldots$$

This can be written as the geometric progression

$$V, \quad 0.8V, \quad (0.8)^2V, \quad (0.8)^3V, \quad (0.8)^4V, \; \ldots$$

We wish to know the amount of gas remaining after six strokes. This amount is the seventh term, l, of the progression:

$$l = ar^{n-1}$$
$$l = V(0.8)^6$$

Use a calculator to find that approximately 26% of the gas remains after six strokes of the pump.   ■

EXERCISE 9.5

*Decide whether each of the following exercises involves an arithmetic or geometric progression, and then solve each problem. **You may use a calculator.***

1. The number of students studying college algebra this year at State College is 623. If a trend has been established that the following year's enrollment is always 10% higher than that of the preceding year, how many professors will be needed in 8 years to teach college algebra if one professor can handle 60 students?

2. If Amelia borrows $5500 interest-free from her mother to buy a new car and agrees to pay her mother back at the rate of $105 per month, how much does she still owe after four years?

3. A Super Ball can always rebound to 95% of the height from which it was dropped. How high will the ball rise after the 13th bounce if it is dropped from a height of 10 meters?

4. If Philip invests $1000 in a 1-year certificate of deposit at $6\frac{3}{4}$% annual interest, compounded daily, how much interest will be earned that year?

5. If a single cell divides into two cells every 30 minutes, how many cells will there be at the end of ten hours?

6. If a lawn tractor, which costs c dollars when new, depreciates 20% of its previous year's value each year, how much is the lawn tractor worth after 5 years?

7. Maria can invest $1000 at $7\frac{1}{2}$%, compounded annually, or at $7\frac{1}{4}$%, compounded daily. If she invests the money for a year, which is the better investment?

8. Find how many feet a brick will travel during the 10th second of its fall.

9. If the population of the world were to double every 30 years, approximately how many people would be on Earth in the year 3000? (Consider the population in 1980 to be 4 billion, and use 1980 as the base year.)

10. If Linda deposits $1300 in a bank at 7% interest, compounded annually, how much will be in the bank 17 years later? (Assume that there are no other transactions on the account.)

11. If a house purchased for $50,000 in 1988 appreciates in value by 6% each year, how much will the house be worth in the year 2010?

12. Calculate the value of $1000 left on deposit for 10 years at an annual rate of 7%, compounded annually.

13. Calculate the value of $1000 left on deposit for 10 years at an annual rate of 7%, compounded quarterly.

14. Calculate the value of $1000 left on deposit for 10 years at an annual rate of 7%, compounded monthly.

15. Calculate the value of $1000 left on deposit for 10 years at an annual rate of 7%, compounded daily.

16. Calculate the value of $1000 left on deposit for 10 years at an annual rate of 7%, compounded hourly.

17. When John was 20 years old, he opened an individual retirement account by investing $2000 that will earn 11% interest, compounded quarterly. How much will his investment be worth when John is 65 years old?

18. One lone bacterium divides to form two bacteria every 5 minutes. If two bacteria multiply enough to fill a petri dish completely in 2 hours, how long will it take one bacterium to fill the dish?

19. A legend tells of an ancient king who was grateful to the inventor of the game of chess and offered to grant him any request. The man was shrewd and he said, "My request is modest, Your Majesty. Simply place one grain of wheat on the first square on the chessboard, two grains on the second, four on the third, and so on, with each square holding double that of the square before. Do this until the board is full." The king, thinking he'd gotten off lightly, readily agreed. How many grains did the king need to fill the chessboard?

20. Estimate the size of the wheat pile in Exercise 19. (*Hint*: There are about one-half million grains of wheat in a bushel.)

21. Does 0.999999 = 1? Explain.   22. Does 0.999 . . . = 1? Explain.

9.6 PERMUTATIONS AND COMBINATIONS

Lydia plans to go to dinner and then attend a movie. If she has a choice of four restaurants and three movies, in how many ways can Lydia spend her evening? There are four choices of restaurants and, for any one of these options, there are three choices of movies. The choices are shown in the tree diagram in Figure 9-1.

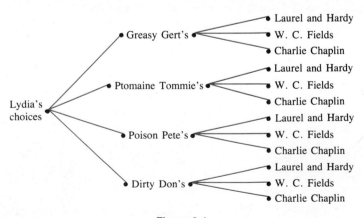

Figure 9-1

This diagram shows that Lydia has 12 ways to spend her evening. One possibility is to visit Ptomaine Tommie's and see W. C. Fields. Another is to dine at Dirty Don's and enjoy Laurel and Hardy.

Any situation that can have several outcomes is called an **event**. Lydia's first event (choosing a restaurant) can occur in 4 ways. Her second event (choosing a movie) can occur in 3 ways. Lydia has 4 times 3, or 12, ways to spend her evening. This example illustrates the **multiplication principle for events.**

The Multiplication Principle for Events. Let E_1 and E_2 be two events. If E_1 can be done in a_1 ways, and if—after E_1 has occurred—E_2 can be done in a_2 ways, then the event "E_1 followed by E_2" can be done in $a_1 \cdot a_2$ ways.

The multiplication principle can be extended to n events.

Example 1 If Frank has four ways to travel from New York to Chicago, three ways to travel from Chicago to Denver, and six ways to travel from Denver to San Francisco, in how many ways can he go from New York to San Francisco?

Solution Let E_1 be the event "going from New York to Chicago," let E_2 be the event "going from Chicago to Denver," and let E_3 be the event "going from Denver to San Francisco." Because there are 4 ways to accomplish E_1, 3 ways to accomplish E_2, and 6 ways to accomplish E_3, the number of routes that Frank can follow is

$$4 \cdot 3 \cdot 6 = 72$$

∎

Permutations

Suppose we wish to arrange 7 books on a shelf. We can fill the first space with any of the 7 books, the second space with any of the remaining 6 books, the third space with any of the remaining 5 books, and so on, until there is only one space to fill with the last book. According to the multiplication principle, the number of ways that we can arrange the books is

$$7 \cdot 6 \cdot 5 \cdot 4 \cdot 3 \cdot 2 \cdot 1 = 5040$$

When computing the number of possible arrangements of the elements in a set such as books on a shelf, we are determining the number of **permutations** of the elements in that set. The number of permutations of 7 books, using all the books, is 5040. The symbol $_nP_r$, which is used in expressing permutation problems, is read as "the number of permutations of n things r at a time."

Example 2 If there are 7 flags of 7 different colors to hang on a flagpole, how many different signals can be sent if only 3 flags are used?

Solution You are asked to find $_7P_3$, the number of permutations of 7 things using only 3 of them. Any one of the 7 flags can hang in the top position on the flagpole. In the middle position, any one of the 6 remaining flags can hang, and in the bottom position can hang any one of the remaining 5 flags. Therefore, according to the multiplication principle,

$$_7P_3 = 7 \cdot 6 \cdot 5 = 210$$

It is possible to send 210 different signals. ∎

Although it is acceptable to write $_7P_3 = 7 \cdot 6 \cdot 5$, there is an advantage in changing the form of this answer to obtain a convenient formula. To get this formula, we will multiply both the numerator and the denominator of $\dfrac{7 \cdot 6 \cdot 5}{1}$ by 4!.

$$_7P_3 = 7 \cdot 6 \cdot 5 = \frac{7 \cdot 6 \cdot 5 \cdot 4 \cdot 3 \cdot 2 \cdot 1}{4 \cdot 3 \cdot 2 \cdot 1} = \frac{7!}{4!} = \frac{7!}{(7 - 3)!}$$

This idea is generalized in the following theorem.

Theorem. The formula for computing the number of permutations of n things r at a time is

$$_nP_r = \frac{n!}{(n - r)!}$$

If $r = n$, then

$$_nP_n = n!$$

The second part of the previous theorem is true because

$$_nP_n = \frac{n!}{(n - n)!} = \frac{n!}{0!} = n! \qquad \text{Remember } 0! = 1.$$

Example 3   In how many ways can a baseball manager arrange the batting order if there are 25 players on the team?

Solution   To find the number of permutations of 25 things 9 at a time ($_{25}P_9$), use the formula $_nP_r = \dfrac{n!}{(n - r)!}$, with $n = 25$ and $r = 9$.

$$_{25}P_9 = \frac{25!}{(25 - 9)!}$$

$$= \frac{25!}{16!}$$

$$= \frac{25 \cdot 24 \cdot 23 \cdot 22 \cdot 21 \cdot 20 \cdot 19 \cdot 18 \cdot 17 \cdot 16!}{16!}$$

$$\approx 741{,}354{,}768{,}000$$

The number of permutations is approximately 741,354,768,000. ∎

Example 4 In how many ways can 5 people stand in a line if 2 people refuse to stand next to each other?

Solution The total number of ways that 5 people can stand in line is

$$_5P_5 = 5! = 5 \cdot 4 \cdot 3 \cdot 2 \cdot 1 = 120$$

To find the number of ways that five people can stand in line if two people *insist* on standing together, consider those two people as a single person. Then there are 4 people to stand in line, and this can be done in $_4P_4 = 4! = 24$ ways. However, there are two arrangements for the pair that insist on standing together, because either could be first. Hence, there are $2 \cdot 4!$ ways that the 5 people can stand in line if 2 people insist on standing together.

The number of ways that 5 people can stand in line if two people *refuse* to stand together is $5! = 120$ (the total number of ways to arrange 5 people) minus $2 \cdot 4! = 48$ (the number of ways to arrange five people if two must stand together):

$$120 - 48 = 72$$

There are 72 ways to stand 5 people in a line if 2 people refuse to stand next to each other. ∎

Example 5 In how many ways can 5 people be seated at a round table?

Solution If you were to seat 5 people in a row, there would be 5! possible arrangements. However, the situation is different when seating people at a round table. At a round table, each person has a neighbor to the left and to the right. If each person moves one place to the left, everyone will still have the same neighbors as before. This same situation applies if everyone moves two, three, four, or five places. Hence, you must divide 5! by 5 to get rid of these duplications. The numbers of ways that 5 people can be seated at a round table is

$$\frac{5!}{5} = 4! = 4 \cdot 3 \cdot 2 \cdot 1 = 24$$

∎

The results of Example 5 can be generalized into the following theorem.

Theorem. There are $(n - 1)!$ ways to place n things in a circle.

Combinations

Suppose a class of 12 students selects a committee of 3 to plan a party. A possible committee is John, Sally, and Joe. In this situation, the order of the 3 is not important, because the committee of John, Sally, and Joe is the same as the committee of Sally, Joe, and John. For the moment, however, we assume that order is important and compute the number of permutations of 12 things 3 at a time:

$$_{12}P_3 = \frac{12!}{(12 - 3)!} = \frac{12 \cdot 11 \cdot 10 \cdot 9!}{9!} = 1320$$

This result indicates the number of ways of arranging 3 people if there are 12 people to choose from. However, with committees we do not care about order. Because there are six ways (3!) of ordering the committee of 3 students, the calculation of $_{12}P_3 = 1320$ provides an answer that is 6 times too big. Actually, the number of possible committees that could plan the party is the number of permutations of 12 things (taken 3 at a time) divided by 6:

$$\frac{_{12}P_3}{6} = \frac{1320}{6} = 220$$

When choosing committee members and in other cases of selection where order is not important, we are interested in **combinations**, not permutations. The symbols $_nC_r$ and $\binom{n}{r}$ both mean the number of combinations of n things r at a time. If a committee of r people is chosen from a total of n people, the number of committees is $_nC_r$, and there will be $r!$ arrangements of each committee. If we consider the committee as an ordered grouping, the number of orderings of r people selected from a group of n people is $_nP_r$. Therefore, the number of *combinations* of n things r at a time multiplied by $r!$ is equal to the number of *permutations* of n things r at a time. This relationship is shown by the equation

$$r!\,_nC_r = \,_nP_r$$

We divide both sides of this equation by $r!$ to obtain the formula for computing $_nC_r$, or $\binom{n}{r}$.

$$_nC_r = \binom{n}{r} = \frac{_nP_r}{r!} = \frac{n!}{r!(n - r)!}$$

This reasoning leads to the following theorem.

Theorem.   The formula for computing the number of combinations of n things r at a time is

$$_nC_r = \binom{n}{r} = \frac{n!}{r!(n - r)!}$$

In the exercises, you will be asked to prove the following theorem.

Theorem.   If n is a whole number, then

$$_nC_n = 1 \qquad \text{and} \qquad _nC_0 = 1$$

Example 6 If Carla must read 4 books from a reading list of 10 books, how many choices does she have?

Solution Because the order in which the books are read is unimportant, calculate the number of combinations of 10 things 4 at a time:

$$_{10}C_4 = \frac{10!}{4!(10-4)!} = \frac{10 \cdot 9 \cdot 8 \cdot 7 \cdot 6!}{4 \cdot 3 \cdot 2 \cdot 1 \cdot 6!}$$

$$= \frac{10 \cdot 9 \cdot 8 \cdot 7}{4 \cdot 3 \cdot 2}$$

$$= 210$$

Carla has 210 options. ∎

Example 7 A class consists of 15 boys and 8 girls. In how many ways can we choose a debate team that will be made up of 3 boys and 3 girls?

Solution There are $_{15}C_3$ ways of choosing the three boys and $_8C_3$ ways of choosing the three girls. By the multiplication principle, there are $_{15}C_3 \cdot _8C_3$ ways of choosing members of the debate team:

$$_{15}C_3 \cdot _8C_3 = \frac{15!}{3!(15-3)!} \cdot \frac{8!}{3!(8-3)!}$$

$$= \frac{15 \cdot 14 \cdot 13 \cdot 12!}{6 \cdot 12!} \cdot \frac{8 \cdot 7 \cdot 6 \cdot 5!}{6 \cdot 5!}$$

$$= \frac{15 \cdot 14 \cdot 13}{6} \cdot \frac{8 \cdot 7 \cdot 6}{6}$$

$$= 25,480$$

There are 25,480 ways to place 3 boys and 3 girls on the debate team. ∎

The formula $_nC_r = n!/[r!(n-r)!]$ gives the coefficient of the $(r+1)$th term of the binomial expansion of $(a+b)^n$. This implies that the coefficients of a binomial expansion can be used to solve problems involving combinations. The binomial theorem is restated below, this time using combination notation.

The Binomial Theorem. If n is any positive integer, then

$$(a+b)^n = \binom{n}{0}a^n + \binom{n}{1}a^{n-1}b + \binom{n}{2}a^{n-2}b^2 + \cdots$$

$$+ \binom{n}{r}a^{n-r}b^r + \cdots + \binom{n}{n}b^n$$

Example 8   Use Pascal's triangle to compute $_7C_5$.

Solution   Consider the eighth row of Pascal's triangle and the corresponding combinations:

$$
\begin{array}{cccccccc}
1 & 7 & 21 & 35 & 35 & 21 & 7 & 1 \\
\dbinom{7}{0} & \dbinom{7}{1} & \dbinom{7}{2} & \dbinom{7}{3} & \dbinom{7}{4} & \dbinom{7}{5} & \dbinom{7}{6} & \dbinom{7}{7}
\end{array}
$$

$$_7C_5 = \dbinom{7}{5} = 21$$

■

When discussing permutations and combinations, a "word" is a distinguishable arrangement of letters. For example, 6 words can be formed with the letters *a*, *b*, and *c* if 3 letters are used exactly once. The six words are *abc*, *acb*, *bac*, *bca*, *cab*, and *cba*. If there are *n* distinct letters and each letter is used once, the number of distinct words that can be formed is $n! = {}_nP_n$. It is more complicated to compute the number of distinguishable words that can be formed with *n* letters if some of the letters appear more than once.

Example 9   Find the number of "words" that can be formed if each of the 6 letters of the word *little* is used once.

Solution   For the moment, assume that all of the letters of the word *little* are distinguishable: "LitTle." The number of words that can be formed using each letter once is $6! = {}_6P_6$. However, in reality you cannot tell the *l*'s or the *t*'s apart. Therefore, divide by an appropriate number to get rid of these duplications; because there are 2! orderings of the two *l*'s and 2! orderings of the two *t*'s, divide by $2! \cdot 2!$. The number of words that can be formed using each letter of the word *little* is

$$\frac{{}_6P_6}{2!2!} = \frac{6!}{2!2!} = \frac{6 \cdot 5 \cdot 4 \cdot 3 \cdot 2 \cdot 1}{2 \cdot 1 \cdot 2 \cdot 1} = 180$$

■

Example 9 illustrates the following general principle.

Theorem.   If a word with *n* letters has *a* of one letter, *b* of another letter, and so on, then the number of distinguishable words that can be formed using each letter of the *n*-letter word exactly once is

$$\frac{n!}{a!b! \cdots}$$

EXERCISE 9.6

1. A lunchroom has a machine with eight kinds of sandwiches, a machine with four kinds of soda, a machine with both white and chocolate milk, and a machine with three kinds of ice cream. How many different lunches can be chosen? (Consider a lunch to be one sandwich, one drink, and one ice cream.)

2. How many six-digit license plates can be manufactured if no license plate number begins with 0?

3. How many different seven-digit phone numbers can be used in one area code if no phone number begins with 0 or 1?

4. In how many ways can the letters of the word *number* be arranged?

5. In how many ways can the letters of the word *number* be arranged if the *e* and *r* must remain next to each other?

6. In how many ways can the letters of the word *number* be arranged if the *e* and *r* cannot be side by side?

7. How many ways can five Scrabble tiles bearing the letters, *F, F, F, L,* and *U* be arranged to spell the word *fluff*?

8. How many ways can six Scrabble tiles bearing the letters *B, E, E, E, F,* and *L* be arranged to spell the word *feeble*?

In Exercises 9–24, evaluate each expression.

9. $_7P_4$

10. $_8P_3$

11. $_7C_4$

12. $_8C_3$

13. $_5P_5$

14. $_5P_0$

15. $\binom{5}{4}$

16. $\binom{8}{4}$

17. $\binom{5}{0}$

18. $\binom{5}{5}$

19. $_5P_4 \cdot _5C_3$

20. $_3P_2 \cdot _4C_3$

21. $\binom{5}{3}\binom{4}{3}\binom{3}{3}$

22. $\binom{5}{5}\binom{6}{6}\binom{7}{7}\binom{8}{8}$

23. $\binom{68}{66}$

24. $\binom{100}{99}$

25. In how many arrangements can 8 girls be placed in a line?

26. In how many arrangements can 5 girls and 5 boys be placed in a line if the girls and boys alternate?

27. In how many arrangements can 5 girls and 5 boys be placed in a line if all the boys line up first?

28. In how many arrangements can 5 girls and 5 boys be placed in a line if all the girls line up first?

29. How many permutations does a combination lock have if each combination has 3 numbers, no two numbers of the combination are the same, and the lock dial has 30 notches? Wouldn't it be better to call these locks "permutation locks"? Explain.

30. How many permutations does a combination lock have if each combination has 3 numbers, no two numbers of the combination are the same, and the lock has 100 notches?

31. In how many ways can 8 people be seated at a round table?

32. In how many ways can 7 people be seated at a round table?

33. In how many ways can 6 people be seated at a round table if 2 of the people insist on sitting together?

34. In how many ways can 6 people be seated at a round table if 2 of the people refuse to sit together?

35. In how many ways can 7 children be arranged in a circle if Sally and John want to sit together and Martha and Peter want to sit together?

36. In how many ways can 8 children be arranged in a circle if Laura, Scott, and Paula want to sit together?

37. In how many ways can 4 candy bars be selected from 10 different candy bars?

38. In how many ways can a hand of 5 cards be selected from a standard deck of 52 playing cards?

39. How many possible bridge hands are there? (*Hint:* There are 13 cards in a bridge hand and 52 cards in a deck.)

40. How many words can be formed from the letters of the word *igloo* if each of the 5 letters is to be used once?

41. How many words can be formed from the letters of the word *parallel* if each letter is to be used once?

42. How many words can be formed from the letters of the word *banana* if each letter is to be used once?

43. How many license plates can be made using two different letters followed by four different digits if the first digit cannot be 0 and the letter O is not used?

44. If there are seven class periods in a school day and a typical student takes 5 classes, how many different time patterns are possible for the student?

45. From a bucket containing 6 red and 8 white golf balls, in how many ways can we draw 6 golf balls of which 3 are red and 3 are white?

46. In how many ways can you select a committee of 3 Republicans and 3 Democrats from a group containing 18 Democrats and 11 Republicans?

47. In how many ways can you select a committee of 4 Democrats and 3 Republicans from a group containing 12 Democrats and 10 Republicans?

48. In how many ways can you select a group of 5 red cards and 2 black cards from a deck containing 10 red cards and 8 black cards?

49. In how many ways can a husband and wife choose 2 different dinners from a menu of 17 dinners?

50. In how many ways can 7 people stand in a row if 2 of the people refuse to stand together?

51. How many lines are determined by 8 points if no 3 points lie on a straight line?

52. How many lines are determined by 10 points if no 3 points lie on a straight line?

53. Use Pascal's triangle to find $_8C_5$.

54. Use Pascal's triangle to find $_{10}C_8$.

55. How many teams can a baseball manager put on the field if the entire squad consists of 25 players? (There are 9 players on the field at a time. Assume that all players can play all positions.)

56. Prove that $_nC_n = 1$ and that $_nC_0 = 1$.

57. Prove that $\binom{n}{r} = \binom{n}{n-r}$.

58. Show that the binomial theorem can be expressed in the form

$$(a + b)^n = \sum_{k=0}^{n} \binom{n}{k} a^{n-k} b^k$$

9.7 PROBABILITY

The probability that a certain event will occur is a measure of the likelihood of that event. A tossed coin, for example, can land in two equally likely ways, either heads or tails. Because one of these two outcomes is heads, we expect that out of several tosses about half will be heads. We say that the probability of obtaining heads in a single toss of the coin is $\frac{1}{2}$.

If a multiple-choice question has five possible answers and only one is correct, we expect to guess the correct answer one time in five. Thus, the probability of guessing the correct answer is $\frac{1}{5}$. If records show that out of 100 days with weather conditions like today's, 30 have received rain, the weather service reports "Today, there is a $\frac{30}{100}$ or 30% probability of rain."

In the language of statistics, an **experiment** is any process for which the outcome is uncertain. Thus, tossing a coin, drawing a card, guessing on a multiple-choice question, and predicting rain are examples of experiments. We define the probability of a favorable outcome of an experiment as follows:

Definition.   If an experiment can have n distinct and equally likely outcomes and E is an event that can occur in s of these ways, the probability of E is

$$P(E) = \frac{s}{n}$$

Because $0 \leq s \leq n$, it follows that $0 \leq \frac{s}{n} \leq 1$ and that all probabilities must have values from 0 to 1. An event that cannot happen has probability 0, and an event that is certain to happen has probability 1.

Saying that the probability of tossing heads on a single toss of a coin is $\frac{1}{2}$ means that if a fair coin is tossed a very large number of times, the ratio of the number of heads to the total number of tosses is very nearly $\frac{1}{2}$. Saying that the probability of a student guessing the one correct answer from 5 choices is $\frac{1}{5}$ means that the student, guessing on an exam consisting of many such questions, can expect a score of about 20%. To say that the probability of rolling a 5 on a single roll of a die is $\frac{1}{6}$ is to say that as the number of rolls approaches infinity, the ratio of the number of favorable outcomes (rolling a 5) to the total number of outcomes (rolling a 1, 2, 3, 4, 5, or 6) approaches $\frac{1}{6}$.

For each experiment, the total list of possible outcomes is called a **sample space**. For example, the sample space of the experiment of tossing two coins is found by listing all possible outcomes. Let H represent "heads," and T, "tails," and use ordered-pair notation to indicate the possible results. Thus, (H, T) represents the outcome "heads on the first coin and tails on the second." The sample space contains the four possible outcomes

$$\text{(H, H)} \qquad \text{(H, T)} \qquad \text{(T, H)} \qquad \text{(T, T)}$$

Because there are two possible outcomes for the first coin and two for the second, we know by the multiplication principle for events that there are $2 \cdot 2$, or 4 total possible outcomes. To find the probability of getting at least one heads on a toss of two coins, we let E be the event "getting at least one heads." Three outcomes are favorable to this event:

$$\text{(H, H)} \qquad \text{(H, T)} \qquad \text{(T, H)}$$

Because there are three favorable outcomes among the four possible outcomes, and all outcomes are equally likely,

$$P(\text{at least one heads}) = \frac{3}{4}$$

Example 1   Exhibit the sample space of the event "rolling two dice a single time."

Solution   The sample space is the listing of all possible outcomes. Use ordered-pair notation, and let the first number of the pair be the result on the first die and the second number the result on the second die:

$$(1, 1) \quad (1, 2) \quad (1, 3) \quad (1, 4) \quad (1, 5) \quad (1, 6)$$
$$(2, 1) \quad (2, 2) \quad (2, 3) \quad (2, 4) \quad (2, 5) \quad (2, 6)$$
$$(3, 1) \quad (3, 2) \quad (3, 3) \quad (3, 4) \quad (3, 5) \quad (3, 6)$$
$$(4, 1) \quad (4, 2) \quad (4, 3) \quad (4, 4) \quad (4, 5) \quad (4, 6)$$
$$(5, 1) \quad (5, 2) \quad (5, 3) \quad (5, 4) \quad (5, 5) \quad (5, 6)$$
$$(6, 1) \quad (6, 2) \quad (6, 3) \quad (6, 4) \quad (6, 5) \quad (6, 6)$$

This sample space contains 36 ordered pairs. Because there are 6 possible outcomes with the first die and 6 possible outcomes with the second die, the multiplication principle for events tells you to expect $6 \cdot 6$, or 36 equally likely possible outcomes. ∎

Example 2   On a single toss of two dice, what is the probability of tossing a sum of seven?

Solution   The sample space for this event is listed in Example 1. The favorable outcomes are the ones that give a sum of 7: (1, 6), (2, 5), (3, 4), (4, 3), (5, 2), and (6, 1). Because there are 6 favorable outcomes among the 36 equally likely outcomes,

$$P(\text{tossing a } 7) = \frac{6}{36} = \frac{1}{6}$$

∎

Example 3   What is the probability of drawing 5 cards, all hearts, from an ordinary deck of cards?

Solution   The number of ways to draw 5 hearts from the 13 hearts is $_{13}C_5$. The number of ways of drawing 5 cards from the complete deck is $_{52}C_5$. The desired probability is the ratio of the number of favorable outcomes to the number of possible outcomes.

$$P(5 \text{ hearts}) = \frac{_{13}C_5}{_{52}C_5}$$

$$P(5 \text{ hearts}) = \frac{\dfrac{13!}{5!8!}}{\dfrac{52!}{5!47!}}$$

$$= \frac{13!}{5!8!} \cdot \frac{5!47!}{52!}$$

$$= \frac{13 \cdot 12 \cdot 11 \cdot 10 \cdot 9 \cdot 8!}{8!} \cdot \frac{47!}{52 \cdot 51 \cdot 50 \cdot 49 \cdot 48 \cdot 47!}$$

$$= \frac{13 \cdot 12 \cdot 11 \cdot 10 \cdot 9}{52 \cdot 51 \cdot 50 \cdot 49 \cdot 48}$$

$$= \frac{33}{66,640}$$

The probability of drawing 5 hearts is $\frac{33}{66,640}$. ∎

There is a multiplication property for probabilities that is similar to the multiplication principle for events that occur in succession.

Multiplication Property of Probabilities. If $P(A)$ represents the probability of event A, and $P(B|A)$ represents the probability that event B will occur after event A, then $P(A \text{ and } B) = P(A) \cdot P(B|A)$.

Example 4 A box contains 40 cubes of the same size. Of these cubes, 17 are red, 13 are blue, and the rest are yellow. If 2 cubes are drawn at random, without replacement, what is the probability that 2 yellow cubes will be drawn?

Solution Of the 40 cubes in the box, 10 are yellow. Thus, the probability of getting a yellow cube on the first draw is

$$P(\text{yellow cube on the first draw}) = \frac{10}{40} = \frac{1}{4}$$

Because there is no replacement after the first draw, 39 cubes remain in the box, and 9 of these are yellow. The probability of getting a yellow cube on the second draw is

$$P(\text{yellow cube on the second draw}) = \frac{9}{39} = \frac{3}{13}$$

The probability of drawing 2 yellow cubes in succession is the product of the probability of drawing a yellow cube on the first draw and the probability of drawing a yellow cube on the second draw.

$$P(\text{drawing two yellow cubes}) = \frac{1}{4} \cdot \frac{3}{13} = \frac{3}{52}$$

■

Example 5 Repeat Example 3 using the multiplication property of probabilities.

Solution The probability of drawing a heart on the first draw is $\frac{13}{52}$, on the second draw $\frac{12}{51}$, on the third draw $\frac{11}{50}$, on the fourth draw $\frac{10}{49}$, and on the fifth draw $\frac{9}{48}$. By the multiplication property of probabilities,

$$P(\text{5 hearts in a row}) = \frac{13}{52} \cdot \frac{12}{51} \cdot \frac{11}{50} \cdot \frac{10}{49} \cdot \frac{9}{48}$$

$$= \frac{33}{66,640}$$

■

Example 6   In a school, 30% of the students are gifted in mathematics, and 10% are gifted in both art and mathematics. If a student earns an A in an advanced mathematics class, what is the probability that he or she is also gifted in art?

Solution   Let $P(M)$ be the probability that a student chosen at random will be gifted in mathematics, and let $P(M \text{ and } A)$ be the probability of such a student being gifted in both art and mathematics. You must determine $P(A|M)$, the probability that the student is gifted in art, given that he or she is gifted in mathematics. To do so, substitute the given values

$$P(M) = 0.3$$
$$P(M \text{ and } A) = 0.1$$

in the formula for multiplication of probabilities, and solve for $P(A|M)$:

$$P(M \text{ and } A) = P(M) \cdot P(A|M)$$
$$0.1 = (0.3)P(A|M)$$
$$P(A|M) = \frac{0.1}{0.3}$$
$$= \frac{1}{3}$$

If a student is already gifted in mathematics, there is a probability of $\frac{1}{3}$ that he or she is also gifted in art.   ■

EXERCISE 9.7

In Exercises 1–4, an ordinary die is tossed. Find the probability of each event.

1. Tossing a 2

2. Tossing a number greater than 4

3. Tossing a number larger than 1 but less than 6

4. Tossing an odd number

In Exercises 5–8, balls numbered from 1 to 42 are placed in a container and stirred. If one is drawn at random, find the probability of each event.

5. The number is less than 20.

6. The number is less than 50.

7. The number is a prime number.

8. The number is less than 10 or greater than 40.

In Exercises 9–12, refer to the spinner in Illustration 1. If the spinner is spun, find the probability of each event. Assume that the spinner never stops on a line.

9. The spinner stops on red.

10. The spinner stops on green.

11. The spinner stops on orange.

12. The spinner stops on yellow.

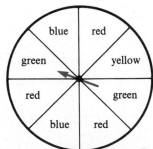

Illustration 1

In Exercises 13–30, find the probability of each given event.

13. Rolling a sum of 4 on one roll of two dice

14. Drawing a diamond on one draw from a card deck

15. Drawing two aces in succession from a card deck if the card is replaced and the deck is shuffled after the first draw

16. Drawing two aces from a card deck without replacing the card after the first draw

17. Drawing a red egg from a basket containing 5 red eggs and 7 blue eggs

18. Getting 2 red eggs in a single scoop from a bucket containing 5 red eggs and 7 yellow eggs

19. Drawing a bridge hand of 13 cards, all of one suit

20. Drawing 6 diamonds from a card deck without replacing the cards after each draw

21. Drawing 5 aces from a card deck without replacing the cards after each draw

22. Drawing 5 clubs from the black cards in a card deck

23. Drawing a face card from a card deck

24. Drawing 6 face cards in a row from a card deck without replacing the cards after each draw

25. Drawing 5 orange cubes from a bowl containing 5 orange cubes and 1 beige cube

26. Rolling a sum of 4 with one roll of three dice

27. Rolling a sum of 11 with one roll of three dice

28. Picking, at random, 5 Republicans from a group containing 8 Republicans and 10 Democrats

29. Tossing 3 heads in 5 tosses of a fair coin

30. Tossing 5 heads in 5 tosses of a fair coin

In Exercises 31–37, assume that the probability that an airplane engine will fail during a torture test is $\frac{1}{2}$ and that the aircraft in question has 4 engines.

31. Construct a sample space showing each of the possible outcomes after the torture test.

32. Find the probability that all of the engines will survive the test.

33. Find the probability that exactly 1 engine will survive.

34. Find the probability that exactly 2 engines will survive.

35. Find the probability that exactly 3 engines will survive.

36. Find the probability that no engines will survive.

37. Find the sum of the probabilities in Exercises 32 through 36.

In Exercises 38–40, assume that a survey of 282 people is taken to determine the opinions of doctors, teachers, and lawyers on a proposed piece of legislation, with the following results:

	Number that favor	Number that oppose	Number with no opinion	Total
Doctors	70	32	17	119
Teachers	83	24	10	117
Lawyers	23	15	8	46
Total	176	71	35	282

A person is chosen at random from those surveyed. Refer to the chart to find each probability.

38.   What is the probability that the person favors the legislation?

39.   If a doctor is chosen, what is the probability that he or she opposes the legislation?

40.   If the person opposes the legislation, what is the probability that he or she is a lawyer?

41.   In a batch of 10 tires, 2 are known to be defective. If 4 tires are chosen at random, find the probability that all 4 tires are good.

42.   Out of a group of 9 patients treated with a new drug, 4 will suffer a relapse. Find the probability that 3 patients of this group, chosen at random, will remain disease-free.

In Exercises 43–50, use the multiplication property of probabilities.

43.   If $P(A) = 0.3$ and $P(B|A) = 0.6$, find $P(A \text{ and } B)$.

44.   If $P(A \text{ and } B) = 0.3$ and $P(B|A) = 0.6$, find $P(A)$.

45.   The probability that a person owns a luxury car is 0.2, and the probability that the owner of such a car also owns a personal computer is 0.7. Find the probability that a person, chosen at random, owns both a luxury car and a personal computer.

46.   If 40% of the population have completed college, and 85% of college graduates are registered to vote, what percent of the population are college graduates and registered voters?

47.   About 25% of the population watches the evening television news coverage as well as the morning soap operas. If 75% of the population watches the news, what percentage of those who watch the news also watch the soaps?

48.   The probability of rain today is 0.40. If it rains, the probability that Bill will forget his raincoat is 0.70. Find the probability that Bill will get wet.

49.   If $P(A \text{ and } B) = 0.7$, is it possible that $P(B|A) = 0.6$? Explain.

50.   Is it possible that $P(A \text{ and } B) = P(A)$? Explain.

9.8   COMPUTATION OF COMPOUND PROBABILITIES

Sometimes we need to compute the probability of one event *or* another, or one event *and* another. Such events are called **compound events**.

Suppose we wish to find the probability of drawing an ace *or* a heart from an ordinary card deck. The probability of drawing an ace is $\frac{4}{52}$, and the probability of drawing a heart is $\frac{13}{52}$. However, the probability of drawing an ace *or* a heart is not the sum of these two probabilities. Because the ace of hearts was counted twice, once as an ace and once as a heart, and the probability of drawing the ace of hearts is $\frac{1}{52}$, we must subtract $\frac{1}{52}$ from the sum of $\frac{4}{52}$ and $\frac{13}{52}$. Thus,

$$P(\text{ace } or \text{ heart}) = P(\text{ace}) + P(\text{heart}) - P(\text{ace of hearts})$$

$$= \frac{4}{52} + \frac{13}{52} - \frac{1}{52}$$

$$= \frac{16}{52}$$

$$= \frac{4}{13}$$

This example suggests the following theorem.

Theorem. If A and B are two events, then

$$P(A \text{ or } B) = P(A) + P(B) - P(A \text{ and } B)$$

If event A and event B are **mutually exclusive** (that is, if one event occurs, the other cannot), then $P(A \text{ and } B) = 0$ and we have another theorem.

Theorem. If A and B cannot occur simultaneously, then

$$P(A \text{ or } B) = P(A) + P(B)$$

Because the events A and $\bar{A}$ (read as "not A") are mutually exclusive,

$$P(A \text{ or } \bar{A}) = P(A) + P(\bar{A})$$

Because either event A or event $\bar{A}$ must happen, $P(A \text{ or } \bar{A}) = 1$. Thus,

$$P(A \text{ or } \bar{A}) = 1$$
$$P(A) + P(\bar{A}) = 1$$
$$P(\bar{A}) = 1 - P(A) \qquad \text{Add} - P(A) \text{ to both sides.}$$

This result gives a third theorem about compound probabilities.

Theorem. If A is any event, then

$$P(\bar{A}) = 1 - P(A)$$

Example 1 A guidance counselor tells a student that his probability of earning a grade of D in algebra is $\frac{1}{5}$, and his probability of failing is $\frac{1}{25}$. What is the probability that the student earns a C or better?

Solution The probability of earning a D or F is given by

$$P(D \text{ or } F) = P(D) + P(F) \qquad \text{Note that } P(D \text{ and } F) = 0.$$

$$= \frac{1}{5} + \frac{1}{25}$$

$$= \frac{6}{25}$$

The probability that the student will receive a C or better is given by

$$P(C \text{ or better}) = 1 - P(D \text{ or } F)$$

$$= 1 - \frac{6}{25}$$

$$= \frac{19}{25}$$

The student's probability of earning a C or better is $\frac{19}{25}$.   ■

If two events do not influence each other, they are called **independent events**.

Definition.   The events A and B are said to be **independent events** if and only if $P(B) = P(B|A)$.

Substituting $P(B)$ for $P(B|A)$ in the multiplication property for probabilities gives a formula for computing probabilities of compound independent events.

Theorem.   If A and B are independent events, then

$$P(A \text{ and } B) = P(A) \cdot P(B)$$

For example, $P(\text{drawing an ace and tossing a heads}) = P(\text{drawing an ace}) \cdot P(\text{tossing a heads})$. This is true because neither event influences the other. Consequently,

$$P(\text{drawing an ace and tossing a heads}) = \frac{4}{52} \cdot \frac{1}{2} = \frac{1}{26}$$

Example 2   The probability that a softball player can get a hit is $\frac{1}{3}$. What is the probability that she will bat safely three times in a row?

Solution   Assume that the three events (each time at bat) are independent; one time at bat does not influence her chances of getting a hit on another turn at bat. Because $P(E_1) = \frac{1}{3}$, $P(E_2) = \frac{1}{3}$, and $P(E_3) = \frac{1}{3}$,

$$P(E_1 \text{ and } E_2 \text{ and } E_3) = \frac{1}{3} \cdot \frac{1}{3} \cdot \frac{1}{3} = \frac{1}{27}$$

The probability that she bats safely three times in a row is $\frac{1}{27}$.   ■

Example 3   A die is tossed three times. What is the probability that the outcome is a six on the first toss, an even number on the second toss, and an odd prime number on the third toss?

Solution The probability of a six on any toss is $P(\text{six}) = \frac{1}{6}$. Because there are three even integers represented on the faces of a die, the probability of tossing an even number is $P(\text{even number}) = \frac{3}{6} = \frac{1}{2}$. The numbers 3 and 5 are the only odd prime numbers on the faces of a die, so the probability of tossing an odd prime is $P(\text{odd prime}) = \frac{2}{6} = \frac{1}{3}$. Because these three events are independent, the probability of the three events happening in succession is the product of the probabilities:

$$P(\text{six and even number and odd prime}) = \frac{1}{6} \cdot \frac{1}{2} \cdot \frac{1}{3} = \frac{1}{36}$$ ∎

Example 4 The probability that the drug Flake Off will cure dandruff is $\frac{1}{8}$. If the drug is used, the probability that the patient will have side effects is $\frac{1}{6}$. What is the probability that a patient who uses the drug will be cured and will suffer no side effects?

Solution The probability that Flake Off will work is $P(\text{works}) = \frac{1}{8}$, and the probability that the patient will have no side effects is $P(\text{no side effects}) = 1 - P(\text{side effects}) = 1 - \frac{1}{6} = \frac{5}{6}$. These events are independent, so $P(\text{cure and no side effects}) = \frac{1}{8} \cdot \frac{5}{6} = \frac{5}{48}$. ∎

EXERCISE 9.8

In Exercises 1–4, assume that you draw one card from a card deck. Find the probability of each event.

1. Drawing a black card

2. Drawing a jack

3. Drawing a black card or an ace

4. Drawing a red card or a face card

In Exercises 5–8, assume that you draw two cards from a card deck without replacement. Find the probability of each event.

5. Drawing two aces

6. Drawing three aces

7. Drawing a club and then drawing another black card

8. Drawing a heart and then drawing a spade

In Exercises 9–12, assume that you roll two dice once.

9. What is the probability of rolling a sum of 7 or 6?

10. What is the probability of rolling a sum of 5 or an even sum?

11. What is the probability of rolling a sum of 10 or an odd sum?

12. What is the probability of rolling a sum of 12 or 1?

In Exercises 13–16, assume that you are dealing with a bucket that contains 7 beige capsules, 3 blue capsules, and 6 green capsules. You make a single draw from the bucket, taking one capsule.

13. What is the probability of drawing a beige or a blue capsule?

14. What is the probability of drawing a green capsule?

15. What is the probability of not drawing a blue capsule?

16. What is the probability of not drawing either a beige or a blue capsule?

In Exercises 17–19, assume that you are using the same bucket of capsules as in Exercises 13–16.

17. On two draws from the bucket, what is the probability of drawing a beige capsule followed by a green capsule? (Assume that the capsule is returned to the bucket after the first draw.)

18. On two draws from the bucket, what is the probability of drawing one blue and one green capsule? (Assume that the capsule is not returned to the bucket after the first draw.)

19. On three successive draws from the bucket (without replacement), what is the probability of failing to draw a beige capsule?

20. Jeff rolls a die and draws one card from a card deck. What is the probability of his rolling a four and drawing a four?

21. Three people are in an elevator together. What is the probability that all three were born on the same day of the week?

22. Three people are on a bus together. What is the probability that at least one was born on a different day of the week than the others?

23. Five people are in a room together. What is the probability that all five were born on a different day of the year?

24. Five people are on a bus together. What is the probability that at least two of them were born on the same day of the year?

25. If the probability that Rick will solve a problem is $\frac{1}{4}$ and the probability that Dinah will solve it is $\frac{2}{5}$, what is the probability that at least one of them will solve the problem?

26. A certain bugle call is based on four pitches and is five notes long. If a child can play these four pitches on a bugle, what is the probability that the first five notes that the child plays will be the bugle call? (Assume that the child is equally likely to play any of the four pitches each time a note is blown.)

27. Valerie visits her cabin in Canada. The probability that her lawn mower will start is $\frac{1}{2}$, the probability that her gas power saw will start is $\frac{1}{3}$, and the probability that her outboard motor will start is $\frac{3}{4}$. What is the probability that all three will start so that Valerie will have a nice vacation? What is the probability that none will start? That exactly one will start? That exactly two will start? What is the sum of your answers?

28. Three children will leave Thailand to start a new life in either the United States or France. The probability that May Xao will go to France is $\frac{1}{3}$, that Tou Lia will go to France is $\frac{1}{2}$, and that May Moua will go to France is $\frac{1}{6}$. What is the probability that exactly two of them will end up in the United States?

9.9 ODDS AND MATHEMATICAL EXPECTATION

There is a concept known as **mathematical odds** that is related to probability.

> **Definition.**   The **odds for an event** is the probability of a favorable outcome divided by the probability of an unfavorable outcome.

> **Definition.**   The **odds against an event** is the probability of an unfavorable outcome divided by the probability of a favorable outcome.

Example 1   The probability that a horse will win a race is $\frac{1}{4}$. What are the odds for and the odds against the horse?

Solution   Because the probability that the horse will win is $\frac{1}{4}$, the probability that the horse will not win is $\frac{3}{4}$. Therefore, the odds for the horse are

$$\frac{\text{probability of a win}}{\text{probability of a loss}} = \frac{\dfrac{1}{4}}{\dfrac{3}{4}} = \frac{1}{3}$$

or 1 to 3, and the odds against the horse are

$$\frac{\text{probability of a loss}}{\text{probability of a win}} = \frac{\dfrac{3}{4}}{\dfrac{1}{4}} = 3$$

or 3 to 1. Note that the odds for an event is the reciprocal of the odds against the event.   ∎

Suppose we have a chance to play a simple game with the following rules:

1. Roll a single die once.
2. If a six appears, win $3.
3. If a five appears, win $1.
4. If any other number appears, win 50¢.
5. The cost to play (one roll of the die) is $1.

Whether it would be advisable to play this game depends on what we can expect to win. In this case the probability of any of the six outcomes—rolling a 6, 5, 4, 3, 2, or 1—is $\frac{1}{6}$, and the winnings are $3, $1, and 50¢. The expected winnings can be found by using the following equation and simplifying the right side:

$$E = \frac{1}{6}(3) + \frac{1}{6}(1) + \frac{1}{6}(0.50) + \frac{1}{6}(0.50) + \frac{1}{6}(0.50) + \frac{1}{6}(0.50)$$

$$= \frac{1}{6}(3 + 1 + 0.50 + 0.50 + 0.50 + 0.50)$$

$$= \frac{1}{6}(6)$$

$$= 1$$

Over the long run, we could expect to win $1 with every play of the game. However, it does cost $1 to play the game, so the expected net gain or loss of playing the game is 0. Because the expected winnings equal the admission price, the game is said to be *fair*.

> **Definition.**   If a certain event has n different outcomes with probabilities $p_1, p_2, p_3, \ldots, p_n$ and the winnings assigned to each outcome are $x_1, x_2, x_3, \ldots, x_n$, the expected winnings, or **mathematical expectation**, E, is given by
>
> $$E = p_1x_1 + p_2x_2 + p_3x_3 + \cdots + p_nx_n$$

Example 2   It costs \$1 to play the following game: Roll two dice, collect \$5 if you roll a sum of 7, and collect \$2 if you roll a sum of 11. All other numbers pay nothing. Is it wise to play this game?

Solution   The probability of rolling a 7 on a single roll of two dice is $\frac{6}{36}$, the probability of rolling an 11 is $\frac{2}{36}$, and the probability of rolling something else is $\frac{28}{36}$. The mathematical expectation is

$$E = \frac{6}{36}(5) + \frac{2}{36}(2) + \frac{28}{36}(0)$$

$$= \frac{17}{18}$$

$$\approx \$0.944$$

By playing the game over and over for a long period of time, you can expect to retrieve a little less than 95¢ for every dollar spent. For the fun of playing, the cost is about 5¢ a game. If the game is enjoyable, it might be worth the expected loss of a nickel. However, the game is slightly unfair. ∎

EXERCISE 9.9

In Exercises 1–6, assume a single roll of a die.

1. What is the probability of rolling a 6?

2. What are the odds in favor of rolling a 6?

3. What are the odds against rolling a 6?

4. What is the probability of rolling an even number?

5. What are the odds in favor of rolling an even number?

6. What are the odds against rolling an even number?

In Exercises 7–12, assume a single roll of two dice.

7. What is the probability of rolling a sum of 6?

8. What are the odds in favor of rolling a sum of 6?

9. What are the odds against rolling a sum of 6?

10. What is the probability of rolling an even sum?

11. What are the odds in favor of rolling an even sum?

12. What are the odds against rolling an even sum?

In Exercises 13–16, assume that you are drawing one card from a card deck.

13. What are the odds in favor of drawing a queen?

14. What are the odds against drawing a black card?

15. What are the odds in favor of drawing a face card?

16. What are the odds against drawing a diamond?

17. If the odds in favor of victory are 5 to 2, what is the probability of victory?

18. If the odds in favor of victory are 5 to 2, what are the odds against victory?

19. If the odds against winning are 90 to 1, what are the odds in favor of winning?

20. What are the odds in favor of rolling a 7 on a single toss of two dice?

21. What are the odds against tossing four heads in a row with a fair nickel?

22. What are the odds in favor of a couple having four girl babies in succession? (Assume $P(\text{boy}) = \frac{1}{2}$.)

23. The odds against a horse are 8 to 1. What is the probability that the horse will win?

24. The odds against a horse are 1 to 1. What is the probability that the horse will lose?

25. It costs $2 to play the following game:
   a. Draw one card from a card deck.   **b.** Collect $5 if an ace is drawn.
   c. Collect $4 if a king is drawn.   **d.** Collect nothing for all other cards drawn.
   Is it wise to play this game? Explain.

26. One thousand tickets are sold for a lottery with two grand prizes of $800. What is a "fair" price for the tickets?

27. What are the odds against a couple having three baby boys in a row? (Assume $P(\text{boy}) = \frac{1}{2}$.)

28. What are the odds in favor of tossing at least three heads in five tosses of a fair coin?

29. Suppose you toss a coin five times and collect $5 if you toss five heads, $4 if you toss four heads, $3 if you toss three heads, and no money for any other combinations. How much should you pay to play the game if the game is to be fair?

30. If you toss two dice one time and collect $10 for double sixes and $1 for double ones, what is a fair price for playing the game?

31. Counting an ace as 1, a face card as 10, and all others at their numerical value, what is the expected value if you draw one card from a card deck?

32. What is the expected sum of one roll of two dice?

33. A multiple-choice test of eight questions gives five possible answers for each question. Only one of the answers for each question is right. What is the probability of getting seven right answers by simple guessing?

34. In the situation described in Exercise 33, what are the odds in favor of getting seven answers right?

CHAPTER SUMMARY

Key Words

arithmetic means (9.4)	*multiplication principle for events* (9.6)
arithmetic progression (9.4)	*multiplication property of*
binomial theorem (9.2)	*probabilities* (9.7)
combinations (9.6)	*odds against an event* (9.9)
compound events (9.8)	*odds for an event* (9.9)
event (9.6)	*Pascal's triangle* (9.2)
factorial notation (9.2)	*permutations* (9.6)
general term of a sequence (9.3)	*probability* (9.7)
geometric means (9.4)	*progression* (9.4)
geometric progression (9.4)	*sample space* (9.7)
independent events (9.8)	*sequence* (9.3)
induction hypothesis (9.1)	*series* (9.3)
mathematical expectation (9.9)	*summation notation* (9.3)
mathematical induction (9.1)	

Key Ideas

(9.1) **The Axiom of Mathematical Induction.** If a statement involving the natural number n has the two properties that

1. the statement is true for $n = 1$ and
2. the statement is true for $n = k + 1$ whenever it is true for $n = k$,

then the statement is true for all natural numbers.

(9.2) $n! = n(n - 1)(n - 2) \cdots 3 \cdot 2 \cdot 1$ $0! = 1$ $n(n - 1)! = n!$

The Binomial Theorem. If n is any positive integer, then

$$(a + b)^n = a^n + \frac{n!}{1!(n - 1)!}a^{n-1}b + \frac{n!}{2!(n - 2)!}a^{n-2}b^2$$

$$+ \frac{n!}{3!(n - 3)!}a^{n-3}b^3 + \cdots + \frac{n!}{r!(n - r)!}a^{n-r}b^r + \cdots + b^n$$

(9.3) If c is a constant, then $\sum_{k=1}^{n} c = nc$.

If c is a constant, then $\sum_{k=1}^{n} cf(k) = c \sum_{k=1}^{n} f(k)$.

$$\sum_{k=1}^{n} [f(k) + g(k)] = \sum_{k=1}^{n} f(k) + \sum_{k=1}^{n} g(k)$$

(9.4) The formula $S_n = \dfrac{n(a + l)}{2}$ gives the sum of the first n terms of an arithmetic progression. In this formula, S_n is the sum, a is the first term, l is the last (or nth) term, and n is the number of terms.

The formula $S_n = \dfrac{a - ar^n}{1 - r}$ gives the sum of the first n terms of a geometric progression. In this formula, S_n is the sum, a is the first term, r is the common ratio, and n is the number of terms. Assume that $r \neq 1$.

If $|r| < 1$, the sum of the terms of an infinite geometric progression is given by the formula $S = \dfrac{a}{1 - r}$, where S is the sum, a is the first term, and r is the common ratio.

(9.6) The formula for computing the number of permutations of n things r at a time is

$$_nP_r = \frac{n!}{(n - r)!}$$

There are $(n - 1)!$ ways to place n things in a circle.

The formula for computing the number of combinations of n things r at a time is

$$_nC_r = \binom{n}{r} = \frac{n!}{r!(n - r)!}$$

If an n-letter word has a of one letter, b of another letter, and so on, then the number of distinguishable words that can be formed using each letter exactly once is

$$\frac{n!}{a!b! \cdots}$$

(9.7)   An event that cannot happen has a probability of 0. An event that is certain to happen has a probability of 1. All other events have probabilities between 0 and 1.

(9.8)   If A and B are two events, then

$$P(A \text{ or } B) = P(A) + P(B) - P(A \text{ and } B)$$

If A and B cannot occur simultaneously, then

$$P(A \text{ or } B) = P(A) + P(B)$$

If A is any event, then

$$P(\bar{A}) = 1 - P(A)$$

If A and B are independent events, then

$$P(A \text{ and } B) = P(A) \cdot P(B)$$

(9.9)   The concepts of probabilities and odds are related.

A game is fair if its cost to play equals the expected winnings.

REVIEW EXERCISES

1. Verify the following formula for $n = 1$, $n = 2$, $n = 3$, and $n = 4$:

$$1^3 + 2^3 + 3^3 + \cdots + n^3 = \frac{n^2(n + 1)^2}{4}$$

2. Prove the formula given in Exercise 1 by mathematical induction.

In Review Exercises 3–6, use the binomial theorem to find the expansion of each expression.

3. $(x - y)^3$

4. $(u + 2v)^3$

5. $(4a - 5b)^5$

6. $(\sqrt{7}r + \sqrt{3}s)^4$

In Review Exercises 7–10, find the required term of each expansion.

7. $(a + b)^8$; fourth term

8. $(2x - y)^5$; third term

9. $(x - y)^9$; seventh term

10. $(4x + 7)^6$; fourth term

In Review Exercises 11–14, evaluate each expression.

11. $\displaystyle\sum_{k=1}^{4} 3k^2$

12. $\displaystyle\sum_{k=1}^{10} 6$

13. $\displaystyle\sum_{k=5}^{8} (k^3 + 3k^2)$

14. $\displaystyle\sum_{k=1}^{30} \left(\frac{3}{2}k - 12\right) - \frac{3}{2}\sum_{k=1}^{30} k$

In Review Exercises 15–18, find the required term of each arithmetic progression.

15. $5, 9, 13, \ldots$; 29th term

16. $8, 15, 22, \ldots$; 40th term

17. $6, -1, -8, \ldots$; 15th term

18. $\dfrac{1}{2}, -\dfrac{3}{2}, -\dfrac{7}{2}, \ldots$; 35th term

In Review Exercises 19–22, find the required term of each geometric progression.

19. 81, 27, 9, . . . ; 11th term

20. 2, 6, 18, . . . ; 9th term

21. $9, \dfrac{9}{2}, \dfrac{9}{4}, \ldots$; 15th term

22. $8, -\dfrac{8}{5}, \dfrac{8}{25}, \ldots$; 7th term

In Review Exercises 23–26, find the sum of the first 40 terms in each progression.

23. 5, 9, 13, . . .

24. 8, 15, 22, . . .

25. 6, −1, −8, . . .

26. $\dfrac{1}{2}, -\dfrac{3}{2}, -\dfrac{7}{2}, \ldots$

In Review Exercises 27–30, find the sum of the first eight terms in each progression.

27. 81, 27, 9, . . .

28. 2, 6, 18, . . .

29. $9, \dfrac{9}{2}, \dfrac{9}{4}, \ldots$

30. $8, -\dfrac{8}{5}, \dfrac{8}{25}, \ldots$

In Review Exercises 31–34, find the sum of each infinite progression, if possible.

31. $\dfrac{1}{3}, \dfrac{1}{6}, \dfrac{1}{12}, \ldots$

32. $\dfrac{1}{5}, -\dfrac{2}{15}, \dfrac{4}{45}, \ldots$

33. $1, \dfrac{3}{2}, \dfrac{9}{4}, \ldots$

34. 0.5, 0.25, 0.125, . . .

In Review Exercises 35–38, use the formula for the sum of the terms of an infinite geometric progression to change each decimal into a common fraction.

35. $0.\overline{3}$

36. $0.\overline{9}$

37. $0.\overline{17}$

38. $0.\overline{45}$

39. Insert three arithmetic means between 2 and 8.

40. Insert five arithmetic means between 10 and 100.

41. Insert three geometric means between 2 and 8.

42. Insert four geometric means between −2 and 64.

43. Find the sum of the first 8 terms of the progression $\frac{1}{3}$, 1, 3, ...

44. Find the seventh term of the progression $2\sqrt{2}$, 4, $4\sqrt{2}$, ...

45. Find the positive geometric mean between 4 and 64.

46. If Leonard invests $3000 in a 6-year certificate of deposit at the annual rate of 7.75%, compounded daily, how much money will be in the account when it matures?

47. The enrollment at Hometown College is growing at the rate of 5% over each previous year's enrollment. If the enrollment is currently 4000 students, what will the enrollment be 10 years from now? What was it 5 years ago?

48. A house trailer that originally cost $10,000 depreciates in value at the rate of 10% per year. How much will the trailer be worth after 10 years?

In Review Exercises 49–60, evaluate each expression.

49. $_6P_6$

50. $\dbinom{7}{4}$

51. $0!$

52. $_{10}P_2 \cdot {}_{10}C_2$

53. $_8P_6 \cdot {}_8C_6$

54. $\dbinom{8}{5}\dbinom{6}{2}$

55. $_7C_5 \cdot {}_4P_0$

56. $_{12}C_0 \cdot {}_{11}C_0$

57. $\dfrac{_8P_5}{_8C_5}$

58. $\dfrac{_8C_5}{_{13}C_5}$

59. $\dfrac{_6C_3}{_{10}C_3}$

60. $\dfrac{_{13}C_5}{_{52}C_5}$

61. Make a tree diagram to illustrate the possible results of tossing a coin four times.

62. State the multiplication principle for events.

63. State the multiplication property for probabilities.

64. In how many ways can you draw a five-card poker hand of 3 aces and 2 kings?

65. What is the probability of drawing the hand described in Review Exercise 64?

66. What is the probability of *not* drawing the hand described in Review Exercise 64?

67. In how many ways can 10 teenagers be seated at a round table if 2 girls wish to sit with their boyfriends?

68. How many distinguishable words can be formed from the letters of the word *casserole* if each letter is used exactly once?

69. What is the probability of having a 13-card bridge hand consisting of 4 aces, 4 kings, 4 queens, and 1 jack?

70. Find the probability of choosing a committee of 3 boys and 2 girls from a group of 8 boys and 6 girls.

71. Find the probability of drawing a club or a spade on one draw from a card deck.

72. Find the probability of drawing a black card or a king on one draw from a card deck.

73. What is the probability of getting an ace-high royal flush (ace, king, queen, jack, and ten of hearts) in poker?

74. What is the probability of being dealt 13 cards of one suit in a bridge hand?

75. What is the probability of getting 3 heads or fewer on 4 tosses of a fair coin?

76. What are the odds against a horse if the probability that the horse will win is $\frac{7}{8}$?

77. What are the odds in favor of a couple having 4 baby girls in a row?

78. What are the expected earnings if you collect $1 for every heads you get when you toss a fair coin 4 times?

79. If the probability that Joe will marry is $\frac{5}{8}$ and the probability that John will marry is $\frac{3}{4}$, what are the odds against either one becoming a husband?

80. If the odds against Priscilla's graduation from college are $\frac{10}{11}$, what is the probability that she will graduate?

81. If the probability that a drug cures a certain disease is 0.83 and we give the drug to 800 people with the disease, what is the number of people that we can expect to be cured?

82. If the total number of subsets that a set with n elements can have is 2^n, explain why

$$\binom{n}{0} + \binom{n}{1} + \binom{n}{2} + \cdots + \binom{n}{n} = 2^n$$

APPENDIX I
A Proof of the Binomial Theorem

The binomial theorem can be proved for positive integral exponents by using mathematical induction.

The Binomial Theorem.   If n is a positive integer, then

$$(a + b)^n = a^n + \frac{n!}{1!(n - 1)!} a^{n-1}b + \frac{n!}{2!(n - 2)!} a^{n-2}b^2 + \cdots$$

$$+ \frac{n!}{r!(n - r)!} a^{n-r}b^r + \cdots + b^n$$

Proof   As in all induction proofs, there are two parts.

Part 1.   Substituting the number 1 for n on both sides of the equation, we have

$$(a + b)^1 = a^1 + \frac{1!}{1!(1 - 1)!} a^{1-1}b^1$$

$$a + b = a + a^0b$$

$$a + b = a + b$$

and the theorem is true when $n = 1$. Part 1 is complete.

Part 2.   We write expressions for two general terms in the statement of the induction hypothesis. We assume that the theorem is true for $n = k$:

$$(a + b)^k = a^k + \frac{k!}{1!(k - 1)!} a^{k-1}b + \frac{k!}{2!(k - 2)!} a^{k-2}b^2 + \cdots$$

$$+ \frac{k!}{(r - 1)!(k - r + 1)!} a^{k-r+1}b^{r-1}$$

$$+ \frac{k!}{r!(k - r)!} a^{k-r}b^r + \cdots + b^k$$

We multiply both sides of the equation above by $a + b$ and hope to obtain a similar equation in which the quantity $k + 1$ replaces all of the n values in the binomial theorem:

$(a + b)^k(a + b)$

$$= (a + b)\left[a^k + \frac{k!}{1!(k - 1)!} a^{k-1}b + \frac{k!}{2!(k - 2)!} a^{k-2}b^2 + \cdots \right.$$

$$\left. + \frac{k!}{(r - 1)!(k - r + 1)!} a^{k-r+1}b^{r-1} + \frac{k!}{r!(k - r)!} a^{k-r}b^r + \cdots + b^k \right]$$

We distribute the multiplication first by a and then by b:

$$(a + b)^{k+1} = \left[a^{k+1} + \frac{k!}{1!(k - 1)!} a^k b + \frac{k!}{2!(k - 2)!} a^{k-1}b^2 + \cdots \right.$$

$$\left. + \frac{k!}{(r - 1)!(k - r + 1)!} a^{k-r+2}b^{r-1} + \frac{k!}{r!(k - r)!} a^{k-r+1}b^r + \cdots + ab^k \right]$$

$$+ \left[a^k b + \frac{k!}{1!(k - 1)!} a^{k-1}b^2 + \frac{k!}{2!(k - 2)!} a^{k-2}b^3 + \cdots \right.$$

$$\left. + \frac{k!}{(r - 1)!(k - r + 1)!} a^{k-r+1}b^r + \frac{k!}{r!(k - r)!} a^{k-r}b^{r+1} + \cdots + b^{k+1} \right]$$

Combining like terms, we have

$$(a + b)^{k+1} = a^{k+1} + \left[\frac{k!}{1!(k - 1)!} + 1 \right] a^k b$$

$$+ \left[\frac{k!}{2!(k - 2)!} + \frac{k!}{1!(k - 1)!} \right] a^{k-1}b^2 + \cdots$$

$$+ \left[\frac{k!}{r!(k - r)!} + \frac{k!}{(r - 1)!(k - r + 1)!} \right] a^{k-r+1}b^r + \cdots + b^{k+1}$$

These results may be written as

$$(a + b)^{k+1} = a^{k+1} + \frac{(k + 1)!}{1!(k + 1 - 1)!} a^{(k+1)-1}b + \frac{(k + 1)!}{2!(k + 1 - 2)!} a^{(k+1)-2}b^2$$

$$+ \cdots + \frac{(k + 1)!}{r!(k + 1 - r)!} a^{(k+1)-r}b^r + \cdots + b^{k+1}$$

This formula has precisely the same form as the binomial theorem, with the quantity $k + 1$ replacing all of the original n values. Therefore, the truth of the theorem for $n = k$ implies the truth of the theorem for $n = k + 1$. Because both parts of the axiom of mathematical induction are verified, the theorem is proved.   □

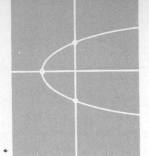

APPENDIX II
Answers to Selected Exercises

Exercise 1.1 (Page 7)

1. ⊆   3. ∈   5. ⊆   7. {3, 5, 7}   9. {1, 2, 3, 4, 5, 6, 7, 8, 9, 10}   11. {1, 3, 5, 7, 9}
13. {2, 3, 4, 5, 6, 7, 8, 10}   15. 10, 12, 14, 15, 16, 18   17. 2   19. infinite   21. finite
23. finite   25. 53   27. false; 27 can be divided by 3 and 9   29. true
31. false; 0 can be written in the form $\frac{a}{b}$. For example, $0 = \frac{0}{5}$ 33. true
35. false; a prime number is a natural number greater than 1
37. false; the product of two primes such as 3 and 5 is a composite: $3 \cdot 5 = 15$
39. false; the square of a prime number such as 7 is a composite: $7^2 = 49$
41. false; the sum of two composites can be composite. For example, $6 + 8 = 14$
43. false; 2 is an even integer that is prime   45. true   47. true
49. false; if two of the integers are odd and one is even, their sum is even: $3 + 5 + 10 = 18$ 51. 0.25
53. $0.\overline{2}$ 55. $-0.41\overline{6}$ 57. $\frac{3}{10}$ 59. $-\frac{25}{33}$ 61. $\frac{41}{333}$ 63. $\frac{1151}{3330}$ 65. $-\frac{853}{99}$
67. $\frac{1601}{990}$ 71. no; $0.999 = \frac{999}{1000}$ 73.

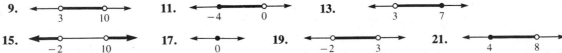

75.
77.
79. Because no numbers are both even and odd, there is no solution set. The graph is the number line with no dots on it.

Exercise 1.2 (Page 15)

1. true   3. false; all integers are rational numbers
5. false; the union of the set of rationals and the set of irrationals is the set of real numbers   7. true
9.   11.   13.
15.   17.   19.   21.
23.   25.   27.   29.
31. 7   33. 1   35. $\sqrt{2} - 1$ 37. 10   39. -30 41. 9   43. -1 45. 5
47. $-\frac{3}{7}$ 49. 3 units   51. transitive   53. reflexive and transitive
55. reflexive, symmetric, and transitive   57. reflexive, symmetric, and transitive

59. commutative property of addition   **61.** closure property   **63.** transitive property of equality
65. reflexive property of equality   **67.** closure property   **69.** commutative property of addition
71. symmetric property of equality   **73.** associative property of multiplication

Exercise 1.3   (Page 23)

1. 169   **3.** 1   **5.** 4   **7.** -1 **9.** x^5 **11.** z^6 **13.** y^{21} **15.** z^{26}

17. $27x^3$ **19.** x^6y^3 **21.** $\dfrac{a^6}{b^3}$ **23.** $\dfrac{1}{z^4}$ **25.** $\dfrac{1}{y^5}$ **27.** x^2 **29.** $\dfrac{1}{a^9}$ **31.** $\dfrac{a^{12}}{b^4}$

33. $\dfrac{1}{r^4}$ **35.** $\dfrac{x^{32}}{y^{16}}$ **37.** $\dfrac{9x^{10}}{25}$ **39.** 1   **41.** $\dfrac{4y^{12}}{x^{20}}$ **43.** $-x^{10}y^{5x}$ **45.** $-x^{8xy}y^{12xy}$

47. x^4 **49.** x^{5n^2} **51.** x^5 **53.** $\dfrac{64z^7}{25y^6}$ **55.** $\dfrac{1}{m^{14}n^{16}p^{12}}$ **57.** 4   **59.** -8

61. 216   **63.** -12 **65.** 20   **67.** 0   **69.** $\dfrac{3}{64}$ **71.** $-\dfrac{8}{9}$ **73.** $-\dfrac{3}{16}$

75. 3.72×10^5 **77.** 1.77×10^8 **79.** 7×10^{-3} **81.** 6.93×10^{-7} **83.** 1×10^{12}
85. 1×10^{-12} **87.** 9.97×10^{-3} **89.** 937,000   **91.** 0.0000221   **93.** 3.2
95. 1.17×10^4 **97.** 7×10^4 **99.** 1.986×10^4 meters per minute   **101.** 1.67248×10^{-15} gram

Exercise 1.4   (Page 35)

1. 3   **3.** 0   **5.** $\frac{1}{5}$ **7.** 8   **9.** -100 **11.** $\frac{1}{8}$ **13.** $\frac{1}{512}$ **15.** $-\frac{1}{27}$ **17.** $\frac{32}{243}$

19. $\frac{16}{9}$ **21.** $10s^2$ **23.** $\dfrac{1}{2y^2z}$ **25.** x^6y^3 **27.** $\dfrac{1}{r^6s^{12}}$ **29.** $\dfrac{4a^4}{25b^6}$ **31.** $\dfrac{100s^8}{9r^4}$

33. $\dfrac{a^2c^6}{b^5}$ **35.** a **37.** $\dfrac{2s}{a^3}$ **39.** $b^{11x/24}$ **41.** 7   **43.** 5   **45.** -3 **47.** $-\dfrac{1}{5}$

49. $6x$ **51.** xy^2 **53.** $2y$ **55.** $\dfrac{xy^2}{z^3}$ **57.** $4|a|$ **59.** $2a^2|a|$ **61.** $-2a$

63. $4b^2$ **65.** $3x$ **67.** $|x|y^2$ **69.** $\sqrt{2}$ **71.** $17x\sqrt{2}$ **73.** $2y^2\sqrt{3y}$ **75.** $12\sqrt[3]{3}$

77. $6z\sqrt[4]{3z}$ **79.** $6x\sqrt{2y}$ **81.** 0   **83.** $\dfrac{x}{2}$ **85.** $\sqrt{3}$ **87.** $\dfrac{2\sqrt{x}}{x}$ **89.** $\sqrt[3]{4}$

91. $\sqrt[3]{5a^2}$ **93.** $\dfrac{2b\sqrt[4]{27a^2}}{3a}$ **95.** $\dfrac{\sqrt{2xy}}{2y}$ **97.** $\dfrac{u\sqrt[3]{6uv^2}}{3v}$ **99.** $\dfrac{2\sqrt{3}}{9}$ **101.** $-\dfrac{\sqrt{2x}}{8}$

103. $\dfrac{x\sqrt{y}}{3}$ **105.** $\dfrac{3y}{x}$ **107.** $\dfrac{x^4y^3\sqrt[5]{x^4y^4}}{2}$ **109.** $\sqrt{3}$ **111.** $\sqrt[5]{4x^3}$ **113.** $\sqrt[6]{72}$

115. $\dfrac{\sqrt[4]{12}}{2}$ **117.** $x \geq 0$ **119.** $x \leq 0$

Exercise 1.5   (Page 45)

1. polynomial, 2, trinomial   **3.** not a polynomial   **5.** polynomial, 3, binomial
7. polynomial, 0, monomial   **9.** polynomial, no degree, monomial   **11.** not a polynomial
13. $6x^3 - 3x^2 - 8x$ **15.** $4y^3 + 14$ **17.** $-x^2 + 14$ **19.** $-28t + 96$ **21.** $-4y^2 + y$
23. $-4x^2 + x$ **25.** $2x^2y - 4xy^2$ **27.** $8x^3y^7$ **29.** $6m^4n^4$ **31.** $-4r^3s - 4rs^3$
33. $12a^2b^2c^2 + 18ab^3c^3 - 24a^2b^4c^2$ **35.** $a^2 + 4a + 4$ **37.** $a^2 - 12a + 36$ **39.** $x^2 - 16$
41. $a^2 - 4ab + 4b^2$ **43.** $9m^2 - 16n^2$ **45.** $x^2 + 2x - 15$ **47.** $3u^2 + 4u - 4$
49. $6x^2 - x - 15$ **51.** $10x^2 + 13x - 3$ **53.** $12v^2 - 7v - 12$ **55.** $6y^2 - 16xy + 8x^2$
57. $9x^3 - 27xy - yx^2 + 3y^2$ **59.** $5z^3 - 5tz + 2tz^2 - 2t^2$ **61.** $27x^3 - 27x^2 + 9x - 1$
63. $x^3y^3 + 2xy^2 - x^2y - 2$ **65.** $6x^3 + 14x^2 - 5x - 3$ **67.** $6x^3 - 5x^2y + 6xy^2 + 8y^3$
69. $x^4 + 2x^3 + x^2 - 1$ **71.** $6y^{2n} - 4y^{n+2} + 2$
73. $-10x^{4n} - 15y^{2n}$ **75.** $x^{2n} - x^n - 12$ **77.** $6r^{2n} - 25r^n + 14$ **79.** $xy + x^{3/2}y^{1/2}$

81. $a - b$ **83.** $\sqrt{3} + 1$ **85.** $x(\sqrt{7} - 2)$ **87.** $2(\sqrt{5} + \sqrt{2})$ **89.** $\dfrac{x(x + \sqrt{3})}{x^2 - 3}$

91. $\dfrac{(y + \sqrt{2})^2}{y^2 - 2}$ **93.** $9 - 4\sqrt{5}$ **95.** $\dfrac{\sqrt{3} + 3 - \sqrt{2} - \sqrt{6}}{2}$ **97.** $\dfrac{x - 2\sqrt{xy} + y}{x - y}$

99. $\dfrac{1}{\sqrt{x + 3} + \sqrt{x}}$ **101.** $\dfrac{2a}{b^3}$ **103.** $\dfrac{x}{2y} + \dfrac{3xy}{2}$ **105.** $\dfrac{2y^3}{5} - \dfrac{3y}{5x^3} + \dfrac{1}{5x^4y^3}$ **107.** $3x + 2$

109. $x - 7 + \dfrac{2}{2x - 5}$ **111.** $x - 3$ **113.** $x^3 - 3 + \dfrac{3}{x^2 - 2}$ **115.** $x^4 + 2x^3 + 4x^2 + 8x + 16$

117. $6x^2 + x - 12$ **119.** $3x^2 - x + 2$ **121.** x^{5a-1} **123.** b^{3a^2+3a} **125.** a^{4x-1} **127.** r^s

Exercise 1.6   (Page 54)

1. $3(x - 2)$ **3.** $4x^2(2 + x)$ **5.** $7x^2y^2(1 + 2x)$ **7.** $3abc(a + 2b + 3c)$ **9.** $(x + y)(b - a)$
11. $(4a + b)(1 - 3a)$ **13.** $(x + 1)(3x^2 - 1)$ **15.** $t(y + c)(2x - 3)$ **17.** $(a + b)(x + y + z)$
19. $(2x + y)(3c + d)$ **21.** $(2x + 3)(2x - 3)$ **23.** $(2 + 3r)(2 - 3r)$ **25.** $(x + z + 5)(x + z - 5)$
27. not factorable over the integers.   **29.** $(x + y - z)(x - y + z)$ **31.** $-4xy$
33. $(x^2 + y^2)(x + y)(x - y)$ **35.** $(x^4 + 8z^2)(x^4 - 8z^2)$ **37.** $3(x + 2)(x - 2)$
39. $2x(3y + 2)(3y - 2)$ **41.** $(x + 4)(x + 4)$ **43.** $(b - 5)(b - 5)$ **45.** $(m + 2n)(m + 2n)$
47. $(x + 7)(x + 3)$ **49.** $(x - 6)(x + 2)$ **51.** not factorable over the integers
53. $(4x - 3y)(3x + 2y)$ **55.** $(6y - 5)(4y - 3)$ **57.** $(6a + 5)(4a - 3)$ **59.** $(3x + 7y)(2x + 5y)$
61. $(2x - 5)(3x + 7)$ **63.** $2(6y - 35)(y + 1)$ **65.** $-x(6x + 7)(x - 5)$ **67.** $x^2(2x - 7)(3x + 5)$
69. $-(6x - 5)(x - 7)$ **71.** $(x^2 + 5)(x^2 - 3)$ **73.** $2(x^2y^2z^2 - 8xyz + 40)$ **75.** $(a^n - 3)(a^n + 1)$
77. $(3x^n - 2)(2x^n - 1)$ **79.** $(2x^n + 3y^n)(2x^n - 3y^n)$ **81.** $(5y^n + 2)(2y^n - 3)$
83. $(2z - 3)(4z^2 + 6z + 9)$ **85.** $2(x + 10)(x^2 - 10x + 100)$
87. $(x + y - 4)(x^2 + 2xy + y^2 + 4x + 4y + 16)$ **89.** $-x(x^2 + 3x + 3)$
91. $(2a + y)(2a - y)(4a^2 - 2ay + y^2)(4a^2 + 2ay + y^2)$ **93.** $(a - b)(a^2 + ab + b^2 + 1)$
95. $(4x^2 + y^2)(16x^4 - 4x^2y^2 + y^4)$ **97.** $(x - 3 + 12y)(x - 3 - 12y)$ **99.** $(a + b - 5)(a + b + 2)$
101. $(3u + 3v + 4)(2u + 2v + 1)$ **103.** $(x + 2)(x^2 - 2x + 4)(x - 1)(x^2 + x + 1)$
105. $(a + b)(c + d + 1)$ **107.** $(x^2 + x + 2)(x^2 - x + 2)$ **109.** $(x^2 + x + 4)(x^2 - x + 4)$
111. $(2a^2 + a + 1)(2a^2 - a + 1)$ **113.** $2(x^2 + 2x + 2)(x^2 - 2x + 2)$ **115.** $2(\frac{3}{2}x + 1)$
117. $x^{1/2}(x^{1/2} + 1)$ **119.** $ab(b^{1/2} - a^{1/2})$ **121.** $a^2(a^n + a^{n+1})$

Exercise 1.7   (Page 61)

1. equal   **3.** not equal   **5.** $\dfrac{8x}{35a}$ **7.** $\dfrac{16}{3}$ **9.** $\dfrac{z}{c}$ **11.** $\dfrac{2x^2y}{a^2b^3}$ **13.** $\dfrac{2}{x + 2}$

15. $-\dfrac{x - 5}{x + 5}$ **17.** $\dfrac{3x - 4}{2x - 1}$ **19.** $\dfrac{x^2 + 2x + 4}{x + a}$ **21.** $\dfrac{x(x - 1)}{x + 1}$ **23.** $\frac{1}{2}$

25. $\dfrac{z - 4}{(z + 2)(z - 2)}$ **27.** $\dfrac{x - 1}{x}$ **29.** $\dfrac{x(x + 1)^2}{x + 2}$ **31.** 1 **33.** $\dfrac{(x + 12)(2x - 1)}{(2x + 1)(2x - 3)}$

35. $\dfrac{x(x + 3)}{x + 1}$ **37.** $\dfrac{x + 5}{x + 3}$ **39.** $\dfrac{-3x^2 + 7x + 4}{(x - 1)(x + 1)}$ **41.** $\dfrac{2a - 4}{(a + 4)(a - 4)}$ **43.** $\dfrac{3y - 2}{y - 1}$

45. $\dfrac{1}{x + 2}$ **47.** $\dfrac{2x - 5}{2x(x - 2)}$ **49.** $\dfrac{2x^2 + 19x + 1}{(x - 4)(x + 4)}$ **51.** 0 **53.** $\dfrac{-x^4 + 3x^3 - 18x^2 - 58x + 72}{(x + 5)(x - 5)(x - 4)(x + 4)}$

55. $\dfrac{b}{2c}$ **57.** $81a$ **59.** -1 **61.** $\dfrac{y + x}{x^2y^2}$ **63.** $\dfrac{y + x}{y - x}$ **65.** $\dfrac{a^2(3x - 4ab)}{ax + b}$ **67.** $\dfrac{x - 2}{x + 2}$

69. $\dfrac{3x^2y^2}{xy - 1}$ **71.** $\dfrac{3x^2}{x^2 + 1}$ **73.** $\dfrac{x^2 - 3x - 4}{x^2 + 5x - 3}$ **75.** $\dfrac{x}{x + 1}$ **77.** $\dfrac{5x + 1}{x - 1}$ **79.** $\dfrac{3x}{3 + x}$

81. $\dfrac{x + 1}{2x + 1}$ **83.** $\dfrac{(x - 1)\sqrt{x^2 + 4}}{x^2 + 4}$ **85.** $\dfrac{(z - 2)\sqrt{z^2 - 7}}{z^2 - 7}$

Exercise 1.8   (Page 70)

1. i **3.** -1 **5.** $x = 3; y = 5$ **7.** $x = \frac{2}{3}; y = -\frac{2}{9}$ **9.** $5 - 6i$ **11.** $-2 - 10i$
13. $4 + 10i$ **15.** $1 - i$ **17.** $-9 + 19i$ **19.** $-5 + 12i$ **21.** $52 - 56i$
23. $-6 + 17i$ **25.** $2 - 11i$ **27.** $0 + i$ **29.** $0 + 7i$ **31.** $\frac{2}{5} - \frac{1}{5}i$ **33.** $\frac{1}{25} + \frac{7}{25}i$
35. $\frac{1}{2} + \frac{1}{2}i$ **37.** $-\frac{7}{13} - \frac{22}{13}i$ **39.** $\frac{-12}{17} + \frac{11}{34}i$ **41.** $\frac{6 + \sqrt{3}}{10} + \frac{3\sqrt{3} - 2}{10}i$ **43.** $\sqrt{13}$
45. $7\sqrt{2}$ **47.** 6 **49.** $\frac{3\sqrt{5}}{5}$ **51.** $34.87 + 32.69i$ **53.** approximately $-6.92 + 9.26i$
55. $(x + 2i)(x - 2i)$ **57.** $2(5m + ni)(5m - ni)$

REVIEW EXERCISES   (Page 73)

1. $\{2\}$ **3.** $\{2\}$ **5.** $0.\overline{925}$ **7.**   **9.**

11.   **13.** 6 **15.** associative property of addition   **17.** additive identity property

19. multiplicative inverse property   **21.** $\frac{x^9}{y^6}$ **23.** $\frac{y^6}{8x^6}$ **25.** $\frac{y^8}{9}$ **27.** $\frac{y^4}{9x^4}$ **29.** 0

31. $\frac{1}{4}$ **33.** 2 **35.** 1 **37.** 2 **39.** -4 **41.** $\frac{8}{27}$ **43.** $\frac{9}{4}$ **45.** 6 **47.** $\frac{3}{5}$

49. $x^6 y$ **51.** $-c$ **53.** xy^2 **55.** $\frac{m^2 n}{p^3}$ **57.** $x^8 y^2 |c|$ **59.** $x^2 y^4$ **61.** $\frac{2\sqrt{5}}{5}$

63 $\frac{\sqrt[3]{4}}{2}$ **65.** $\sqrt{3} + 1$ **67.** $\frac{2x(\sqrt{x} + 2)}{x - 4}$ **69.** $7\sqrt{2}$ **71.** $5 + 2\sqrt{6}$

73. $\sqrt{6} + \sqrt{2} + \sqrt{3} + 1$ **75.** a polynomial, degree of 3, binomial
77. a polynomial, degree of 2, monomial   **79.** $2x^3 - 7x^2 - 6x$ **81.** $8a^2 - 8ab - 6b^2$
83. $\frac{2b}{a^3} + \frac{3}{a}$ **85.** $x^2 + 3x - 2$ **87.** $3x - 3 + \frac{4}{x + 1}$ **89.** $3x(x + 1)(x - 1)$
91. $(3x + 8)(2x - 3)$ **93.** $(2x - 5)(4x^2 + 10x + 25)$ **95.** $(x + 3 + 2t)(x + 3 - 2t)$
97. $(2z + 7)(4z^2 - 14z + 49)$ **99.** $(11z - 2)(11z - 2)$ **101.** $(y - 2z)(2x - w)$
103. $(x - 2)(x + 3)$ **105.** $\frac{x + 1}{5}$ **107.** $\frac{(x - 2)(x + 3)(x - 3)}{(x - 1)(x + 2)^2}$ **109.** $\frac{3x^2 - 10x + 10}{(x - 4)(x + 5)}$
111. $\frac{3x^3 - 12x^2 + 11x}{(x - 1)(x - 2)(x - 3)}$ **113.** $\frac{-x^3 + x^2 - 12x - 15}{x^2 (x + 1)}$ **115.** $\frac{20}{3x}$ **117.** $\frac{y + x}{xy(x - y)}$
119. $0 + i$ **121.** $6 + i$ **123.** $11 - 2i$ **125.** $0 - 3i$ **127.** $\frac{3}{2} - \frac{3}{2}i$ **129.** $1 + 2i$
131. $\sqrt{53}$ **133.** $\frac{3\sqrt{5}}{5}$

Exercise 2.1   (Page 82)

1. all real numbers   **3.** all real numbers except 0   **5.** all real numbers greater than or equal to 0
7. all real numbers except 3 and -2 **9.** $x = 5$, conditional equation   **11.** no solution
13. $x = 7$, conditional equation   **15.** no solution   **17.** identity   **19.** $b = 6$, conditional equation
21. identity   **23.** $x = 1$ **25.** $z = 9$ **27.** $z = 10$ **29.** $x = -3$ **31.** $x = 6$
33. $x = -2$ **35.** $x = \frac{5}{2}$ **37.** $x = 1$ **39.** $x = -\frac{14}{11}$ **41.** identity   **43.** $y = 2$
45. $s = 4$ **47.** $x = -4$ **49.** no solution   **51.** $n = 17$ **53.** $x = -\frac{2}{5}$ **55.** $x = \frac{2}{3}$
57. $n = 3$ **59.** $y = 5$ **61.** no solutions   **63.** $a = 2$

Exercise 2.2   (Page 88)

1. 7   **3.** 17 and 37   **5.** 84   **7.** $10,000   **9.** 327   **11.** $\frac{4}{3}$ hours   **13.** $\frac{190}{9}$ hours
15. 1 liter   **17.** 600 cubic centimeters   **19.** about 45.5%
21. 39 miles per hour going; 65 miles per hour returning   **23.** $2\frac{1}{2}$ hours   **25.** 50 seconds
27. 12 miles per hour   **29.** 8 of each
31. 600 pounds of barley, 1637 pounds of oats, 163 pounds of soybean meal

Exercise 2.3   (Page 99)

1. 3, -2 **3.** 12, -12 **5.** 2, $-\frac{5}{2}$ **7.** 2, $\frac{3}{5}$ **9.** $\frac{3}{5}$, $-\frac{5}{3}$ **11.** $\frac{3}{2}$, $\frac{1}{2}$ **13.** 0, -5, -4
15. 0, $\frac{4}{3}$, $-\frac{1}{2}$ **17.** 5, -5, 1, -1 **19.** 6, -6, 1, -1 **21.** $3\sqrt{2}$, $-3\sqrt{2}$, $\sqrt{5}$, $-\sqrt{5}$
23. 0, 1   **25.** $\frac{1}{8}$, -8 **27.** 1, 144   **29.** $\frac{1}{64}$ **31.** 3, -3 **33.** $5\sqrt{2}$, $-5\sqrt{2}$
35. 3, -1 **37.** 2, -4 **39.** 5, 3   **41.** 2, -3 **43.** 0, 25   **45.** $\frac{2}{3}$, -2
47. $\dfrac{-5 + \sqrt{5}}{2}, \dfrac{-5 - \sqrt{5}}{2}$ **49.** $\dfrac{-2 + \sqrt{7}}{3}, \dfrac{-2 - \sqrt{7}}{3}$ **51.** $2\sqrt{3}$, $-2\sqrt{3}$ **53.** 3, $-\frac{5}{2}$
55. 2, $-\frac{1}{5}$ **57.** 1, -2 **59.** $\dfrac{-5 + \sqrt{13}}{6}, \dfrac{-5 - \sqrt{13}}{6}$ **61.** $\dfrac{-1 + \sqrt{61}}{10}, \dfrac{-1 - \sqrt{61}}{10}$
63. 3, -4 **65.** $\frac{3}{2}$, $-\frac{1}{4}$ **67.** $\frac{1}{2}$, $-\frac{4}{3}$ **69.** $\frac{5}{6}$, $-\frac{2}{5}$ **71.** $4 + 2\sqrt{2}$, $4 - 2\sqrt{2}$
73. $-1 + i$, $-1 - i$ **75.** $-2 + i$, $-2 - i$ **77.** $\frac{1}{3} + \frac{1}{3}i$, $\frac{1}{3} - \frac{1}{3}i$

Exercise 2.4   (Page 105)

1. rational and equal   **3.** not real numbers   **5.** rational and unequal   **7.** 2, 10   **9.** yes
11. 1, -1 **13.** $-\frac{1}{2}$, 5   **15.** -2 **17.** 3, $-\frac{8}{11}$ **19.** $\frac{1}{2}$, $-\frac{1}{2}$, $\frac{1}{3}$, $-\frac{1}{3}$ **21.** 27   **23.** 1
25. $-\frac{5}{2}$ **27.** 9   **29.** 20   **31.** 2   **33.** 2   **35.** 3, 4   **37.** $\frac{1}{5}$, -1 **39.** 3, 5
41. -2, 1   **43.** 2, $-\frac{5}{2}$ **45.** -2 **47.** no solution   **49.** 3   **51.** 1
53. $r_1 + r_2 = -\frac{b}{a}$; $r_1 r_2 = \frac{c}{a}$

Exercise 2.5   (Page 108)

1. 6 and 8   **3.** 4 feet by 8 feet   **5.** 9 centimeters
7. 20 miles per hour going, 10 miles per hour returning   **9.** 7 hours   **11.** 25 seconds
13. about 9.5 seconds   **15.** 4 hours   **17.** about 9.5 hours   **19.** no
21. Matilda at 8%, Maude at 7%   **23.** 10   **25.** 20 spokes   **27.** 10 meters and 24 meters

Exercise 2.6   (Page 111)

1. $p = \dfrac{k}{2.2}$ **3.** $b_2 = \dfrac{2A}{h} - b_1$ **5.** $r = \sqrt{\dfrac{3V}{\pi h}}$ **7.** $s = \dfrac{f(P_n - L)}{i}$ **9.** $r = \dfrac{r_1 r_2}{r_1 + r_2}$

11. $n = \dfrac{l - a + d}{d}$ **13.** $F = \dfrac{9}{5}C + 32$ **15.** $y = \dfrac{-3x^2 + 1}{3x - 7}$ **17.** $a = S - Sr + lr$

19. $r_1 = \dfrac{-Rr_2 r_3}{Rr_3 + Rr_2 - r_2 r_3}$ **21.** $x = \dfrac{y^2}{a - y}$ **23.** $n = \dfrac{360}{180 - a}$ **25.** $h = \dfrac{V}{B_1 + B_2 + \sqrt{B_1 B_2}}$

27. $\theta = \dfrac{2A}{r^2} + \phi$ **29.** $y = \dfrac{r - x}{1 + rx}$ **31.** $y = \dfrac{y_2 - y_1}{x_2 - x_1}(x - x_1) + y_1$ **33.** 0, $3 - 2y$

35. $y = \dfrac{x \pm \sqrt{4x^3 - 7x^2 + 4x}}{2(x - 1)}$ **37.** $n = \dfrac{C - 8}{0.10}$; 200, 500, 700   **39.** 150 newspapers

REVIEW EXERCISES   (Page 113)

1. all real numbers   **3.** all real numbers   **5.** all x such that $x \geq 0$ **7.** $\frac{16}{3}$, conditional equation
9. 4, conditional equation   **11.** 7, conditional equation   **13.** 7, conditional equation   **15.** $2, -\frac{3}{2}$
17. $0, \frac{8}{5}$ **19.** 3, 5   **21.** $\dfrac{1 + \sqrt{21}}{10}, \dfrac{1 - \sqrt{21}}{10}$ **23.** $2, -7$ **25.** $\dfrac{-1 + \sqrt{21}}{10}, \dfrac{-1 - \sqrt{21}}{10}$
27. $2, -3$ **29.** $1, 1, -1, -1$ **31.** 5   **33.** 0   **35.** $\frac{1}{3}$ **37.** 1.5 liters   **39.** $5\frac{1}{7}$ hours
41. \$4500 at 11%, \$5500 at 14%   **43.** either 95 by 110 yards or 55 by 190 yards   **45.** $f_1 = \dfrac{ff_2}{f_2 - f}$
47. $l = \dfrac{a + Sr - S}{r}$ **49.** $y = \dfrac{x^2}{2 - x}$

Exercise 3.1   (Page 122)

1. I   **3.** III   **5.** I   **7.** positive x-axis   **9.** positive y-axis   **11.** negative x-axis
13.

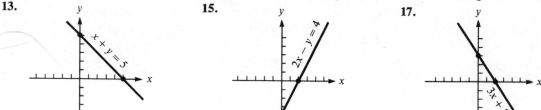

19.

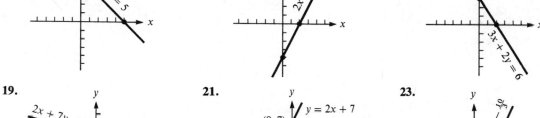

25.

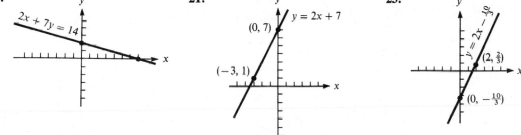

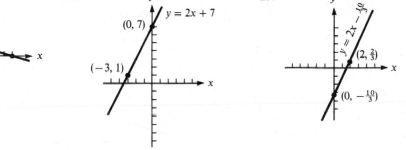

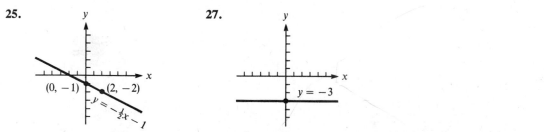

29. 5   **31.** $\sqrt{13}$ **33.** $\sqrt{a^2 + b^2}$ **35.** 5   **37.** 10   **39.** $2\sqrt{2}$ **41.** 2   **43.** 3
45. $\sqrt{x^2 + y^2}$ **47.** $(4, 6)$ **49.** $(0, 1)$ **51.** $\left(\dfrac{\sqrt{5}}{2}, \dfrac{\sqrt{5}}{2}\right)$ **53.** $\left(\dfrac{a + 2}{2}, \dfrac{2 + a}{2}\right)$
55. $(5, 6)$ **57.** $(5, 15)$ **63.** $(-4, -3)$ **65.** $(5, 0), (0, 5), (-5, 0), (0, -5)$ **67.** $\sqrt{2}$ units

Exercise 3.2   (Page 128)

1. 5   **3.** $-\frac{7}{4}$ **5.** -2 **7.** no defined slope   **9.** -1 **11.** 3   **13.** $\frac{1}{2}$ **15.** $\frac{2}{3}$
17. 0   **19.** perpendicular   **21.** parallel   **23.** perpendicular   **25.** parallel
27. perpendicular   **29.** neither   **31.** They do not lie on a straight line.
33. No two are perpendicular   **35.** PQ and PR are perpendicular

Exercise 3.3   (Page 135)

1. $y = 2x$ **3.** $y = 2x + \frac{7}{2}$ **5.** $y = -5x + 20$ **7.** $y = 0$ **9.** $y = 3x - 2$
11. $y = \sqrt{2}x + \sqrt{2}$ **13.** $y = ax + \frac{1}{a}$ **15.** $y = ax + a$ **17.** $3x - 2y = 0$
19. $3x + y = -4$ **21.** $\sqrt{2}x - y = -\sqrt{2}$ **23.** $sx - ry = 0$ **25.** $x + y = 5$
27. $6x + y = 16$ **29.** $13x + 9y = 0$ **31.** $m = \frac{42}{5}, b = \frac{21}{5}$ **33.** no defined slope; no y-intercept
35. $m = -1; b = 2$ **37.** $m = 3; b = 11$ **39.** $m = \frac{1}{3}; b = \frac{7}{3}$ **41.** $x - 2y = -5$
43. $8x + 7y = 12$ **45.** $11x + 2y = 0$ **47.** $x - y = -7$ **49.** $3x - 4y = -12$
51. $2x + 3y = 30$ **53.** $y = -5$ **55.** $x + 3y = -15$ **57.** $5x - 3y = 15$
59. $y = -\frac{3}{2}x$ **61.** $y = \frac{5}{3}x + \frac{19}{3}$ **63.** $y = \frac{s}{r}x$ **65.** $C = \frac{5}{9}(F - 32)$
67. The temperature at sea level is $70°$.   **75.** $y = \frac{1}{4}x$ **77.** $x = 0$

Exercise 3.4   (Page 145)

1. function   **3.** not a function   **5.** function   **7.** not a function   **9.** not a function
11. function   **13.** domain is the set of real numbers, range is the set of real numbers
15. domain is the set of real numbers, range is the set of nonnegative real numbers
17. domain is the set of real numbers except -1, range is the set of real numbers except 0
19. domain is the set of nonnegative real numbers, range is the set of nonnegative real numbers
21. $f(2) = 4, f(-3) = -11, f(k) = 3k - 2, f(k^2 - 1) = 3k^2 - 5$
23. $f(2) = 4, f(-3) = \frac{3}{2}, f(k) = \frac{1}{2}k + 3, f(k^2 - 1) = \frac{1}{2}k^2 + \frac{5}{2}$
25. $f(2) = 4, f(-3) = 9, f(k) = k^2, f(k^2 - 1) = k^4 - 2k^2 + 1$
27. $f(2) = \frac{1}{3}, f(-3) = 2, f(k) = \frac{2}{k + 4}, f(k^2 - 1) = \frac{2}{k^2 + 3}$
29. $f(2) = \frac{1}{3}, f(-3) = \frac{1}{8}, f(k) = \frac{1}{k^2 - 1}, f(k^2 - 1) = \frac{1}{k^4 - 2k^2}$
31. $f(2) = \sqrt{5}, f(-3) = \sqrt{10}, f(k) = \sqrt{k^2 + 1}, f(k^2 - 1) = \sqrt{k^4 - 2k^2 + 2}$ **33.** function
35. function   **37.** not a function
39.

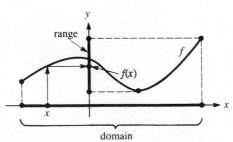

41.

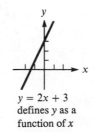

$y = 2x + 3$
defines y as a
function of x

43.

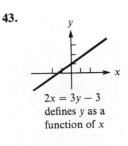

$2x = 3y - 3$
defines y as a
function of x

45.

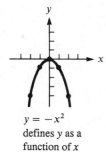

$y = -x^2$
defines y as a
function of x

47.

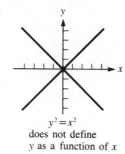

$y^2 = x^2$
does not define
y as a function of x

49.

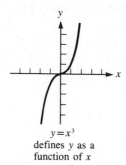

$y = x^3$
defines y as a
function of x

51.

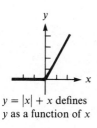

$y = |x| + x$ defines
y as a function of x

53.

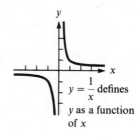

$y = \dfrac{1}{x}$ defines
y as a function
of x

55.

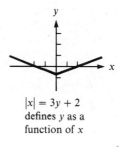

$|x| = 3y + 2$
defines y as a
function of x

57.

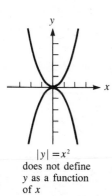

$|y| = x^2$
does not define
y as a function
of x

59.

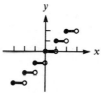

The phrase "y is the
greatest integer less than
or equal to x" defines y
as a function of x

61.

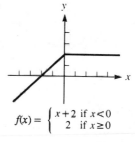

$f(x) = \begin{cases} x+2 & \text{if } x<0 \\ 2 & \text{if } x\geq 0 \end{cases}$

63.

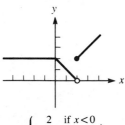

$f(x) = \begin{cases} 2 & \text{if } x<0 \\ 2-x & \text{if } 0\leq x<2 \\ x & \text{if } x\geq 2 \end{cases}$

65. $r = \dfrac{C}{2\pi}$ **67.** $p = 4\sqrt{A}$ **69.** $V = A^{3/2}$ **71.** $A = \dfrac{1}{2}d^2$ **73.** $F = \dfrac{9}{5}C + 32$

75. $c = \dfrac{3}{2500}n + \dfrac{7}{2}$

Exercise 3.5   (Page 158)

1.
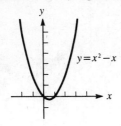
$y = x^2 - x$

3.
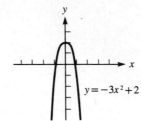
$y = -3x^2 + 2$

5.
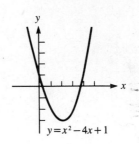
$y = x^2 - 4x + 1$

7. (2, 0)   **9.** (−3, −12)   **11.** (3, 1)   **13.** symmetric about the origin; odd function
15. symmetric about the x-axis   **17.** neither even nor odd function
19. symmetric about the y-axis; even function

21.
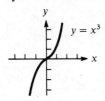
$y = x^3$

23.
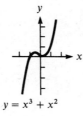
$y = x^3 + x^2$

25.
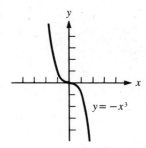
$y = -x^3$

27.

$y = x^4 - 2x^2 + 1$

29. decreasing for $x < 1$; increasing for $x > 1$
31. decreasing for $x < 1$; increasing for $x > 1$

33.
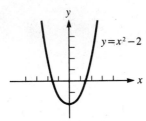
$y = x^2 - 2$

35.
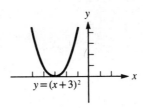
$y = (x+3)^2$

37.
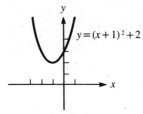
$y = (x+1)^2 + 2$

39.
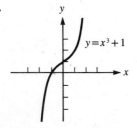
$y = x^3 + 1$

41.
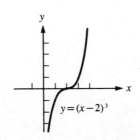
$y = (x-2)^3$

43.
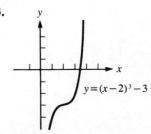
$y = (x-2)^3 - 3$

45.

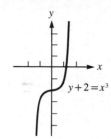

$y+2=x^3$

47.

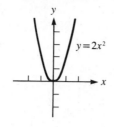

$y=2x^2$

49.

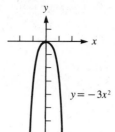

$y=-3x^2$

51.

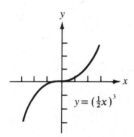

$y=(\tfrac{1}{2}x)^3$

53.

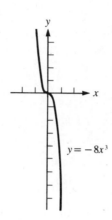

$y=-8x^3$

55.

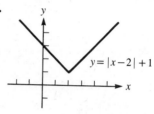

$y=|x-2|+1$

57.

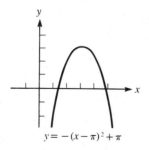

$y=\sqrt{x-2}+1$

59.

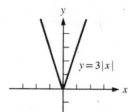

$y=3|x|$

61.

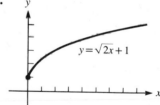

$y=\sqrt{2x}+1$

63.

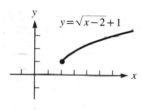

$y=-(x-\pi)^2+\pi$

65. 20 units   **67.** $a=\dfrac{100}{\sqrt{2}}+1=50\sqrt{2}+1$ **69.** 25 by 25 feet

71. $w=12$ inches, $d=6$ inches   **73.** both numbers are 3   **75.** 6 by $4\tfrac{1}{2}$ units   **77.** $\tfrac{5}{2}$ seconds

79. 100 feet

Exercise 3.6 (Page 170)

1. all real numbers except 2   3. all real numbers except 5 and −5
5. all real numbers except 0, 1, and −1

7.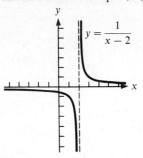
$$y = \frac{1}{x-2}$$

9.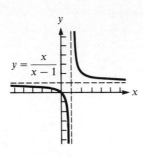
$$y = \frac{x}{x-1}$$

11.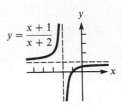
$$y = \frac{x+1}{x+2}$$

13.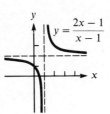
$$y = \frac{2x-1}{x-1}$$

15.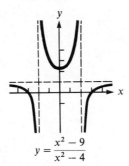
$$y = \frac{x^2-9}{x^2-4}$$

17.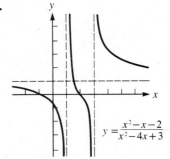
$$y = \frac{x^2-x-2}{x^2-4x+3}$$

19.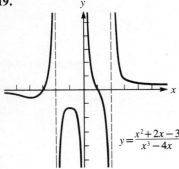
$$y = \frac{x^2+2x-3}{x^3-4x}$$

21.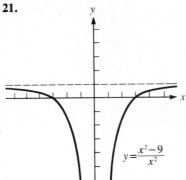
$$y = \frac{x^2-9}{x^2}$$

23.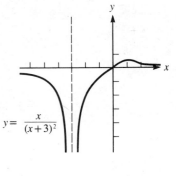
$$y = \frac{x}{(x+3)^2}$$

25.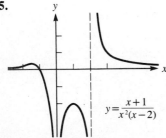
$$y = \frac{x+1}{x^2(x-2)}$$

27.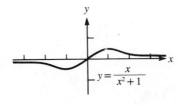
$$y = \frac{x}{x^2+1}$$

29.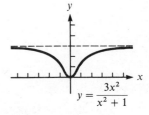
$$y = \frac{3x^2}{x^2+1}$$

31.

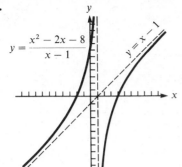

$$y = \frac{x^2 - 2x - 8}{x - 1}$$

$y = x - 1$

33.

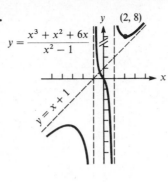

$$y = \frac{x^3 + x^2 + 6x}{x^2 - 1}$$

$(2, 8)$

$y = x + 1$

35.

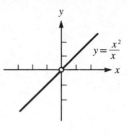

$$y = \frac{x^2}{x}$$

37.

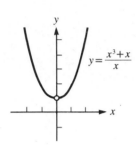

$$y = \frac{x^3 + x}{x}$$

39.

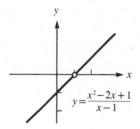

$$y = \frac{x^2 - 2x + 1}{x - 1}$$

41. no

Exercise 3.7   (Page 177)

1. $(f + g)(x) = 5x - 1$; all real numbers   **3.** $(f \cdot g)(x) = 6x^2 - x - 2$; all real numbers

5. $(f - g)(x) = x + 1$; all real numbers

7. $(f/g)(x) = \dfrac{x^2 + x}{x^2 - 1} = \dfrac{x(x + 1)}{(x + 1)(x - 1)} = \dfrac{x}{x - 1}$; all real numbers except 1 and -1 **9.** 7   **11.** 1

13. 12   **15.** no value   **17.** $f(x) = 3x^2$, $g(x) = 2x$ **19.** $f(x) = 3x^2$, $g(x) = x^2 - 1$

21. $f(x) = 3x^3$, $g(x) = -x$ **23.** $f(x) = x + 9$, $g(x) = x - 2$ **25.** 11   **27.** -17 **29.** 190

31. 145   **33.** $(f \circ g)(x) = 3x + 3$; all real numbers   **35.** $(f \circ f)(x) = 9x$; all real numbers

37. $(g \circ f)(x) = 2x^2$; all real numbers   **39.** $(f \circ f)(x) = x^4$; all real numbers

41. $(f \circ g)(x) = \sqrt{x^2 + 1}$; all real numbers   **43.** $(f \circ f)(x) = \sqrt{\sqrt{x}}$ or $\sqrt[4]{x}$; all nonnegative numbers

45. $(g \circ f)(x) = x$; all $x \geq -1$ **47.** $(g \circ g)(x) = x^4 - 2x^2$; all real numbers

49. $f(x) = x - 2$, $g(x) = 3x$ **51.** $f(x) = x - 2$, $g(x) = x^2$ **53.** $f(x) = x^2$, $g(x) = x - 2$

55. $f(x) = \sqrt{x}$, $g(x) = x + 2$ **57.** $f(x) = x + 2$, $g(x) = \sqrt{x}$ **59.** $f(x) = x$, $g(x) = x$

63. $(f \circ f)(x) = -\dfrac{1}{x}$

Exercise 3.8   (Page 184)

1. one-to-one   **3.** not one-to-one

5. not one-to-one (there are at least 3 values of x (1, -1, 0) such that $y = 0$)   **7.** not one-to-one

9. not one-to-one   **11.** one-to-one   **13.** not one-to-one

15. $(f \circ g)(x) = f(g(x)) = 5(\frac{1}{5}x) = x$

$(g \circ f)(x) = g(f(x)) = \frac{1}{5}(5x) = x$

17. $(f \circ g)(x) = f(g(x)) = \dfrac{\dfrac{1}{x-1}+1}{\dfrac{1}{x-1}} = \dfrac{\dfrac{x}{x-1}}{\dfrac{1}{x-1}} = x$

$(g \circ f)(x) = g(f(x)) = \dfrac{1}{\dfrac{x+1}{x}-1} = \dfrac{1}{\dfrac{1}{x}} = x$

19. $y = \frac{1}{3}x$ **21.** $y = \dfrac{x-2}{3}$ **23.** $y = \dfrac{1-3x}{x}$ **25.** $y = \dfrac{1}{2x}$

27.

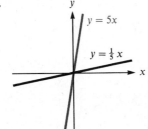

29.

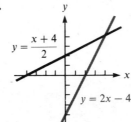

31.

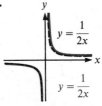

33.

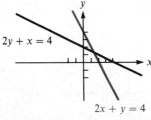

35.

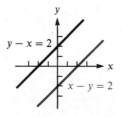

37. $f^{-1}(x) = -\sqrt{x+3}$ $(y \le 0)$ **39.** $f^{-1}(x) = \sqrt[4]{x+8}$ $(y \ge 0)$
41. $f^{-1}(x) = \sqrt{4-x^2}$ $(0 \le y \le 2)$
43. domain is the set of all real numbers except 1, range is the set of all real numbers except 1
45. domain is the set of all real numbers except 0, range is the set of all real numbers except -2

Exercise 3.9   (Page 189)

1. $x = 14$ **3.** $x = -3$ or $x = 2$ **5.** 18 girls   **7.** $\frac{1}{2}$ **9.** 1000   **11.** 1   **13.** $\frac{21}{4}$

15. -8 **17.** $\frac{160}{11}$ cubic feet   **19.** 3 seconds   **21.** 20 volts   **23.** 432 hertz   **25.** $\dfrac{\sqrt{3}}{4}$

REVIEW EXERCISES   (Page 193)

1. 10; (0, 3)   **3.** 16; $(-\sqrt{3}, 1)$ **5.** -6 **7.** -1 **9.** $7x + 5y = 0$ **11.** $2x + y = 9$
13. $y = 17$ **15.** $7x + y = 54$ **17.** $3x - 4y = 6$
19. a function; both domain and range are the set of real numbers

21. a function; domain is the set of all x such that $x \geq 1$, range is the set of nonnegative real numbers

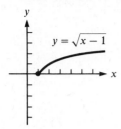

23. a function; domain is the set of all real numbers, range is the set of all real numbers greater than or equal to $\frac{3}{4}$

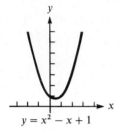

25.

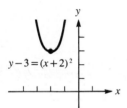

27.

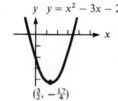

29.

31.

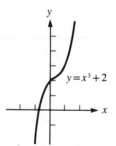

33.

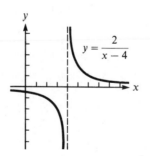

35.

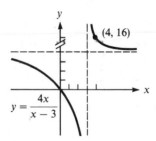

37.

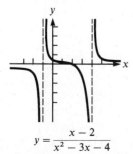

39. $f(2) = 6$ **41.** $(f + g)(2) = 11$ **43.** $(f \cdot g)(0) = 0$ **45.** $(f \circ g)(-2) = 1$
47. $(f + g)(x) = \sqrt{x + 1} + x^2 - 1$ **49.** $(f \circ g)(x) = \sqrt{x^2} = |x|$

51. all x such that $x > -1$ and $x \neq 1$ **53.** one-to-one, $f^{-1}(x) = \frac{1}{7}x$ **55.** one-to-one, $f^{-1}(x) = \frac{1 + x}{x}$

57. one-to-one, $f^{-1}(x) = \frac{2 + 3x}{x - 1}$

59. domain is the set of real numbers except $\frac{1}{2}$; range is the set of real numbers except 1   **61.** $\frac{9}{5}$ pounds
63. about 2.8

Exercise 4.1   (Page 199)

1. $(-\infty, 1)$ **3.** $[1, \infty)$ **5.** $(-\infty, 1)$ **7.** $[1, \infty)$

9. $(-\infty, 3]$ **11.** $(-\frac{10}{3}, \infty)$ **13.** $(5, \infty)$ **15.** $[14, \infty)$

17. $(-\infty, \frac{15}{4}]$ **19.** $(-\frac{44}{41}, \infty)$ **21.** $(6, 9]$ **23.** $(8, 22]$

25. $[-11, 4]$ **27.** $[5, 21]$ **29.** $(-\infty, 0)$ **31.** $(0, \infty)$

33. $(3, \frac{14}{3})$ **35.** $(2, \infty)$ **37.** $(-4, \frac{5}{6})$ **39.** $[-2, \infty)$

41. all x such that $x > 1$ **43.** all x such that $x < -4$ **45.** all x such that $x > -4$
47. all x such that $x < 3$ **49.** from 5.25 to 6.25 miles   **51.** between $16\frac{2}{3}$ and 20 centimeters.
53. $40 + 2w < P < 60 + 2w$ **55.** 12   **57.** $\{x : x \geq \frac{4}{3}\}$ **59.** $\{x : x > -\frac{2}{3}\}$

Exercise 4.2   (Page 205)

1. 7   **3.** 0   **5.** 2   **7.** $\pi - 2$ **9.** x if $x \geq 0$; $-x$ if $x < 0$ **11.** $x - 3$ **13.** 0, -4
15. 2, $-\frac{4}{3}$ **17.** $\frac{14}{3}$, -2 **19.** 7, -3 **21.** no values of x **23.** $\frac{2}{7}$, 2   **25.** $x \geq 0$
27. $-\frac{3}{2}$ **29.** 0, -6 **31.** 0   **33.** $\frac{3}{5}$, 3   **35.** $-\frac{3}{13}$, $\frac{9}{5}$
37. $(-3, 9)$ **39.** $(-\infty, -9) \cup (3, \infty)$ **41.** $(-\infty, -7] \cup [3, \infty)$ **43.** $[-\frac{13}{3}, 1]$

45. $(-\infty, -3) \cup (-3, \infty)$ **47.** $(-1, \frac{1}{5})$ **49.** $(-\infty, -\frac{7}{9}) \cup (\frac{13}{9}, \infty)$ **51.** $(-5, 7)$

53. $(-2, -\frac{1}{2}) \cup (-\frac{1}{2}, 1)$ **55.** $(-\frac{7}{3}, -\frac{2}{3}) \cup (\frac{4}{3}, 3)$ **57.** $(-7, -1] \cup [11, 17)$

59. $(-18, -6) \cup (10, 22)$ **61.** $(-10, -7] \cup [5, 8)$

63. $|x - 5| \leq 2$ **65.** $|x - 2| \leq \frac{5}{2}$ **67.** $0 < |x - 2| < 2$ **69.** $0 < |x| < 5$ **71.** $[-\frac{1}{2}, \infty)$
73. $(-\infty, 0)$ **75.** $(-\infty, -\frac{1}{2})$ **77.** $[0, \infty)$

Exercise 4.3   (Page 209)

1.

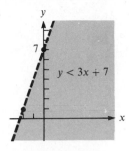

$y < 3x + 7$

3.

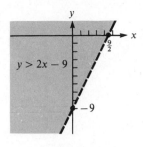

$y > 2x - 9$

5.

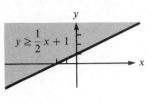

$y \geq \frac{1}{2}x + 1$

7.

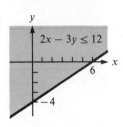

$2x - 3y \leq 12$

9.

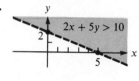

$2x + 5y > 10$

11.

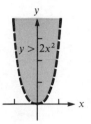

$y > 2x^2$

13.

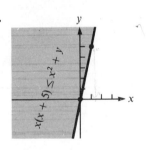

$x(x + 5) \leq x^2 + y$

15.

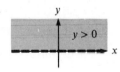

$y > 0$

17.

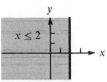

$x \leq 2$

19.

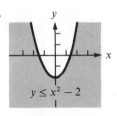

$y \leq x^2 - 2$

21.

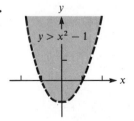

$y > x^2 - 1$

23.

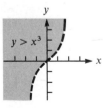

$y > x^3$

25.

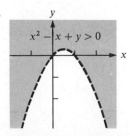

$x^2 - x + y > 0$

27.

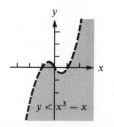

$y < x^3 - x$

29.

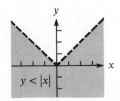

$y < |x|$

31.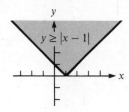

33.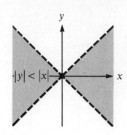

35.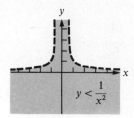

37.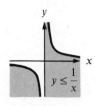

39.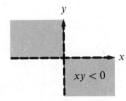

Exercise 4.4   (Page 215)

1. ○—○ $-4\ -3$

3. ●—● $2\quad 3$

5. ○—○ $-3\ -2$

7. ●—● $-\frac{1}{2}\ -\frac{1}{3}$

9. ○—○ $\frac{1}{3}\quad \frac{1}{2}$

11. ●—● $-\frac{3}{2}\quad 1$

13. ●—○ $-3\quad 2$

15. ○——○ $-1\quad 0\quad 1$

17. ○—●—○ $-3\ -2\quad 2$

19. ●—○——○ $-6\ -5\ -3\ -1$

21. ○——○ $-9\quad -4\ -2$

23. ○—○ $0\quad \frac{3}{2}$

25. ○—○ $0\quad \frac{3}{2}$

27. ○—● $2\quad \frac{13}{5}$

29. ○——● $-2\ -\frac{5}{4}$

31. ○—○——○—○ $-\sqrt{7}\ -1\quad 1\quad \sqrt{7}$

33. ○—○——○ $-3\quad 0\quad 1$

35. ○ 0

37. ○—○——○ $0\quad 1\quad 2$

39. ●—●—● $-1\quad 0\quad 1$

41. ●—● $2\quad \frac{14}{3}$

43. $\{x : x \le -3 \text{ or } x \ge 3\}$

45. $\{x : x \le -3 \text{ or } x \ge -1\}$

REVIEW EXERCISES   (Page 216)

1. $(-\infty, 7)$
○ 7

3. $(\frac{35}{3}, \infty)$
○ $\frac{35}{3}$

5. $(-\infty, \frac{5}{3})$
○ $\frac{5}{3}$

7. $(-\infty, 7)$
○ 7

9. $[-\frac{9}{2}, \frac{15}{2})$
●—○ $-\frac{9}{2}\quad \frac{15}{2}$

11. 8

13. -9

15. $(-6, 0)$
○—○ $-6\quad 0$

17. 5, -7

19. $(-5, 1)$
○—○ $-5\quad 1$

21.

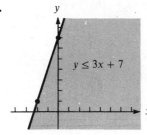

$y \le 3x + 7$

23.

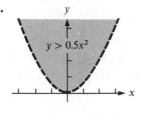

$y > 0.5x^2$

25.

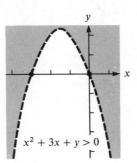

$x^2 + 3x + y > 0$

27. $(-1, 3)$

$-1 \quad 3$

29. $(-\infty, -7] \cup [\frac{3}{5}, \infty)$.

$-7 \quad \frac{3}{5}$

31. $(-\frac{5}{2}, -\frac{2}{3})$

$-\frac{5}{2} \quad -\frac{2}{3}$

33. $[-2, 1] \cup (3, \infty)$

$-2 \quad 1 \quad 3$

35. $(-1, -\frac{1}{2}] \cup [3, \infty)$

$-1 \; -\frac{1}{2} \quad 3$

37. $(-\infty, -2) \cup (-1, 1)$

$-2 \; -1 \quad 1$

Exercise 5.1   (Page 224)

1.

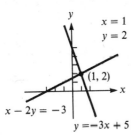

$x = 1$
$y = 2$
$(1, 2)$
$x - 2y = -3$
$y = -3x + 5$

3.

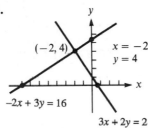

$(-2, 4)$
$x = -2$
$y = 4$
$-2x + 3y = 16$
$3x + 2y = 2$

5. $x = 0, y = 0$ **7.** $x = 3, y = -2$ **9.** $x = \frac{1}{2}, y = \frac{1}{3}$ **11.** no solution   **13.** $x = 3, y = 1$
15. $x = 3, y = 2$ **17.** $x = -3, y = 0$ **19.** $x = 1, y = -\frac{1}{2}$ **21.** all (x, y) such that $x = 2y$
23. $x = 2, y = 2$ **25.** $x = 1, y = 2, z = 0$ **27.** $x = 0, y = -\frac{1}{3}, z = -\frac{1}{3}$
29. $x = 1, y = 2, z = -1$ **31.** $x = 1, y = 0, z = 5$ **33.** no solution
35. $x = 2 - y, y = $ any number, $z = 2 - y$ **37.** $x = \frac{2}{3}, y = \frac{1}{4}, z = \frac{1}{2}$
39. $x = 2 - y, y = $ any number, $z = 1$ **41.** a table costs \$112, a chair costs \$45   **43.** 2100 sq cm
45. 3, 6, 10   **47.** 15 hours cooking hamburgers, 10 hours pumping gas, and 5 hours as a janitor
49. 14 nickels, 12 dimes, 6 quarters   **51.** 50 units from A, the rest from B and C in any combination

Exercise 5.2   (Page 234)

1. $x = 2, y = -1$ **3.** $x = -2, y = 0$ **5.** $x = 3, y = 1$ **7.** no solution
9. $x = 1, y = 0, z = 2$ **11.** $x = 2, y = -2, z = 1$ **13.** $x = 1, y = 1, z = 2$
15. $x = -1, y = 3, z = 1$ **17.** $x = 1, y = -3$ **19.** $x = \frac{8}{7} + \frac{1}{7}z, y = \frac{10}{7} - \frac{4}{7}z, z = $ any number
21. $w = y, x = 1, y = $ any number, $z = 1$ **23.** no solution   **25.** $x = 1, y = 2, z = 1, t = 1$
27. $x = 1, y = 2, z = 0, t = 1$ **29.** $x = \pm 2, y = \pm 1, z = \pm 3$ **31.** $x = 16, y = 1, z = 0$
33. $x = \frac{9}{4}, y = -3, z = \frac{3}{4}$ **35.** $x = -\frac{1}{2} - \frac{3}{2}z, y = 9 + 7z, z = $ any number

Exercise 5.3   (Page 243)

1. $x = 2, y = 5$ **3.** no values   **5.** $x = 1, y = 2$ **7.** $x = 2, y = 2$ **9.** $\begin{bmatrix} -1 & 2 & 1 \\ -6 & 0 & 0 \end{bmatrix}$

11. $\begin{bmatrix} 5 & -4 & 3 \\ -7 & -4 & -2 \\ 0 & 4 & -8 \end{bmatrix}$ **13.** not possible   **15.** $[-3, -3, -3]$ **17.** $\begin{bmatrix} 3 & 2 & -7 \\ 9 & 19 & 15 \\ 0 & -4 & -1 \end{bmatrix}$

19. $\begin{bmatrix} 2 & -2 \\ 3 & 10 \end{bmatrix}$ **21.** $\begin{bmatrix} -22 & -22 \\ -105 & 126 \end{bmatrix}$ **23.** $\begin{bmatrix} 4 & 2 & 10 \\ 5 & -2 & 4 \\ 2 & -2 & 1 \end{bmatrix}$ **25.** $\begin{bmatrix} 32 \\ 2 \end{bmatrix}$

27. not possible   **29.** $\begin{bmatrix} 19 & 23 & 27 \\ 17 & 22 & 27 \end{bmatrix}$ **31.** $\begin{bmatrix} 2 & 0 & 3 \\ 3 & 0 & 5 \end{bmatrix}$ **33.** $\begin{bmatrix} 12 & -4 & 9 \\ 16 & -2 & 2 \\ 10 & -2 & 1 \end{bmatrix}$

35. $\begin{bmatrix} 7 \\ 6 \end{bmatrix}$ **37.** $\begin{bmatrix} 4 & 5 \\ -7 & -1 \end{bmatrix}$ **39.** not possible   **41.** $\begin{bmatrix} 24 & 16 \\ 39 & 26 \end{bmatrix}$ **43.** $\begin{bmatrix} 47 \\ 81 \end{bmatrix}$

45. $\begin{bmatrix} 64 & 64 \\ 64 & 64 \end{bmatrix}$ **47.** $A^2 = \begin{bmatrix} 1 & 0 \\ 2 & 1 \end{bmatrix}, A^3 = \begin{bmatrix} 1 & 0 \\ 3 & 1 \end{bmatrix}, A^n = \begin{bmatrix} 1 & 0 \\ n & 1 \end{bmatrix}$

49. No. Let $A = \begin{bmatrix} 1 & 1 \\ 1 & 1 \end{bmatrix}$ and $B = \begin{bmatrix} 1 & 0 \\ 0 & 0 \end{bmatrix}$, for example. Then

$$(AB)^2 = \left(\begin{bmatrix} 1 & 1 \\ 1 & 1 \end{bmatrix} \begin{bmatrix} 1 & 0 \\ 0 & 0 \end{bmatrix}\right)^2 = \begin{bmatrix} 1 & 0 \\ 1 & 0 \end{bmatrix}^2 = \begin{bmatrix} 1 & 0 \\ 1 & 0 \end{bmatrix} \begin{bmatrix} 1 & 0 \\ 1 & 0 \end{bmatrix} = \begin{bmatrix} 1 & 0 \\ 1 & 0 \end{bmatrix}$$

$$A^2 B^2 = \left(\begin{bmatrix} 1 & 1 \\ 1 & 1 \end{bmatrix} \begin{bmatrix} 1 & 1 \\ 1 & 1 \end{bmatrix}\right) \left(\begin{bmatrix} 1 & 0 \\ 0 & 0 \end{bmatrix} \begin{bmatrix} 1 & 0 \\ 0 & 0 \end{bmatrix}\right) = \begin{bmatrix} 2 & 2 \\ 2 & 2 \end{bmatrix} \begin{bmatrix} 1 & 0 \\ 0 & 0 \end{bmatrix} = \begin{bmatrix} 2 & 0 \\ 2 & 0 \end{bmatrix}$$

The answers are different.

51. One example is $\begin{bmatrix} 1 & 0 \\ 0 & 0 \end{bmatrix}$.

Exercise 5.4   (Page 249)

1. $\begin{bmatrix} 3 & 4 \\ 2 & 3 \end{bmatrix}$ **3.** $\begin{bmatrix} 5 & -7 \\ -2 & 3 \end{bmatrix}$ **5.** $\begin{bmatrix} -40 & 16 & 9 \\ 13 & -5 & -3 \\ 5 & -2 & -1 \end{bmatrix}$ **7.** $\begin{bmatrix} 4 & 1 & -3 \\ -5 & -1 & 4 \\ -1 & -1 & 1 \end{bmatrix}$

9. no inverse   **11.** $\begin{bmatrix} 1 & -2 & 1 \\ 0 & 1 & -2 \\ 0 & 0 & 1 \end{bmatrix}$ **13.** no inverse   **15.** $\begin{bmatrix} 1 & -2 & 1 & 0 \\ 0 & 1 & -2 & 1 \\ 0 & 0 & 1 & -2 \\ 0 & 0 & 0 & 1 \end{bmatrix}$

17. $x = 23, y = 17$ **19.** $x = 70, y = -30$ **21.** $x = -10, y = 4, z = 1$
23. $x = 7, y = -9, z = -1$ **25.** $x = 4, y = -6, z = 3$ **31.** $x = \pm 4$
35. $E = \begin{bmatrix} 1 & 0 & 0 \\ 0 & 1 & 0 \\ 3 & 0 & 1 \end{bmatrix}, E^{-1} = \begin{bmatrix} 1 & 0 & 0 \\ 0 & 1 & 0 \\ -3 & 0 & 1 \end{bmatrix}$ **37.** $\begin{bmatrix} -6 & -30 & -3 & 7 \\ -5 & -20 & 0 & 5 \\ 8 & 35 & 3 & -8 \\ -1 & -5 & 0 & 1 \end{bmatrix}$

Exercise 5.5   (Page 260)

1. 8   **3.** 1   **5.** −54   **7.** −7   **9.** 86   **11.** −2   **13.** 12   **15.** 1
17. $x = 1, y = 2$ **19.** $x = 3, y = 0$ **21.** $x = 1, y = 0, z = 1$ **23.** $x = 1, y = -1, z = 2$
25. $x = 6, y = 6, z = 12$ **27.** $p = 1, q = 1, r = 1, s = 1$
31. For example, $\begin{vmatrix} 1 & 0 \\ 0 & 1 \end{vmatrix} + \begin{vmatrix} 1 & 0 \\ 0 & 1 \end{vmatrix} = 1 + 1 = 2;$

however $\left| \begin{bmatrix} 1 & 0 \\ 0 & 1 \end{bmatrix} + \begin{bmatrix} 1 & 0 \\ 0 & 1 \end{bmatrix} \right| = \begin{vmatrix} 2 & 0 \\ 0 & 2 \end{vmatrix} = 4.$

33. $\begin{vmatrix} a & b & c \\ 0 & d & e \\ 0 & 0 & f \end{vmatrix} = f \begin{vmatrix} a & b \\ 0 & d \end{vmatrix} = adf$ **39.** $x = 8$ **41.** $x = -1$

43. domain is the set of $n \times n$ matrices, range is the set of real numbers
45. Yes, because $|AB| = |A| \, |B|$, and $|AB| = 0$ implies that $|A| \, |B| = 0$, and therefore $|A| = 0$ or $|B| = 0$.

Exercise 5.6   (Page 266)

1. $\dfrac{1}{x + 1} + \dfrac{2}{x - 1}$ **3.** $\dfrac{1}{x^2 + 2} - \dfrac{3}{x + 1}$ **5.** $\dfrac{1}{x} + \dfrac{2}{x^2} - \dfrac{3}{x - 1}$ **7.** $\dfrac{1}{x^2} + \dfrac{1}{x^2 + 1}$

9. $\dfrac{2}{x} + \dfrac{3x + 2}{x^2 + 1}$ **11.** $\dfrac{1}{x} + \dfrac{1}{x^2} + \dfrac{2}{x^2 + x + 1}$ **13.** $\dfrac{1}{x^2 + x + 5} + \dfrac{x + 1}{x^2 + 1}$

15. $\dfrac{-1}{x^2 + 1} - \dfrac{x}{(x^2 + 1)^2} + \dfrac{1}{x}$ **17.** $\dfrac{1}{x^2 + 1} + \dfrac{x + 2}{x^2 + x + 2}$ **19.** $\dfrac{1}{x} + \dfrac{x}{x^2 + 2x + 5} + \dfrac{x + 2}{(x^2 + 2x + 5)^2}$

21. $2 + \dfrac{1}{x} + \dfrac{2}{x^2} + \dfrac{3}{x + 1}$ **23.** $1 + \dfrac{x + 1}{x^2 + 1} + \dfrac{2}{(x^2 + 1)^2}$

Exercise 5.7   (Page 272)

1. $y = -2x + 3$ $y = 3x + 2$

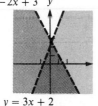

3. $3x + 2y = 6$ $x + 3y = 2$

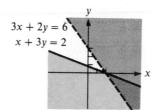

5. $-x + 2y = 9$ $3x + y = 1$

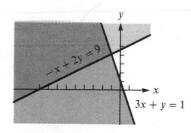

7. $2x - y = 4$ $y = -x^2 + 2$

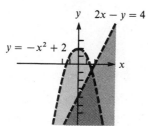

9. $y = x^2 - 4$ $y = -x^2 + 4$

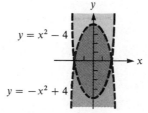

11. $x = 0$ $2x + 3y = 5$ $3x + y = 1$

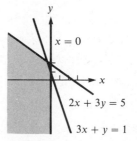

13.

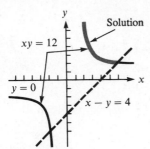

$xy = 12$

$y = 0$

Solution

$x - y = 4$

15.

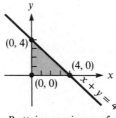

$3x + 6y = 18$

$9x + 3y = 18$

17.

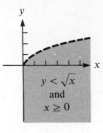

$y < \sqrt{x}$
and
$x \geq 0$

19.

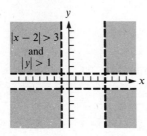

$|x - 2| > 3$
and
$|y| > 1$

21.

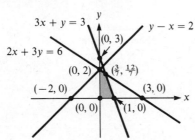

$(0, 4)$

$(4, 0)$

$(0, 0)$

$x + y = 4$

P attains maximum of
12 at $(0, 4)$

23.

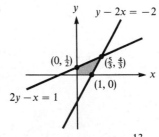

$y - 2x = -2$

$(0, \frac{1}{2})$

$(\frac{5}{3}, \frac{4}{3})$

$(1, 0)$

$2y - x = 1$

P attains maximum of $\dfrac{13}{6}$

at $\left(\dfrac{5}{3}, \dfrac{4}{3}\right)$

25.

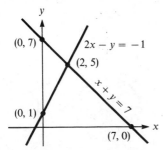

$3x + y = 3$

$y - x = 2$

$(0, 3)$

$2x + 3y = 6$

$(0, 2)$ $(\frac{3}{7}, \frac{12}{7})$

$(-2, 0)$ $(3, 0)$

$(0, 0)$ $(1, 0)$

P attains maximum value

of $\dfrac{18}{7}$ at $\left(\dfrac{3}{7}, \dfrac{12}{7}\right)$

27.   2 tables and no chairs   **29.**   30 of each
31.   10 square meters of strawberries, 30 square meters of pumpkins
33.   10 ounces of X and 25 ounces of Y per day

REVIEW EXERCISES   (Page 273)

1.

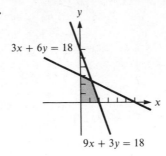

$(0, 7)$

$2x - y = -1$

$(2, 5)$

$x + y = 7$

$(0, 1)$

$(7, 0)$

3.

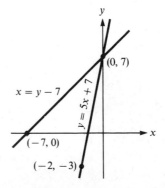

$(0, 7)$

$x = y - 7$

$y = 5x + 7$

$(-7, 0)$

$(-2, -3)$

5. $x = 1, y = 5$ **7.** $x = 0, y = -3$ **9.** $x = -3, y = 2$ **11.** $x = 2, y = -1$

13. $x = 1, y = 0, z = 1$ **15.** $x = 0, y = 1, z = 2$ **17.** $x = 1, y = 1$

19. $x = 1 - \frac{4}{7}z, y = 2 - \frac{1}{7}z, z = $ any number   **21.** $w = 1, x = 2, y = 0, z = -1$

23. $\begin{bmatrix} 1 & 3 & 4 \\ 4 & 0 & 2 \end{bmatrix}$ **25.** $\begin{bmatrix} -7 & -1 \\ -7 & 4 \end{bmatrix}$ **27.** $[5]$ **29.** $\begin{bmatrix} 2 & -1 & 1 & 3 \\ 4 & -2 & 2 & 6 \\ 2 & -1 & 1 & 3 \\ 10 & -5 & 5 & 15 \end{bmatrix}$

31. $[-18]$ **33.** $\begin{bmatrix} \frac{5}{14} & -\frac{3}{14} \\ \frac{3}{14} & \frac{1}{14} \end{bmatrix}$ **35.** $\begin{bmatrix} 1 & -3 & 32 \\ 0 & 1 & -9 \\ 0 & 0 & 1 \end{bmatrix}$ **37.** $\begin{bmatrix} 9 & 16 & -56 \\ -3 & -5 & 18 \\ -1 & -2 & 7 \end{bmatrix}$

39. $x = 1, y = 2, z = -1$ **41.** -7 **43.** -6 **45.** $x = 1, y = -2$

47. $x = 1, y = -1, z = 3$ **49.** $\dfrac{1}{x} + \dfrac{3x + 4}{x^2 + 1}$ **51.** $\dfrac{1}{x} + \dfrac{-1}{x^2 + x + 5}$

53.

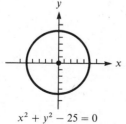

P attains maximum
of 6 at (3, 0)

55.

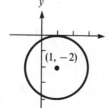

P attains maximum
of 2 at (1, 1)

57. 1000 bags of X, 1400 bags of Y

Exercise 6.1   (Page 282)

1. $x^2 + y^2 = 1$ **3.** $(x - 6)^2 + (y - 8)^2 = 16$ **5.** $(x + 5)^2 + (y - 3)^2 = 25$

7. $(x - 3)^2 + (y + 4)^2 = 2$ **9.** $(x - 3)^2 + (y - 3)^2 = 25$ **11.** $(x + 5)^2 + (y - 1)^2 = 65$

13. $(x + 3)^2 + (y - 4)^2 = 25$ **15.** $(x + 2)^2 + (y + 6)^2 = 40$ **17.** $x^2 + (y + 3)^2 = 157$

19. $(x - 5)^2 + (y - 8)^2 = 338$ **21.** $(x + 4)^2 + (y + 2)^2 = 98$ **23.** $(x - 1)^2 + (y + 2)^2 = 36$

25. $x^2 + (y + 12)^2 = 10$ **27.** no

29.

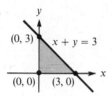

$x^2 + y^2 - 25 = 0$

31.

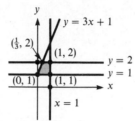

$(x - 1)^2 + (y + 2)^2 = 4$

33.

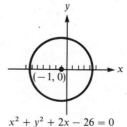

$x^2 + y^2 + 2x - 26 = 0$

35.

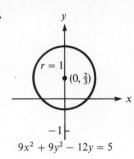

$r = 1$

$(0, \frac{2}{3})$

$9x^2 + 9y^2 - 12y = 5$

37.

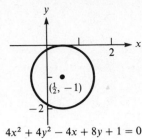

$(\frac{1}{2}, -1)$

$4x^2 + 4y^2 - 4x + 8y + 1 = 0$

39. $x^2 + (y - 3)^2 = 25$
41. $A = 5\pi$

Exercise 6.2 (Page 288)

1. $x^2 = 12y$ **3.** $y^2 = 12x$ **5.** $(x - 3)^2 = -12(y - 5)$ **7.** $(x - 3)^2 = -28(y - 5)$
9. $(x - 2)^2 = -2(y - 2)$ or $(y - 2)^2 = -2(x - 2)$ **11.** $(x + 4)^2 = -\frac{16}{3}(y - 6)$ or $(y - 6)^2 = \frac{9}{4}(x + 4)$
13. $(y - 8)^2 = -4(x - 6)$ **15.** $(x - 3)^2 = \frac{1}{2}(y - 1)$

17.

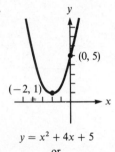

$(0, 5)$

$(-2, 1)$

$y = x^2 + 4x + 5$
or
$y - 1 = (x + 2)^2$

19.

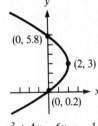

$(0, 5.8)$

$(2, 3)$

$(0, 0.2)$

$y^2 + 4x - 6y = -1$
or
$(y - 3)^2 = -4(x - 2)$

21.

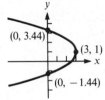

$(0, 3.44)$

$(3, 1)$

$(0, -1.44)$

$y^2 + 2x - 2y = 5$
or
$(y - 1)^2 = -2(x - 3)$

23.

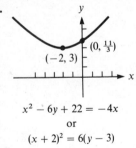

$(0, \frac{11}{3})$

$(-2, 3)$

$x^2 - 6y + 22 = -4x$
or
$(x + 2)^2 = 6(y - 3)$

25.

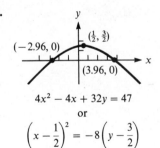

$(-2.96, 0)$

$(\frac{1}{2}, \frac{3}{2})$

$(3.96, 0)$

$4x^2 - 4x + 32y = 47$
or
$\left(x - \frac{1}{2}\right)^2 = -8\left(y - \frac{3}{2}\right)$

27. $x^2 = -\frac{45}{2}y$ **29.** 8 cabins   **31.** $\frac{80}{7}$ meters   **33.** $\frac{9}{4}$ feet   **37.** $y = x^2 + 4x + 3$

Exercise 6.3 (Page 296)

1. $\frac{x^2}{25} + \frac{y^2}{16} = 1$ **3.** $\frac{9x^2}{16} + \frac{9y^2}{25} = 1$ **5.** $\frac{x^2}{7} + \frac{y^2}{16} = 1$ **7.** $\frac{(x - 3)^2}{4} + \frac{(y - 4)^2}{9} = 1$

9. $\frac{(x - 3)^2}{9} + \frac{(y - 4)^2}{4} = 1$ **11.** $\frac{(x - 3)^2}{41} + \frac{(y - 4)^2}{16} = 1$ **13.** $\frac{x^2}{36} + \frac{(y - 4)^2}{20} = 1$

15. $\frac{x^2}{100} + \frac{y^2}{64} = 1$

17.

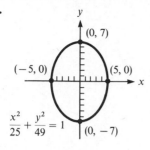

$$\frac{x^2}{25} + \frac{y^2}{49} = 1$$

19.

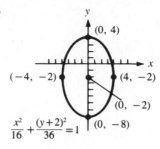

$$\frac{x^2}{16} + \frac{(y+2)^2}{36} = 1$$

21.

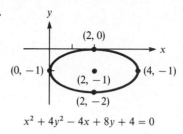

$$x^2 + 4y^2 - 4x + 8y + 4 = 0$$

23.

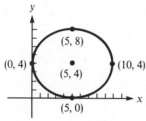

$$16x^2 + 25y^2 - 160x - 200y + 400 = 0$$

25. 199,395 miles   **27.** $\dfrac{x^2}{2500} + \dfrac{y^2}{900} = 1$; 36 meters

29. In the ellipse, $b^2 = a^2 - c^2$, or $a = \sqrt{b^2 + c^2}$. Using the distance formula, you have $d(FB) = \sqrt{b^2 + c^2}$.

31. Substitute c for x in the equation $\dfrac{x^2}{a^2} + \dfrac{y^2}{b^2} = 1$. Use the fact that $c^2 = a^2 - b^2$, and solve for y to determine FA'. The focal width is $2y$.

33. The thumbtacks are the foci at $(\pm 1, 0)$. Hence, $c = 1$. The string is 6 meters, so $2a = 6$, or $a^2 = 9$. Because $b^2 = a^2 - c^2$, you have $b^2 = 8$. The equation is $\dfrac{x^2}{9} + \dfrac{y^2}{8} = 1$.

Exercise 6.4   (Page 303)

1. $\dfrac{x^2}{25} - \dfrac{y^2}{24} = 1$ **3.** $\dfrac{(x-2)^2}{4} - \dfrac{(y-4)^2}{9} = 1$ **5.** $\dfrac{(y-3)^2}{9} - \dfrac{(x-5)^2}{9} = 1$ **7.** $\dfrac{y^2}{9} - \dfrac{x^2}{16} = 1$

9. $\dfrac{(x-1)^2}{4} - \dfrac{(y+3)^2}{16} = 1$ or $\dfrac{(y+3)^2}{4} - \dfrac{(x-1)^2}{16} = 1$ **11.** $\dfrac{x^2}{10} - \dfrac{3y^2}{20} = 1$ **13.** 24 square units

15. 12 square units   **17.** $\dfrac{(x+2)^2}{4} - \dfrac{4(y+4)^2}{81} = 1$ and $\dfrac{(y+4)^2}{4} - \dfrac{4(x+2)^2}{81} = 1$ **19.** $\dfrac{x^2}{36} - \dfrac{16y^2}{25} = 1$

21.

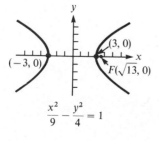

$$\frac{x^2}{9} - \frac{y^2}{4} = 1$$

23.

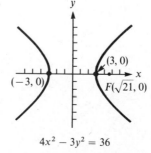

$$4x^2 - 3y^2 = 36$$

25.

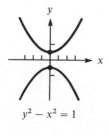

$$y^2 - x^2 = 1$$

27.

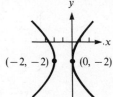

$$4x^2 - 2y^2 + 8x - 8y = 8$$
or
$$\frac{(x + 1)^2}{1} - \frac{(y + 2)^2}{2} = 1$$

29.

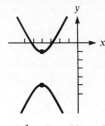

$$y^2 - 4x^2 + 6y + 32x = 59$$

31.

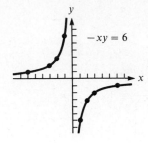

$-xy = 6$

33. $\dfrac{(x - 3)^2}{9} - \dfrac{(y - 1)^2}{16} = 1$ **35.** $4x^2 - 5y^2 - 60y = 0$

37. From the geometry of Figure 6-17, $PF + F'F > F'P$. This is equivalent to $F'P - PF < F'F$ or $2a < 2c$, which implies that $c > a$.

Exercise 6.5   (Page 308)

1.

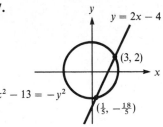

$x = 2y$
$8x^2 + 32y^2 = 256$

3.

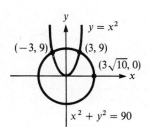

$y = x^2$
$(-3, 9)$ $(3, 9)$
$(3\sqrt{10}, 0)$
$x^2 + y^2 = 90$

5.

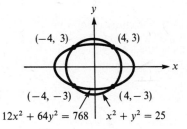

$(-4, 3)$ $(4, 3)$
$(-4, -3)$ $(4, -3)$
$12x^2 + 64y^2 = 768$ $x^2 + y^2 = 25$

7.

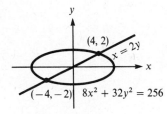

$y = 2x - 4$
$(3, 2)$
$x^2 - 13 = -y^2$
$(\frac{1}{5}, -\frac{18}{5})$

9.

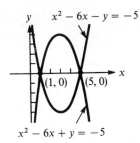

$x^2 - 6x - y = -5$
$(1, 0)$ $(5, 0)$
$x^2 - 6x + y = -5$

11. $(3, 0), (0, 5)$ **13.** $(1, 1)$ **15.** $(1, 2), (2, 1)$ **17.** $(-2, 3), (2, 3)$
19. $(\sqrt{5}, 5), (-\sqrt{5}, 5)$ **21.** $(3, 2), (3, -2), (-3, 2), (-3, -2)$
23. $(2, 4), (2, -4), (-2, 4), (-2, -4)$ **25.** $(-\sqrt{15}, 5), (\sqrt{15}, 5), (-2, -6), (2, -6)$
27. $(0, -4), (-3, 5), (3, 5)$ **29.** $(-2, 3), (2, 3), (-2, -3), (2, -3)$ **31.** $(3, 3)$
33. $(6, 2), (-6, -2), (\sqrt{42}, 0), (-\sqrt{42}, 0)$ **35.** $(\frac{1}{2}, \frac{1}{3}), (\frac{1}{3}, \frac{1}{2})$ **37.** 7 by 9 centimeters
39. 14 and -5 **41.** Either \$750 at 9% or \$900 at 7.5%

Exercise 6.6   (Page 314)

1. $P(3, 1)$ **3.** $R(1, -3)$ **5.** $P(4, -8)$ **7.** $R(2, -4)$

9.

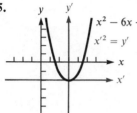

$y = 3x - 2$
$y' = 3x' + 3$
$(0, -5)$

11.

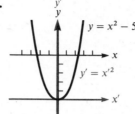

$y = x^2 - 5$
$y' = x'^2$

13.

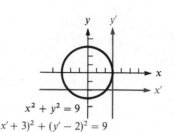

$x^2 + y^2 = 9$
$(x' + 3)^2 + (y' - 2)^2 = 9$

15.

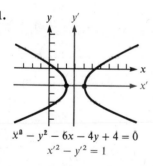

$x^2 - 6x + 7 = y$
$x'^2 = y'$

17.

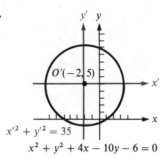

$O'(-2, 5)$
$x'^2 + y'^2 = 35$
$x^2 + y^2 + 4x - 10y - 6 = 0$

19.

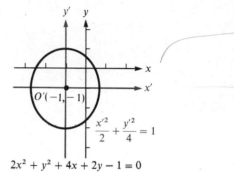

$O'(-1, -1)$
$\dfrac{x'^2}{2} + \dfrac{y'^2}{4} = 1$
$2x^2 + y^2 + 4x + 2y - 1 = 0$

21.

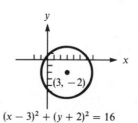

$x^2 - y^2 - 6x - 4y + 4 = 0$
$x'^2 - y'^2 = 1$

REVIEW EXERCISES   (Page 316)

1. $x^2 + y^2 = 50$ **3.** $(x - 5)^2 + (y - 10)^2 = 85$ **5.**

$(3, -2)$

$(x - 3)^2 + (y + 2)^2 = 16$

7. $y^2 = -2x$ **9.** $\left(x + \dfrac{b}{2a}\right)^2 = \dfrac{1}{a}\left(y + \dfrac{b^2 - 4ac}{4a}\right)$; vertex at $\left(-\dfrac{b}{2a}, \dfrac{4ac - b^2}{4a}\right)$

11.

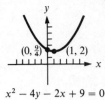

$(0, \frac{9}{4})$ $(1, 2)$

$x^2 - 4y - 2x + 9 = 0$

13.

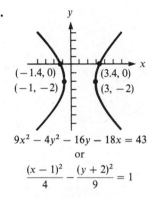

$(4, 3)$

$(3, 1)$

$(4, -1)$

$y^2 - 4x - 2y + 13 = 0$

15. $\dfrac{(x + 2)^2}{16} + \dfrac{(y - 3)^2}{9} = 1$

17.

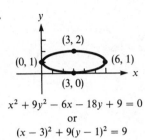

$(3, 2)$

$(0, 1)$ $(6, 1)$

$(3, 0)$

$x^2 + 9y^2 - 6x - 18y + 9 = 0$
or
$(x - 3)^2 + 9(y - 1)^2 = 9$

19.

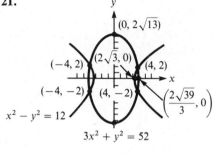

$(-1.4, 0)$ $(3.4, 0)$
$(-1, -2)$ $(3, -2)$

$9x^2 - 4y^2 - 16y - 18x = 43$
or
$\dfrac{(x - 1)^2}{4} - \dfrac{(y + 2)^2}{9} = 1$

21.

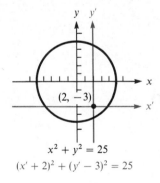

$(0, 2\sqrt{13})$

$(-4, 2)$ $(2\sqrt{3}, 0)$ $(4, 2)$

$(-4, -2)$ $(4, -2)$

$\left(\dfrac{2\sqrt{39}}{3}, 0\right)$

$x^2 - y^2 = 12$

$3x^2 + y^2 = 52$

23.

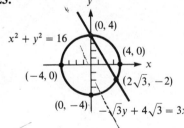

$x^2 + y^2 = 16$

$(0, 4)$

$(4, 0)$

$(-4, 0)$

$(2\sqrt{3}, -2)$

$(0, -4)$ $-\sqrt{3}y + 4\sqrt{3} = 3x$

25.

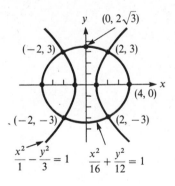

$(0, 2\sqrt{3})$

$(-2, 3)$ $(2, 3)$

$(4, 0)$

$(-2, -3)$ $(2, -3)$

$\dfrac{x^2}{1} - \dfrac{y^2}{3} = 1$ $\dfrac{x^2}{16} + \dfrac{y^2}{12} = 1$

27.

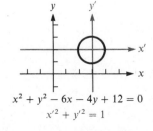

$(2, -3)$

$x^2 + y^2 = 25$
$(x' + 2)^2 + (y' - 3)^2 = 25$

29.

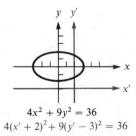

$4x^2 + 9y^2 = 36$
$4(x' + 2)^2 + 9(y' - 3)^2 = 36$

31.

$x^2 + y^2 - 6x - 4y + 12 = 0$
$x'^2 + y'^2 = 1$

Exercise 7.1 (Page 322)

1. $P(2) = 35$ **3.** $P(0) = -1$ **5.** $P(-4) = 719$ **7.** $P(2) = 31$ **9.** $P(-3) = 142$
11. true **13.** true **15.** false **17.** true **19.** $3, -1, -5$ **21.** $1, 1, \sqrt{3}, -\sqrt{3}$
23. $x^3 - 3x^2 + 3x - 1$ **25.** $x^3 - 11x^2 + 38x - 40$ **27.** $x^4 - 3x^2 + 2$
29. $x^3 - \sqrt{2}x^2 + x - \sqrt{2}$ **31.** $x^3 - 2x^2 + 2x$ **33.** $1, -\dfrac{1}{2} + \dfrac{\sqrt{3}}{2}i, -\dfrac{1}{2} - \dfrac{\sqrt{3}}{2}i$
35. $-5, \dfrac{5}{2} + \dfrac{5\sqrt{3}}{2}i, 5 - \dfrac{5\sqrt{3}}{2}i$ **37.** $3, 2, i, -i$ **39.** $4, -1, 1, -2$ **41.** $a_1 = 0$

Exercise 7.2 (Page 327)

1. $P(2) = 47$ **3.** $P(-5) = -569$ **5.** $P(0) = 1$ **7.** $P(-i) = -1 + 6i$ **9.** $P(1) = 3$
11. $P(-2) = 30$ **13.** $P(\frac{1}{2}) = \frac{15}{8}$ **15.** $P(i) = 5$ **17.** $P(1) = 0$ **19.** $P(-3) = 384$
21. $P(3) = 0$ **23.** $P(2) = 9$ **25.** $P(0) = 8$ **27.** $P(2) = 0$ **29.** $P(i) = 16 + 2i$
31. $P(-2i) = 40 - 40i$ **33.** $(x + 1)(3x^2 - 5x - 1) - 3$ **35.** $(x - 2)(3x^2 + 4x + 2) + 0$
37. $x(3x^2 - 2x - 6) - 4$ **39.** $7x^2 - 10x + 5 + \dfrac{-4}{x + 1}$ **41.** $4x^3 + 9x^2 + 27x + 80 + \dfrac{245}{x - 3}$
43. $3x^4 + 12x^3 + 48x^2 + 192x$ **45.** $3^5 = 243$ **47.** $2^7 = 128$
49.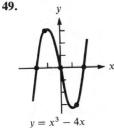
$y = x^3 - 4x$
51.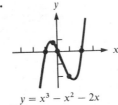
$y = x^3 - x^2 - 2x$
53.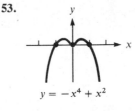
$y = -x^4 + x^2$
55.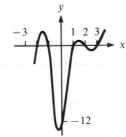
$y = x^5 - 3x^4 - 5x^3 + 15x^2 + 4x - 12$
57. $k = -12$ **59.** $P(1.3) = 4.9643$

Exercise 7.3 (Page 333)

1. 10 **3.** 4 **5.** 0 or 2 positive; 1 negative; 0 or 2 nonreal
7. 0 positive; 1 or 3 negative; 0 or 2 nonreal **9.** 0 positive; 0 negative; 4 nonreal
11. 1 positive; 1 negative; 2 nonreal **13.** 0 positive; 0 negative; 10 nonreal
15. 0 positive; 0 negative; 8 nonreal; 1 root of 0 **17.** 1 positive; 1 negative; 2 nonreal **19.** $-1, 6$
21. $-4, 6$ **23.** $-4, 3$ **25.** $-5, 2$
27. An odd-degree polynomial equation must have an odd number of roots. Since complex roots occur in conjugate pairs, one root must be left over, and it is real.

Exercise 7.4   (Page 339)

1. $1, -1, 5$ **3.** $3, -3, 2$ **5.** $1, -1, 2$ **7.** $1, 2, 3, 4$ **9.** $2, -5, \sqrt{3}, -\sqrt{3}$
11. $1, -1, 2, -2, -3$ **13.** $0, 2, 2, 2, -2, -2, -2$ **15.** $\frac{2}{3}, 2i, -2i$ **17.** $-1, \frac{2}{3}, 2, 3$
19. $-3, \frac{1}{3}, \frac{1}{2}, \frac{1}{2}$ **21.** $-3, 2, 2, -\frac{1}{3}, \frac{1}{2}$ **23.** $-\frac{3}{5}, \frac{2}{3}, \frac{3}{2}$ **25.** $\frac{2}{3}, -\frac{3}{5}, 4$ **27.** $-\frac{2}{3}, 3, -1$
29. $\frac{1}{2}, \frac{1}{2}, \frac{1}{2}, 1, 1$
31. Because there is no change of sign in $P(x)$ or in $P(-x)$, there can be no positive or negative roots. 0 is not a root. Hence, all roots are nonreal.

Exercise 7.5   (Page 342)

1. $P(-2) = 3; P(-1) = -2$ **3.** $P(4) = -40; P(5) = 30$ **5.** $P(1) = 8; P(2) = -1$
7. $P(2) = -72; P(3) = 154$ **9.** $P(0) = 10; P(1) = -60$ **11.** $\sqrt{3} \approx 1.73$ **13.** $x \approx 1.7$
15. $x \approx -1.2$ **17.** $1.732, -1.732$ **19.** $0.5, 2.236, -2.236$
21. $1, -1, 1.414, -1.414, 1.732, -1.732$

REVIEW EXERCISES   (Page 343)

1. $P(0) = -2$ **3.** $P(-3) = 277$ **5.** false   **7.** true   **9.** $-4, 2 + 2\sqrt{3}i, 2 - 2\sqrt{3}i$
11. $2x^3 - 5x^2 - x + 6$ **13.** $3x^3 + 9x^2 + 29x + 90$ with remainder 277   **15.** 6
17. 2 or 0 positive; 2 or 0 negative; 0, 2, or 4 nonreal   **19.** 0 positive; 0 negative; 4 nonreal
21. $-5, -\frac{3}{2}, -2$ **23.** $P(0) = 18; P(-1) = -9$ **25.** $\sqrt{7} \approx 2.65$

Exercise 8.1   (Page 351)

1.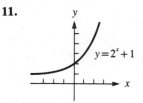

3.

5.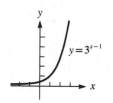

7.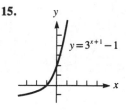

9.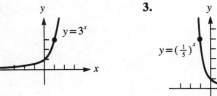

11.

13.

15.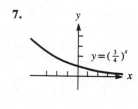

17.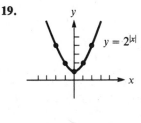

19.

21. $b = 5$ **23.** no value of b **25.** $b = \frac{1}{2}$ **27.** $\frac{32}{243}A_0$ **29.** $A_0 2^{-3000/5700} \approx 0.6943A_0$
31. $\$1342.53$ **33.** $\$2,273,996.13$ **35.** 1.68×10^8 **37.** 2.83

Exercise 8.2   (Page 356)

1.

3.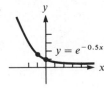

5.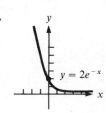

7.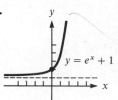

9. yes   **11.** no   **13.** no   **15.** no   **17.** $19,171.75   **19.** $4500   **21.** $8753.36
23. 9.44×10^5 **25.** 2.6   **27.** 202   **29.** e^3 **31.**

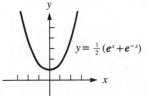

Exercise 8.3   (Page 365)

1. $3^4 = 81$ **3.** $(\frac{1}{2})^3 = \frac{1}{8}$ **5.** $4^{-3} = \frac{1}{64}$ **7.** $x^z = y$ **9.** $\log_8 64 = 2$ **11.** $\log_4 \frac{1}{16} = -2$
13. $\log_{1/2} 32 = -5$ **15.** $\log_x z = y$ **17.** 3   **19.** -3 **21.** 2   **23.** 2   **25.** 7
27. 4   **29.** $-\frac{3}{2}$ **31.** $\frac{2}{3}$ **33.** 5   **35.** $\frac{3}{2}$ **37.** $\frac{1}{9}$ **39.** 8
41.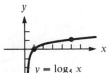

43.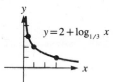

45.

47.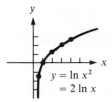

49.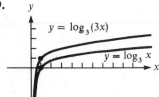

51.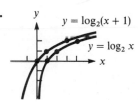

53.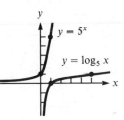

55. no value of b **57.** 3   **59.** no value of b **61.** yes   **63.** no

65. $\log_b xyz = \log_b x + \log_b y + \log_b z$ **67.** $\log_b \left(\frac{x}{y}\right)^2 = 2\log_b x - 2\log_b y$

69. $\log_b x\sqrt{z} = \log_b x + \frac{1}{2}\log_b z$ **71.** $\log_b \sqrt[4]{\frac{x^3 y^2}{z^4}} = \frac{3}{4}\log_b x + \frac{1}{2}\log_b y - \log_b z$

73. $\log_b(x+1) - \log_b x = \log_b \frac{x+1}{x}$ **75.** $2\log_b x + \frac{1}{3}\log_b y = \log_b x^2 y^{1/3}$

77. $-3\log_b x - 2\log_b y + \frac{1}{2}\log_b z = \log_b \frac{z^{1/2}}{x^3 y^2}$ **79.** $\log_b\left(\frac{x}{z}+x\right) - \log_b\left(\frac{y}{z}+y\right) = \log_b \frac{\frac{x}{z}+x}{\frac{y}{z}+y} = \log_b \frac{x}{y}$

81. true   **83.** false   **85.** true   **87.** false   **89.** true   **91.** false   **93.** false
95. true   **97.** true   **99.** false   **101.** true   **103.** 1.4472   **105.** 0.3521
107. 1.1972   **109.** 2.4014   **111.** 2.0493   **119.** Because log 0.9 is a negative number

Exercise 8.4   (Page 372)

1. 0.5119   **3.** −0.0726   **5.** −0.2752   **7.** 69.4079   **9.** 0.0002   **11.** 120.0719
19. 4.77   **21.** from 0.000501 to 0.00126   **23.** 0.71 V

25. $10 \log \dfrac{P_O}{P_I} = 10 \log \dfrac{kE_O^2}{kE_I^2} = 10 \log \left(\dfrac{E_O}{E_I}\right)^2 = 20 \log \dfrac{E_O}{E_I}$ **27.** 4.4   **29.** 2500 μm   **31.** 19.9 h

33. $L = L_0 + k \ln 2$ where $L_0 = k \ln I$ **35.** about 5.8 years   **37.** 3 years old
39. about 10.8 years

Exercise 8.5   (Page 378)

1. $x = \dfrac{\log 5}{\log 4} \approx 1.16$ **3.** $x = \dfrac{\log 2}{\log 13} + 1 \approx 1.27$ **5.** $x = \dfrac{\log 2}{\log 3 - \log 2} \approx 1.71$ **7.** $x = 0$

9. $x = \sqrt{\dfrac{1}{\log 7}} \approx 1.09$ **11.** $x = 0$ or $x = \dfrac{\log 9}{\log 8} \approx 1.06$ **13.** $x = 0$

15. $x = \dfrac{\log 2}{2 \log 6 - \log 3} \approx 0.2789$ **17.** $x = 1$ or $x = 3$ **19.** $x = 0$ **21.** $x = 7$

23. $x = 4$ **25.** $x = 10$ or $x = -10$ **27.** $x = 50$ **29.** $x = 20$ **31.** $x = 10$
33. $x = 3$ or $x = 4$ **35.** $x = 1$ or $x = 100$ **37.** $x = 9$ **39.** $x = 4$
41. $y = 7$ or $y = 1$ **43.** 1.771   **45.** 2.322   **47.** about 5.146 years   **49.** about 42.7 days
51. about 4200 years old   **53.** 4.03 years; 3.92 years   **55.** about 6.96%   **57.** about 3.15 days

Exercise 8.6   (Page 384)

1. 0.7760   **3.** 0.6263   **5.** 5.6355   **7.** −2.8633   **9.** 3.14   **11.** 8380
13. 0.00284   **15.** 0.8384   **17.** −2.4614   **19.** 4.281   **21.** 1598   **23.** 0.007702
25. 0.1913   **27.** 3.388   **29.** 4.354   **31.** 15.33   **33.** 2.022   **35.** 1.5369   **37.** 7.73
39. 3.3810   **41.** −4.8000

REVIEW EXERCISES   (Page 386)

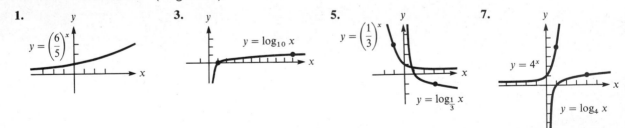

1. $y = \left(\dfrac{6}{5}\right)^x$ **3.** $y = \log_{10} x$ **5.** $y = \left(\dfrac{1}{3}\right)^x$; $y = \log_{\frac{1}{3}} x$ **7.** $y = 4^x$; $y = \log_4 x$

9. 8   **11.** 3   **13.** 1   **15.** $\frac{1}{6}$ **17.** −2   **19.** 27   **21.** 32   **23.** 27   **25.** $\frac{1}{8}$

27. 4   **29.** 10   **31.** $\frac{1}{2} \log_b x - \frac{1}{2} \log_b y - \log_b z$ **33.** 2.64   **35.** $x = \dfrac{\log 7}{\log 3}$

37. $x = \dfrac{\log 3}{\log 3 - \log 2}$ **39.** $x = 4$ **41.** $x = \dfrac{\ln 9}{\ln 2}$ **43.** $x = \dfrac{e}{e - 1}$

45. approximately 3300 years old   **47.** $pH = \log_{10} \dfrac{1}{[H^+]} = \log_{10}[H^+]^{-1} = -\log_{10}[H^+]$ **49.** 8.141

51. 0.0001617

Exercise 9.1   (Page 393)

25. no

Exercise 9.2   (Page 398)

1. 24   **3.** 720   **5.** 1440   **7.** $\frac{1}{1320}$ **9.** 18,564   **11.** $a^4 + 4a^3b + 6a^2b^2 + 4ab^3 + b^4$

13. $a^5 - 5a^4b + 10a^3b^2 - 10a^2b^3 + 5ab^4 - b^5$ **15.** $8x^3 - 12x^2y + 6xy^2 - y^3$

17. $16x^4 + 32x^3y + 24x^2y^2 + 8xy^3 + y^4$ **19.** $256x^4 - 768x^3y + 864x^2y^2 - 432xy^3 + 81y^4$

21. $216x^3 - 540x^2 + 450x - 125$ **23.** $6a^2b^2$ **25.** $35a^3b^4$ **27.** $-b^5$ **29.** $28x^2y^6$

31. $20\sqrt{2}x^3y^2$ **33.** $4608x^2y^7$ **35.** $\frac{r!}{3!(r-3)!}\, a^{r-3}b^3$ **37.** $\frac{n!}{(r-1)!(n-r+1)!}\, a^{n-r+1}b^{r-1}$

39. 1, 2, 4, 8, 16, 32, 64, 128, 256, 512. They are sequential powers of 2.

41. The sixth term is $\dfrac{10!}{5!(10-5)!}\, a^5\left(-\dfrac{1}{a}\right)^5$, or -252.

43. $(x^{1/2})^{25-r}\left(\dfrac{1}{x}\right)^r$ must be x^8. Hence, $r = 3$. Coefficient is $\dfrac{25!}{3!(25-3)!} \cdot \dfrac{1}{2^3} = \dfrac{575}{2}$.

Exercise 9.3   (Page 404)

1. 0, 10, 30, 60, 100, 150, 210, 280   **3.** 21   **5.** $a + 4d$ **7.** 15   **9.** 15   **11.** 15

13. $\frac{242}{243}$ **15.** 35   **17.** 3, 7, 15, 31   **19.** $-4, -2, -1, -\frac{1}{2}$ **21.** k, k^2, k^4, k^8

23. $8, \dfrac{16}{k}, \dfrac{32}{k^2}, \dfrac{64}{k^3}$ **25.** an alternating infinite series   **27.** not an alternating infinite series   **29.** 30

31. -50 **33.** 40   **35.** 500   **37.** $\frac{7}{12}$ **39.** 160

Exercise 9.4   (Page 411)

1. 1, 3, 5, 7, 9, 11   **3.** $5, \frac{7}{2}, 2, \frac{1}{2}, -1, -\frac{5}{2}$ **5.** $9, \frac{23}{2}, 14, \frac{33}{2}, 19, \frac{43}{2}$ **7.** 285   **9.** 555

11. $157\frac{1}{2}$ **13.** 44   **15.** $\frac{25}{2}, 15, \frac{35}{2}$ **17.** $-\frac{82}{15}, -\frac{59}{15}, -\frac{36}{15}, -\frac{13}{15}$ **19.** 10, 20, 40, 80

21. $-2, -6, -18, -54$ **23.** $3, 3\sqrt{2}, 6, 6\sqrt{2}$ **25.** 2, 6, 18, 54   **27.** 124   **29.** $-29, 524$

31. $\frac{1995}{32}$ **33.** 18   **35.** 8   **37.** $10\sqrt[4]{2}, 10\sqrt{2}, 10\sqrt[4]{8}$ **39.** 8, 32, 128, 512   **41.** $\frac{5}{9}$

43. $\frac{25}{99}$ **47.** \$21,474,836.47

49. arithmetic mean is $\frac{11}{16}$; geometric mean is $\sqrt{7}/4$; the arithmetic mean is larger   **51.** $1 - \frac{1}{101} = \frac{100}{101}$

Exercise 9.5   (Page 415)

1. 23   **3.** 5.13 meters   **5.** 1,048,576   **7.** She will earn \$0.19 more on the $7\frac{1}{4}\%$ investment.

9. approximately 6.87×10^{19} **11.** \$180,176.87   **13.** \$2001.60   **15.** \$2013.62

17. \$264,094.58   **19.** 1.8447×10^{19} grains

21. no, because $0.999999 = \dfrac{999999}{1000000}$, which does not equal 1

Exercise 9.6   (Page 422)

1. 144   **3.** 8,000,000   **5.** 240   **7.** 6   **9.** 840   **11.** 35   **13.** 120   **15.** 5

17. 1   **19.** 1200   **21.** 40   **23.** 2278   **25.** 40,320   **27.** 14,400   **29.** 24,360

31. 5040   **33.** 48   **35.** 96   **37.** 210   **39.** approximately 6.35×10^{11} **41.** 3360

43. 2,721,600   **45.** 1120   **47.** 59,400   **49.** 272   **51.** 28   **53.** 56

55. 2,042,975 (ignoring the order of the players)

Exercise 9.7   (Page 428)

1. $\frac{1}{6}$ **3.** $\frac{2}{3}$ **5.** $\frac{19}{42}$ **7.** $\frac{13}{42}$ **9.** $\frac{3}{8}$ **11.** 0   **13.** $\frac{1}{12}$ **15.** $\frac{1}{169}$ **17.** $\frac{5}{12}$
19. approximately 6.3×10^{-12} **21.** 0   **23.** $\frac{3}{13}$ **25.** $\frac{1}{6}$ **27.** $\frac{1}{8}$ **29.** $\frac{5}{16}$
31. (S-survive; F-fail) SSSS, SSSF, SSFS, SFSS, FSSS, SSFF, SFSF, FSSF, SFFS, FSFS, FFSS, SFFF, FSFF,
FFSF, FFFS, FFFF   **33.** $\frac{1}{4}$ **35.** $\frac{1}{4}$
37. 1; Exercises 32 through 36 exhaust all possibilities and are mutually exclusive   **39.** $\frac{32}{119}$ **41.** $\frac{1}{3}$
43. 0.18   **45.** 0.14   **47.** about 33%
49. no, for then $P(A)$ would be greater than 1, which is impossible

Exercise 9.8   (Page 433)

1. $\frac{1}{2}$ **3.** $\frac{7}{13}$ **5.** $\frac{1}{221}$ **7.** $\frac{25}{204}$ **9.** $\frac{11}{36}$ **11.** $\frac{7}{12}$ **13.** $\frac{5}{8}$ **15.** $\frac{13}{16}$ **17.** $\frac{21}{128}$
19. $\frac{3}{20}$ **21.** $\frac{1}{49}$ **23.** 0.973   **25.** $\frac{11}{20}$ **27.** $\frac{1}{8}, \frac{1}{12}, \frac{3}{8}, \frac{5}{12}, 1$

Exercise 9.9   (Page 436)

1. $\frac{1}{6}$ **3.** 5 to 1   **5.** 1 to 1   **7.** $\frac{5}{36}$ **9.** 31 to 5   **11.** 1 to 1   **13.** 1 to 12
15. 3 to 10   **17.** $\frac{5}{7}$ **19.** 1 to 90   **21.** 15 to 1   **23.** $\frac{1}{9}$
25. no; the expected winnings are $\$\frac{9}{13}$ **27.** 7 to 1   **29.** \$1.72   **31.** 6.54   **33.** $\frac{32}{390,625}$

REVIEW EXERCISES   (Page 439)

3. $x^3 - 3x^2y + 3xy^2 - y^3$ **5.** $1024a^5 - 6400a^4b + 16,000a^3b^2 - 20,000a^2b^3 + 12,500ab^4 - 3125b^5$

7. $56a^5b^3$ **9.** $84x^3y^6$ **11.** 90   **13.** 1718   **15.** 117   **17.** -92 **19.** $\frac{1}{3^6}$

21. $\frac{9}{2^{14}}$ **23.** 3320   **25.** -5220 **27.** $\frac{3280}{27}$ **29.** $\frac{2295}{128}$ **31.** $\frac{2}{3}$ **33.** no sum

35. $\frac{1}{3}$ **37.** $\frac{17}{99}$ **39.** $\frac{7}{2}, 5, \frac{13}{2}$ **41.** $2\sqrt{2}, 4, 4\sqrt{2}$ **43.** $\frac{3280}{3}$ **45.** 16
47. 6516, 3134   **49.** 720   **51.** 1   **53.** 564,480   **55.** 21   **57.** 120   **59.** $\frac{1}{6}$
65. 0.00000923   **67.** 20,160   **69.** about 6.3×10^{-12} **71.** $\frac{1}{2}$ **73.** $\frac{1}{2,598,960}$ **75.** $\frac{15}{16}$
77. $\frac{1}{15}$ **79.** 23 to 1   **81.** 664

APPENDIX III
Tables

Table A Powers and Roots

n	n^2	$\sqrt{n}$	n^3	$\sqrt[3]{n}$	n	n^2	$\sqrt{n}$	n^3	$\sqrt[3]{n}$
1	1	1.000	1	1.000	51	2,601	7.141	132,651	3.708
2	4	1.414	8	1.260	52	2,704	7.211	140,608	3.733
3	9	1.732	27	1.442	53	2,809	7.280	148,877	3.756
4	16	2.000	64	1.587	54	2,916	7.348	157,464	3.780
5	25	2.236	125	1.710	55	3,025	7.416	166,375	3.803
6	36	2.449	216	1.817	56	3,136	7.483	175,616	3.826
7	49	2.646	343	1.913	57	3,249	7.550	185,193	3.849
8	64	2.828	512	2.000	58	3,364	7.616	195,112	3.871
9	81	3.000	729	2.080	59	3,481	7.681	205,379	3.893
10	100	3.162	1,000	2.154	60	3,600	7.746	216,000	3.915
11	121	3.317	1,331	2.224	61	3,721	7.810	226,981	3.936
12	144	3.464	1,728	2.289	62	3,844	7.874	238,328	3.958
13	169	3.606	2,197	2.351	63	3,969	7.937	250,047	3.979
14	196	3.742	2,744	2.410	64	4,096	8.000	262,144	4.000
15	225	3.873	3,375	2.466	65	4,225	8.062	274,625	4.021
16	256	4.000	4,096	2.520	66	4,356	8.124	287,496	4.041
17	289	4.123	4,913	2.571	67	4,489	8.185	300,763	4.062
18	324	4.243	5,832	2.621	68	4,624	8.246	314,432	4.082
19	361	4.359	6,859	2.668	69	4,761	8.307	328,509	4.102
20	400	4.472	8,000	2.714	70	4,900	8.367	343,000	4.121
21	441	4.583	9,261	2.759	71	5,041	8.426	357,911	4.141
22	484	4.690	10,648	2.802	72	5,184	8.485	373,248	4.160
23	529	4.796	12,167	2.844	73	5,329	8.544	389,017	4.179
24	576	4.899	13,824	2.884	74	5,476	8.602	405,224	4.198
25	625	5.000	15,625	2.924	75	5,625	8.660	421,875	4.217
26	676	5.099	17,576	2.962	76	5,776	8.718	438,976	4.236
27	729	5.196	19,683	3.000	77	5,929	8.775	456,533	4.254
28	784	5.292	21,952	3.037	78	6,084	8.832	474,552	4.273
29	841	5.385	24,389	3.072	79	6,241	8.888	493,039	4.291
30	900	5.477	27,000	3.107	80	6,400	8.944	512,000	4.309
31	961	5.568	29,791	3.141	81	6,561	9.000	531,441	4.327
32	1,024	5.657	32,768	3.175	82	6,724	9.055	551,368	4.344
33	1,089	5.745	35,937	3.208	83	6,889	9.110	571,787	4.362
34	1,156	5.831	39,304	3.240	84	7,056	9.165	592,704	4.380
35	1,225	5.916	42,875	3.271	85	7,225	9.220	614,125	4.397
36	1,296	6.000	46,656	3.302	86	7,396	9.274	636,056	4.414
37	1,369	6.083	50,653	3.332	87	7,569	9.327	658,503	4.431
38	1,444	6.164	54,872	3.362	88	7,744	9.381	681,472	4.448
39	1,521	6.245	59,319	3.391	89	7,921	9.434	704,969	4.465
40	1,600	6.325	64,000	3.420	90	8,100	9.487	729,000	4.481
41	1,681	6.403	68,921	3.448	91	8,281	9.539	753,571	4.498
42	1,764	6.481	74,088	3.476	92	8,464	9.592	778,688	4.514
43	1,849	6.557	79,507	3.503	93	8,649	9.644	804,357	4.531
44	1,936	6.633	85,184	3.530	94	8,836	9.695	830,584	4.547
45	2,025	6.708	91,125	3.557	95	9,025	9.747	857,375	4.563
46	2,116	6.782	97,336	3.583	96	9,216	9.798	884,736	4.579
47	2,209	6.856	103,823	3.609	97	9,409	9.849	912,673	4.595
48	2,304	6.928	110,592	3.634	98	9,604	9.899	941,192	4.610
49	2,401	7.000	117,649	3.659	99	9,801	9.950	970,299	4.626
50	2,500	7.071	125,000	3.684	100	10,000	10.000	1,000,000	4.642

Table B   Base-10 Logarithms

N	0	1	2	3	4	5	6	7	8	9
1.0	.0000	.0043	.0086	.0128	.0170	.0212	.0253	.0294	.0334	.0374
1.1	.0414	.0453	.0492	.0531	.0569	.0607	.0645	.0682	.0719	.0755
1.2	.0792	.0828	.0864	.0899	.0934	.0969	.1004	.1038	.1072	.1106
1.3	.1139	.1173	.1206	.1239	.1271	.1303	.1335	.1367	.1399	.1430
1.4	.1461	.1492	.1523	.1553	.1584	.1614	.1644	.1673	.1703	.1732
1.5	.1761	.1790	.1818	.1847	.1875	.1903	.1931	.1959	.1987	.2014
1.6	.2041	.2068	.2095	.2122	.2148	.2175	.2201	.2227	.2253	.2279
1.7	.2304	.2330	.2355	.2380	.2405	.2430	.2455	.2480	.2504	.2529
1.8	.2553	.2577	.2601	.2625	.2648	.2672	.2695	.2718	.2742	.2765
1.9	.2788	.2810	.2833	.2856	.2878	.2900	.2923	.2945	.2967	.2989
2.0	.3010	.3032	.3054	.3075	.3096	.3118	.3139	.3160	.3181	.3201
2.1	.3222	.3243	.3263	.3284	.3304	.3324	.3345	.3365	.3385	.3404
2.2	.3424	.3444	.3464	.3483	.3502	.3522	.3541	.3560	.3579	.3598
2.3	.3617	.3636	.3655	.3674	.3692	.3711	.3729	.3747	.3766	.3784
2.4	.3802	.3820	.3838	.3856	.3874	.3892	.3909	.3927	.3945	.3962
2.5	.3979	.3997	.4014	.4031	.4048	.4065	.4082	.4099	.4116	.4133
2.6	.4150	.4166	.4183	.4200	.4216	.4232	.4249	.4265	.4281	.4298
2.7	.4314	.4330	.4346	.4362	.4378	.4393	.4409	.4425	.4440	.4456
2.8	.4472	.4487	.4502	.4518	.4533	.4548	.4564	.4579	.4594	.4609
2.9	.4624	.4639	.4654	.4669	.4683	.4698	.4713	.4728	.4742	.4757
3.0	.4771	.4786	.4800	.4814	.4829	.4843	.4857	.4871	.4886	.4900
3.1	.4914	.4928	.4942	.4955	.4969	.4983	.4997	.5011	.5024	.5038
3.2	.5051	.5065	.5079	.5092	.5105	.5119	.5132	.5145	.5159	.5172
3.3	.5185	.5198	.5211	.5224	.5237	.5250	.5263	.5276	.5289	.5302
3.4	.5315	.5328	.5340	.5353	.5366	.5378	.5391	.5403	.5416	.5428
3.5	.5441	.5453	.5465	.5478	.5490	.5502	.5514	.5527	.5539	.5551
3.6	.5563	.5575	.5587	.5599	.5611	.5623	.5635	.5647	.5658	.5670
3.7	.5682	.5694	.5705	.5717	.5729	.5740	.5752	.5763	.5775	.5786
3.8	.5798	.5809	.5821	.5832	.5843	.5855	.5866	.5877	.5888	.5899
3.9	.5911	.5922	.5933	.5944	.5955	.5966	.5977	.5988	.5999	.6010
4.0	.6021	.6031	.6042	.6053	.6064	.6075	.6085	.6096	.6107	.6117
4.1	.6128	.6138	.6149	.6160	.6170	.6180	.6191	.6201	.6212	.6222
4.2	.6232	.6243	.6253	.6263	.6274	.6284	.6294	.6304	.6314	.6325
4.3	.6335	.6345	.6355	.6365	.6375	.6385	.6395	.6405	.6415	.6425
4.4	.6435	.6444	.6454	.6464	.6474	.6484	.6493	.6503	.6513	.6522
4.5	.6532	.6542	.6551	.6561	.6571	.6580	.6590	.6599	.6609	.6618
4.6	.6628	.6637	.6646	.6656	.6665	.6675	.6684	.6693	.6702	.6712
4.7	.6721	.6730	.6739	.6749	.6758	.6767	.6776	.6785	.6794	.6803
4.8	.6812	.6821	.6830	.6839	.6848	.6857	.6866	.6875	.6884	.6893
4.9	.6902	.6911	.6920	.6928	.6937	.6946	.6955	.6964	.6972	.6981
5.0	.6990	.6998	.7007	.7016	.7024	.7033	.7042	.7050	.7059	.7067
5.1	.7076	.7084	.7093	.7101	.7110	.7118	.7126	.7135	.7143	.7152
5.2	.7160	.7168	.7177	.7185	.7193	.7202	.7210	.7218	.7226	.7235
5.3	.7243	.7251	.7259	.7267	.7275	.7284	.7292	.7300	.7308	.7316
5.4	.7324	.7332	.7340	.7348	.7356	.7364	.7372	.7380	.7388	.7396

Table B *(continued)*

N	0	1	2	3	4	5	6	7	8	9
5.5	.7404	.7412	.7419	.7427	.7435	.7443	.7451	.7459	.7466	.7474
5.6	.7482	.7490	.7497	.7505	.7513	.7520	.7528	.7536	.7543	.7551
5.7	.7559	.7566	.7574	.7582	.7589	.7597	.7604	.7612	.7619	.7627
5.8	.7634	.7642	.7649	.7657	.7664	.7672	.7679	.7686	.7694	.7701
5.9	.7709	.7716	.7723	.7731	.7738	.7745	.7752	.7760	.7767	.7774
6.0	.7782	.7789	.7796	.7803	.7810	.7818	.7825	.7832	.7839	.7846
6.1	.7853	.7860	.7868	.7875	.7882	.7889	.7896	.7903	.7910	.7917
6.2	.7924	.7931	.7938	.7945	.7952	.7959	.7966	.7973	.7980	.7987
6.3	.7993	.8000	.8007	.8014	.8021	.8028	.8035	.8041	.8048	.8055
6.4	.8062	.8069	.8075	.8082	.8089	.8096	.8102	.8109	.8116	.8122
6.5	.8129	.8136	.8142	.8149	.8156	.8162	.8169	.8176	.8182	.8189
6.6	.8195	.8202	.8209	.8215	.8222	.8228	.8235	.8241	.8248	.8254
6.7	.8261	.8267	.8274	.8280	.8287	.8293	.8299	.8306	.8312	.8319
6.8	.8325	.8331	.8338	.8344	.8351	.8357	.8363	.8370	.8376	.8382
6.9	.8388	.8395	.8401	.8407	.8414	.8420	.8426	.8432	.8439	.8445
7.0	.8451	.8457	.8463	.8470	.8476	.8482	.8488	.8494	.8500	.8506
7.1	.8513	.8519	.8525	.8531	.8537	.8543	.8549	.8555	.8561	.8567
7.2	.8573	.8579	.8585	.8591	.8597	.8603	.8609	.8615	.8621	.8627
7.3	.8633	.8639	.8645	.8651	.8657	.8663	.8669	.8675	.8681	.8686
7.4	.8692	.8698	.8704	.8710	.8716	.8722	.8727	.8733	.8739	.8745
7.5	.8751	.8756	.8762	.8768	.8774	.8779	.8785	.8791	.8797	.8802
7.6	.8808	.8814	.8820	.8825	.8831	.8837	.8842	.8848	.8854	.8859
7.7	.8865	.8871	.8876	.8882	.8887	.8893	.8899	.8904	.8910	.8915
7.8	.8921	.8927	.8932	.8938	.8943	.8949	.8954	.8960	.8965	.8971
7.9	.8976	.8982	.8987	.8993	.8998	.9004	.9009	.9015	.9020	.9025
8.0	.9031	.9036	.9042	.9047	.9053	.9058	.9063	.9069	.9074	.9079
8.1	.9085	.9090	.9096	.9101	.9106	.9112	.9117	.9122	.9128	.9133
8.2	.9138	.9143	.9149	.9154	.9159	.9165	.9170	.9175	.9180	.9186
8.3	.9191	.9196	.9201	.9206	.9212	.9217	.9222	.9227	.9232	.9238
8.4	.9243	.9248	.9253	.9258	.9263	.9269	.9274	.9279	.9284	.9289
8.5	.9294	.9299	.9304	.9309	.9315	.9320	.9325	.9330	.9335	.9340
8.6	.9345	.9350	.9355	.9360	.9365	.9370	.9375	.9380	.9385	.9390
8.7	.9395	.9400	.9405	.9410	.9415	.9420	.9425	.9430	.9435	.9440
8.8	.9445	.9450	.9455	.9460	.9465	.9469	.9474	.9479	.9484	.9489
8.9	.9494	.9499	.9504	.9509	.9513	.9518	.9523	.9528	.9533	.9538
9.0	.9542	.9547	.9552	.9557	.9562	.9566	.9571	.9576	.9581	.9586
9.1	.9590	.9595	.9600	.9605	.9609	.9614	.9619	.9624	.9628	.9633
9.2	.9638	.9643	.9647	.9652	.9657	.9661	.9666	.9671	.9675	.9680
9.3	.9685	.9689	.9694	.9699	.9703	.9708	.9713	.9717	.9722	.9727
9.4	.9731	.9736	.9741	.9745	.9750	.9754	.9759	.9763	.9768	.9773
9.5	.9777	.9782	.9786	.9791	.9795	.9800	.9805	.9809	.9814	.9818
9.6	.9823	.9827	.9832	.9836	.9841	.9845	.9850	.9854	.9859	.9863
9.7	.9868	.9872	.9877	.9881	.9886	.9890	.9894	.9899	.9903	.9908
9.8	.9912	.9917	.9921	.9926	.9930	.9934	.9939	.9943	.9948	.9952
9.9	.9956	.9961	.9965	.9969	.9974	.9978	.9983	.9987	.9991	.9996

Table C   Base-*e* Logarithms

N	0	1	2	3	4	5	6	7	8	9
1.0	.0000	.0100	.0198	.0296	.0392	.0488	.0583	.0677	.0770	.0862
1.1	.0953	.1044	.1133	.1222	.1310	.1398	.1484	.1570	.1655	.1740
1.2	.1823	.1906	.1989	.2070	.2151	.2231	.2311	.2390	.2469	.2546
1.3	.2624	.2700	.2776	.2852	.2927	.3001	.3075	.3148	.3221	.3293
1.4	.3365	.3436	.3507	.3577	.3646	.3716	.3784	.3853	.3920	.3988
1.5	.4055	.4121	.4187	.4253	.4318	.4383	.4447	.4511	.4574	.4637
1.6	.4700	.4762	.4824	.4886	.4947	.5008	.5068	.5128	.5188	.5247
1.7	.5306	.5365	.5423	.5481	.5539	.5596	.5653	.5710	.5766	.5822
1.8	.5878	.5933	.5988	.6043	.6098	.6152	.6206	.6259	.6313	.6366
1.9	.6419	.6471	.6523	.6575	.6627	.6678	.6729	.6780	.6831	.6881
2.0	.6931	.6981	.7031	.7080	.7129	.7178	.7227	.7275	.7324	.7372
2.1	.7419	.7467	.7514	.7561	.7608	.7655	.7701	.7747	.7793	.7839
2.2	.7885	.7930	.7975	.8020	.8065	.8109	.8154	.8198	.8242	.8286
2.3	.8329	.8372	.8416	.8459	.8502	.8544	.8587	.8629	.8671	.8713
2.4	.8755	.8796	.8838	.8879	.8920	.8961	.9002	.9042	.9083	.9123
2.5	.9163	.9203	.9243	.9282	.9322	.9361	.9400	.9439	.9478	.9517
2.6	.9555	.9594	.9632	.9670	.9708	.9746	.9783	.9821	.9858	.9895
2.7	.9933	.9969	1.0006	.0043	.0080	.0116	.0152	.0188	.0225	.0260
2.8	1.0296	.0332	.0367	.0403	.0438	.0473	.0508	.0543	.0578	.0613
2.9	.0647	.0682	.0716	.0750	.0784	.0818	.0852	.0886	.0919	.0953
3.0	1.0986	.1019	.1053	.1086	.1119	.1151	.1184	.1217	.1249	.1282
3.1	.1314	.1346	.1378	.1410	.1442	.1474	.1506	.1537	.1569	.1600
3.2	.1632	.1663	.1694	.1725	.1756	.1787	.1817	.1848	.1878	.1909
3.3	.1939	.1969	.2000	.2030	.2060	.2090	.2119	.2149	.2179	.2208
3.4	.2238	.2267	.2296	.2326	.2355	.2384	.2413	.2442	.2470	.2499
3.5	1.2528	.2556	.2585	.2613	.2641	.2669	.2698	.2726	.2754	.2782
3.6	.2809	.2837	.2865	.2892	.2920	.2947	.2975	.3002	.3029	.3056
3.7	.3083	.3110	.3137	.3164	.3191	.3218	.3244	.3271	.3297	.3324
3.8	.3350	.3376	.3403	.3429	.3455	.3481	.3507	.3533	.3558	.3584
3.9	.3610	.3635	.3661	.3686	.3712	.3737	.3762	.3788	.3813	.3838
4.0	1.3863	.3888	.3913	.3938	.3962	.3987	.4012	.4036	.4061	.4085
4.1	.4110	.4134	.4159	.4183	.4207	.4231	.4255	.4279	.4303	.4327
4.2	.4351	.4375	.4398	.4422	.4446	.4469	.4493	.4516	.4540	.4563
4.3	.4586	.4609	.4633	.4656	.4679	.4702	.4725	.4748	.4770	.4793
4.4	.4816	.4839	.4861	.4884	.4907	.4929	.4951	.4974	.4996	.5019
4.5	1.5041	.5063	.5085	.5107	.5129	.5151	.5173	.5195	.5217	.5239
4.6	.5261	.5282	.5304	.5326	.5347	.5369	.5390	.5412	.5433	.5454
4.7	.5476	.5497	.5518	.5539	.5560	.5581	.5602	.5623	.5644	.5665
4.8	.5686	.5707	.5728	.5748	.5769	.5790	.5810	.5831	.5851	.5872
4.9	.5892	.5913	.5933	.5953	.5974	.5994	.6014	.6034	.6054	.6074
5.0	1.6094	.6114	.6134	.6154	.6174	.6194	.6214	.6233	.6253	.6273
5.1	.6292	.6312	.6332	.6351	.6371	.6390	.6409	.6429	.6448	.6467
5.2	.6487	.6506	.6525	.6544	.6563	.6582	.6601	.6620	.6639	.6658
5.3	.6677	.6696	.6715	.6734	.6752	.6771	.6790	.6808	.6827	.6845
5.4	.6864	.6882	.6901	.6919	.6938	.6956	.6974	.6993	.7011	.7029

Use the properties of logarithms and $\ln 10 \approx 2.3026$ to find logarithms of numbers less than 1 or greater than 10.

Table C   *(continued)*

N	0	1	2	3	4	5	6	7	8	9
5.5	1.7047	.7066	.7084	.7102	.7120	.7138	.7156	.7174	.7192	.7210
5.6	.7228	.7246	.7263	.7281	.7299	.7317	.7334	.7352	.7370	.7387
5.7	.7405	.7422	.7440	.7457	.7475	.7492	.7509	.7527	.7544	.7561
5.8	.7579	.7596	.7613	.7630	.7647	.7664	.7681	.7699	.7716	.7733
5.9	.7750	.7766	.7783	.7800	.7817	.7834	.7851	.7867	.7884	.7901
6.0	1.7918	.7934	.7951	.7967	.7984	.8001	.8017	.8034	.8050	.8066
6.1	.8083	.8099	.8116	.8132	.8148	.8165	.8181	.8197	.8213	.8229
6.2	.8245	.8262	.8278	.8294	.8310	.8326	.8342	.8358	.8374	.8390
6.3	.8405	.8421	.8437	.8453	.8469	.8485	.8500	.8516	.8532	.8547
6.4	.8563	.8579	.8594	.8610	.8625	.8641	.8656	.8672	.8687	.8703
6.5	1.8718	.8733	.8749	.8764	.8779	.8795	.8810	.8825	.8840	.8856
6.6	.8871	.8886	.8901	.8916	.8931	.8946	.8961	.8976	.8991	.9006
6.7	.9021	.9036	.9051	.9066	.9081	.9095	.9110	.9125	.9140	.9155
6.8	.9169	.9184	.9199	.9213	.9228	.9242	.9257	.9272	.9286	.9301
6.9	.9315	.9330	.9344	.9359	.9373	.9387	.9402	.9416	.9430	.9445
7.0	1.9459	.9473	.9488	.9502	.9516	.9530	.9544	.9559	.9573	.9587
7.1	.9601	.9615	.9629	.9643	.9657	.9671	.9685	.9699	.9713	.9727
7.2	.9741	.9755	.9769	.9782	.9796	.9810	.9824	.9838	.9851	.9865
7.3	.9879	.9892	.9906	.9920	.9933	.9947	.9961	.9974	.9988	2.0001
7.4	2.0015	.0028	.0042	.0055	.0069	.0082	.0096	.0109	.0122	.0136
7.5	2.0149	.0162	.0176	.0189	.0202	.0215	.0229	.0242	.0255	.0268
7.6	.0281	.0295	.0308	.0321	.0334	.0347	.0360	.0373	.0386	.0399
7.7	.0412	.0425	.0438	.0451	.0464	.0477	.0490	.0503	.0516	.0528
7.8	.0541	.0554	.0567	.0580	.0592	.0605	.0618	.0631	.0643	.0656
7.9	.0669	.0681	.0694	.0707	.0719	.0732	.0744	.0757	.0769	.0782
8.0	2.0794	.0807	.0819	.0832	.0844	.0857	.0869	.0882	.0894	.0906
8.1	.0919	.0931	.0943	.0956	.0968	.0980	.0992	.1005	.1017	.1029
8.2	.1041	.1054	.1066	.1078	.1090	.1102	.1114	.1126	.1138	.1150
8.3	.1163	.1175	.1187	.1199	.1211	.1223	.1235	.1247	.1258	.1270
8.4	.1282	.1294	.1306	.1318	.1330	.1342	.1353	.1365	.1377	.1389
8.5	2.1401	.1412	.1424	.1436	.1448	.1459	.1471	.1483	.1494	.1506
8.6	.1518	.1529	.1541	.1552	.1564	.1576	.1587	.1599	.1610	.1622
8.7	.1633	.1645	.1656	.1668	.1679	.1691	.1702	.1713	.1725	.1736
8.8	.1748	.1759	.1770	.1782	.1793	.1804	.1815	.1827	.1838	.1849
8.9	.1861	.1872	.1883	.1894	.1905	.1917	.1928	.1939	.1950	.1961
9.0	2.1972	.1983	.1994	.2006	.2017	.2028	.2039	.2050	.2061	.2072
9.1	.2083	.2094	.2105	.2116	.2127	.2138	.2148	.2159	.2170	.2181
9.2	.2192	.2203	.2214	.2225	.2235	.2246	.2257	.2268	.2279	.2289
9.3	.2300	.2311	.2322	.2332	.2343	.2354	.2364	.2375	.2386	.2396
9.4	.2407	.2418	.2428	.2439	.2450	.2460	.2471	.2481	.2492	.2502
9.5	2.2513	.2523	.2534	.2544	.2555	.2565	.2576	.2586	.2597	.2607
9.6	.2618	.2628	.2638	.2649	.2659	.2670	.2680	.2690	.2701	.2711
9.7	.2721	.2732	.2742	.2752	.2762	.2773	.2783	.2793	.2803	.2814
9.8	.2824	.2834	.2844	.2854	.2865	.2875	.2885	.2895	.2905	.2915
9.9	.2925	.2935	.2946	.2956	.2966	.2976	.2986	.2996	.3006	.3016

Use the properties of logarithms and ln 10 ≈ 2.3026 to find logarithms of numbers less than 1 or greater than 10.

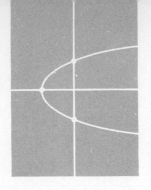

Index